The extreme properties of pulsars demand a broad range of physics: the magnetosphere requires plasma physics and electrodynamics whilst the structure and dynamics of the very dense interior call upon nuclear and condensed matter physics. At The Los Alamos Workshop on Isolated Pulsars recent advances were presented by international specialists; from this, contributions have been selected to encompass all aspects of pulsar detection and understanding.
Observational highlights include results from the Compton Gamma-Ray Observatory, ROSAT, the ASGAT and Whipple Observatory TeV experiments, and the Hubble Space Telescope. Theoretical contributions embody thermal and non-thermal emission mechanisms, properties of dense matter and the generation and evolution of magnetic fields. Together these form a comprehensive and up-to-date guide to the observation and theory of isolated neutron stars.

Isolated Pulsars

Isolated Pulsars

Proceedings of the Los Alamos Workshop,
held in Taos, New Mexico,
February 23–28, 1992

Edited by

K. A. VAN RIPER, R. EPSTEIN, and C. HO
Los Alamos National Laboratory, New Mexico

CAMBRIDGE
UNIVERSITY PRESS

Published by the Press Syndicate of the University of Cambridge
The Pitt Building, Trumpington Street, Cambridge CB2 1RP
40 West 20th Street, New York, NY 10022–4211, USA
10 Stamford Road, Oakleigh, Melbourne 3166, Australia

First published 1993

Printed in Great Britain at the University Press, Cambridge

A catalogue record for this book is available from the British Library

Library of Congress cataloguing in publication data available

ISBN 0 521 434297 hardback

Table of Contents

Pulsar Magnetosphere and Radio Emission

Optical to Gamma-Ray Pulsar Emission

VHE/UHE Emission

Particle Acceleration

Pulsar Timing Analysis

Pulsar Evolution and Related Phenomena

Preface

The goal of the 1992 Los Alamos Workshop on Isolated Pulsars was to bring together scientists working on different facets of pulsar research, to report observational and theoretical results, and to stimulate new directions of pulsar research. A special emphasis of the Workshop was to study pulsars through the high-energy phenomena illuminated by the new data. The Workshop was held February 23 to 28, 1992, at the Sagebrush Inn in Taos, New Mexico. Nearly 100 scientists participated in the Workshop sessions and took part in lively discussions in the Workshop and on the ski slope. In this Workshop, participants reported on new theoretical and observational results. These proceedings contain detailed reports of the contributions to the Workshop. They are grouped into sections which reflect the content of the sessions of the Workshop.

The physics of pulsars, magnetized spinning neutron stars, spans a wide range of disciplines, from nuclear and condensed matter physics of very dense matter in neutron star interiors, to plasma physics and electrodynamics of the magnetospheres, to relativistic magnetohydrodynamics of electron-positron pulsar winds interacting with the surrounding ambient medium. Observationally, pulsar research is advancing rapidly. A great array of new space instruments were recently launched: the Hubble Space Telescope, German ROSAT satellite, the Compton Gamma-Ray Observatory, and the Russian satellite GRANAT. These instruments are opening new windows in pulsar research with high quality data in energy bands from optical to gamma-rays. The impressive new X-ray measurements with the ROSAT satellite have enabled comparison of thermal emission from neutron star surfaces with atmospheric models having extreme gravity and magnetic fields. The discovery of new pulsars by the Compton Gamma-Ray Observatory, discussed at and reported soon after the Workshop, is enlarging the class of high-energy pulsars to permit meaningful studies of their general properties. The observation of the onset and real-time evolution of a glitch occurring in the Crab pulsar could prove invaluable in our study of the intriguing glitch phenomena. The confirmation of the TeV emission from the Crab nebula further solidifies the powerful nature of the pulsar/nebula interaction which provides important clues to the overall energetics of the pulsar phenomena.

The Space Astronomy and Astrophysics Group at Los Alamos hosted the workshop. The sponsors were the Space Science and Technology Division and the Institute of Geophysics and Planetary Physics of Los Alamos, and the Astrophysics Division of NASA. Jon Arons, Jim Cordes, David Helfand, Richard Lamb, Hakki Ögelman, Curt Michel, Mal Ruderman, and Mel Ulmer worked with us on the Scientific Organizing committee. The local organizing committee consisted of the editors, John Middleditch, and Darragh Nagle. We are deeply indebted to the Workshop staff, Elaine Ruhe, Margaret Flaugh, Deanna Davis, and Robin Naffziger, for taking care of the myriad details necessary for the smooth running of the meeting. The staff at the Sagebrush Inn provided a pleasant environment for the Workshop and superb refreshments and lunches. Patricia Mendius provided valuable assistance with copy editing, Pam Paine helped with the graphics, and Ann Nagy organized the final copy. We thank Ruth

Robichaud for her help in putting together the Workshop logo, which incorporates the light curve of the Crab pulsar, the Zia sun symbol of New Mexico, and the adobe pueblo structure for which Taos is famous. This Workshop was organized under the auspices of the US Department of Energy.

Ken Van Riper, Richard Epstein, Cheng Ho
Los Alamos, New Mexico
September, 1992

ISOLATED PULSARS
Taos, New Mexico February 1992

Colophon

The camera-ready copy for this book was prepared with the TEX formatting program, using macros based on those prepared by Christopher Mauche for the 1989 CV Workshop and hacked on by the editors. The TEX program and postprocessors are from Arbortext. The fonts are LucidaBright and LucidaNewMath PostScript outline fonts supplied by Y&Y of Carlisle MA ("Lucida" is a registered trademark of Bigelow and Holmes, Inc.).

List of Participants

Carl Akerlof	U. Michigan (U.S.)
Ali Alpar	M.E.T.U. (Turkey)
Jonathon Arons	UC Berkeley (U.S.)
Estelle Asseo	Ecole Polytech (France)
Matthew Bailes	Nuffield Radio Astr. Lab. (U.K.)
Menashe Banit	Columbia U. (U.S.)
Gordon Baym	U. Illinois (U.S.)
Werner Becker	Max Planck (Germany)
George Berzins	Los Alamos National Lab. (U.S.)
James Biggs	NASA/GSFC (U.S.)
Donald Casperson	Los Alamos National Lab. (U.S.)
Hoi Fung Chau	U. Hong Kong (Hong Kong)
Kwong-Sang Cheng	U. Hong Kong (Hong Kong)
Anton Chernenko	Inst. Spa. Res. (Russia)
James Chiang	Stanford U. (U.S.)
James Cordes	Cornell U. (U.S.)
Kwan-Ying Ding	U. of Hong Kong (Hong Kong)
Vladimir Dolidze	Inst. Spa. Res. (Russia)
Darrell Drake	Los Alamos National Lab. (U.S.)
Richard Epstein	Los Alamos National Lab. (U.S.)
Doyle Evans	Los Alamos National Lab. (U.S.)
Edward Fenimore	Los Alamos National Lab. (U.S.)
John Finley	U. Wisconsin (U.S.)
Roger Foster	NRL (U.S.)
Jack Fry	U. Wisconsin (U.S.)
Yves Gallant	UC Berkeley (U.S.)
Urlich Geppert	Spa. Res. Inst. (Germany)
Janusz Gil	Max Planck (Germany)
Peter Goldreich	Caltech (U.S.)
Isabelle Grenier	Saclay (France)
Timothy Hankins	NRAO (U.S.)
Alice Harding	NASA/GSFC (U.S.)
Dieter Hartmann	Clemson U. (U.S.)
David Helfand	Columbia U. (U.S.)
Heinz Herold	U. Tubingen (Germany)
Cheng Ho	Los Alamos National Lab. (U.S.)
Gottfried Kanbach	Max Planck (Germany)
Nobuyuki Kawai	Riken (Japan)
Alan Kerrick	UC Riverside (U.S.)
Ray Klebesadel	Los Alamos National Lab. (U.S.)
Valerij Kostenko	Inst. Spa. Res. (Russia)
Dong Lai	Cornell U. (U.S.)
Frederick Lamb	U. Illinois (U.S.)
Richard Lamb	Iowa State U. (U.S.)

Bennett Link	Los Alamos National Lab. (U.S.)
Andrew Lyne	Nuffield Radio Astr. Lab (U.K.)
Jerzy Madej	Warsaw U. (Poland)
Mark McConnell	UNH (U.S.)
Mike Meier	Los Alamos National Lab. (U.S.)
F. Curtis Michel	Rice U. (U.S.)
John Middleditch	Los Alamos National Lab. (U.S.)
M. Coleman Miller	U. Illinois (U.S.)
Igor G. Mitrofanov	Inst. Space Res. (Russia)
Cal Moss	Los Alamos National Lab. (U.S.)
Darragh Nagle	Los Alamos National Lab. (U.S.)
Takashi Nakamura	Kyoto U. (Japan)
George Nakano	Los Altos (U.S.)
Robert Nelson	Cornell U. (U.S.)
David Nice	Princeton U. (U.S.)
Hakki Ögelman	U. Wisconsin (U.S.)
Franco Pacini	Oss. Astr., Firenze (Italy)
David Pines	U. Illinois (U.S.)
Bill Priedhorsky	Los Alamos National Lab. (U.S.)
Michael Punch	Whipple Obs. (U.S.)
Roger Romani	Stanford U. (U.S.)
Diane Roussel-Dupre	Los Alamos National Lab. (U.S.)
Malvin Ruderman	Columbia U. (U.S.)
Frederick Seward	SAO (U.S.)
Maurice Shapiro	U. Maryland (U.S.)
Yurii Shibanov	Ioffe Phy.-Tech. Inst. (Russia)
Noriaki Shibazaki	Rikkyo Univ. (Japan)
David Simons	Los Alamos National Lab. (U.S.)
Ian Smith	U. Chicago (U.S.)
Sergey Solodukhin	Moscow U. (Russia)
Tod Strohmayer	U. Rochester (U.S.)
David Thompson	NASA/GSFC (U.S.)
Sachiko Tsuruta	Montana State (U.S.)
Takahiro Tsutsumi	U. N. M. (U.S.)
Melville Ulmer	Northwestern U. (U.S.)
Kenneth Van Riper	Los Alamos National Lab. (U.S.)
Joseph Ventura	U. Crete (Greece)
Zhenru Wang	Nanjing U. (China)
Trevor Weekes	Whipple Obs. (U.S.)
Daniel Weeks	UNM (U.S.)
Ralph Wijers	Princeton U. (U.S.)
Robert Wilson	NASA/MSFC (U.S.)
Tser-Yuan Yang	UC Berkeley (U.S.)
Gaurang Yodh	UC Irvine (U.S.)

Neutron Star Interiors

Gordon Baym

Department of Physics, University of Illinois at Urbana/Champaign

1 Abstract

The outer lower density material in a neutron star forms a solid crust, whereas the denser liquid interior is composed primarily of neutron-rich nuclear matter. Various exotic states - ranging from pion or kaon condensed matter to normal or strange quark matter - have been proposed as candidates for the matter deep in the core. We review possible states of matter in neutron stars, and their effects on neutron star structure and cooling.

2 Introduction

Beneath a solid crust a neutron star has a liquid core. The gross structure of neutron stars, *e.g.*, their mass density profiles $\epsilon(r)/c^2$, radii R, and moments of inertia I, as functions of the stellar mass M - as well as the maximum possible neutron star mass M_{max} - are determined by the equation of state of matter in the liquid interior. On the other hand, the crust is ultimately responsible for the remarkable timing stability of pulsars and is, in addition, the most likely site for the mechanism of timing irregularities, including sudden pulsar speedups or glitches (as discussed in the following papers in these Proceedings).

The transition from crust to core occurs when the nuclei in the crust become so closely packed that they merge together, at a density of order that of the matter in the interior of a large laboratory nucleus, $\rho_0 \simeq 2.8 \times 10^{14}$ g/cm^3. When looked at in detail, the nuclei in the transition appear to proceed through highly distorted "pasta" states, reminiscent of lasagna, spaghetti, meatballs, etc. (Ravenhall *et al.* 1983). As a substantial fraction of the mass of the crust is in such nuclei (Lorentz *et al.* 1992), their effects, *e.g.*, on glitch mechanisms, pose an important unsolved question.

In the core near the transition, matter is essentially nuclear matter liquid, or "dissolved nuclei." However, the nature of the matter in the innermost core of neutron stars remains considerably uncertain. A number of exotic states, including pion and kaon condensed matter, and normal and strange quark matter, have all been studied. The reason for much of the interest in exotic states of matter in the interior is their enhanced neutrino emissivity; the rapid early cooling of neutron stars that would result is a possible signal of such matter in the interior. But, as we shall see, cooling via neutrino emission from normal nuclear matter may in fact proceed as rapidly as that in exotic states. Having recently reviewed in some detail the nature of the matter in the interior (Baym 1991, 1992), I will briefly summarize the interior states, and focus on more recent developments.

3 Nuclear Matter in the Interior

At densities near ρ_0, matter in the liquid interior consists primarily of degenerate neutrons with an admixture of degenerate protons and electrons to maintain equilibrium under the weak interactions

$$n \to p + e^- + \bar{\nu}_e, \quad e^- + p \to n + \nu_e. \tag{1}$$

The neutron fluid is most likely superfluid and the proton fluid superconducting (see Ainsworth *et al.* (1989) and Alpar and Pines (1992) for recent discussions of the states of superfluidity in neutron stars.) The superfluidity of the neutron fluid extends outward into the inner crust, where it is believed to be the key to glitches (discussed in the several following papers in these Proceedings).

The equation of state of matter in the liquid near ρ_0 can be obtained by extrapolation from laboratory nuclear physics. The first step is to extract the nucleon-nucleon interaction from pp and pn scattering experiments at energies below $\sim$ 300 MeV, constrained by fitting the properties of the deuteron (reviewed in Wiringa *et al.* 1988, hereafter WFF), and then to solve numerically the many-body Schrödinger equation, *e.g.*, by variational techniques (WWF, Pandharipande and Wiringa 1989), to find the energy density as a function of baryon number. Modern two-body potentials include the Paris, the Urbana-v_{14} (UV14), and the Argonne-v_{14} (AV14) potentials, described in WWF, where the 14's refer to the number of different components, such as central, spin-orbit, etc., included in the interactions.

Calculations with two-body forces alone, however, predict too high a saturation density of symmetric nuclear matter, as well as too small a binding of light nuclei ($A \leq 4$). These problems can be cured by including phenomenological intrinsic three-body forces between nucleons [arising from, *e.g.*, two nucleons interacting while one or both are internally excited from interactions with a third nucleon, processes not included in the two-body interactions]. The three-body forces must increase the binding of symmetric nuclear matter in the neighborhood of ρ_0, but, to avoid overbinding, they must become repulsive at higher densities. This repulsion leads to a stiffening of the equation of state of neutron star matter at higher densities over that computed from two-body forces alone. The stiffening effects can be seen in Fig. 1a, the energy per nucleon, $E = \epsilon/\rho$, of neutron matter with inclusion of the phenomenological Urbana "Three-Nucleon" (TNI) and three-body Urbana VII (UVII) interactions. Uncertainties in the three-body interaction are reflected in the equation of state, particularly at higher densities.

Despite these uncertainties, the modern potentials yield very similar models of neutron stars. Figure 1b shows the mass density profile of a $1.4M_\odot$ star, calculated by WFF with the UV14 and AV14 two-body, and UVII three-body potentials; and, for comparison, the density profiles for the very soft PΛ equation of state (Friedman and Pandharipande 1981), and the very stiff tensor interaction model (TI) equation of state (Pandharipande and Smith 1975). The $1.4M_\odot$ models calculated with the modern

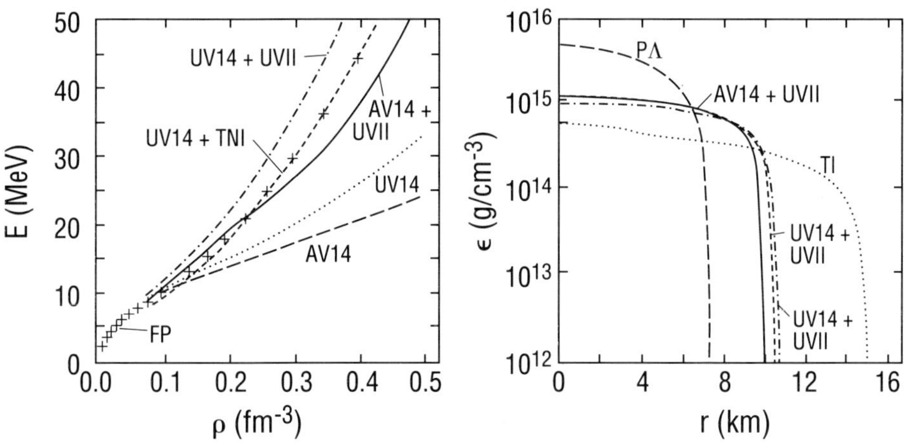

Fig. 1—Energy per nucleon vs. baryon density with and without three-body forces in pure neutron matter (left panel), and density profiles of a $1.4M_{\odot}$ neutron star calculated for various equations of state of neutron matter in beta equilibrium, with three-body interactions included (right panel). From WFF.

potentials have a radius $\simeq$ 10.4-11.2 km, and a central density $\sim 6\rho_0$. This density is sufficiently low to give one confidence in the extrapolations from low-energy nuclear experiments. The crusts of these stars are only $\sim$ 1 km thick, very thin compared with that of the stiff TI model. Figure 2a shows the gravitational mass vs. central density for families of stars calculated with the same equations of state as in Fig. 1 (with beta equilibrium included). The maximum mass for the modern equations of state is $\sim 2M_{\odot}$. The corresponding mass vs. radius of the families of models is shown in Fig. 2b; the radii vary only slowly with mass, except at the extremes.

A major uncertainty, arising in part from the not-well-determined form of the three-body interaction, particularly its isospin dependence, is the fraction, x_p, of the nucleons that are protons. In a non-interacting gas of nucleons and electrons in beta equilibrium at $\rho \sim \rho_0$, $x_p \simeq 5\%$; however, in matter interacting by means of the modern potentials, x_p could reach 15% or higher.

Such a large proton fraction would have a striking effect on the cooling rate of stars through neutrino emission (Lattimer *et al.* 1991). For $x_p < 1/9$ the normal Urca process (1) cannot, at low temperatures, conserve energy and momentum, since the neutron Fermi momentum is much larger than that of the protons and electrons, and also much larger than that available to the emitted low-energy neutrino; only by transferring the excess momentum to a bystander nucleon can the normal process work. The resulting cooling timescale for this modified Urca process is $\tau_{Urca'} \sim 1y/T_9^6$, where T_9 is the temperature in units of 10^9K. This rate is down by a factor $\sim (T/T_f)^2$ (where T is the temperature, and T_f is the bystander Fermi momentum) compared with that for the normal Urca process, were it allowed, because of the Pauli-suppressed density of states available to the bystander in degenerate matter. On the other hand for $x_p > 1/9$ (rising after muons make their appearance at densities slightly above ρ_0 as the electron chemical potential begins to exceed the mass of the muon,

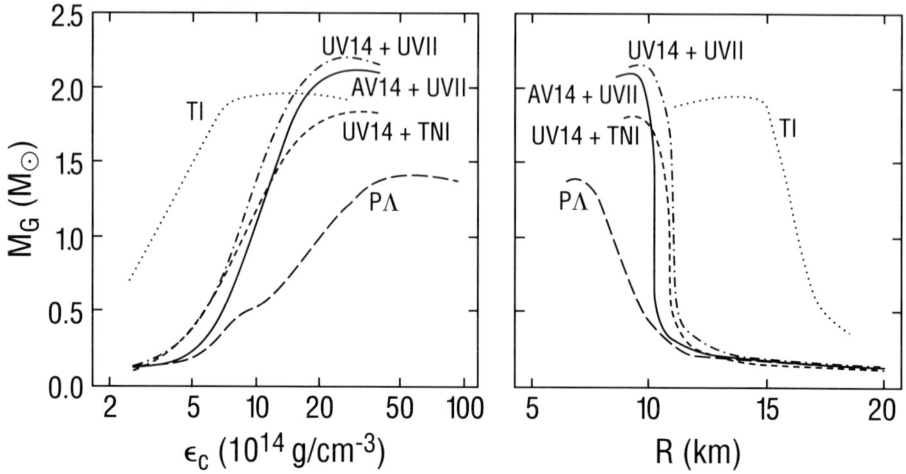

Fig. 2—Mass vs. central density (left panel), and mass vs. radius, for the same equations of state as in Fig. 1 (right panel). From WFF.

eventually to 0.148), the normal Urca process (1) is allowed, and the cooling timescale falls to $\tau_{Urca} \sim 10^2 s/T_9^4$ (Lattimer *et al.* 1991; Pethick 1992).

Although the recent nuclear equations of state undoubtedly give an accurate description of neutron star matter in the neighborhood of ρ_0, they fail to take into account the rich variety of hadronic and quark degrees of freedom in the nuclear system that become important with increasing density. Further hadronic degrees of freedom include Δ-isobars, hyperons, and higher baryon resonances, as well as mesonic degrees of freedom. The Δ's can both be excited as real states in the matter and contribute indirectly through the three-body forces among nucleons. The major uncertainties in including them in the equation of state are their interactions with nucleons and with other Δ's, interactions that become increasing important with increasing density.

Strange baryons, *e.g.*, Λ and Σ, can similarly be excited via weak conversions of nucleons, or by electron capture processes, $e^- + p \rightarrow \Lambda + \bar{\nu}_e$, $e^- + n \rightarrow \Sigma^- + \bar{\nu}_e$ (the latter process occurs at lower density than the former when the electron Fermi energy exceeds the Σ - Λ mass difference). Again, lack of detailed knowledge of the interactions of hyperons prevents an accurate description of their effect on the equation of state. However, even a small concentration of hyperons could lead to rapid cooling via direct Urca processes, $\Lambda \rightarrow e^- + p + \nu_e$, $\Sigma^- \rightarrow e^- + n + \nu_e$, and the corresponding electron capture processes, above (Prakash *et al.* 1992; Pethick, 1992).

Beyond a few times ρ_0 one can no longer expect the forces between particles to be describable in terms of static few-body potentials; nor can one continue to assume that at higher densities the system can even be described in terms of well-defined "asymptotic" laboratory particles. For example, the meson fields that mediate the low-energy interactions between baryons also become dynamical degrees of freedom with increasing density; in dense matter the meson fields can become macroscopically excited, leading to pion or kaon condensation. At sufficiently high densities, quark degrees of freedom are expected to become dominant. Let us overview these possibilities.

4 Pion and Kaon Condensation

Mesons, as bosons, when present, would macroscopically occupy the lowest available mode, *i.e.*, form a condensate, as in ordinary Bose-Einstein condensation. The state of matter with condensed charged pions or kaons would be superconducting. [The theory of condensation is reviewed in Baym and Campbell (1979).] Such states in neutron stars could be astrophysically important since their neutrino emissivity is comparable to the non-suppressed Urca process and hence could account for more rapid stellar cooling.

Essentially, condensation would occur when the collective mode of nuclear matter with the quantum numbers of the charged pion - an oscillation of the matter with spatially-varying nucleon spin (S=1) and isospin (I=1) - becomes "soft" at a critical density, ρ_π. In this softening the mode falls to zero frequency at a critical wavevector, k_c, causing the nucleon eigenstates to rotate in isospin space; instead of being pure neutron $|n\rangle$ and proton $|p\rangle$, the eigenstates spontaneously undergo a [chiral $SU(2)\otimes SU(2)$] rotation to become quantum-mechanical linear superpositions of neutron and proton states of the form: $|N'\rangle = \cos\theta|n\rangle + \sin\theta|p\rangle$, $|P'\rangle = \cos\theta|p\rangle - \sin\theta|n\rangle$, where the condensation angle θ grows from zero as the density increases above ρ_π. To conserve charge as the nucleon eigenstates are rotated, the system develops a macroscopic spatially-varying pion field $\langle\pi\rangle$ - the condensate - of net negative charge. An analogous neutral pion condensed state can also be formed through softening of the neutron particle-hole collective mode; the neutral condensed state is characterized by a spatially varying finite expectation value of the neutral pion field.

Pion condensation would enhance the cooling rates of neutron stars by allowing a much larger phase space for the Urca process (1) than is available in the modified Urca process. In charged-pion condensed matter, the rotated nucleon states, N' and P', can directly participate in the Urca process, effectively β-decaying, $N' \to N' + e^- + \bar{\nu}_e$, and electron-capturing, $e^- + N' \to N' + \nu_e$, into themselves. The spatially-varying condensate serves to absorb the excess momentum without the need for a bystander nucleon; the nucleons are Bragg-scattered by the condensate. The resulting cooling timescale for small θ is $\tau_\pi \sim 10s/\theta^2 T_9^4$, longer than that for the normal Urca at high proton concentration but significantly shorter than that for modified Urca.

Early estimates predicted the onset of charged pion condensation at $\rho_\pi \sim 2\rho_0$. However, recent measurements in nuclei tend to indicate that the effective neutron-hole proton repulsion in the I=1, S=1 channel is too strong to permit condensation (Brown *et al.* 1988). An alternative, possibly more likely, form of condensation involves spontaneous formation of K mesons (kaons). The underlying chiral SU(3)⊗SU(3) symmetry of strong interactions, exact in the limit that the up (u), down (d) and strange (s) quark masses vanish, leads to a density-dependent kaon effective mass which lowers the kaon energy in the matter. The energy of a K^+ falls below $\mu_p - \mu_n$, the chemical potential for K^+, at a critical density $\rho_K \sim 2.5\text{-}3\ \rho_0$ (Kaplan and Nelson 1986); above this density the system should form a kaon or strangeness condensate, with a macroscopic expectation value of the charged kaon field.

Analogous to the rotation of the nucleon eigenstates in pion condensation, the nucleons in kaon condensation undergo a chiral SU(3)⊗SU(3) rotation in which a neutron state becomes a linear superposition of a neutron and a Σ^-, $|N'\rangle = \cos\theta_K|n\rangle + \sin\theta_K|\Sigma^-\rangle$, while a proton state is rotated into a linear superposition of a proton, Σ^0 and Λ. From the point of view of the underlying quark structure of the baryons (n = udd, p = uud, Σ^- = dds, Σ^0 = uds, Λ = uds), the u and s components are rotated into each other by the angle θ_K. The rotation leads to a non-zero field expectation value in matter, $\langle\bar{s}u\rangle$, with the quantum numbers of the K^+ (where u and s are here the quark fields). The matter can also form an η meson condensate, *i.e.*, a state with a non-vanishing expectation value, $\langle\bar{s}s\rangle$, with the quantum numbers of the η meson.

Kaon condensation would also enhance the neutrino luminosity compared with the modified Urca process in normal matter. Even though the condensate is expected to be spatially uniform, the rotated nucleons have phase space to β-decay into themselves and thus undergo an Urca cycle. The cooling timescale for small θ_K is $\tau_K \sim 10^4 s/\theta_K^2 T_9^4$ (Brown *et al.* 1988), not as large as the enhanced cooling for pion-condensed matter but still significantly larger than the normal neutrino luminosity. Detailed calculations of neutron star cooling in the presence of a kaon condensate are given by Page and Baron (1990). The uncertainties in the parameters of the effective interaction of kaons and the effective strange-baryon nucleon-hole couplings in neutron star matter lead, however, to large uncertainties in the condensation angle expected at given density.

5 Normal and Strange Quark Matter

Nuclear matter, when squeezed to sufficiently high density, turns into a liquid of uniform quark matter in which the quarks are no longer confined in individual hadrons but are free to roam over macroscopic distances in the core. Quark matter in neutron stars would consist of Fermi seas of u and d quarks.

The transition from hadronic to quark matter is uncertain. The only reliable approach at present to determining the transition is through Monte-Carlo calculations of lattice gauge theory (reviewed by Gottlieb 1991). At zero baryon density these calculations indicate that the transition to a quark-gluon plasma is likely first order, at a critical temperature $\sim$ 200 MeV (or 10^{12}K), with a latent heat of order a few GeV/fm^3. However, lattice gauge calculations at non-zero baryon density are beset by technical problems; to date one does not have a reliable estimate of the transition density at zero temperature from nuclear to quark matter or even compelling evidence that there is a sharp phase transition.

One can roughly estimate the point at which quarks become deconfined by asking whether nuclear matter or quark matter has a lower energy per baryon at given baryon density; such comparisons (based on inequivalent physical descriptions of the two phases, and thus not expected to give an accurate description of the transition) imply a first-order transition with a discontinuity in the baryon density at $\rho \sim 5-10\rho_0$. This density range is possibly above the central density found in models of neutron stars with $M \sim 1.4M_\odot$ based on nuclear equations of state. However, in the absence of a good theory of the equation of state at very high densities, the question of whether neutron stars can have quark matter cores remains open, as is the issue of whether a distinct family of quark stars with higher central densities than neutron stars can exist.

Quark matter, in common with pion and kaon condensed matter, should have a much higher neutrino luminosity than ordinary neutron star matter at low proton concentration. The fundamental Urca processes are $d \to u+e^-+\bar{\nu}_e$ and $e^-+u \to d+\nu_e$. Massless non-interacting u and d quarks have essentially no phase space for these reactions. However, the strong interactions increase the energy of the individual u and d quarks [in lowest order perturbation theory, $E_p = (1+(2/3\pi)\alpha_c)cp$, where p is the quark momentum, and α_c is the quantum chromodynamic fine structure constant], thus allowing phase space for the Urca process (Iwamoto 1980). The resulting cooling timescale is $\tau_{qm} \sim 10^2 s/\alpha_c T_9^4$; within the uncertainties in parameters, the effect of a quark core on the cooling is roughly that of a pion-condensed core.

A second intriguing form of quark matter is one containing a substantial fraction of strange quarks, *strange quark matter* - a state distinct from the strangeness-condensed hadronic matter discussed above. At sufficiently high densities, strange quark matter should have lowest energy per baryon. Consider a high density at which normal quark matter is preferred over hadronic matter. By converting a u or d quark in the quark matter to a strange quark s, through the interactions $d+u \to s+u$, $e^- + u \to s + \nu_e$, one creates further Fermi seas for the same number of fermions, thus lowering the Fermi energy of the system. The cost of the conversion, neglecting interaction effects for the moment, is the increased mass of the s quark, $m_s \sim 150$ MeV. If the gain in Fermi energy exceeds the mass cost, such conversions would lower the energy of the matter.

One cannot even rule out the possibility that self-bound strange quark matter is the absolute ground state of matter and that ordinary nuclear matter is metastable (Witten 1984). This possibility depends critically on the interactions among the constituents of the matter (Farhi and Jaffe 1984), an issue that is far from settled theoretically.

If the baryon density ρ_s above which strange quark matter becomes the most favorable state is less than the central density of a neutron star made of hadronic or normal quark matter, then the central regions of neutron stars would have strange cores at densities above ρ_s. Were strange quark matter the absolute ground state, entirely *strange stars* could form (reviewed in Alcock and Olinto 1988; Baym 1988). A strange star would be made in a supernova explosion through the initial formation of a normal neutron star. At the very high central densities in the neutron star, a seed of strange quark matter can form through a fluctuation. As long as strange matter is more favorable than normal hadronic matter, then once formed the seed would "burn" the matter around it into strange matter. The burning front would first convert the liquid core of the neutron star to exotic matter; the heat ahead of the front would melt the crust of the neutron star, as well as melt the nuclei in the crust into normal fluid nuclear matter, and, if strange matter is the absolute ground state, within $\sim 10^3$ s, the entire star would be converted into a strange neutron star (Heiselberg *et al.* 1991; Olinto 1987). If, on the other hand, normal matter is preferable at lower densities, the core would become strange with the outer regions remaining normal. If strange stars can be made, many, if not all, neutron stars should be strange.

The gross structure of stars made entirely of strange quark matter is not very different from that of normal neutron stars. Whereas the radius of a normal neutron star generally decreases with increasing mass, that of a strange quark star, made of self-bound matter, increases with increasing mass. The crucial difference is the absence of a normal solid crust, which would prevent strange stars from acting as pulsars with good timing stability (Alpar 1987), unless either a) lower density strange quark matter is unstable against breaking into droplets, as normal nuclear matter at low pressure breaks up into nuclei, in which case a strange star could have a crust made of strange quark droplets arranged in a lattice, or b) strange quark matter is not self-bound, in which case strange stars would have normal crusts. Strange stars are promising candidates for gamma-ray burst sources (cf. the model of Alcock *et al.* 1986); they also provide the basis of an explanation for the reported deep underground muons with source in the direction of, and having the 4.79 hour periodicity of Cygnus X-3 (Baym *et al.* 1985).

In common with other exotic states of matter, strange matter should also have a high neutrino luminosity via Urca cycles $d \to u + e^- + \bar{\nu}_e$, $e^- + u \to d + \nu_e$, as well as $s \to u + e^- + \bar{\nu}_e$, $e^- + u \to s + \nu_e$. Stars with strange cores would cool faster than would normal neutron stars via the modified Urca process.

In summary, measurements of cooling of neutron stars could yield evidence on the nature of matter in their interiors; observations of early rapid cooling would indicate that more than modified Urca is at work, whether due to exotic matter in their interiors, or nuclear matter with sufficiently large proton fraction. However, our knowledge of neutron stars remains limited by the lack of experimentally-based information on the equation of state of matter in their interiors beyond a few times ρ_0. In the laboratory the only way to learn about high density matter is to collide large atomic nuclei together at ultrarelativistic energies. Such a program to study experimentally the properties of nuclear matter with extreme energy and baryon density is currently underway at CERN and Brookhaven National Laboratory (Awes *et al.* 1992) and should begin to provide insights into neutron stars.

This work was supported in part by National Science Foundation Grant DMR 91-22385.

References

Alcock, C., Farhi, E., and Olinto, A. 1986, *Phys. Rev. Lett.* **57**, 2088.
Alcock, C. and Olinto, A. 1988 *Ann. Rev. Nucl. Part. Sci.* **38**, 161.
Ainsworth, T., Pines, D., and Wambach, J. 1989, *Phys. Lett.*, **222B**, 173.
Alpar, M. A. 1987, *Phys. Rev. Lett.* **58**, 2152.
Alpar, M. A. and Pines, D. 1992, these Proceedings.
Awes, T. *et al.* eds. 1992, *Quark Matter '91, Proc. 9th Intl. Conf. on Ultra-relativistic Nucleus-Nucleus Collisions, Nucl. Phys.* **A544**.
Baym, G. 1988, *Nucl. Phys.* **A479**, 27c.
Baym, G. 1991, in *Neutron Stars: Theory and Observation*, D. Pines and J. Ventura, eds., Kluwer Academic Publ. Dordrecht, p. 21.
Baym, G. 1992, in *The Structure and Evolution of Neutron Stars*, D. Pines, R. Tamagaki, and S. Tsuruta, eds., Addison-Wesley, Menlo Park, p. 188.
Baym, G. and Campbell, D. K. 1979, in *Mesons in Nuclei*, v. 3, M. Rho and D. Wilkinson, eds., North-Holland Publ. Co., Amsterdam, p. 1031.
Baym, G., Jaffe, R., Kolb, E. W., McLerran, L., and Walker, T. P. 1985, *Phys. Lett.* **160B**, 181.
Brown, G.E., Kubodera, K., Page, D. and Pizzochero, P. 1988, *Phys. Rev.***D37**, 2042.
Farhi, E. and Jaffe, R. L. 1984, *Phys. Rev.* **D30**, 2379.
Friedman, B. and Pandharipande, V. R. 1981, *Nucl. Phys.* **A361**, 502.
Gottlieb, S. 1991, *Nucl. Phys.* **B** (Proc. Suppl.) **20**, 247.
Heiselberg, H., Baym, G., and Pethick, C. J. 1991, *Nucl. Phys.* **B** (Proc. Suppl.) **24**, 144.
Iwamoto, N. 1980, *Phys. Rev. Lett.* **44**, 1637.
Kaplan, D. B. and Nelson, A. E. 1986, *Phys. Lett.* **175B**, 57.
Lattimer, J. M., Pethick, C. J., Prakash, M. and Haensel, P. 1991, *Phys. Rev. Lett.* **66**, 2701.
Lorentz, C., Ravenhall, D. G. and Pethick, C. J. 1992, to be published.
Olinto, A. V. 1987, *Phys. Lett.* **192B**, 71.
Pandharipande, V. R. and Smith, R. A. 1975, *Nucl. Phys.* **A237**, 507.
Pandharipande, V. R. and Wiringa, R. B. 1989, in *The Nuclear Equation of State*, W. Greiner and H. Stöcker, eds., Plenum Press, New York, p. 585.
Page, D., and Baron, E. 1990 *Ap. J.* **354**, L17.
Pethick, C. J. 1992, *Rev. Mod. Phys.* **64**, No. 3.
Prakash, M., Prakash, M., Lattimer, J. M., and Pethick, C. J. 1992, *Ap. J. Lett.* **390**, L77.
Ravenhall, D. G., Pethick, C. J., and Wilson, J. 1983, *Phys. Rev. Lett.* **50**, 2066.
Wiringa, R. B., Fiks, V., and Fabrocini, A. 1988, *Phys. Rev.* **C38**, 1010.
Witten, E. 1984, *Phys. Rev.* **D30**, 272.

Neutron Superfluid Dynamics

Richard I. Epstein,[1] Gordon Baym,[1,2] and Bennett Link[1]

[1]Los Alamos National Laboratory
[2]Physics Department, University of Illinois at Urbana/Champaign

1 Abstract

Neutron superfluid dynamics and the coupling between the superfluid and the solid crust play important roles in the rotational, thermal, and magnetic evolution of neutron stars. Here we describe the physical processes by which the neutron superfluid and the crust interact.

2 Introduction

If neutron stars were composed solely of normal matter, they would be dense, uniformly rotating objects, and pulsars would not exhibit much of the behavior that has intrigued astrophysicists for years. However, neutron stars likely contain several superfluids. In their dense interiors the electron Fermi energy is so high that free neutrons are stable, and the attractive nucleon interaction pairs them to produce a superfluid. In the inner crust, where the free neutrons coexist with a lattice of neutron rich nuclei, the neutrons form an isotropic s-wave superfluid. At higher densities nuclei dissolve, and the stellar matter is thought to consist of an anisotropic $^3P_2$ neutron superfluid, protons in a $^1S_0$, type II superconductor, and normal electrons. The densest part of the star may contain exotic matter, such as a pion or kaon condensate, or a quark fluid.

Differential rotation between the superfluids and the crust is a source of free energy for driving glitches, sudden increases in the rotation rate of the stellar crust. In this picture, glitches are attributed to the rapid transfer of angular momentum to the crust from the more rapidly rotating superfluid (Anderson and Itoh 1975; Ruderman 1976). The post-glitch behavior, during which much of the excess rotation rate created by a glitch decays, occurs through the gradual re-establishment of rotational equilibrium between the crust and superfluid (Alpar *et al.* 1984; Link, Epstein, and Baym 1993). Perturbations in the rotation rate of the star are damped by the coupling between the superfluid and the crust. For example, neutron star precession, as possibly exhibited by Her X-1, must be continuously driven to compensate for crust-superfluid dissipation (Alpar and Ögelman 1987); general relativistic rotational instabilities are suppressed by this dissipation (Lindblom and Mendell 1992). The neutron superfluids may tie together the spin and magnetic evolution of neutron stars in two ways. In the core the interaction of neutron vortices with magnetic flux tubes couples these two distributions (Śrinivasan *et al.* 1990; Ding, Cheng, and Chau 1992). In the crust superfluid stresses tend to move plates of solid matter in which the magnetic fields are entrained (Ruderman 1991). The existence of superfluids also has significant consequences for the thermal evolution of neutron stars. Frictional heating from the drag between the superfluid and crust significantly alters the observable surface temperatures of older stars (Alpar *et al.* 1984; Shibazaki and Lamb 1989; Van Riper 1991) and may lead to thermal and rotational instabilities (Shibazaki 1992). Superfluidity also changes the stellar heat capacity and thermal conductivity.

In the inner crust the coexistence of nuclei and a neutron superfluid creates the possibility of vortex pinning, vortex creep, and highly dissipative vortex flow, phenomena with important and direct observational consequences. This report focuses on some of our recent progress in understanding superfluid dynamics in the inner crust region and how the angular velocities of the crust and superfluid evolve prior to, during, and following a glitch.

3 Physical processes

A neutron superfluid, as an inviscid quantum liquid, is described by a macroscopic wave function $\psi(r,t)$; the fluid velocity is $\mathbf{v}_s = (\hbar/2m_n)\nabla S$, where S is the phase of the wave function, m_n is the neutron mass, and $2m_n$ is the mass of the Cooper pair. Because the velocity is proportional to a gradient, the superfluid is irrotational; *i.e.*, $\nabla\times\mathbf{v}_s = 0$. The superfluid thus cannot rotate uniformly as a normal fluid. Nevertheless, it can carry angular momentum and approximate normal rotation by forming an array of vortex lines, thin filamentary regions in which the superfluidity vanishes and vorticity is concentrated. The circulation around each vortex line is quantized, $\oint \mathbf{v}_s\cdot d\mathbf{l} = \kappa \equiv \pi\hbar/m_n$. The mean superfluid angular velocity $\Omega_s(r)$ at a distance r from the rotation axis is proportional to the number of vortex lines $N_v(r)$ interior to r:

$$\Omega_s(r) = \frac{\kappa N_v(r)}{2\pi r^2}. \tag{1}$$

The distribution of vortex lines completely specifies $\Omega_s(r)$, and the superfluid angular velocity changes only as the vortex lines move radially.

The rotational evolution of the superfluid is determined by the motion of its vortex lines. The lines move according to a first-order equation of motion

$$\rho_s\vec{\kappa}\times(\mathbf{v}_v - \mathbf{v}_s) + \mathbf{f}_{\rm other} = 0, \tag{2}$$

where ρ_s is the superfluid density, $\vec{\kappa}$, whose magnitude is κ, is aligned with the vortex line, and $\mathbf{v}_v$ is the velocity of the vortex line. The first term, the Magnus force, arises from the relative velocity of the vortex line with respect to the superfluid. This classical force can be understood in terms of the interaction of the average superfluid velocity and the velocity induced by an individual vortex line. On one side of the vortex, the velocity field of the vortex adds constructively with the flow past the line. The increased velocity results, by Bernoulli's law, in a lower than average pressure. In contrast, on the other side the velocities partially cancel, and the pressure is higher. The net force due to the pressure differential is the Magnus force. The second term in the equation of motion includes all other forces such as those due to vortex interactions with nuclei, electrons, magnetic flux tubes or the vortex self-energy (tension).

If there were no other forces (*i.e.*, if $\mathbf{f}_{\rm other} = 0$), vortex lines would be swept along with the superfluid, and $N_\nu(r)$ would remain unchanged. Only when azimuthal forces act on the vortex lines, thereby applying a torque on the superfluid, do the lines move in the radial direction and change the angular momentum of the superfluid.

To illustrate the effects of dissipative forces acting on a vortex line, consider viscous drag between the vortex line and the normal component of the star:

$$\mathbf{f}_{\rm other} \to \mathbf{f}_{\rm drag} = -\eta \Delta \mathbf{v}, \qquad (3)$$

where η is a drag coefficient and $\Delta\mathbf{v}$ is the relative velocity of the vortex line through the crust. Substitution of equation (3) in the dynamical equation (2) shows that the radial component of the vortex velocity is $v_r = \Delta v \sin\theta_d$, where $\theta_d \equiv \tan^{-1}(\eta/\rho_s\kappa)$ is the *dissipation angle*. If only the Magnus and drag forces act on the vortex lines, the lines spiral outward or inward to bring the superfluid into corotation with the rest of the star on a time scale $\sim [\Omega_s \sin 2\theta_d]^{-1}$ (Bildsten and Epstein 1989; Epstein and Baym 1992).

In the liquid core of the star, Fermi liquid effects magnetize the neutron vortices by coupling the motions of the superfluid protons and neutrons. Electron scattering from these magnetized vortex lines creates a strong drag force which damps any differential rotation between the core superfluid and the crust within minutes (Alpar, Langer, and Sauls 1984; Alpar and Sauls 1988). However, the dynamics of the core is complicated by vortex interactions with magnetic flux tubes which may prevent the superfluid from achieving complete corotation (Ding, Cheng, and Chau 1992).

In the inner crust a superfluid vortex line interacts with the lattice of nuclei - being either attracted to the nuclei or repelled by them, depending on the local matter density. In either case the vortex line tends to bind to the crystal lattice by pinning to the nuclei or in the interstices between the nuclei. The interaction energy for a nucleus near a vortex line was calculated by Alpar (1977) and by Epstein and Baym (1988). In the former treatment nuclei were treated as hard spheres where the superfluid properties change abruptly. In the improved latter treatment, gradients in the superfluid density and nuclear structure were accommodated by minimizing a Ginzburg-Landau free energy functional; this approach is not completely adequate, since the dimensions of the vortex lines and the nuclei are smaller than or comparable to that of the superfluid coherence length ξ. Full BCS calculations, currently being pursued by De Blasio *et al.* (1992), should produce more reliable results.

The vortex tension $\hat{T}$, or self-energy, corresponding primarily to the kinetic energy of the flow around a vortex, is significant in superfluid dynamics. If a perturbation of wavenumber k is made in the line, the excess energy per unit length is of the form $\hat{T} = \rho_s\kappa^2\Lambda/4\pi$, where $\Lambda \simeq 0.116 - \ln k\xi$. The importance of tension is characterized by the dimensionless *stiffness*, $\tau \equiv \hat{T}r_0/F_{\rm max}l$ (r_0 is the range of the pinning potential, typically tens of fm; $F_{\rm max}$ is the maximum value of the pinning force and l is the distance between pinning sites). A flexible vortex has $\tau \ll 1$ whereas a stiff vortex has $\tau \gg 1$ (Link and Epstein 1991). At stellar densities near 10^{13} g cm^{-3} where the pinning forces are the greatest, vortex lines may be flexible; elsewhere they are at least moderately stiff ($\tau \gtrsim 1$).

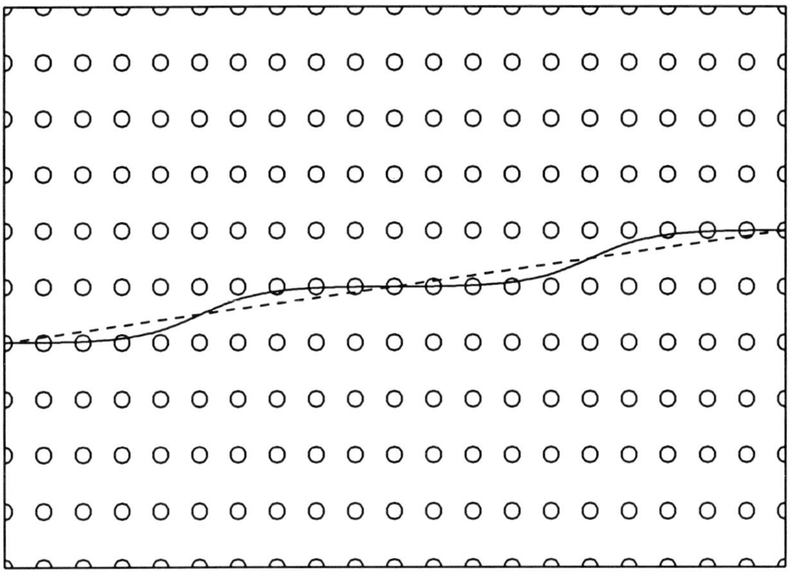

Fig. 1— Path of a vortex line through a crystal. The solid line shows a vortex of stiffness $\tau = 1$. The dashed line shows an infinitely stiff line.

A completely rigid vortex line would not intercept many pinning sites unless it happened to be oriented along a major axis of the crystal. This situation is illustrated by the dashed line in Figure 1. The net pinning forces on a randomly oriented rigid line largely cancel since, on average, there are as many nearby pinning sites on either side of the line (Jones 1992). However, for a vortex line of finite tension, pinning is much more efficient because the increase in the self-energy caused by slight bending is compensated by the large change in the interaction energy when the vortex line passes through nearly all of the nearby pinning sites (Link, Epstein, and Baym 1993; Link and Epstein 1992).

4 Superfluid dynamics in the inner crust

In a young, hot neutron star, the newly formed superfluid initially corotates with the normal matter. As the spin frequency of the crust decreases because of the magnetic torque, the drag forces on the vortex lines tend to move the vortex lines radially outward to maintain the neutron superfluid in corotation with the normal matter. The pinning forces, on the other hand, freeze the lines in the crust and preventing corotation. An angular velocity lag ω between the superfluid and crust builds up until the Magnus force pulls the vortex lines free from their pinning sites. The critical lag ω_{cr}, the largest differential rotation that can be maintained by pinning, is obtained by equating the Magnus force to $F_{\max}/l$, the maximum value of the pinning force per unit length; this calculation yields $\omega_{cr}(r) = F_{\max}/\rho_s \kappa l r$. The thick solid lines in Figure 2 show estimates of ω_{cr} throughout the inner crust based on the Ginzburg-Landau (*GL*) approach. For comparison we show the older, hard-sphere (*HS*) approach with

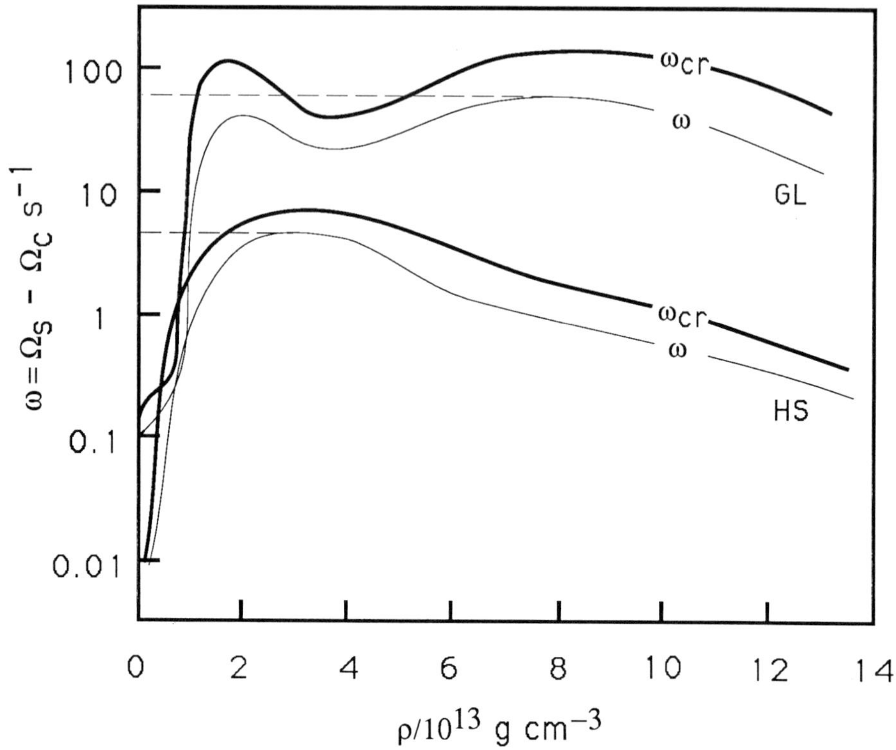

Fig. 2—The thick solid lines give the critical angular velocity lag ω_{cr} as a function of the stellar mass density for pinning forces based on the Ginzburg-Landau (*GL*) and hard-sphere (*HS*) calculations. The thin solid lines illustrate steady-state lag and the dashed lines give the low-density steady-state lag if the vortex lines are restricted to have only one orientation.

the superfluid gaps calculated by Ainsworth *et al.* (1989). In regions of vortex pinning, vortex lines move by thermal activation or quantum tunneling (Link, Epstein, and Baym 1993), and steady state is achieved for $\omega(r) < \omega_{cr}(r)$.

As one sees in Figure 2, ω_{cr} falls off precipitously in the lower density part of the inner crust. The steady-state configuration, as a consequence, *develops regions of negative vorticity, i.e.*, vortex circulation opposite to the rotation of the crust - exactly as is observed in laboratory Taylor-Couette flow of superfluid HeII between concentric rotating cylinders at sufficiently rapid rotation of the inner cylinder (Swanson and Donnelly 1987). To understand this behavior, we note that the steady-state configuration that results if all lines in the inner crust are oriented only positively is at best metastable. When all lines have positive orientation, the net number of lines $N_v(r)$ interior to cylindrical radius r increases monotonically with r, and thus according to equation (1) the superfluid angular velocity $\Omega_s(r)$ cannot decrease faster than r^{-2}. In the higher density part of the inner crust, where ω_{cr} increases with radius, the steady state lag ω tracks ω_{cr}, as illustrated by the thin solid lines in Figure 2. However, in the lower density part of the inner crust, ω can no longer follow ω_{cr}, but must remain effectively constant, as shown by the dashed lines (the r^{-2} dependence yields only a slight decrease in ω because the thickness of this part of the stellar crust is at most several percent of the stellar radius); the vortex density is zero in this region. This configuration does not have the lowest free energy. The free energy associated with superfluid flow is

$$F_{\rm rot} = E_{\rm kin} - L_s\Omega_{crust} = \frac{1}{2}\int \rho_s r^2\omega^2 d^3r - \frac{1}{2}I_s\Omega_{crust}^2, \tag{4}$$

where $E_{\rm kin}$ is the kinetic energy of the superfluid, L_s its angular momentum, and I_s its moment of inertia. In regions where $\omega < \omega_{cr}$ pinning forces can stabilize the flow even though $F_{\rm rot}$ is not a minimum. Were, however, ω greater than the local ω_{cr}, as, *e.g.*, shown by the dashed lines in Figure 2, the superfluid would lower the free energy by forming vortex lines of circulation opposite to the overall circulation of the superfluid, so as to bring ω below ω_{cr} (Epstein, Baym, and Link 1992). In the evolution of the star, the system remains in the lowest free-energy configuration by creating lines of negative vorticity, a situation analogous to creation of polarization pairs in a dielectric by an external electric field.

A negative vortex line can be created at the outer edge of the inner crust where the superfluid energy gap vanishes, or be nucleated from a small vortex ring in the interior. If the superfluid velocity were given by the dashed line in Figure 2, creation of a negative vortex line would lower the rotational energy of the superfluid by $\sim (M_{ld}/2\,m_n)\hbar\omega \gtrsim 4\times 10^{33}\,(M_{ld}/10^{-4}\,M_\odot)\,(\omega/100\ {\rm s}^{-1})$ MeV, where M_{ld} is the mass of the low-density portion of the inner crust (the vortex self-energy is insignificant). The energy required to nucleate a small vortex ring is $\sim \rho_s\kappa^3/(16\pi\Delta\nu) \sim 10\,(\rho_s/10^{13}\ {\rm g\ cm^{-3}})\,(\omega/100\ {\rm s}^{-1})^{-1}$ MeV. A plausible steady-state lag is illustrated by the thin solid lines in Figure 2. For $\omega(r)$ as sketched, the vortices have positive orientation for $d(\log\omega)/d(\log r) > -2$ and negative orientation for logarithmic slopes less than -2. As the superfluid slows its rotation, the negatively-oriented vortex lines creep *inward*, since they experience a Magnus force of opposite sign, and annihilate with positive lines moving outward. The evolution of the rotation of the inner-crust superfluid is thus more complex than hitherto assumed, involving the formation and annihilation of lines as well as the more familiar creep process, to which we now turn.

Vortex creep involves transitions in which a vortex line in some initial pinned configuration (state *i*) hops to a final state f due to thermal or quantum fluctuations. If we take $R_{i\to f}$ to be the transition rate from state i to state f and $\overline{\Delta r}_{i\to f}$ to be the mean radial shift of the line in this jump, then the radial component of the creep velocity is

$$v_{\rm cr} = \sum_f \overline{\Delta r}_{i\to f} R_{i\to f}. \tag{5}$$

The vortex creep velocity based on equation (5) is calculated in Link, Epstein, and Baym (1993). Since stiff vortex lines cannot bend sharply, vortex creep for stiffness $\tau \gg 1$ proceeds by large segments unpinning and repinning in concert. In the regions where vortex lines are flexible, nearby pinning bonds are weakly coupled, and the vortex line moves mostly by short segments breaking free and repinning. Link, Epstein, and Baym (1993) find that the creep rate for stiff lines ($\tau \gg 1$) is

$$v_{\rm cr} \simeq v_{\rm eff}\, l \left[\frac{2\pi k_B T_{\rm eff}}{A}\right]^{1/2} {\rm e}^{-A/k_B T_{\rm eff}}, \tag{6}$$

and for flexible vortices ($\tau \ll 1$) the creep rate is

$$\nu_{\rm cr} \simeq \nu_{\rm eff}\, l\, {\rm e}^{-A/k_B T_{\rm eff}}. \tag{7}$$

In these expressions $\nu_{\rm eff}$, the effective attack frequency, takes into account the oscillation frequency of the vortex line as well as statistical weight factors; $k_B T_{\rm eff}$, the effective thermal energy, equals the thermal energy at high temperatures and approaches the vortex zero-point energy at low temperatures; and $A(\omega)$ are the activation energies which were estimated in Link and Epstein (1991). For a path integral approach to vortex creep see Chau and Cheng (1991, 1992).

Although considerable progress has been made in understanding the dynamics of the neutron superfluid in the inner crusts of neutron stars, many perplexing astrophysics and physics issues remain. The greatest unresolved question is the origin of the glitches observed in at least nine pulsars (nine times in the Vela pulsar alone). It has been proposed that glitches are caused by the sudden unpinning of many vortex lines. After unpinning, the vortex lines rapidly flow outward, sharply reducing the angular momentum of the superfluid and spinning up the stellar crust (Anderson and Itoh 1975). Although this mechanism is appealing, the mechanism for the sudden unpinning has yet to be elucidated. Alternatively, glitches may be a result of crust cracking driven by superfluid stresses (Ruderman 1991). The observations of two spin-up time scales in a glitch in the Crab pulsar (Lyne 1992) pose a new challenge to explanations of glitches. Our understanding of the pulsar spin behavior after a glitch seems to be in better shape. The general theoretical picture, in which this behavior is ascribed to a gradual return to a rotational equilibrium between the inner crust superfluid and the rest of the star (Alpar *et al.* 1984), has stood up well to almost a decade of observations. Nevertheless, to relate the observed spin rates to specific physical processes in the star requires a much better understanding of microscopic physical processes. In particular, improved calculations of the interaction energy between a superfluid vortex and a nucleus are needed. We must also understand how a vortex pins to the nuclear lattice when the vortex core dimension is comparable to the lattice spacing. Recent results suggest that this poorly-understood regime is the most important one for explaining the observed post-glitch phenomena (Link, Epstein, and Baym 1993; Alpar and Pines 1992). Finally, we mention that the core neutron superfluid and its interaction with the proton superconductor are not understood nearly well enough and may hold some marvelous surprises.

This work was carried out under the auspices of the Department of Energy and supported in part by National Science Foundation Grant DMR 91-22385 and NASA's High Energy Astrophysics Theory and Data Analysis Program.

References

Ainsworth, T., Pines, D., and Wambach, J. 1989, *Phys. Lett.*, **B222**, 173.
Alpar, M. A. 1977, *Ap. J.*, **213**, 527.
Alpar, M. A., Anderson, P.W., Pines, D. and Shaham, J. 1984, *Ap. J.*, **276**, 325.
Alpar, M. A., Langer, S. A. and Sauls, J. A. 1984, *Ap. J.*, **282**, 533.
Alpar, M. A. and Pines, D. 1992, these Proceedings.
Alpar, M. A. and Ögelman, H. 1987, *Ap. J.*, **185**, 196.
Alpar, M. A. and Sauls, J. A. 1988, *Ap. J.*, **327**, 723.
Anderson, P. W. and Itoh, N. 1975, *Nature*, **256**, 25.
Bildsten L. and Epstein, R. I. 1989, *Ap. J.*, **342**, 951.
Chau, H. F and Cheng, K. S. 1991, *Phys. Rev. A*, **44**, 3478.
Chau, H. F and Cheng, K. S. 1992, these Proceedings.
De Blasio, F. V., Lazzari, G., Lazzari, C., and Broglia, R. A. 1992, in preparation.
Ding, K. Y., Cheng, K. S., and Chau, H. F. 1992, these Proceedings.
Epstein, R. I. and Baym, G. 1988, *Ap. J.*, **328**, 680.
Epstein, R. I. and Baym, G. 1992, *Ap. J.*, **387**, 276.
Epstein, R. I., Baym, G. and Link, B. 1992, in preparation.
Jones, P. B. 1992, *MNRAS*, in press.
Lindblom, L. and Mendell, G. 1992, in *The Structure and Evolution of Neutron Stars*, eds. D. Pines, R. Tamagaki, and S. Tsuruta (Addison-Wesley).
Link, B. and Epstein, R. I. 1991, *Ap. J.*, **373**, 592.
Link, B. and Epstein, R. I. 1992, in preparation.
Link, B., Epstein, R. I. and Baym, G. 1993, *Ap. J.*, in press.
Lyne, A. G. 1992, this Workshop.
Ruderman, M. 1976, *Ap. J.*, **203**, 213.
Ruderman, M. 1991, *Ap. J.*, **366**, 261.
Shibazaki, N. and Lamb, F. K. 1989, *Ap. J.*, **346**, 308.
Shibazaki, N. 1992, this Workshop.
Srinivasan, G., Bhattacharya, D. Muslimov, A. G., and Tsygan, A. I. 1990, *Current Sci.*, **59**, 31.
Swanson, C. E. and Donnelly 1987, *J. Low Temp. Phys.*, **67**, 185.
Van Riper, K. A. 1991, *Ap. J.*, **372**, 251.

Pulsar Glitches and Neutron Star Superfluidity - A 1992 Perspective

M. Ali Alpar[1] and David Pines[2]

[1]Physics Department, Middle East Technical University, Ankara
[2]Physics Department, University of Illinois

1 Introduction

The transfer of angular momentum from the pinned crustal superfluid in a neutron star to the observed crust proceeds by a radially outward current of quantized vortices. A detailed analysis of glitches and postglitch relaxation from the Vela pulsar indicates that continuous vortex flow (creep) and sudden discharges of pinned vorticity (glitches) are the two modes of angular momentum transfer, analogous to continuous dissipative currents in resistors and to sudden discharges in capacitors which do not transmit current otherwise. In each Vela pulsar glitch a large part of the spinup does not relax back: it constitutes the discrete part of the vortex current angular momentum transfer that couples the pinned superfluid to the external torque on the star. The Vela pulsar, which exhibits the same glitch behavior repeatedly, is in a steady state dynamical mode: vortex trap/discharge regions (capacitors) have already been established to near saturation. In a younger pulsar, such as the Crab, glitches indicate that the vortex capacitors are still being formed. We present this interpretation of the Crab and Vela pulsar glitch signatures in Section 2. In section 3, we discuss older pulsars, which are expected to be like Vela in their glitch signature, but experience glitches less frequently. The rates of dynamical energy dissipation in older pulsars are also of observational interest. In Section 4 we return to the Vela pulsar to discuss the implications of our model fits to its detailed postglitch behavior including the transients. These fits lead to interesting constraints on the neutron superfluid gaps and on the equation of state.

2 Postglitch Behavior of the Crab and Vela Pulsars

The glitches and postglitch behavior of the Crab and Vela pulsars have been studied in detail (for the Crab glitches see Lyne and Pritchard 1987, and Lyne 1992; for a discussion of the first six Vela glitches and references to the observations see Cordes *et al.* 1987; for model fits to the first eight glitches and references to observations

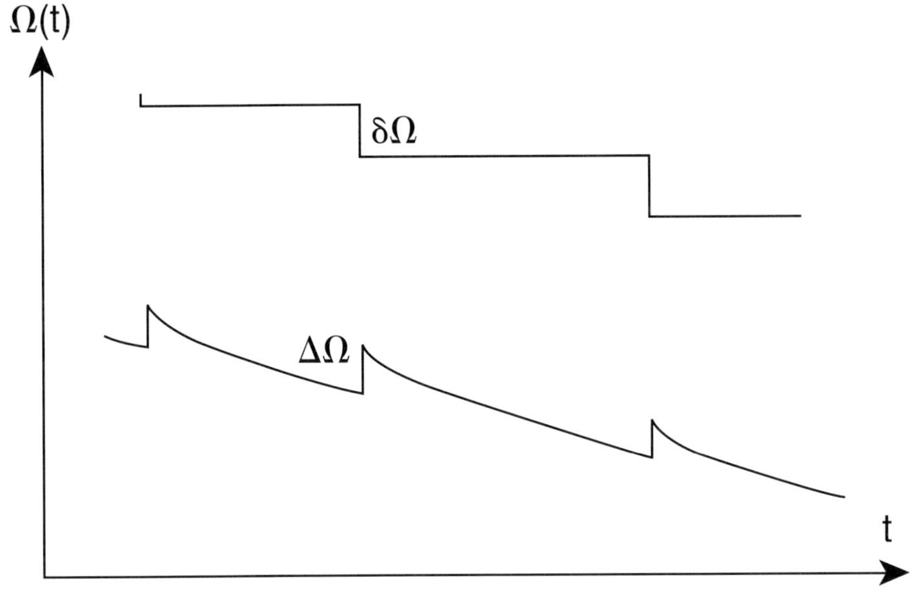

Fig. 1—A qualitative sketch of the Vela pulsar glitches showing the frequency Ω, of the observed outer crust as a function of time. The postglitch transients are not shown here in order to highlight the long term postglitch relaxation, which is exaggerated out of scale here. After each glitch there is an increase $\Delta\dot{\Omega}$ in the spindown rate, which subsequently heals at a constant rate $\Delta\ddot{\Omega}$ seen as a parabolic $\Omega(t)$. Note that when the next glitch arrives, there is a net remnant spinup with respect to the extrapolation from preglitch timing. The upper curve depicts the rotation rate in those regions B which relax only at glitches, by amounts $\delta\Omega$. Note that because there is steady state dynamical coupling between the two components, the average spindown rate is the same for both curves.

through the latest, ninth glitch see Alpar *et al.* 1992). The characteristic glitch behavior pattern in Crab is very different from that of Vela; we explain the difference below in terms of an evolutionary model.

Figure 1 sketches the long term glitch signature of the Vela pulsar. Part of the initial jumps in frequency and in the spindown rate decays promptly and exponentially following a glitch. In discussing the long term dynamics of the pulsar, we ignore these transients and concentrate on the remaining offsets, $\Delta\Omega$ and $\Delta\dot{\Omega}$. The observed long term features are:

1. Although the pulsar has a net spindown, the spinup introduced by each glitch mostly does not relax back to the extrapolation of preglitch behavior. Indeed after eight glitches there is a net unrelaxed remnant spinup from glitches: the glitches constitute a discrete spinup torque on the pulsar's observed outer crust (against the dominant external spindown torque) As there is no evidence that the external (electromagnetic) torque on the neutron star changes at glitches, an internal component of the star must be spinning *down* at glitches; the part of the glitch which does not relax is then just a discrete internal torque allowing some part of the neutron star interior to spin down in response to the external torque on the star.

2. There is a long-term offset $\Delta\dot{\Omega}$ in the spindown rate. This offset recovers as a linear function of time.

The three observed long-term parameters are the frequency jump, the (long-term) jump in spindown rate, and the second derivative of the rotation frequency, which appears as the slope of the recovery in spindown rate that proceeds as a linear function of time. Equations (1), (2), and (3) relate these observed quantities to the parameters of a simple model:

$$I\Delta\Omega = (I_B + I_A/2)\delta\Omega_s. \tag{1}$$

Equation (1) states that at the time of the glitch angular momentum is transferred to the crust as suddenly unpinned vortices move outward. The effective moment of inertia of the crust is practically equal to the total moment of inertia I of the star as the core superfluid is coupled rigidly to the observed crust on time scales of interest (Alpar, Langer and Sauls 1985). The superfluid rotation rate is thereby reduced by $\delta\Omega_s$ in regions with total moment of inertia I_B. $\delta\Omega_s$ is uniform in these regions because vortices move through but no vortex unpinning or repinning occurs. There are, in addition, regions I_A, which experience vortex unpinning and repinning. The average decrease in superfluid rotation rate is $\delta\Omega/2$ in such regions (Alpar *et al.* 1992).

$$\Delta\dot{\Omega}/\dot{\Omega} = I_A/I. \tag{2}$$

Equation (2) states that the glitch stops the continuous spindown of the superfluid in a part of the superfluid, the regions A. Before the glitch there is a continuous vortex creep process driven by the lag between the superfluid and crust rotation rates. The sudden vortex motion at the glitch reduces the superfluid rotation rate and the lag, and stops the vortex creep which depends exponentially on the lag. As the regions A thus get decoupled from the crust, the external torque acts on less moment of inertia and the spindown rate increases by a fraction I_A/I. In principle, the regions B responsible for the angular momentum transfer could also be involved in vortex creep and in changes in the spindown rate. Including I_B in Equation (2) leads to overestimation of the times between glitches by large factors, and other signatures of such a model are not present in the data (Alpar *et al.* 1992). Including only the regions A as regions whose vortex creep coupling is stopped at glitches leads to equation (2). Vortex creep then gradually restarts in the successive parts of the regions A at times $t = \delta\Omega_s(r)/|\dot{\Omega}|$; a spatially variable $\delta\Omega_s$ with constant $dI/d(\delta\Omega_s)$ corresponds to a constant density of unpinned vortices and translates to a healing of the initial long-term offset in the spindown rate linearly in time. The rate of this healing, measured as the second derivative of the rotation rate, is then related to the model parameters through

$$\Delta\ddot{\Omega}/|\dot{\Omega}| = I_A/(It_g). \tag{3}$$

Here $t_g = \delta\Omega_s/|\dot{\Omega}|$, where $\delta\Omega_s$ is uniform in the regions B and is also the maximum $\delta\Omega(r)$ for the regions A. This time scale t_g is an estimate of the time to refill the vortex traps; so it is also the estimated time between glitches. Table 1 gives t_g obtained from the solution of equations (1),(2) and (3) after each of the first 8 Vela pulsar glitches. The agreement with the observed times between the glitches is encouraging, being about

TABLE 1

Date	2/28/69	8/29/71	9/28/75	7/3/78	11/11/81	8/10/82	7/12/85	12/24/88
$t_g\,(d)$	1624	1375	1036	1371	616	1485	972	1422
$t_{obs}\,(d)$	912	1491	1009	1227	272	1067	1261	907

20 per cent on the average, whereas the scatter in the observed interglitch times is 33 per cent.

The basic feature of these fits is that there is a component of the neutron star interior, with moment of inertia I_B, which transfers angular momentum to the crust at glitches (eq. [1]) but does not contribute to the continuous creep and the spindown rate between glitches (Alpar 1992, Pines and Alpar 1992). These regions are therefore not involved either in the changes induced on the spindown rate by glitches (eq. (2)) or in the subsequent relaxation (eq.(3)). The moments of inertia obtained from postglitch fits are consistent with this model. I_B/I is between 8×10^{-3} and 1.2×10^{-2}, and I_A/I is between 4.7×10^{-3} and 7.2×10^{-3} for the first eight glitches (Alpar *et al.* 1992).

To reiterate, the pinned superfluid couples to the external torque through a radially outward current of vortex lines. This vortex current transfers angular momentum to the crust and dissipates energy. Regions of superfluid, through which there is a radially outward vortex motion, spin down and thereby exert a spinup torque on the crust. The regions A where there is a continuous vortex current (vortex creep) are analogous to resistances in a circuit, whereas the regions B are analogous to capacitors. There is no vortex transport through such regions except at glitches (vortex discharges). Inward of such regions there are vortex traps, which we label collectively as the regions A, that withhold vortices from the regions B until a critical number of stored vortices is reached. Between glitches, the regions A fill in gradually by the continuous vortex creep process, which is stopped temporarily by the vortex discharge at a glitch. The stopping of creep decouples the regions A from the crust and hence effects a sudden increase in the spindown rate as the external torque now acts on less moment of inertia. The increase $\Delta\dot{\Omega}$ in the spindown rate heals gradually as the vortices start to move in and fill the trap. The observed nature of this healing, at a constant rate, is expressed in equation (3). Outward of the regions B there are vortex depletion regions from which vortex creep zones further out derive their current. These are the opposite plates of the capacitor, and their contribution to the spindown rate has the same signature as the vortex traps: they are also parts of the regions A in our model.

After each of the first eight glitches analyzed to date, the Vela pulsar exhibits the same behavior. The parameters of the fits are quite similar. The Vela pulsar seems to have settled into a steady mode of dynamical behavior, of which the glitches form a major part. A key point is that the regions B exchange angular momentum only in glitches, do not have a continuous spindown, and therefore do not contribute to the jumps in the spindown rate of the crust. These regions are already well established vortex capacitors in the present era. Further, the moment of inertia I_B is comparable to the overall moment of inertia of the superfluid crust, so the process of formation of the vortex traps/capacitors is practically over and saturated in the Vela pulsar. The

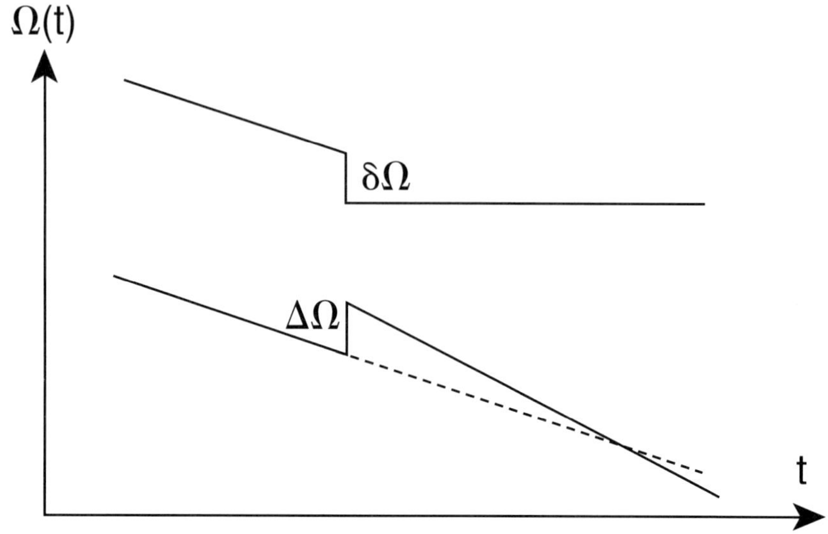

Fig. 2—A qualitative sketch of $\Omega(t)$ for the Crab pulsar glitches. The postglitch transients are not shown, and the scale is exaggerated. The increase $\Delta\dot{\Omega}$ in the spindown rate does not relax. Note the overshooting with respect to the extrapolation from preglitch timing. The upper curve depicts the rotation rate in the regions B, which spin down by creep before the glitch but become a "newly formed capacitor" after the glitch, so that there is no vortex creep and spindown after the glitch. Note that the overshooting occurs here within an internal torque model, but requires that the coupling between the two components changes at the glitch. As this occurs in several Crab glitches, the dynamical coupling between the two components is not (yet) in a steady state.

only exceptions to this picture are two small disturbances $2'$ and $9'$ which occurred after the large glitches 2 and 9, and are similar in size to the Crab pulsar glitches. These small glitches probably signal the last bits of vortex trap being formed. It will be interesting to test whether a persistent shift in the spindown rate associated with the small glitches $2'$ and $9'$ can be resolved in the Vela timing data.

By contrast, the Crab glitches leave a long term offset in the spindown rate that seems to be permanent (see Fig. 2). The spindown rate becomes more negative at the glitch, signaling that the external spindown torque acts on less moment of inertia after the glitch than it did before. In contrast with the situation in the Vela pulsar, the change in the spindown rate does not relax back. This "persistent shift" in the spindown rate causes the pulsar crust's rotation frequency to decrease to values less than the extrapolation from the preglitch behavior. In terms of internal torque models, in which the crust is coupled to an interior component of the neutron star (the pinned crust superfluid in the present models), such overshooting is impossible unless the parameters of the coupling have changed at the glitch (Link, Epstein, and Baym 1992). *Such a change is the signature of a newly formed vortex capacitor*: a region through which there was a vortex current prior to the glitch is transformed during the glitch to a region controlled by a newly formed vortex trap which holds and stores the incoming vortices. The times between glitches in the Crab pulsar do not agree with the estimates of the vortex unpinning model alone (Alpar *et al.* 1981). Thus, the glitches may be triggered by crustquakes which lead to sudden vortex motion (Ruderman 1991). We conjecture that the crustquakes form new vortex traps by introducing a new arrangement of pinning centers in the lattice. The absence of a $\Delta\ddot{\Omega}$ was given in equation (3) for Vela indicates that the moment of inertia I_A of the newly formed vortex trap region ("capacitor plates") is negligible. The persistent shift in the spindown rate gives the moment of inertia I_B in the newly formed vortex free region controlled by this trap. I_B/I, labeled $I_{n\ell}/I$ in earlier work was found to be 2×10^{-4} and 4×10^{-4} in the 1975 and 1989 Crab pulsar glitches (Pines and Alpar, 1992), along with $\delta\Omega/\Omega \cong 3.3 \times 10^{-5}$ and 2.1×10^{-5} (Pines and Alpar, 1992).

Comparing this I_B with the moment of inertia I_B in the *established* capacitor regions in the Vela pulsar, we find that it will take about 10^2 Crab glitches, or 10^3 years at the rate of a Crab glitch every 10 years, to jointly form the established capacitor regions with moment of inertia I_B well before the 10^4 years age of the Vela pulsar. Once formed, a capacitor region does not allow vortex currents and does not contribute to the persistent shifts in subsequent Crab glitches which indicate the formation of new vortex capacitors at other locations. Thus the evolution of dynamical behavior from Crab's age to Vela's reflects a gradual formation and linking of vortex capacitor regions percolating to a final large capacitor network with effective moment of inertia I_B.

The number of vortices N involved in a glitch is related to the decrease $\delta\Omega$ in the superfluid angular velocity by

$$N\kappa = 2\pi R^2 \delta\Omega, \tag{4}$$

where $\kappa = h/2m_n \cong 2 \times 10^{-3} cm^2 s^{-1}$ is the quantum of vorticity (h is the Planck constant and m_n the neutron mass) and R is the star's radius. $N \sim 1.6 - 4 \times 10^{13}$ for the Vela pulsar glitches and a similar number $N \sim 1.2 - 2 \times 10^{13}$ is obtained for the Crab pulsar glitches. This may be a critical number of vortices accommodated in a single trap before a discharge. The Crab pulsar glitches reflect the discharge of this critical number of vortices through one newly established vortex capacitor, whereas in the Vela pulsar the same critical number of vortices percolates through a network of vortex capacitors that collectively make up the regions B.

The Vela pulsar microglitches $2'$ and $9'$ may well indicate that occasional Crab-like glitches are still taking place. In this connection it will be very interesting to see if there are persistent shifts associated with these glitches. It is interesting that the distribution of glitch sizes from the entire pulsar sample seems to follow a power law, whereas individual pulsars with repeated glitches tend to exhibit glitches of the same order of magnitude in a given pulsar. The present statistics are consistent with the glitch size saturating at the size $\Delta\Omega/\Omega \sim 10^{-6}$ typical of the Vela pulsar, with most older pulsars exhibiting Vela-size glitches. We can then predict that both the postglitch behavior and the rate of energy dissipation expected from the dynamical activity of the Vela pulsar extend to old pulsars with appropriate scaling of their interglitch intervals and energy dissipation rates as well as their postglitch relaxation time scales with their spindown rates and temperatures.

3 Glitch Activity and Energy Dissipation in Old Pulsars

Taking the viewpoint that the Vela pulsar has reached dynamical maturity implies that similar dynamics applies in the older pulsars too, with appropriate scaling. The extent of vortex creep regions, vortex traps and vortex free regions, the numbers of vortices discharged in each glitch, and therefore the glitch magnitudes in older pulsars should be similar to those in Vela. The critical conditions for vortex discharge are achieved as a result of spindown. The time between successive glitches is therefore longer in inverse proportion as the spindown rate. It is unlikely that one will be able to see successive glitches from an old pulsar and establish its systematics. Pulsars with spindown rates comparable to Vela's, say within an order of magnitude, present a much better chance, with interglitch intervals of a few years to a few decades. Indeed pulsars with high glitch activity, like PSR 1737-30, have spindown rates comparable to that of Vela.

The response of the regions A gives an anomalous second derivative and braking index. If the vortex traps are already saturated in Vela, we expect older pulsars to have similar I_A and $\delta\Omega$ in these regions; as in the Vela pulsar the "anomalous" braking index ν is always larger than the braking index ν_0 because of the pulsar braking torque so that this typical postglitch signature of Vela may be observable in older pulsars between glitches. The glitches are less frequent in older pulsars, but if they are like Vela, the $\Delta\ddot{\Omega}$ will be present all the time from one glitch to the next. Indeed, it may have been observed as an anomalous braking index from certain old pulsars (Boynton 1992). Such observations provide an important test of the hypothesis that the dynamics of the Vela pulsar is typical of older pulsars when appropriately scaled. $\Delta\ddot{\Omega}$, like the second derivative $\ddot{\Omega}$ from the dipole torque on the pulsar, is proportional to $(\dot{\Omega})^2$, so eventually this effect will become unobservable, though being of order $I_A/I\Omega/\delta\Omega$ times $\ddot{\Omega}$, the anomalous $\Delta\ddot{\Omega}$ should dominate in measurements of the second derivative.

Postglitch relaxation times, which depend on the temperature as well as the spindown rate, also become longer as a pulsar ages. The postglitch relaxation of an old pulsar in the observable time scales of hours, days and months will display the response of superfluid regions whose relaxation time scales in the Vela pulsar are too short to be observable, whereas the regions whose postglitch relaxation is observed in the Vela pulsar will have very long relaxation times in the postglitch data of an old pulsar. This makes it difficult to make a direct comparison of postglitch fits from pulsars of different ages. We shall not give the detailed expressions for the postglitch relaxation times here (Alpar *et al.* 1992). The main result of vortex creep theory for the evolution of postglitch response is that, as a pulsar ages, there should be less moment of inertia involved in linear response (simple exponential relaxation) (Alpar *et al.* 1989). An old pulsar with moment of inertia 10^{-2} involved in exponential relaxation would be incompatible with the predictions of the vortex creep model.

An important consequence of the vortex creep theory is that in the presence of regions of nonlinear response there will be an enhanced rate of energy dissipation, $\dot{E} = I_p\omega_{cr}|\dot{\Omega}|$, instead of the much smaller $\dot{E} = I_p\dot{\Omega}^2\tau$ typical of linear response with relaxation time τ. Here $\omega_{cr} \sim 10^{-1} - 10^{-2}$rad s^{-1} is a critical lag between superfluid and crust rotation rates and I_p, the moment of inertia of the pinned superfluid involved in the dissipation. Upper limits (Alpar *et al.* 1987) and possible detections of this energy dissipation as soft X-rays emitted from the neutron star surface give information on the neutron star structure. This emission is detectable in pulsars which have already cooled off their initial heat content. Such pulsars must be older than Vela, but if they are too old, the energy dissipation rate is too low for detection. In any case the pulsar must be relatively nearby and must have a small column density of neutral hydrogen. With likely values of $I_p = 10^{43} gm\ cm^2$ and $\omega_{cr} = 10^{-1}$rad s^{-1}, the luminosity expected from a pulsar with $\dot{\Omega} = 10^{-11}$rad s^{-1} is 10^{31}erg s^{-1} with a corresponding surface temperature of 0.3 kev on a neutron star of 10 Km radius. This dissipation rate is a possible observational test of the theory, but once again it is applicable to relatively young pulsars, with spindown ages of order 10^5 yrs.

Next we turn to the energy dissipated in the glitch itself. Associated with the angular momentum transfer, equation (1), an energy $I_B \delta\Omega^2$ must be dissipated. In the Vela pulsar, emerging from the surface on thermal diffusion time scales that are comparable to the interglitch time, this gives an average luminosity of $I_B \delta\Omega\ |\dot{\Omega}|$, of order 10^{31}erg s^{-1}, which is smaller than the standard cooling luminosity of the Vela pulsar and therefore not discernible in the observed thermal luminosity. In older pulsars the cooling luminosity is much less, whereas the average dissipation luminosity associated with each glitch is of order $I_B \delta\Omega\ |\dot{\Omega}| t_g / t_{diff}$ where t_{diff} is the thermal diffusion time for the energy dissipated in the glitch. t_{diff} does not depend on the pulsar's age; for t_{diff} of the order of a few years, the thermal luminosity after a glitch, though being on only t_{diff}/t_g of the time, would be enhanced by a factor of a hundred with respect to the continuous rate of energy dissipation in a pulsar with t_g of the order of a hundred years. Calculations of the thermal diffusion time scale through the neutron star crust and observational searches for a postglitch thermal signal in soft X rays are of interest in this connection (Van Riper, Epstein, and Miller 1991).

4 Constraints on the Equation of State and on Superfluidity

Detailed fits to postglitch relaxation yield information on the moments of inertia in various components of the pinned crust superfluid, on the pinning parameters, and on the neutron superfluid gap function. This fitting has been done most extensively for the Vela pulsar, where independent fits to the data following the first eight glitches are consistent with the same or similar values of the parameters (Alpar *et al.* 1992). The most important results are constraints on the equation of state and on the neutron superfluid gap.

The postglitch dynamics of the Vela pulsar involves several different components of exponential relaxation as well as the long-term response of the vortex traps, with moment of inertia I_A, and the inference, from the angular momentum transfer at glitches, of the vortex free regions with moment of inertia I_B. Adding to I_A and I_B the moments of inertia involved in the exponential relaxation components, we obtain a lower bound on the total moment of inertia of the pinned superfluid in the neutron star crust. This quantity is also a lower bound on the fractional moment of inertia in the entire crust, which must be at least 2.9×10^{-2} of the total moment of inertia in the star (Alpar *et al.* 1992). For a given mass, soft equations of state give neutron star models with thinner crusts. Assuming that the Vela pulsar has a mass of $1.4M_{\odot}$, one finds that soft equations of state like those based on the Reid potential are ruled out (Pines 1991, Datta and Alpar 1992). As a large part of the moment of inertia in the crust superfluid is associated with the regions I_B, this inference essentially follows from the premise that the Vela pulsar is in dynamical steady state, with part of the pinned crust superfluid transferring angular momentum only at glitches. We thus obtain an important constraint on the equation of state of dense matter at neutron star core densities, above the density of nuclear matter, from a simple observation.

Constraints on the neutron superfluid gap function are more model dependent. In the vortex creep model there are two different pinning regimes, the weak and superweak pinning regimes; there is also an independent classification of vortex creep, as linear and nonlinear creep. A detailed description of the model for the pinning strengths and the creep regimes is given in earlier work (Alpar *et al.* 1992). The model fits to the Vela postglitch data require that the transition from weak to superweak pinning occurs at a lower density in the crust than the transition from nonlinear to linear response. Expressing this requirement with the criteria for both transitions as functions of the superfluid energy gap implies that the gap function must have a peak value of at least 1 MeV, in agreement with the recent family of gap functions calculated by Ainsworth, Pines and Wambach (1989), and ruling out gap functions that are significantly weaker. Gap functions that are much stronger are indicated only if superweak pinning is characterized by vortex core diameters that are larger than two lattice spacings in the crust crystal, rather than the more natural scale of a single lattice spacing. Such gaps would lead to high-energy dissipation rates which are inconsistent with observational upper limits (Alpar *et al.* 1987). In any case it is remarkable that the requirements of consistent timing fits lead to gap functions in exactly the range of the most sophisticated theoretical calculations.

5 Conclusions

The vortex creep model applied to postglitch timing data from the Vela pulsar yields important information on the equation of state, that it cannot be soft, and on the neutron superfluid gap, that the current calculations are in the right range.

The comparison of the Crab and Vela pulsars' glitches and postglitch behavior suggests that the dynamical behavior of pulsars evolves, from juvenile behavior exemplified by Crab to mature coupling exemplified by Vela. The evolution involves gradual establishment of a steady state vortex transfer circuit through the establishment of vortex traps (capacitors) to saturation. The conclusion that Vela is already in a dynamical steady state leads to several predictions for older pulsars.

This research was supported by a NATO collaborative research grant CRG 900074, by NSF grant INT89-18665, and NSF THY86-00377, and by the Scientific and Technical Research Council of Turkey grant TBAG1049.

References

Ainsworth, T. L., Wambach, J., and Pines, D. 1989, *Phys. Lett. B*, **222**,173.
Alpar, M. A. 1992, in *The Structure and Evolution of Neutron Stars*, D. Dines, R. Tamagaki, and S. Tsuruta, eds. (Redwood City: Addison-Wesley)
Alpar, M. A., Chau, H. F., Cheng, K. S., and Pines, D. 1992, *Ap. J.*, submitted.
Alpar, M. A., Langer, S. A., and Sauls, J. A. 1984, *Ap. J.*, **282**, 533.
Alpar, M. A., Kiziloğlu, Ü., Ögelman, H. and Pines, D. 1987, *Astr. Ap.*,**177**,101.
Boynton, P. E. 1992, Private Communication
Cheng, K. S., Alpar, M. A., Pines, D., and Shaham, J. 1988, *Ap. J.*, **330**, 835.
Cordes, J. M., Downs, G. S., and Krause-Polstorff, J. 1988, *Ap. J.*, **330**, 841.
Datta, B. and Alpar, M. A. 1992, to be submitted.
Link, B., Epstein, R. I. & Baym, G. 1992, *Ap. J. Lett.*, **390**, L21
Lyne, A. G. 1992, this Workshop
Lyne, A. G. and Pritchard, R. S. 1987, *M. N. R. A. S.*, **229**, 223.
Pines, D. 1991, in *Neutron Stars: Theory and Observation, Proc. NATO ASI, Agia Pelagia, Greece, 1990*, Pines, D. and Alpar, M. A. 1992, in *The Structure and Evolution of Neutron Stars*, D. Dines, R. Tamagaki, and S. Tsuruta, eds. (Redwood City: Addison-Wesley)
Ruderman, M. 1991, *Ap. J.*, **382**, 587 (RIII).
Van Riper, K. A., Epstein, R. I., and Miller, G. S. 1991, *Ap. J. Lett.*, **381**, L47.

Constraints on the Nuclear Matter Equation of State from Pulsar Glitches

Bennett Link, Richard I. Epstein, and Kenneth A. Van Riper

Los Alamos National Laboratory

Abstract

We study the post-glitch response of four pulsars to obtain lower limits on the total moment of inertia of the inner crust superfluid. In contrast to previous work, our constraints are independent of the form of the crust-superfluid coupling. We conclude that the superfluid must comprise $\gtrsim 0.8\%$ of the total moment of inertia of the star. This constraint rules out the softest equations of state.

1 Introduction

Substantial progress has been made in determining the properties of matter near nuclear density by extrapolation from experimental physics, yet many questions remain regarding the state of the denser matter deep in the neutron star interior (see Baym 1992). The equation of state of matter above nuclear density determines to a large extent the overall properties of a neutron star, such as the average mass density, moment of inertia, radius, and the maximum stable mass. The thermal evolution and associated surface emission of a neutron star are very sensitive to the equation of state.

An interesting question is to what extent can we use studies of pulsar timing irregularities to probe the liquid interior? In the inner crust and core the neutrons are believed to form a superfluid (Migdal 1959; Hoffberg *et al.* 1970). Sudden transfers of angular momentum from the superfluid to the crust are likely responsible for glitches (Anderson and Itoh 1975), while post-glitch recoveries, occurring over days to years, have been explained as the gradual return to rotational equilibrium between the superfluid and the rest of the star (Alpar *et al.* 1984). General properties of the superfluid can be deduced directly from observations. Here we use pulsar timing data to set a lower limit on the moment of inertia of the superfluid in the inner crust and discuss the implications for the properties of matter above nuclear density.

Most of the moment of inertia of a neutron star is contained in the superfluid neutrons of the inner crust and core. The superfluid velocity field is characterized by singular regions of vorticity, or *vortex lines.* Frictional coupling between the crust and superfluid, in which angular momentum is exchanged, is accompanied by the radial motion of vortices. The time scale for momentum transfer between the two components varies enormously with stellar density. The portion of the superfluid that responds to changes in the crust rotation rate on time scales of days to years, the "moderately-coupled superfluid", probably resides in the inner crust where vortex interactions with nuclei inhibit radial vortex motion. In the core, however, where superfluid neutrons co-exist with superconducting protons, entrainment of protons by the flowing neutrons induces magnetic moments about the vortices; the magnetized vortices scatter electrons efficiently. Following a glitch, the core superfluid achieves corotation with the crust within minutes (Alpar *et al.* 1985); over longer time scales, the solid crust and core behave as a single rigid body.

The frictional coupling between the inner crust superfluid and the crust is poorly understood, and depends on the manner in which vortices move through the inhomogeneous environment of the nuclear lattice. Interactions between vortex lines and nuclei can pin the lines to the nuclear lattice (Alpar 1977; Alpar *et al.* 1984; Epstein and Baym 1988; Link and Epstein 1991) thereby impeding vortex motion. Nevertheless, pinned vortex lines can move by the process of vortex creep, whereby they unpin from one configuration and repin in another (Alpar *et al.* 1984; Chau and Cheng 1991; Link, Epstein, and Baym 1992). On the other hand, if vortex pinning is sufficiently strong, forces on pinned vortices lead to crust cracking, and the movement of fragments or plates of the crust changes the vortex distribution (Ruderman 1991). At the opposite extreme, in some regions of the inner crust vortex pinning may be very weak, and vortex lines move in near corotation with the crust (Jones 1990).

Our goal is to constrain the properties of the moderately-coupled superfluid, whose response we observe following glitches, without making special assumptions about the nature of the crust-superfluid coupling.

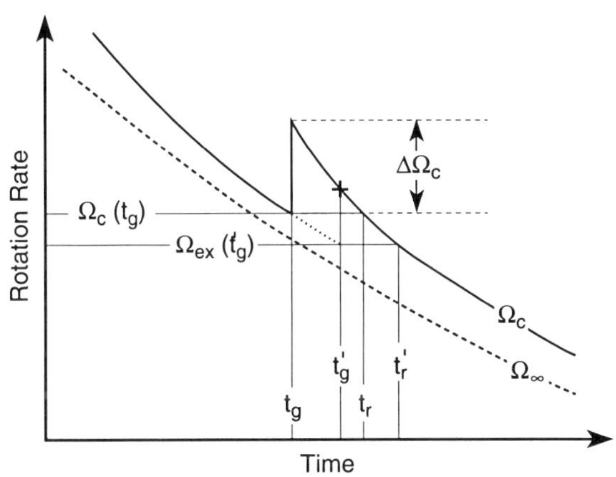

Fig. 1—Schematic evolution of the crust angular velocity Ω_c through a glitch. After a glitch at time t_g, the pulsar recovers its pre-glitch rotation rate at t_r. The first post-glitch measurements (shown by the +) are at t_g' and the corresponding recovery time is t_r'. The dashed line represents the asymptotic spin rate Ω_∞.

2 Approach

Typical spin behavior of a glitching pulsar is illustrated in Figure 1. In the absence of glitches or other violent changes, the deceleration of the crust $\dot{\Omega}_c(t)$ and the superfluid $\dot{\Omega}_s(t,\vec{r})$ approach an asymptotic rate $\dot{\Omega}_\infty(t)$. The moderately-coupled superfluid slows by virtue of its drag with the crust, and hence, $\Omega_s(t,\vec{r}) \geq \Omega_c(t)$. When there is no glitch activity, all of the superfluid is slowing and transferring angular momentum to the crust, *i.e.*, $\dot{\Omega}_s(t,\vec{r}) \leq 0$. During a glitch at time t_g (see Fig. 1), the crust increases its angular velocity by an amount $\Delta\Omega_c$, and then decelerates more rapidly than before the glitch.

After time t_r, the *recovery time*, the crust falls below its pre-glitch rotation rate. When $t_g < t < t_r$, the crust spins faster than before the glitch and possibly spins faster than some of the superfluid. During this interval, the crust may impart angular momentum to the superfluid [Alpar *et al.* (1990) argue that such a situation occurred during the 0.4 d response following the December 1988 glitch in the Vela pulsar (Flanagan 1990)]. After t_r, however, the crust must once again have a lower angular velocity than all of the superfluid so that $\dot{\Omega}_s(> t_r) < 0$ everywhere. These ideas are illustrated in Fig. 2.

We define $I \equiv I_c + I_s$ to be the total effective moment of inertia of the star, where I_s is the moment of inertia of the moderately-coupled superfluid, and I_c is that of the normal matter of the crust plus that of the liquid core. Since the superfluid is losing angular momentum for times $t > t_r$, it follows from angular momentum conservation that (Link, Epstein, and Van Riper 1992)

$$\frac{I_s}{I} > 1 - \frac{\dot{\Omega}_c(> t_r)}{\dot{\Omega}_c(< t_g)}, \tag{1}$$

independent of the details of the crust-superfluid frictional coupling, as long as the external torque does not increase. A lower limit on I_s can be obtained for any time after t_r prior to the next glitch; however the strongest constraint is obtained for t very close to t_r, when $|\dot{\Omega}_c|$ is still relatively large. The lower limit on I_s/I derived from eq. [2] is less than the fractional change in $\dot{\Omega}_c$ across the glitch, as some of the discontinuity in the spin-down rate decays by $t > t_r$.

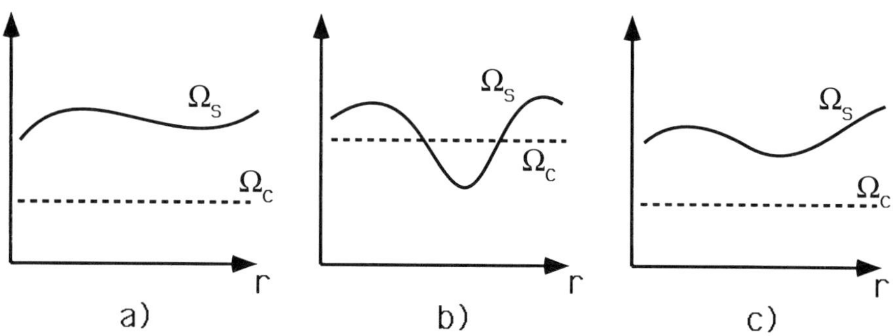

Fig. 2—An example of the superfluid rotational velocity distribution in the star a) just before a glitch, b) just after the glitch, and c) at t_r. The superfluid is losing angular momentum in a), may be either gaining or losing angular momentum in b), and is again losing angular momentum in c).

To account for the uncertainty in the glitch time, we use the following procedure. We extrapolate the fit of $\Omega_c(t)$ for the pre-glitch data to the time of the first post-glitch measurement at time t'_g; we denote this quantity by $\Omega_{\rm ex}(t'_g)$ (see Fig. 1). The corresponding recovery time, t'_r, is found from $\Omega_c(t'_r) = \Omega_{\rm ex}(t'_g)$. Since t'_r always exceeds t_r, we use $\dot{\Omega}_c(t'_r)$ for $\dot{\Omega}_c(> t_r)$ in eq. [1]. For the quantity $\dot{\Omega}_c(< t_g)$ we

TABLE 1
Glitch Parameters

PSR	$\Omega/2\dot{\Omega}$ yr	date	$10^8(\Delta\Omega/\Omega)_g$	$10^2(\Delta\dot{\Omega}/\dot{\Omega})_g$	$t'_r - t'_g$ d	$10^2 I_s/I >$
0531+21	1.2 K	2/75[a]	3.8	0.25	0.04	0.25
(Crab)		8/86[b]	0.92	0.25	0.008	0.23
0833-45	11 K	2/69[c]	234	1.0	19	0.63
(Vela)		8/71[c]	205	1.5	17	0.72
		9/75[c]	199	1.1	16	0.81
		7/78[c]	306	1.8	25	0.70
		10/81[c]	114	0.84	9	0.67
		10/81[d]	115	4.9	9	0.70
		8/82[c]	205	2.3	17	0.65
		8/82[d]	206	6.3	17	0.74
		7/85[d]	160	1.4	13	0.68
		12/88[e]	181	7.7	15	0.76
0355+54	0.6 M	1/85[f]	0.56	0.18	2.3	0.16
0525+21	1.5 M	1/74[g]	0.13	0.46	1.4	0.37

[a]Lohsen (1981); [b]Lyne and Pritchard (1987); [c]Cordes *et al.* (1988); [d]McCulloch *et al.* (1987); [e]McCulloch *et al.* (1990); [f]Lyne (1987); [g]Downs (1982). In using the fitting formulae of Cordes *et al.* (1988), the pre-glitch $\dot{\Omega}_c$ extrapolated to t'_g was used in eq. (1).

evaluate $\dot{\Omega}_c(t)$ at the time of the last pre-glitch observation. In practice $\dot{\Omega}_c(< t_g)$ varies so slowly that the precise time used is unimportant.

Table 1 shows lower limits on I_s/I obtained from data on selected glitches in PSRs 0531+21 (Crab), 0833-45 (Vela), 0355+54, and 0525+21. We also tabulate the approximate glitch date (which is $\le t'_g$), the discontinuities in Ω_c and $\dot{\Omega}_c$ across each glitch (indicated with the subscript g), and the interval $t'_r - t'_g$. In the cases of the Crab pulsar and PSR 0525+21, recovery occurs so soon after the glitch (as reflected in the small values for $t'_r - t'_g$) that little of the initial discontinuity in $\dot{\Omega}_c$ decays, and the limit on I_s is $I_s/I \gtrsim (\Delta\dot{\Omega}_c/\dot{\Omega}_c)_g$. In the Vela pulsar, on the other hand, much of the discontinuity in $\dot{\Omega}_c$ decays before t'_r. Our limits on I_s/I account for the uncertainties in the glitch parameters. For the Vela pulsar, we experimented with the errors given with the fitting parameters and found that the corresponding changes in the limits on I_s were less than $\sim 10\%$.

While PSR 1737-30 is one of the most active glitching pulsars known, we have not included it in this study. Glitches are so frequent that values of $t_r - t_g$ typically exceed the average time between glitches (Michel *et al.* 1990). Neither has the large 1986 glitch of PSR 0355+54 been included in Table 1 since the published data on its spin rate do not include the entire recovery period up to t_r.

The 1975 giant glitch of the Vela pulsar yields the most stringent constraint on the moment of inertia of the moderately-coupled superfluid: $I_s/I \ge 0.0081$. Although there is a large range in $(\Delta\dot{\Omega}_c/\dot{\Omega}_c)_g$ observed for glitches in the Vela pulsar, the constraints we obtain for I_s all lie within $\simeq 25\%$ of one another. Moreover, all glitches listed in Table 1 give limits on I_s within a factor of ~ 5 of each other. This striking similarity among pulsars of considerably different ages (> 1000-fold variation) with glitches of greatly varying amplitudes (> 2000-fold variation), provides compelling evidence that post-glitch behavior is due to physical processes common to all glitching neutron stars. We now compare these observational limits on I_s to those implied by calculations of neutron star structure and inner crust superfluidity.

In principle, the moderately-coupled superfluid can exist anywhere in the neutron star inner crust where there are superfluid neutrons. Throughout most of the inner crust, extending from the density of neutron drip (4.3×10^{11} g cm^{-3}) to the density of nuclear matter ($\rho_0 = 2.8 \times 10^{14}$ g cm^{-3}), the free neutrons form an isotropic $^1S_0$ superfluid, while near nuclear density and above, the neutrons form a $^3P_2$ superfluid (Hoffberg *et al.* 1970). The $^1S_0$ neutron superfluidity vanishes at densities near ρ_0, but uncertainties in the 3-body nucleon interaction potential as well as calculational complexities make a precise determination of the maximum density difficult (see, *e.g.*, Ainsworth *et al.* 1989). Within current theoretical uncertainties, $^1S_0$ superfluidity may extend up to ρ_0 (Ainsworth *et al.* 1989), or may persist to densities of only $\simeq 0.27\rho_0$ (Chen *et al.* 1986). The situation is also uncertain for the $^3P_2$ superfluid (Hoffberg *et al.* 1970), which may not penetrate the inner crust at all. Furthermore, if there are regions in the inner crust where vortex lines are effectively immobilized, for example by strong pinning to the nuclear lattice, the superfluid in these regions would not exchange angular momentum with the crust. To allow for the possibility that only a fraction of the superfluid of the inner crust contributes to I_s, we treat the maximum density to which superfluidity extends as a parameter ρ_{max}.

In general, for a given mass stellar models with softer equations of state have thinner inner crusts with smaller moments of inertia. In Figure 3, we show the total moment of inertia of the crustal superfluid as a function of neutron star mass for selected maximum densities for superfluidity ($\rho_{max}/\rho_0 = 1.0$, 0.7 and 0.5) and the following equations of state: BPS (Reid soft-core nucleon potential and hyperons; Baym *et al.* 1971), P (Reid potential and pure neutron matter; Pandharipande 1971), FP (two- and three-nucleon interactions; Friedman and Pandharipande 1981), BJ (Reid potential with short-range repulsion and hyperons; Bethe and Johnson 1976), CCLR (two-body Levinger and Simmons potential; Cohen *et al.* 1970), and PS (nucleon coupling to a mean scalar field; Pandharipande and Smith 1975).

In Figure 3 our lower limit $I_s/I \geq 0.0081$ is shown. Pines and Alpar (1991) have obtained estimates of I_s using their multi-zone model for neutron star vortex creep. They derive $I_s/I \simeq 0.028$ from the 1978 Vela glitch; this estimate is shown in Figure 3 for comparison. We see that for $\rho_{max}/\rho_0 = 1.0$, the model-independent constraint on I_s is barely consistent with the Vela pulsar being a 1.4 $M_\odot$ neutron star with a soft equation of state (*e.g.*, BPS), and the model-dependent estimate excludes this possibility. For the more plausible case of $\rho_{max}/\rho_0 = 0.7$, soft equations of state are ruled out for any reasonable mass of the Vela pulsar, whereas a moderately stiff equation of state (*e.g.*, FP) is consistent with the model-dependent estimate. For superfluidity extending to $0.5\rho_0$, moderately stiff equations of state are allowed by the model-independent constraint, although not by the model-dependent estimate (BPS, P, and FP all excluded).

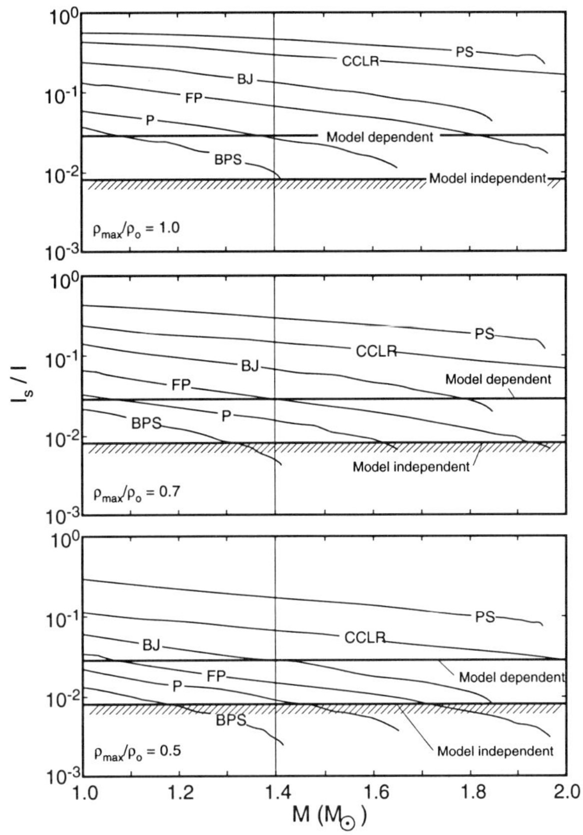

Fig. 3— Total moment of inertia of the inner crust superfluid as a function of neutron star mass for selected equations of state and maximum densities ρ_{max} for superfluidity. The model-independent constraint, obtained here, and the model-dependent estimate, of Pines and Alpar (1991), are shown. Values of I_s that fall in the cross-hatched regions are ruled out.

3 Discussion

Pulsar timing data afford a powerful means of studying the internal structure of neutron stars. We have obtained lower limits on the moment of inertia I_s of the superfluid component decoupled from the crust during a glitch. Aside from direct neutron star mass determinations, this approach provides, by virtue of its model-independence, the strongest astrophysical constraint on the equation of state of matter above nuclear density. We find $I_s/I \gtrsim 0.8\%$, regardless of the detailed form of the crust-superfluid frictional coupling. Moreover, unless the moderately-coupled superfluid consists of virtually all the inner crust free neutrons ($\rho_{max}/\rho_0 \simeq 1$), or the Vela pulsar is significantly less massive than 1.4 $M_\odot$, equations of state for neutron star matter as soft as BPS are ruled out. The BPS equation of state, which has a maximum stable mass of $1.41 M_\odot$, is also inconsistent with the lower mass limit for PSR 1913+16 of $1.439 M_\odot$ (Wolszczan 1991).

The inclusion of three-body nucleon interactions has led overall to a stiffening of the nuclear equation of state. Recent structure calculations yield maximum neutron star masses of 1.8–2.2 $M_\odot$ (see, *e.g.*, Wiringa *et al.* 1988). However, Brown (1988) has argued that the equation of state may be much softer than generally thought and suggests that the maximum stable neutron star mass is near $1.5 M_\odot$. Brown *et al.* (1992) propose that the failure to observe a pulsar in SN 1987A indicates that the core collapsed to form a black hole, as would be the case were the equation of state very soft. The theoretical situation is far from resolved, and we note that the method presented here may be capable of settling this issue. Stronger constraints should become possible as calculations of neutron superfluidity improve and as more timing data becomes available.

This work was carried out under the auspices of the U. S. Department of Energy.

References

Ainsworth, T. L., Wambach, J. and Pines, D. 1989, *Phys. Lett. B*, **222**, 173.
Alpar, M. A. 1977, *Ap. J.*, **213**, 527.
Alpar, M. A., Anderson, P. W., Pines, D. and Shaham, J. 1984, *Ap. J.*, **276**, 325.
Alpar, M. A., Langer, S. A., and Sauls, J. A. 1985, *Ap. J.*, **282**, 533.
Alpar, M. A., Pines, D., and Cheng, K. S. 1990, *Nature*, **348**, 707.
Anderson, P. W. and Itoh, N. 1975, *Nature*, **256**, 25.
Baym, G. 1992, these Proceedings.
Baym, G., Pethick, C., and Sutherland, P. 1971, *Ap. J.*, **170**, 299.
Bethe, H. A. and Johnson M. 1976, *Nucl. Phys. A*, **230**, 1.
Brown, G. E. 1988, *Nature*, **336**, 519.
Brown, G. E., Bruenn, S. W., and Wheeler, J. C. 1992, *to appear in Comments on Astrophysics.*
Cohen, J. M, Langer, W. D., Rosen, L. C., and Cameron, A. G. W. 1970, *Ap. Space Sci.*, **6**, 228.
Chau, H. F. and Cheng, K. S. 1991, *Phys. Rev. A*, **44**, 3478.
Chen, J. M. C., Clark, J. W., Krotscheck, E. and Smith, R. A. 1986, *Nucl. Phys. A.*, **451**, 509.
Cordes, J. M., Downs, G. S. and Krause-Polstorff, J. 1988, *Ap. J.*, **330**, 847.
Downs, G. S. 1982, *Ap. J.*, **257**, L67.
Epstein, R. I., and Baym, G. 1988, *Ap. J.*, **328**, 680.
Flanagan, C. S. 1990, *Nature*, **345**, 416.
Friedman, B. and Pandharipande, V. R. 1981, *Nucl. Phys. A*, **361**, 502.
Hoffberg, M., Glassgold, A. E., Richardson, R. W., and Ruderman, M. 1970, *Phys. Rev. Lett.*, **24**, 175.
Jones, P. B. 1990, *MNRAS*, **243**, 257.
Link, B. and Epstein, R. I. 1991, *Ap. J.*, **373**, 592.
Link, B., Epstein, R. I. and Baym, G. 1992, *Ap. J.*, *in Press.*
Link, B., Epstein, R. I., and Van Riper, K. A. 1992, *submitted.*
Lohsen, E. H. G. 1981, *Astron. Astrophys. Suppl.*, **44**, 1.
Lyne, A. G. 1987, *Nature*, **326**, 569.
Lyne, A. G. and Pritchard, R. S. 1987, *MNRAS*, **229**, 223.
McCulloch, P. M., Hamilton, P. A., McConnell, D. and King, E. A. 1990, *Nature*, **346**, 822.
McCulloch, P. M., Klekociuk, A. R., Hamilton, P. A. and Royle, G. W. R. 1987, *Aust. J. Phys.*, **40**, 725.
Michel, F. C., Bland Hawthorn, J. and Lyne, A. G 1990, *Mon. Not. R. Astr. Soc.*, **246**, 624.
Migdal, A. B. 1959, *Nucl. Phys.*, **13**, 655.
Pandharipande, V. R. 1971, *Nucl. Phys. A*, **174**, 641.
Pandharipande, V. R. and Smith, R. A. 1975, *Phys. Lett. B*, **59**, 15.
Pines, D. and Alpar M. A., in Structure and Evolution of Neutron Stars, eds. Pines, D., Tamagaki R., and Tsuruta, S. (New York: Addison-Wesley, 1991).
Ruderman, M. 1991, *Ap. J.*, **366**, 261.
Wiringa, R. B., Fiks, V., and Fabrocini, A. 1988, *Phys. Rev. C*, **38**, 1010.
Wolszczan, A. 1991, *Nature*, **350**, 688.

Neutron Vortex Creep With Tension In Neutron Star

H. F. Chau and Kwong-Sang Cheng

Department of Physics, University of Hong Kong

Abstract

We have calculated the correlated creep velocity of a vortex line with line tension, which is pinned to a perfect simple cubic lattice and creeps out under the action of a biased potential by thermal activation. Recently we also extended the result to the case where the biased potential, the line tension and the pinning energy of the line are all functions of position. Although evaluating the exact correlated creep velocity is difficult, a good approximate form can be found. We can show that the Anderson and Kim thermal activation formula in the vortex creep model is still valid after suitable modification even though their uncorrelated creeping assumption is incorrect. We apply our formalism to study the motion of vortex lines in the crustal region of neutron stars. The model results suggest that the steady state configuration of vortex lines should form a set of hyperbolic curves because the stronger pinning regions (also higher tension) are located in the inner regions of the crust. Perturbations, like star quakes, can trigger a large number of vortex lines to unpin from the stronger pinning region and the tension forces the vortex lines to move a macroscopic distance. Such mechanism may be responsible for the glitches of pulsars.

1 Introduction

It is well known that the magnetic fluxoids in a type II superconductor can pin to certain locations of the sample (say the lattice imperfection, for example) because it is energetically favorable for the fluxoid to pass through these pinning centers. However, the fluxoids can still hop through the pinning energy barriers by means of thermal assisted flux creeping (Anderson 1963, Anderson and Kim 1964). According to the model of Anderson and Kim, different parts of the flux line are treated as independent and therefore uncorrelated entities. The correlation effect between different parts of the fluxoids is not handled seriously. Let us take a small part of the pinned flux line out. Suppose that the pinning energy of this elementary segment is E_P, and if there is a biased external force whose effect can be represented by means of a potential, is acting on the line. Suppose that the work done by the force when the line is activated and hops to the adjacent site is E_{ext}, then the average creep rate of the line can be shown to be $2w\ \exp(\ E_P/kT)\sinh(E_{ext}/kT)$, where w is the microscopic attempt rate of the line (Anderson 1963, Anderson and Kim 1964). However, there is a great conceptual problem in the derivation: suppose we have a straight flux line at the beginning and we let it creep out of the superconducting sample. The standard derivation between different parts of the line will be ever increasing, and therefore the total length of the flux line will be an ever increasing quantity which is impossible. Energy (called the length energy) is required to create the core of the magnetic flux line and hence a line tension is present whenever the flux line is elongated. We have reexamined the theory of flux creeping under the presence of line tension earlier by means of a simple model using the path integral method (Chau and Cheng 1991). Recently, we have also extended this work to the case where all the physical parameters, such as the pinning energy, are position dependent (Chau and Cheng, 1992). We shall apply the correlated creep formula to study the problem of pulsar glitches.

2 Calculation of the Correlated Creep Velocity

Let us model the system by an infinite perfect simple cubic crystal lattice where the distance between two nearest neighbors is of unit length. A vortex line is allowed to pass through the crystal roughly along the z-direction. An external force (Lorentz force for type II superconductor and Magnus force for superfluid), which can be expressed in terms of a biased potential E_{ext}, is imposed on this system along the x-axis. Moreover, the line is energetically favorable to pass through the lattice points, called pinned to the lattice. The pinning energy per site is denoted by E_P. Finally, the energy required to create a unit length of vortex line is given by E_L. Should E_L be large compared with kT, different parts of the line will tend to correlate with each other and move as a whole. The (correlated) creep velocity of the line is just the sum of the contributions from individual microscopic transition pairs with a proper statistical weight (Chau and Cheng 1991; Epstein, Link, and Baym 1992). In fact, we have shown that under the high tension limit, when E_L is large in comparison with kT, the creep velocity of the line is given by (Chau and Cheng 1991)

$$v_{drift} \approx 8wL \exp(-E_P/kT) \sinh(E_{ext}/2kT), \tag{1}$$

where w is the microscopic attempt rate, T is the temperature of the system, the lattice spacing is set to unit length, and $L = \exp[-(\sqrt{2}-1)E_L/kT]$. So the creep velocity of the line is of the same form as the original one proposed by Anderson and Kim except that the pinning energy E_P is replaced by the sum $E_P + (\sqrt{2}-1)E_L$. We cannot measure the pinning energy directly using the creep formula because the pinning energy is always accompanied by the length energy. Besides, we have also showed that the vortex line will creep out gradually as in Figure 1 (Chau and Cheng 1991).

Moreover, the above result can be generalized to the case where E_{ext}, E_L, E_P, T and w are functions of position. In other words, we consider the case where the vortex line passes through a number of layers of crystal so that in each layer the physical quantities associated with the vortex line may be different. Let us illustrate the idea by considering a two-dimensional lattice. Extension to higher dimension is just straight forward. In the high tension limit, we know that the possible local shapes of the vortex line are given by Figures 2a–2c. The probability of occurrence of the three local shapes

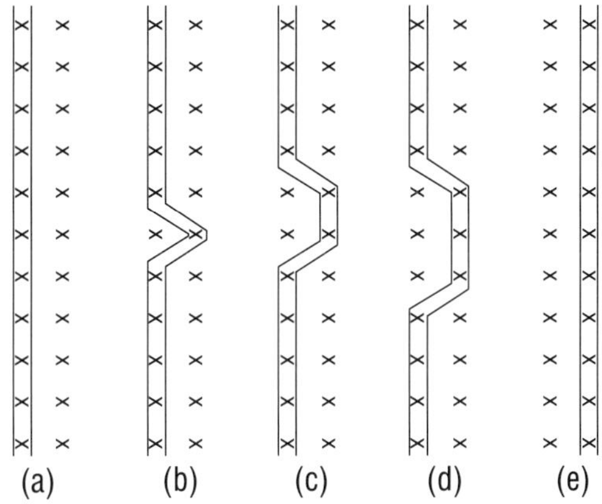

Fig. 1—Schematic representation of the propagation of a vortex line. (a) an arbitrary pinned point of the vortex line jumps to a new pinning site, (b) either one of the two nearer pinning points jumps to the corresponding nearest pinning point, (c)-(e) the process of (b) repeats itself until the entire line propagates one unit length.

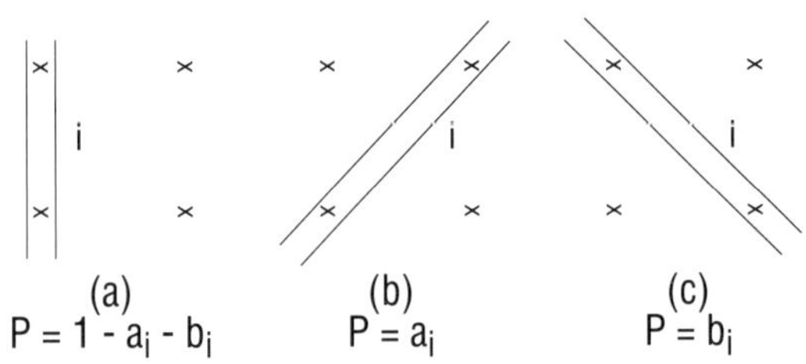

Fig. 2—The possible local shapes of the vortex line between two successive pinning sites in the high vortex tension limit in two dimensions. The probabilities of occurrence of (a), (b) and (c) are $1 - a_i - b_i$, a_i and b_i respectively.

is given by $1 - a - b, a$ and b respectively. Hence, the free energy F of the line is given by (Chau and Cheng, 1992)

$$F = \sum_i E_i - T_i S_i = \sum_i T_i \{ (1 - a_i - b_i) \ln(1 - a_i - b_i) + a_i \ln(a_i) + b_i \ln(b_i) - (a_i + b_i) \ln L_i \} \tag{2}$$

where the first three terms are the contributions from entropy of the lines and the last term is the length energy contribution. Also the sum over i denotes the sum of contributions from each line segment. Note that all the E_i,T_i,S_i and hence a_i, b_i are in general function of position. We have also dropped all the pinning energy contributions because we assume that the line must pass through the same number of lattice points so that every pinning configuration has the same amount of total pinning energy. Moreover, the equilibrium configuration of the line is subject to the constraint that the lines have to creep in a correlated way, so that the mean creep velocity of each part of the line must agree. In the high tension limit, this is given by (Chau and Cheng, 1992)

$$v \equiv v_i \approx \mathrm{w}_i \exp(-Ep_i/kT_i) \left[G_i (\gamma_i + \beta_i)(\gamma_{i+1} + \alpha_{i+1}) - \frac{1}{G_i} (\gamma_i + \alpha_i)(\gamma_{i+1} + \beta_{i+1}) \right] \quad \forall i, \tag{3}$$

where w_i is the microscopic attempt rate of the line at pinning site i, and v is the correlated (mean) creep velocity of the vortex line. Also $G_i = \exp[(F_{Ei} + F_{Ei+1})/4kT_i]$,

$\alpha_i = a_i/L_i^{1/2}$, $\beta_i = b_i/L_i^{1/2}$ and $\gamma_i = (1 - a_i - b_i)L_i^{1/2}$. Therefore, the equilibrium configuration of the vortex line can be obtained by minimizing the free energy F subject to the correlated creep constraints as in equation (3).

Although the exact solution of this constrained minimization problem is almost impossible, some of the particular cases, which are also physically interesting, can be solved readily. Suppose that the system, which is isothermal, is divided into a number of portions labeled from i=1 to n. The total number of pinning sites is in the ratio $\ell(1):\cdots:\ell(n)$. Moreover, region 1 has a much slower locally correlated creep velocity than all the other regions. Then the entire macroscopic line has to move as a whole at the expense of the lines in the regions other than 1 that have to be curved (See Fig. 3). In fact, it can be shown that the correlated creep velocity v of the line is given by

$$\left[\frac{4L_1\chi_1}{v}\right]^{1/2} - 1 \approx \pm\frac{\chi_1}{\ell(1)} \sum_{i>1} \frac{\ell(i)}{\chi_i} \ln\left[\frac{L_i v}{\chi_i}\right] \approx 0, \tag{4}$$

where $\chi_i = w_i \exp(-E_{Pi}/kT)(G_i - 1/G_i)$. In other words, the correlated creep velocity $v \approx 4L_1\chi_1$ is close to that of the locally correlated one of the slowest moving region (Chau and Cheng, 1992).

3 A Possible Picture of Pulsar Glitch

It is commonly accepted that the glitches of pulsars are related to the superfluid unpinning in the inner crust of neutron stars. The angular momentum of the superfluid is thereby transferred to the crust and speed up of rotation of the star. The post-glitch evolution of the rotation of the star reflects how the pinned neutron superfluid responds to such a catastrophic event (Anderson and Itoh 1975; Alpar, Anderson, Pines, and Shaham 1984; Pines and Alpar, 1985). However, there are still some subtle problems in this model. First, the neutron vortices in the inner crust must pass through a number of regions with different densities and distance from the rotational axis. Therefore, different parts of the vortex may experience different pinning forces, Magnus forces, and tension forces. All previous models do not take this into account seriously. The use of the Anderson-Kim formula for thermal vortex (or fluxoid) creep in such a highly complex system where the vortex is likely to pass through a number of different regions is questionable. Second, the method of triggering a glitch is not clear. The conventional explanation of a catastrophic superfluid unpinning caused by the increase of Magnus force does not fit into a self-consistent scheme: once we allow thermal creeping (which is necessary for explaining the post-glitch evolution), the system finds it very difficult to accumulate enough stress to trigger a glitch. On the other hand, some authors used rather exotic mechanisms such as the strong pinning region (Ruderman 1991a, 1991b, 1991c; Link and Epstein 1991). However, there is still not enough observational sign for the strong pinning region to show up in the post-glitch relaxation of pulsars. Third, in the conventional picture, vortices have to unpin and travel a macroscopic distance in order to account for the observed sizes of glitches. However, the energy required to unpin a vortex with macroscopic length almost at the same time may be a difficult task. Moreover, the Kelvin circulation theorem forces the unpinned free vortex to corotate with the crustal superfluid. Thus it is not very clear why it will predominantly move out radially. Now we are going to show that with the introduction of vortex tension by means of the length energy together with a natural triggering mechanism, all the above problems may be solved.

The superfluid gap energy Δ and the Fermi energy E_F are functions of density ρ and hence radial position of the star. The pinning energy of each site is given by (Anderson and Itoh 1975; Alpar, Anderson, Pines, and Shaham 1984; Pines and Alpar, 1985)

$$E_P = \frac{3}{8\pi}\left[\left(\frac{\Delta^2 n}{E_F}\right)_{out} - \left(\frac{\Delta^2 n}{E_F}\right)_{in}\right]V \tag{5}$$

where n is the number density of neutron superfluid, the subscripts "in" and "out" denote the values inside and outside the site, and V is the volume of the pinning site. The site can be nucleus or lattice imperfections (Chau and Cheng, 1991; Anderson and Itoh 1975; Alpar, Anderson, Pines, and Shaham 1984; Pines and Alpar, 1985). In a similar way, the biased potential is just $E_{ext} = \rho r \kappa \omega \xi b$, where κ is the vorticity quantum, r is the distance from the rotational axis, b is the spacing between successive pinning sites, ξ is the coherence length of the neutron superfluid, and ω is the angular velocity difference between the superfluid and the crust (Chau and Cheng, 1992; Anderson and Itoh 1975; Alpar, Anderson, Pines, and Shaham 1984; Pines and Alpar, 1985). Finally, the length energy is $E_L = \alpha\pi\xi^2 \frac{3\Delta^2 n}{8\pi E_F} b$, where α is the geometrical factor which is of order of unity for strong or weak pinning regions whereas $\alpha \ll 1$ for superweak pinning regions (Anderson and Itoh 1975; Alpar, Anderson, Pines, and Shaham 1984; Pines and Alpar, 1985). Therefore the values of E_P and E_L are very sensitive to the density and hence to the radial position of the vortex line (Chau and Cheng, 1992). Also, the typical temperature of the inner crust, which is isothermal, is about 10keV for young pulsars. It should be much less for older ones. But the gap energy is of the order of a few tenths of MeV (Chen, *et al.* 1986; Ainsworth, Wambach, and Pines, 1989). With the typical values of $\xi(\approx 10\text{fm}), n(\approx 10^{35}\text{cm}^{-3}), b(\approx 10\text{fm})$ and $E_F(\approx 100\text{MeV})$, the length energy of a neutron vortex is so high that the high tension limit is a good approximation (Chau and Cheng, 1991, 1992; Anderson and Itoh 1975; Alpar, Anderson, Pines, and Shaham 1984; Pines and Alpar, 1985). Without going into the detailed calculations, which are model and equation of state dependent, we can argue some important features of the vortex lines in the inner crust of the neutron star.

Because E_P is sensitively dependent on the position of the vortex in the star and χ depends exponentially on E_P, whenever the vortex passes through a very slow moving region or what we called a stronger pinning region, equation (4) tells us that the correlated creep velocity of the vortex will be more or less equal to the local creep velocity of this stronger pinning region. Besides, the rest of the vortex is highly curved. In this way, the correlated creep velocity is represented by a single, slow moving region, which is in the form of equation (1). Therefore, the use of creep velocity in the Anderson-Kim form in the derivation of post-glitch relaxation is justified even though the original derivation did not take correlation effect into account. Also the steady state vortex density is given by the continuity equation:

$$\nabla \cdot (n_v v) = 0, \tag{6}$$

where v is the correlated creep velocity of the line and n_v is the number density of neutron vortices. So there is a natural pile up of vortex lines in the stronger pinning regions. Physically, there are a number of cases for the vortices in the inner crust to give rise to such a pile up: if the vortex line passes through both strong and weak pinning regions or similarly passes through weak and superweak pinning regions, the correlated creep velocity of the line will be roughly equal to the local correlated creep velocity in the former region. More important, the vortex is highly curved in the latter regions (see Fig. 3). Therefore, such a configuration is not uncommon in neutron stars. By means of an external triggering mechanism, say a star quake, part of the vortices may unpin. Although vortex lines can be unpinned from both the stronger and weaker pinning regions, the tension of the stronger pinning region will move the line radially outward. On the other hand, the part of the vortex lines in the weaker pinning region remains unchanged because the tension is along the line itself (see Fig. 3). Therefore the vortex tension will drive the vortex to move a macroscopic distance in the radially outward direction before stopping or dissipating in the inner crust boundary (see Fig. 3). In fact, the parts needed to be unpinned are those in the stronger pinning region. Compared with the original picture, the energy required to trigger a glitch is much reduced. Moreover, it thus avoids the problem of simultaneously unpinning a vortex with a total length comparable to the stellar diameter. equation (6) also tells us that the stronger pinning region also acts as a reservoir of vortices. In fact, this observation is also consistent with the recent interpretation of the post-glitch relaxation of the Vela pulsar, especially the region I_B (Alpar, Cheng, and Pines 1989; Alpar, Pines, and Cheng 1990; Alpar, Chau, Cheng, and Pines 1992; Alpar and Pines 1992).

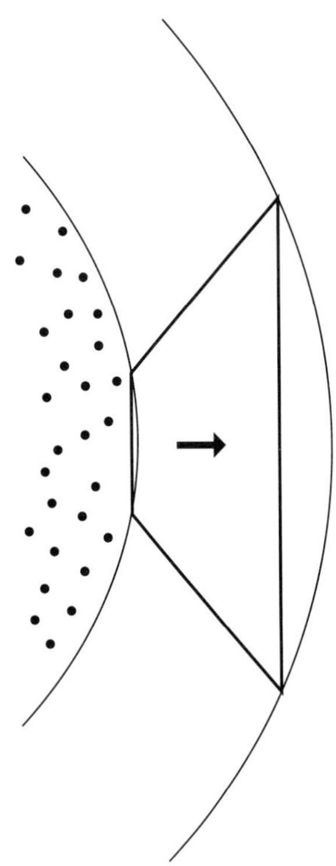

Fig. 3—The equilibrium shape of the vortex line (heavy dark line) before and after unpinning in the inner crust of the neutron star. Note that the vortex is highly curved because of the presence of a stronger pinning region (the dotted region). Hence, once the vortex is unpinned in the stronger pinning region in the neutron star, the vortex tension will drive the line to move through a macroscopic distance before stopping or dissipating in the stellar boundary.

4 Summary

With the help of the correlated creep velocity formula that we have recently derived, we suggested that the correlated creep velocity of the neutron vortex lines is governed by that of the stronger pinning region alone. Moreover, the rest of the vortex is highly curved. The number density of vortices at this region is also high. It is therefore a good candidate for the trigger region of glitch. Moreover, the post-glitch response in such a picture is consistent with the recent Vela post-glitch observations.

We would like to thank Professors M. A. Alpar, G. Baym, and D. Pines for their interesting comments and discussions. This work is supported in part by the UPGC grant of Hong Kong and the LANL conference grant. Moreover, one of the authors (H. F. C.) would also like to thank the Croucher Foundation for the Studentship support.

References

Ainsworth, T. L., Wambach, J., and Pines, D. 1989, *Phys. Lett. B*, **222**, 173.
Alpar, M. A. and Pines, D. 1992, these Proceedings.
Alpar, M. A., Anderson, P. W., Pines, D., and Shaham, J. 1984, *Ap. J.*,**276**, 325.
Alpar, M. A., Chau, H. F., Cheng, K. S., and Pines, D. 1992, *Ap. J.*, submitted.
Alpar, M. A., Cheng, K. S., and Pines, D. 1989, *Ap. J.*, **346**, 823.
Alpar, M. A., Pines, D., and Cheng, K. S. 1990, *Nature*, **348**, 707.
Anderson, P. W. 1963, *Phys. Rev. Lett.*, **9**, 309.
Anderson, P. W., and Itoh, I. 1975, *Nature*, **256**, 25.
Anderson, P. W., and Kim, Y. M. 1964, *Rev. Mod. Phys.*, **36**, 39.
Chau, H. F., and Cheng, K. S. 1991, *Phys. Rev. A*, **44**, 3478.
Chau, H. F., and Cheng, K. S. 1992, *Phys. Rev. B*, submitted.
Chen, J. M. C., *et al.* 1986, *Nucl. Phys. A*, **451**, 509.
Epstein, R. I., Link, B. K., and Baym, G. 1992, these Proceedings.
Link, B. K., and Epstein, R. I. 1991, *Ap. J.*, **373**, 592.
Pines, D. and Alpar, M. A. 1985, *Nature*, **316**, 27.
Ruderman, M. A. 1991a, *Ap. J.*, **366**, 261.
Ruderman, M. A. 1991b, *Ap. J.*, **382**, 576.
Ruderman, M. A. 1991c, *Ap. J.*, **382**, 587.

Effect of Interpinning in Core Field Decay of Neutron Stars

K. Y. Ding, K. S. Cheng, and H. F. Chau

Department of Physics, University of Hong Kong

Abstract

We propose a new mechanism in which the magnetic field in the core of a neutron star can be rapidly driven out from the core on a time scale $<< 10^6$ years but still maintains a residual field $\geq 10^8$ Gauss on the Hubble time scale ($\sim 3 \times 10^{10}$ years). Such a driving force arises from the fact that the magnetized vortex lines of the $^3P_2$ neutron superfluid are effectively interpinned by the magnetic fluxoids of the $^1S_0$ proton superconducting fluid. Nevertheless, the neutron vortex lines and the proton fluxoids are not always moving out with the same speed despite the strong pinning energy, $\sim$10 MeV per intersection.

1 Introduction

The evolution of neutron star magnetic fields has been an interesting topic ever since the discovery of radio pulsars (Hewish *et al.* 1968). Statistical analysis of the observed pulsar characteristics showed evidence for exponential decay of magnetic fields with a time scale of a few million years (Gunn and Ostriker 1970; Lyne, Anderson and Salter 1982; Lyne, Manchester and Taylor 1985). Besides, there are other alternatives to explain the observed pulsar characteristics (Kundt 1981; Blair and Candy 1989; Taam and van den Heuvel 1986; Blondin and Freese 1986; Romani 1990; Narayan and Ostriker 1990).

In view of the recent observation of neutron stars in binary systems, and millisecond pulsars (Kulkarni 1986; Kulkarni, Djorgovski and Klemola 1991), it has been suggested that the magnetic fields of the binary millisecond pulsars ($\sim 10^9$ G) must persist for at least 10^9 years if their progenitors are low mass X-ray binaries (van den Heuvel, van Paradijs and Taam 1986). It comes to the question that IF one accepts the "recycled" scenario of the millisecond pulsars and the current models of neutron stars in binary systems, what are the mechanisms of the field decay which allow the field to decay exponentially but maintain the field strength about $10^8 - 10^{10}$ Gauss for at least 10^9 years? It is not our intention here to clear-up all these models, instead, our goal is to construct a self-consistent model of the magnetic field evolution by assuming that millisecond pulsars are recycled pulsars whose magnetic fields do decay from 10^{12} Gauss to $10^8 - 10^{10}$ Gauss. We shall concentrate on the mechanism of magnetic field decay in the core and assume that the crustal field decays exponentially which is consistent with the observed data. We shall examine the dynamics of the flux tubes and the vortex lines in the core of a neutron star by assuming they are uniformly distributed inside the core for simplicity and adopt the present understanding of the equations of state of nuclear matter. We will show how our model can quantitatively lead to a self-consistent picture of the observed magnetic field decay time scale and the residual fields of millisecond pulsars.

2 Interpinning and Creeping in the Core

The superconducting protons in the core are expected to form a Type II superconductor, the microscopic proton currents circulating flux lines producing the stellar magnetic field. Each flux tube carries a quantized flux, $\Phi_\circ$, confined to the radius of the flux tube which is of order of the London penetration depth, λ. The magnetic field of each proton flux tube is determined by

$$B_p = 1.9 \times 10^{16} x_p \rho_{15} \ln(\lambda/\xi)\ \mathrm{G}, \tag{1}$$

where $\xi \le \lambda/\sqrt{2}$ is the coherence length, ρ_p is the proton density, $x_p = \rho_p/\rho$ is the fractional concentration of proton and ρ_{15} is the core density in units of 10^{15} gcm^{-3}.

The superfluid drag effect induces a charge supercurrent and results in a magnetic field contribution associated with each neutron vortex line of magnitude (Alpar *et al.* 1984)

$$B_n = 3.7 \times 10^{16} \left(\frac{m_p^*}{m_p}\right)^{-1} \left(\frac{\delta m_p^*}{m_p}\right) x_p \rho_{15}\ \mathrm{G}, \tag{2}$$

where m_p^* is the proton effective mass.

Then when these two kinds of lines are to cross over each other, they will encounter a magnetic energy barrier, E_p, given by (Chau, Cheng and Ding 1991)

$$E_p = \frac{2\vec{B_p} \cdot \vec{B_n}}{8\pi}(\alpha_g \lambda)(\pi \lambda_*^2) = 9.1 \times 10^7 \alpha_g \left(\frac{\delta m_p^*}{m_p}\right)(x_p \rho_{15})^{1/2} \ln(\lambda/\xi) \cos\chi\ \mathrm{eV}, \tag{3}$$

where χ is the angle between $\vec{B_n}$ and $\vec{B_p}$, and α_g is a geometrical factor of order of unity. Pinning effects between the two kinds of lines have been discussed by other authors (Sauls 1989; Srinivasan *et al.* 1990; Sedrakyan and Sedrakyan 1991) but these pinning energies are as a result of proton density perturbations in the center of a vortex line and are at least three orders of magnitude smaller than the magnetic pinning energy considered here (Sedrakyan and Sedrakyan 1991 have included the superfluid drag effect to calculate E_p due to proton condensation; their result is an order of magnitude smaller than ours).

3 Motion of the Neutron Vortices in the Core

The neutron vortex lines are interpinned with the proton flux tubes which are coupled to the crust; we expect both kinds of lines to be rotating with the crust with an angular velocity, Ω_c. As a result, any difference between Ω_c and Ω_s will result in a response force, namely the magnus force, acting on the neutron vortices which tends to drive them out. This force is given per unit length by

$$\vec{f_m} = \rho \mathcal{K} r (\Omega_s - \Omega_c) \hat{e_r} = \rho \mathcal{K} r \omega \hat{e_r}, \tag{4}$$

where ω is the angular velocity lag between Ω_s and Ω_c, $\mathcal{K} = \frac{h}{2m_n}$ is the quantized vorticity, and r is the distance of the vortex line from the rotation axis of the star. This force can be effectively treated as a bias potential and reduces the magnetic energy barrier, E_p, by

$$\Delta E = f_m b_p \lambda_*, \tag{5}$$

where $b_p = \frac{1}{\sin\chi}(\frac{\Phi_\circ}{B_c})^{1/2}$ is the mean distance between pinning sites. Hence, the neutron vortices can creep radially outward by thermal activation in response to the dipole radiation torque. Finally, a steady state lag is established (Chau, Cheng and Ding 1991)

$$\omega_\infty = \frac{kT\omega_{cr}}{E_p} \sinh^{-1}\left[\frac{r}{4\Omega_\circ \nu_\circ} \mid \dot{\Omega}_\infty \mid \exp(E_p/kT)\right] \approx \omega_{cr} (\text{ for } kT/E_p << 1), \tag{6}$$

where B_{c12} is the core magnetic field in units of 10^{12} G, $\nu_\circ$ is the microscopic velocity of a vortex line, $\dot{\Omega}_\infty$ is the steady state spin down rate, such that $\dot{\Omega}_s = \dot{\Omega}_c = \dot{\Omega}_\infty$, and ω_{cr} is the maximum lag that can be sustained by the pinning force, defined by

$$\omega_{cr} = \frac{E_p}{\lambda_* \rho \mathcal{K} b_p r} = 8.7 \times 10^{-2} x_p \alpha_g r_6 \left(\frac{\delta m_p^*}{m_p}\right) \left(\frac{m_p^*}{m_p}\right)^{1/2} B_{c12}^{1/2} \ln(\lambda/\xi) \sin 2\chi \text{ rads}^{-1}. \tag{7}$$

The radially outward velocity of the neutron vortices, v_n, at the core-crustal boundary is governed by the spin down rate of the superfluid. Since $\mid \frac{\omega_\infty}{\Omega_c} \mid << 1$, assume the deceleration torque is due completely to dipole radiation, then

$$v_n(t) = \frac{R_c k(t) \Omega_s(t)^2}{2} \tag{8}$$

with $k(t)$ defined as

$$k(t) \equiv \frac{2}{3} \frac{B_s(t)^2 R^6 \sin^2\chi}{Ic^3} \tag{9}$$

where B_s is the surface magnetic field, R_c is the core radius, R is the radius of the star and I is the total moment of inertia of the neutron star.

So, the large energy barrier between the two kinds of lines which result in interpinning and together with the fact that creeping takes place, we would expect no perfect lattice of vortices can be formed and we can assume no collective motion of vortices in our calculation.

4 Motion of the Proton Flux Tubes in the Core

The proton flux tubes in the core are moving radially outward which causes the decay of the core magnetic field. The most important forces that act on them are as follows.

4.1 Vortex Acting Force

Since the neutron vortices will move radially outward with a velocity of v_n, the force acting on the proton flux tubes due to the neutron vortices cannot be neglected. If the velocity of the neutron vortices is faster than that of the proton flux tubes, this driving force will drive the proton flux tubes outward; if the reverse is true, it will act as a resistive force to prevent the flux tubes moving outward. Assuming that the neutron vortices and proton flux tubes are uniformly distributed in the core, together with the fact that the magnus force acting on the neutron vortex lines per unit length is given by f_m in equation (4), then the force acting on unit length of a flux tube due to neutron vortices is given by (Ding, Cheng and Chau 1991)

$$\vec{f_n} = \frac{N_n}{N_p} \rho \mathcal{K} r \omega \hat{e_r}, \tag{10}$$

where N_n, N_p are the total number of neutron vortices and proton flux tubes in the core respectively; the value of ω will depend on whether there is creeping or not.

4.2 Buoyancy Force

The existence of the magnetic stress in the core of a flux tube results in a buoyancy force acting on unit length of a flux tube given by (Muslimov and Tsygan 1985)

$$f_b = \left(\frac{\Phi_\circ}{4\pi\lambda}\right)^2 \frac{1}{R_c} \ln(\lambda/\xi). \qquad ((11)$$

4.3 Drag Force

When a flux tube drifts out of the core through the plasma, it will experience a drag force which arises from the scattering of the degenerate relativistic electrons on the flux tube magnetic field. We treat each flux tube as individual entity drifting radially outward with a velocity v_p, rather than considering a lattice of flux tubes. (This treatment is valid as when there is interpinning of flux tubes and vortices, creeping is present, and when we expect no perfect lattice can be formed.) The drag force (Jones 1987) is given by

$$f_v = -\frac{3\pi}{64} \frac{n_e e^2 \Phi_\circ^2}{E_f(e)\lambda} \frac{v_p}{c}, \tag{12}$$

where n_e is the number density of electrons and $E_f(e)$ is the electron Fermi energy.

4.4 Tension

This consists of two components. One is due to bending of flux tubes, the other is due to the fact that a flux tube would tend to reduce its length (Harvey *et al.* 1986). As the buoyancy force and tension have a similar dependence on the parameters, we choose to group them together as γf_b. Since γ is of the order of unity in most regions of the core, we simply take $\gamma = 1$ in our following calculations. Besides these dominant forces, there are also magnus forces acting on the flux tubes due to their relative motions with the local superfluid, electron-proton scattering (Feibelman 1971; Jones 1987), and the buoyancy force acting on neutron vortices which, in effect, acts on the flux tubes. These forces have magnitudes much smaller than those mentioned above, so we neglect them in our calculation.

Now, the equation of motion of a flux tube can be written as

$$f_n(\omega, B_c, t) + f_b + f_v(v_p, t) = 0 \text{ where } \omega(t) |\leq| \omega_{cr}(t), \tag{13}$$

since the neutron vortices cannot exert a force higher than the pinning force. Using the typical physical parameters of a canonical pulsar, it can be estimated that in the early stage of neutron star evolution, the velocity of neutron vortices is much higher than the velocity of proton flux tubes given by equation (13) with $\omega = \omega_{cr}$. Therefore, the neutron vortices will creep over the proton flux tubes at this stage—we call this *forward creeping.* It can be estimated that the vortex acting force is the dominant driving force that acts on the proton flux tubes when the pulsar is young (age $\leq 10^5$ year). Since v_p is decreasing at a slower rate than v_n, after certain time, t_1, we find $v_p = v_n$(Notice that v_p cannot be greater than v_n at this instant since once v_p is greater than v_n (see Fig. 2), the neutron vortices will no longer push on the proton flux tubes, so that f_n will decrease and the relation $v_n = v_p$ is again established). The neutron vortices now move radially outward together with the proton flux tubes with the same velocity. We call this stage *comoving.* At this comoving stage, we have $v_n = v_p$ and $|\ \omega(t)|<|\ \omega_{cr}(t)|$. As f_n will decrease with v_n since f_b is a constant, until

$$f_b + f_v(v_p = v_n, t) > 0, \tag{14}$$

after which ω will become negative. This means that the magnus force is acting radially inward on the neutron vortices and thus the vortex acting force, f_n, is also acting inward and acts to oppose to the motion of proton flux tubes at this stage. It can be shown that the vortex acting force dominates the drag force as the resistive force. Later on, if $\omega(t) = -\ |\ \omega_{cr}(t)|$, the force balance equation is maintained by having v_p greater than v_n, and the proton flux tubes will now creep over the neutron vortices—we call this stage *reverse creeping.*

5 Model of Magnetic Field Decay

The core magnetic field of a neutron star can be simply expressed as

$$B_c(t) = \frac{\Phi_o}{\pi R_c^2} N_p(r,t)|_{r=R_c}\,. \tag{15}$$

We can interpret $v_p(t)$ as the velocity of the flux tubes moving out of the core to the crust at the core-crustal boundary. Then

$$\frac{dB_c(t)}{dt} = -\frac{2}{R_c} B_c(t) v_p(t). \tag{16}$$

Note that the magnetic field decay time scale is equal to the spin down time scale only at comoving stage. In this paper, the detailed mechanism of crustal field decay has

TABLE 1

Results obtained from exact computation using different equations of state with $B_c(0) = 2 \times 10^{12}$ G, $B_s(0) = 5 \times 10^{12}$ G, $B_{crust}(0) = 3 \times 10^{12}$ G, $\rho_{15} = 0.2$, $\tau_c = 10$ Myr, $\gamma = 1$, $\chi = 45^o$ and $\lambda/\xi = \sqrt{2}$.

MODEL	$\frac{\delta m_p^*}{m_p}$	x_p	$\log[t_1(s)]$	$B_{c12}(t_1)$	$B_{c12}(\tau_c)$	$B_{c12}(10^{18}s)$	FINAL STAGE
$AV14$	0.19	0.025	13.44	1.24	0.0042	0.0041	reverse creeping
$UV14$	0.17	0.027	13.47	1.22	0.0032	0.0032	reverse creeping
PAL	0.30	0.027	13.41	1.20	0.0120	0.0120	reverse creeping
MPA	0.14	0.035	13.50	1.23	0.0021	0.0021	reverse creeping
HS	0.20	0.020	13.44	1.21	0.0047	0.0047	reverse creeping
G	0.25	0.04	13.47	1.20	0.0077	0.0076	reverse creeping

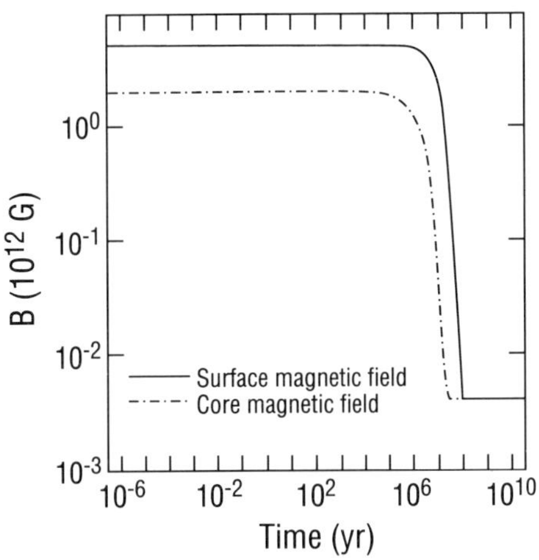

Fig. 1—The evolution of the core and surface fields of a neutron star with EOS, $AV14$. We have used $B_c(0) = 2 \times 10^{12}$ G, $B_{crust} = 3 \times 10^{12}$ G, $\rho_{15} = 0.2$, $\lambda/\xi = \sqrt{2}$, $\tau_c = 10$ Myr.

not been considered, and we simply adopt the observed life time scale, 10 Myr, for τ_c (Narayan and Ostriker 1990). The evolution of the surface field is given by

$$\frac{dB_s(t)}{dt} = -\frac{B_s(t) - B_c(t)}{\tau_c}. \tag{17}$$

By specifying the values of parameters $\delta m_p^*/m_p$ and x_p, and the initial physical parameters of a pulsar, we can deduce the magnetic field at any time by equations (16) and (17) where v_p is given by equation (13). Figure 1 shows the decay curves of the core and the surface fields of EOS, $AV14$ (In calculating m_p^* and x_p, we have used $\rho_{15} \approx 0.2$ where the $^3P_2$ neutron superfluid and $^1S_0$ proton superconducting fluid start to coexist in the liquid core of the neutron star). Figures 2 and 3 are the velocity-time graph and force-time graph respectively. We have computed the evolution of the core and the surface magnetic fields by using various equations of state (Glendenning 1985 Model G; Horowitz and Serot 1987 Model HS; Muether, Prakash and Ainsworth 1987 Model MPA; Prakash, Ainsworth and Lattimer 1988 Model PAL; Winnja, Fisk and Fibrocini 1988 Model $AV14$ and $UV14$), The physical parameters used and results obtained are

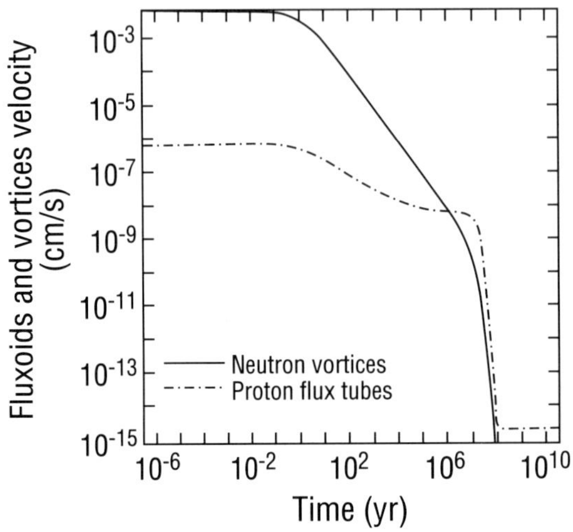

Fig. 2—The velocity time graph of the proton flux tubes and the neutron vortices using Model $AV14$ ($m_p^* = 0.81, x_p = 0.025$) and conditions specified in Fig. 1. With the equation of state chosen, three stages are obtained as discussed in the text.

summarized in Table I. We find that the evolution of fields only weakly depends on the EOS chosen.

First of all, let us summarize some common features of the decay curves obtained. The core field remains unchanged until a time, $\sim t_1$, and then has a decay of about 1-3 orders of magnitude, after that it ceases to decay until the Hubble time scale. Both t_1 and the magnitude of decay depend on m_p^*, x_p, $B_c(0)$, and $B_s(0)$. A quasi-steady state is established around a time t_2 which seems to be independent of those four input parameters and roughly equals τ_c. Qualitatively speaking, the steady state of the core fields results from the fact that the vortex acting force, f_n, changes from a driving force to a resistive force on the proton flux tubes in the late stage of evolution. The detail discussion and the quantitative treatment can be found in Ding, Cheng, Chau 1991.

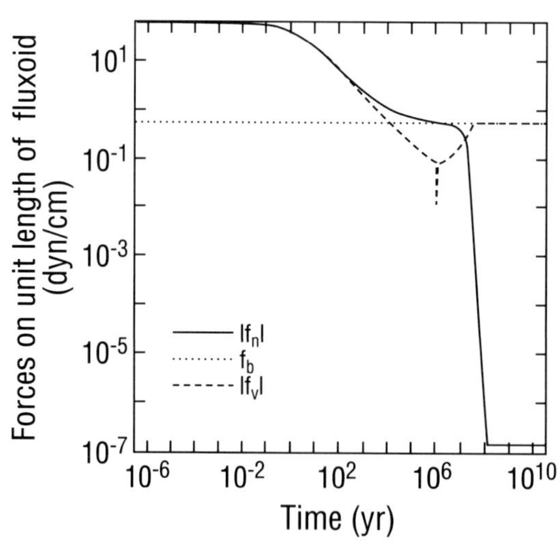

Fig. 3—The absolute values of the forces act on proton flux tubes using Model $AV14$ and conditions specified in Fig. 1. At the beginning, the vortex acting force acts as the dominant driving force, balanced by the drag force. The vortex acting force then drops sharply at the comoving stage; after the turnover, it takes over the drag force to balance against the buoyancy force very rapidly.

In conclusion, if one accepts that the millisecond pulsars are recycled old neutron stars, based on the simple assumptions quoted in the text, our model provides a self-consistent picture for the evolution of neutron star magnetic field.

References

Alpar, M. A., Cheng, A. F., Ruderman, M. A., and Shaham, J. 1982, *Nature*, **300**, 729.
Alpar, M. A., Langer, S. A., and Sauls, J. A. 1984, *Ap. J.*, **282**, 533.
Blair, D. G., and Candy, B. N. 1989, *Timing Neutron Star*, ed. Ögelman, H., and van den Heuvel, E. P. J.
Blondin, J. M., and Freese, K. 1986, *Nature*, **323**, 786.
Chau, H. F., Cheng, K. S., and Ding, K. Y. 1992, *Ap. J.*, in press.
Ding, K. Y., Cheng, K. S., and Chau, H. F., 1991, submitted.
Feibelman, P. 1971, *Phys. Rev.*, **D4**, 1589.
Glendenning, N. K. 1985, *Ap. J.*, **293**, 470.
Gunn, J. E. and Ostriker, J. P. 1970, *Ap. J.*, **160**, 979.
Harvey, J. A., Ruderman, M. A., and Shaham, J. 1986, *Phys. Rev.* D, **33**, 2084.
Hewish, A., Bell, S. J., Pilkington, J. D., Scott, P. F., and Collins, R. A. 1968, *Nature*, **217**, 709.
Horowitz, C. J., and Serot, B. D. 1987, *Nucl. Phys.*, **A464**, 613.
Jones, P. B. 1987, *M. N. R. A. S.*, **228**, 513.
Kulkarni, S. R. 1986, *Ap. J.*, **306**, L85.
Kulkarni, S. R., Djorgovski, S., and Klemola, A. R. 1991, *Ap. J.*, **367**, 221.
Kundt, W. 1981, *Astr. Ap.*, **98**, 207.
Lyne, A. G., Anderson, B., and Salter, M. J. 1982, *M. N. R. A. S.*, **201**, 503.
Lyne, A. G., Manchester, R. N., and Taylor, J. H. 1985, *M. N. R. A. S.*, **213**, 613.
Muether, H., Prakash, M., and Ainsworth, T. L. 1987, *Phys. Lett.*, **199B**, 469.
Muslimov, A. G., and Tsygan, A. I. 1985, *Ap. Sp Sci.*, **115**, 41.
Narayan, R., and Ostriker, J. P. 1990, *Ap. J.*, **352**, 222.
Prakash, M., Ainsworth, T. L., and Lattimer, J. M. 1988, *Phys. Rev. Lett.*, **61**, 2518.
Romani, R. W. 1990, *Nature*, **347**, 741.
Ruderman, M. A. 1991, *Neutron Stars: Theory and Observation*, ed. Ventura, J., and Pines, D.
Sauls, J. A. 1989, *Timing Neutron Star*, ed. Ögelman, H., and van den Heuvel, E. P. J.
Sedrakyan, D. M., and Sedrakyan, A. D. 1991, *Sov. Phys., J. E. T. P.*, **73**, 193.
Srinivasan, G., Bhattacharya, D., Muslimov, A. G., and Tsygan, A. I. 1990, *Current Sci.*, **59**, 31.
Taam, R. E., and van den Heuvel, E. P. J. 1986, *Ap. J.*, **305**, 235.
van den Heuvel, E. P. J., van Paradijs, J. A., and Taam, R. E. 1986, *Nature*, **322**, 153.
Winnja, R. B., Fisk, V., and Fibrocini, A. 1988, *Phys. Rev. C*, **38**, 1010.

Light Curves of Rotating, Oscillating Neutron Stars

Tod E. Strohmayer

Department of Physics and Astronomy, University of Rochester

Abstract

The detection of oscillations of neutron stars would provide astrophysicists with a powerful probe of the neutron star interior. Such oscillations could conceivably be excited in X-ray bursters, X-ray binaries and pulsars. To investigate this possibility we have developed a method for computing the light curve produced by a rotating, oscillating neutron star that emits radiation from a circular polar cap region, as is thought to exist in pulsars, X-ray binaries, and perhaps X-ray bursters. The theory includes the effects of general relativity on the photon trajectories and allows for anisotropic (beamed) emission from the stellar surface. We allow general orientations of the line of sight, neutron star rotation axis, and polar cap emission region. We investigate the light curves produced by individual oscillation modes for various veiwing geometries. To investigate quantitatively the likelihood of detecting such variations, we have simulated observations of X-ray bursts with XTE and AXAF. We find that XTE should be able to probe millisecond scale variations with amplitudes on the order of a few percent of the instantaneous X-ray flux.

1 Introduction

Recently, Schoelkopf and Kelley (1991) reported the presence of a 7.6 Hz oscillation during an X-ray burst from Aql X-1. They suggested that the variation was due to rotational modulation of a thermonuclear detonation which was initially localized on the surface of the star. If this scenario is correct it may be possible to detect and identify intensity variations produced by individual oscillation modes which have been driven to observable amplitudes by the thermonuclear flash (*cf.* McDermott and Taam 1987; and Strohmayer 1992). Further, accreting X-ray pulsars are believed to have thermal "hotspots" produced by accretion of material onto the magnetic poles (Pines 1980, Joss and Rappaport 1984). It is conceivable that the energy released by this accretion flow may also act as a driving mechanism for oscillations.

2 Luminosity of an Oscillating Polar Cap

In computing the light curve produced by rotation and oscillation of a neutron star, we take into account the effects of relativistic gravity on the photon trajectories. We restrict this treatment to stars which are "slowy rotating," in the sense that the

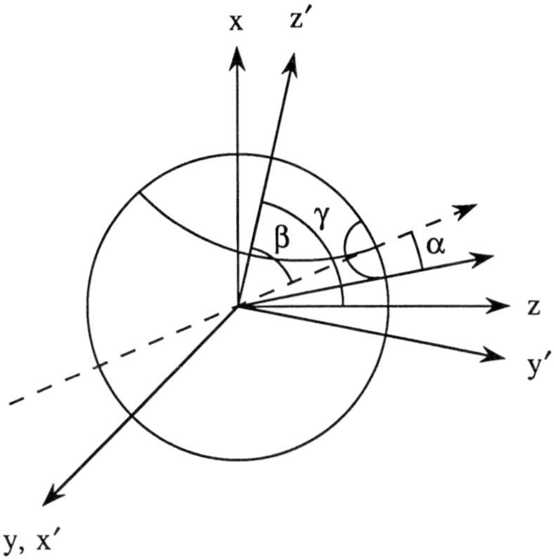

Fig. 1—Geometry used in the calculation of the time-dependent luminosity produced by a polar cap hotspot on a rotating, oscillating neutron star. The parameters are discussed in section 2.2

gravitational acceleration is much greater than the centripetal acceleration due to rotation at the stellar surface. In this limit the Kerr-like metric, which describes the structure of space-time in the vicinity of a rotating neutron star, is well approximated by the Schwarzchild metric. We also neglect the effects of the stellar oscillations on the geometry of space-time, *e.g.*, in producing gravitational radiation. This is entirely consistent with oscillations that produce only small perturbations to the equilibrium structure of the star.

2.1 Photon Orbits in the Schwarzchild Metric

In Schwarzchild geometry, all photons travel along null geodesics. These paths satisfy the equation (Weinberg 1972),

$$\left(\frac{dr}{d\phi}\right)^2 + r^2\,(1 - 2m/r) = \frac{r^4}{c^2 b^2}, \tag{1}$$

where r and ϕ are spherical coordinates, b is the impact parameter of the photon orbit, $m \equiv GM/c^2$, and the motion is confined to a plane through the origin at the center of mass of the star (*cf.* Fig. 1). A convenient change of variables to $u = m/r$ yields the integral

$$\phi = \int_0^{m/R} \left[u_b^2 - u^2(1 - 2u)\right]^{-1/2} du\,, \tag{2}$$

where $u_b = m/b$. Equation (2) maps a given impact parameter, b, onto the surface of the star. This mapping provides a means of relating the integral over the neutron star in the observer's sky to the integral over coordinates centered on the neutron star, as Pechenick *et al.* (1983) have previously demonstrated.

2.2 Luminosity Integral

We assume that all emission comes from the stellar surface and that the radiation is homogeneous but not necessarily isotropic. We investigate the effect of radiation beaming through an intensity dependence on the angle δ between the normal to the stellar surface and the direction to the observer. This choice for the beaming dependence is convenient in that δ is independent of the azimuthal angle, ϕ, thus simplifying the angular integrals. It is not essential to our treatment, however.

The orientation of the system is depicted in Figure 1. The observer is situated on the z-axis a distance $z = r_0 \gg R$ from the neutron star, where R is the stellar radius. The rotation axis of the star (the z'-axis) is inclined at an angle γ relative to the direction to the observer. We consider the emission region to be a circular polar cap with angular radius, α, whose symmetry axis is oriented at an angle β with respect to the rotation axis of the star. To compute the luminosity, we must integrate the intensity over the solid angle subtended by the neutron star in the observer's sky. We thus define a third system of coordinates, (x'', y'', z''), which is centered on the observer's location. This system is oriented with the z''-axis directed along the $-z$ axis toward the neutron star. The element of solid angle in these coordinates is defined to be

$$d\Omega'' = \theta'' d\theta'' d\phi'' , \tag{3}$$

where we have used the small angle approximation in equation (3) since θ'' is infinitesimal for $r_0 \gg R$. In terms of the impact parameter $b = r_0\theta''$, the element of solid angle is

$$d\Omega'' = \frac{1}{r_0^2} b \, db \, d\phi . \tag{4}$$

Here we have replaced $d\phi''$ with $d\phi$, since, by construction of the (x'', y'', z'') coordinate system, the two are equivalent. Changing variables to $x \equiv b/m$, we write equation (4) as

$$d\Omega'' = (m/r_0)^2 x dx d\phi . \tag{5}$$

The luminosity dL originating from a differential element of solid angle $d\Omega''$ at the surface of the neutron star is

$$dL = I(r_0, \Omega'', t) d\Omega'' = (1 - 2m/R)^2 I(R, \Omega'', t) d\Omega'' , \tag{6}$$

where $I(r, \Omega'', t)$ is the time-dependent intensity (Pechenick *et al.* 1983). The quantity $(1 - 2m/R)^2$ is the correction to the intensity due to general relativity. We now write the intensity as

$$I(R, \Omega'', t) = \left[I_0 + \sum_{l,n} I_{ln} Y_{ln}(\theta', \phi') \cos(\omega_{ln} t) \right] f(\delta) , \tag{7}$$

where I_0 is the time-independent component of the intensity, and $f(\delta)$ is the beaming function. The oscillations are assumed to produce time-dependent intensity variations with angular frequency ω_{ln}, and angular dependence given by the spherical harmonics $Y_{ln}(\theta', \phi')$. The maximum intensity variation produced by a given oscillation mode is defined to be I_{ln}. The coordinates (θ', ϕ') are measured with respect to the reference frame (x', y', z'), which rotates with the star. This choice is made for consistency with the analysis of Strohmayer (1991). Substituting equations (5) and (7) into equation (6) yields

$$dL = (m/r_0)^2 (1 - 2m/R)^2 \left[I_0 + \sum_{l,n} I_{ln} Y_{ln}(\theta', \phi') \cos(\omega_{ln} t) \right] f(\delta) x dx d\phi . \tag{8}$$

We now use equation (2) to express θ as a function of $x = b/m$,

$$\theta(x) = \int_0^{m/R} \left[x^{-2} - (1 - 2u)u^2\right]^{-1/2} du \,. \tag{9}$$

It can be shown from equation (1) that there is a maximum impact parameter b_{max} at which a photon can still be traced back to the surface of the star. This sets the upper limit for the integration on x; that is, x ranges from 0 to $x_{max} = b_{max}/m = (R/m)(1 - 2m/R)^{-1/2}$. The angle δ is also easily expressed as a function of x (Pechenick *et al.* 1983), $\sin\delta = x/x_{max}$. Since the oscillation modes are referenced to a coordinate system that rotates with the neutron star, we must be able to transform from these coordinates, (θ', ϕ'), to the integration coordinates $(\theta[x], \phi)$. These transformations are readily derived by performing a pair of rotations that map the integration coordinates onto the rotating coordinates. With the use of these transformations the time-dependent luminosity becomes

$$L = \left(\frac{m}{r_0}\right)^2 \left(1-\frac{2m}{R}\right)^2 \int_0^{x_{max}} \left[I_0 \int_{\phi_-}^{\phi_+} d\phi + \int_{\phi_-}^{\phi_+} \sum_{l,n} I_{ln} Y_{ln}(\theta', \phi') \cos(\omega_{ln} t) d\phi \right] f(\delta) x dx. \tag{10}$$

Here, $\phi_\pm \equiv \phi_0 \pm h(\theta; \alpha, \theta_0)/2$, and

$$\int_{\phi_-}^{\phi_+} d\phi \equiv h(\theta; \alpha, \theta_0) = 2\cos^{-1}\left[\frac{\cos\alpha - \cos\theta\cos\theta_0}{\sin\theta\sin\theta_0}\right]. \tag{11}$$

The function $h(\theta; \alpha, \theta_0)$ defines the angular range in ϕ of the polar cap emission region in a one-dimensional slice at constant θ. Equations (9) - (11) allow us to compute numerically the light curve produced by any linear combination of oscillation modes.

3 Light Curves of Rotating, Oscillating Neutron Stars

In this section we give a brief description of the light curves produced by individual oscillation modes for typical values of m/R, and different viewing geometries. Current measurements of neutron star masses cluster near $1.4 M_\odot$ (Taylor and Weisberg 1982, Joss and Rappaport 1984, and Wolszczan 1991). Theoretical models typically have radii from 9 km to 16 km, depending on the equation of state (Pines 1980). Therefore, we consider values from $m/R = 0$ (Newtonian) to $m/R = .30$ (strongly relativistic).

We begin by investigating the light curve produced by a single dipole ($l = 1$) oscillation mode, with isotropic emission ($f[\delta] = 1$). In Figure 2 we show the light curves for one-half of a rotation with $\alpha = 5°$, $\beta = \gamma = 90°$ for an $l = 1, n = 0$ mode with a period of $P_{1,0} = (1/50)P_{rot}$, where P_{rot} is the rotation period of the neutron star. The legend indicates the value of m/R used for each calculation. The intensity of the oscillating component was $I_{ln} = 1.0$ for this calculation. Notice the flattening of the light curve with increasing m/R. This flattening is the principal effect of general relativity, as the bending of light allows a photon to reach the observer from "behind" the star. Also of interest is the peak near 180° in the $m/R = .30$ calculation. This peak is produced by gravitational lensing of the single polar cap hot spot and was previously discussed by Pechenick *et al.* (1983).

Note that this observational geometry reveals no variations in the light curve due to the oscillations because the polar cap is centered on the equator, which is a node of the $l = 1$ oscillation mode, when $\beta = \gamma = 90°$. The observer sees no variations simply because the oscillations integrate to zero over the polar cap.

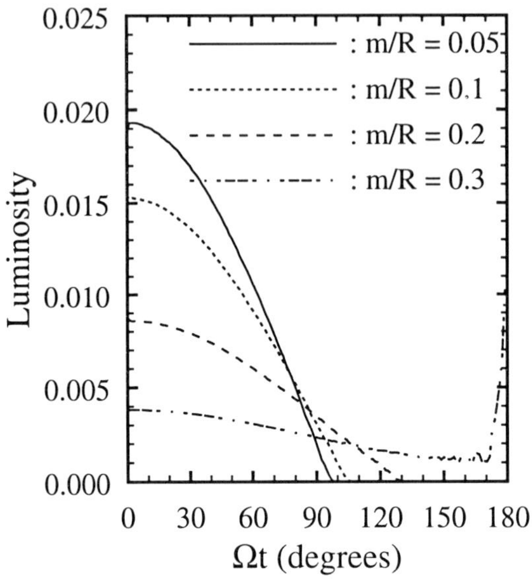

Fig. 2—Light curves from an $l = 1$, $n = 0$ oscillation mode for four different choices of the gravitational field strength m/R. Here $\alpha = 5°$, $\beta = \gamma = 90°$, and $f(\delta) = 1$. The oscillation period is $P_{1,0} = P_{rot}/50.0$, where P_{rot} is the rotation period of the neutron star, with $I_{1,0} = 1$. In this case the oscillations reveal no intensity variations because of the viewing geometry.

In Figures 3a and 3b we show the light curves for the mode with $l = 1, n = 0$, but now for $\beta = 70°$ and $\beta = 50°$ respectively, with the same oscillation period as in the previous calculation. These may be compared with the light curve for $\beta = 90°$ which is shown in Figure 2. For both panels of Figure 3 we have used the value $m/R = 0.2$. Note the relative increase in the periodic intensity variations due to the oscillations as β decreases from 90°. This increase is due to the change in aspect of the polar cap as seen by the observer.

4 Observability of Global Oscillation Modes in X-ray Sources

In this section we simulate observations of type I X-ray bursts using parameters appropriate to the X-ray Timing Explorer (XTE) and the Advanced X-ray Astrophysics Facility (AXAF). Our goal is to assess the chances for detecting global oscillation modes of neutron stars in such systems.

To do this, we have generated a series of simulations of type I X-ray bursts. In these simulations we estimate the X-ray count-rates that would be detected from typical X-ray bursts by XTE and AXAF, two X-ray astronomy satellites scheduled for launch later in this decade. We model the X-ray burst as a thermal source with a temperature T_b. The temporal behavior of the burst is assumed to have the form

$$I(t) = \left\{I_q + I_b e^{-(t-t_b)/\tau_2}\left[1 - e^{-(t-t_b)/\tau_1}\right]\left[1 + a_i \cos(2\pi t/P_i)\right]\right\} , \qquad (12)$$

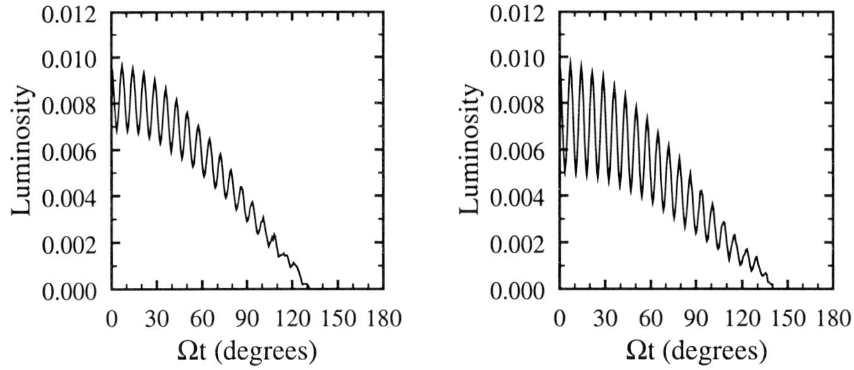

Fig. 3—Light curves from an $l = 1$, $n = 0$ oscillation mode. For these calculations $\alpha = 5°$, $\gamma = 90°$, $f(\delta) = 1$, and $m/R = 0.2$. The left-hand panel shows the light curve for $\beta = 70°$, whereas the right-hand panel was computed with $\beta = 50°$.

where I_q is the quiescent source intensity, I_b the burst intensity, t_b the time at the beginning of the burst, τ_1 the rise-time of the burst, and τ_2 the decay-time of the burst. The quantity a_i is the relative intensity change, expressed as a percentage of the instantaneous burst intensity, produced by an oscillation mode with period P_i. To compute the intensity I_b (counts s^{-1}) for a given simulation, we assume a thermal spectrum is produced by a neutron star with effective temperature T_b and radius R located at a distance d from the observer. In each case the spectrum is integrated over the effective detector areas appropriate for XTE (Bradt *et al.* 1988) and AXAF (Canizares 1988).

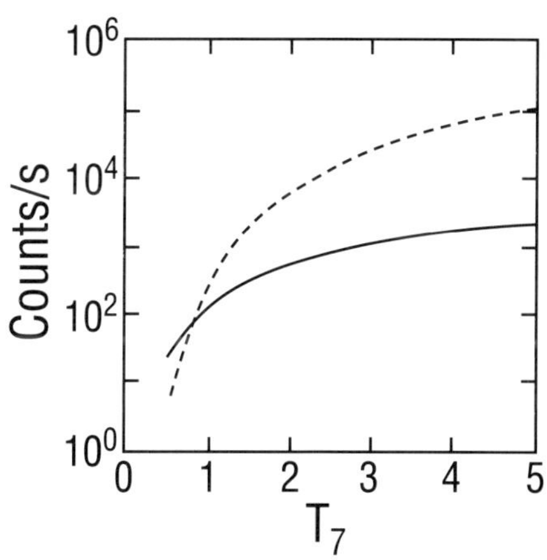

Fig. 4—The X-ray count rates expected in the XTE proportional counter array (PCA) and the HRC detectors of AXAF from a neutron star of radius $R = 16$ km at a distance $d = 5$ kpc are shown as a function of the temperature $T_7 \equiv T_b/10^7$ K. The distance to the object was assumed to be 5 kpc.

The count-rates thus computed from a neutron star of radius $R = 16$ km at a distance $d = 5$ kpc are shown in Figure 4 as a function of temperature T_b. These numbers should be considered upper limits since we have neglected absorption of photons in the interstellar medium. The highest count-rates, especially in XTE, are essential for probing the high-frequency oscillation spectra of neutron stars. To illustrate the type of sensitivity possible with such observations, we have simulated the observation of an X-ray burst by XTE. In modeling each burst, we first generate a series of photon arrival times using the source intensity of equation (12). The time separations between successive photons are distributed according to Poisson statistics. The mean time between photons in the Poisson distribution is taken to be the reciprocal of the instantaneous intensity (equation 12). These photon arrival times can then be binned according to the particular requirements of the simulation. For example, we generated a series of simulations for a burst with $T_b = 1.0 \times 10^7$K, $\tau_1 = 0.15$s, $\tau_2 = 5.0$s, and also with a $P_i = 1.0$ ms oscillation and variable oscillation amplitudes a_i. A total of 20.0 s of data from these simulated bursts was sampled with 0.247 ms integration times and then Fourier analyzed. Figure 5 shows the amplitude of the resulting power spectrum plotted as a function of the oscillation amplitude a_i. In each case the power spectrum was evaluated at a frequency of 1000 Hz ($P_i = 1$ ms). The solid and dashed horizontal lines indicate the mean and 3σ values, respectively, of the power spectrum. From this analysis we can conclude that a 1.0 ms oscillation in a burst of this type can just be detected by XTE if the amplitude is $\geq 5\%$.

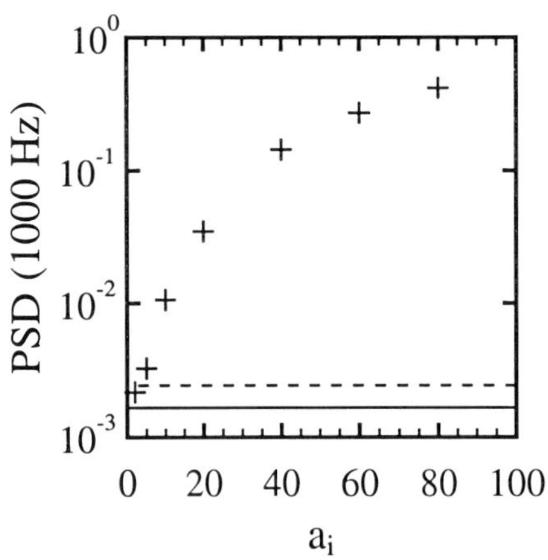

Fig. 5—The amplitude of the power spectrum evaluated at the frequency of a 1 ms oscillation signal (1000 Hz) is shown as a function of oscillation amplitude a_i. These results are from a simulated XTE observation of an X-ray burst.

5 Conclusions

We have described a technique for computing the light curves produced from circular, polar cap hotspots on rotating, oscillating neutron stars. The calculations include the effects of general relativity on the photon trajectories and allow for anisotropic beaming of radiation from the polar caps. We have given several examples of light curves produced by single, dipole oscillation modes.

We have also examined simulations of X-ray observations of X-ray bursts with XTE and AXAF, both of which are scheduled for launch in this decade. We find that XTE should be able to probe the millisecond regime of neutron star oscillation spectra if such oscillations in X-ray bursts can produce X-ray intensity variations on the order of a few percent of the instantaneous X-ray flux. The present model represents the first step toward a more precise description of the light curve from an oscillating neutron star. However, to make precise identifications of oscillation modes, it will be necessary to calculate the beaming function produced by a specific magnetic field, and in the case of accreting neutron stars to take into account the beaming effects which may result from inhomogeneous accretion onto the star.

References

Bradt, H. V., Swank, J. H., and Rothschild, R. E. 1988, in: *Advances in Space Research*, Proceedings of Symposium 14 COSPAR 27th Plenary session, 297.

Canizares, C. R. 1988, in: *Advances in Space Research*, Proceedings of Symposium 14 COSPAR 27th Plenary session, 261.

Joss, P. C. and Rappaport, S. A. 1984, *Ann. Rev. Astron. Astrophys.*, **22**, 537.

McDermott, P. N., and Taam, R. E. 1987, *Ap. J.*, **318**, 278.

Pechenick, K. R., Ftaclas, C., and Cohen, J. M. 1983, *Ap. J.*, **274**, 846.

Pines, D., 1980, *Science*, **207**, 597.

Schoelkopf, R. J. and Kelly, R. L. 1991, *Ap. J.*, **375**, 696.

Strohmayer, T. E., 1991, *Ap. J.*, **372**, 573.

Strohmayer, T. E., 1992, *Ap. J.*, **388**, 138.

Taylor, J. H., and Weisberg, J. M., 1982, *Ap. J.*, **253**, 908.

Weinberg, S. 1972, *Gravitation and Cosmology*, (Wiley and Sons).

Wolszczan, A., 1991, *Nature*, **350**, 688.

How Fast Can Neutron Stars Rotate?

Heinz Herold, Joachim Heimberger, Hanns Ruder, and Xuejun Wu

Theoretische Astrophysik, Universität Tübingen

1 Introduction

The first observation of millisecond pulsars in 1982 has stimulated the research on rapidly rotating neutron stars. Since neutron stars are extremely compact objects with strong gravity, general relativity must be employed in modeling such stars. General relativistic calculations of *rotating* neutron stars are relatively rare. The case of slow rotation has been treated by various authors (see Hartle 1967; Hartle and Thorne 1968; Datta 1988, *e.g.*). The construction of realistic models for *rapidly* rotating neutron stars has been performed by Friedman *et al.* (1986) (see also Friedman *et al.* 1989). The numerical method employed by these authors is based on the idea to discretize directly Einstein's field equations for a chosen parametrization of the metric. To the authors' knowledge, there are no other groups which have developed another independent approach for the problem of rapidly rotating neutron stars.

In this contribution we will show that the minimal surface formalism is well suited for the numerical calculation of the gravitational fields and the structure of rapidly rotating neutron stars, especially since in this method it is not necessary to consider the field equations explicitly. After a short description of the basic formulation of our approach in Section 2, the numerical procedure is explained in Section 3. Typical results of the numerical calculation are presented in Section 4, where global properties of fast rotating neutron stars are discussed. Furthermore, the solutions are visualized by embedding diagrams and 4D ray-tracing pictures.

2 Basic Formulation

In this section we summarize the basic equations which are used for the description of the structure of rapidly rotating neutron stars and their gravitational fields (for more details, see Neugebauer and Herold 1992). The space-time generated by an isolated rotating star admits two commuting Killing vectors: the (at least asymptotically) time-like vector $(\xi^\mu) = \partial/\partial t$ and the space-like (azimuthal) vector $(\eta^\mu) = \partial/\partial\varphi$. Additionally, it is assumed that the matter is a perfect fluid with the energy-momentum tensor

$$T_{\mu\nu} = (\varepsilon + p)\, u_\mu u_\nu + p\, g_{\mu\nu}, \tag{1}$$

and rotates rigidly with the angular velocity Ω. Thus, the four velocity is given by

$$u^\mu = e^{-V}(\xi^\mu + \Omega\eta^\mu). \tag{2}$$

The minimal surface formulation for the stationary and axially symmetric field equations requires that the metric is parametrized in the following form:

$$ds^2 = -e^{2U}(dt + A\,d\varphi)^2 + e^{-2U}\left(e^{2\alpha}(\nabla W)^2(dr^2 + r^2 d\theta^2) + W^2 d\varphi^2\right). \qquad (3)$$

Therein we have specialized the – in principle arbitrary – "meridional" coordinates x^1, x^2 (which are needed additionally to the Killing coordinates t, φ) to spherical coordinates r, θ. It should be noted that the potentials $U(r,\theta)$, $W(r,\theta)$, $A(r,\theta)$, $\alpha(r,\theta)$ are scalar quantities which can invariantly be defined through the Killing vectors.

The normalization of the four vector (2) of the matter yields an expression for the quantity V, which represents in the non-relativistic limit the sum of the Newtonian and the centrifugal potential:

$$V = U + \frac{1}{2}\ln\left((1 + \Omega A)^2 - \Omega^2 W^2 e^{-4U}\right). \qquad (4)$$

A consequence of energy-momentum conservation is that the energy density ε as well as the pressure p only depend on V and are related by

$$\varepsilon(p) + p = -\frac{dp}{dV} \quad . \qquad (5)$$

For a given equation of state (EOS) $\varepsilon = \varepsilon(p)$, the differential equation (5) can be integrated to obtain the function $p = p(V)$. The zero point of V is fixed by the prescription of vanishing pressure on the surface, $p(V_0) = 0$. Thus, the parameters which characterize a rigidly rotating neutron star are the angular velocity Ω and the surface gravity V_0, which describes the compactness of the system.

It can be shown (Neugebauer and Herlt 1984; Neugebauer 1985, 1988) that, for the situation described, Einstein's field equations are equivalent to the minimal surface equations in an abstract Riemannian potential space with a well-defined (indefinite) metric, whose coordinates are the potentials U, W, A, α. One can formulate the problem with the help of an Lagrangian $\mathcal{L}$, whose variation must vanish, $\delta\mathcal{L} = 0$. The Lagrangian is given by

$$\mathcal{L} = \frac{1}{2G}\int_0^{\frac{\pi}{2}} d\theta \int_0^{\infty} dr\, r\left[-\nabla\alpha\nabla W + W(\nabla U)^2 - \frac{e^{4U}}{4W}(\nabla A)^2 \right.$$
$$\left. - 8\pi G\, W e^{2\alpha - 2U} p(V)(\nabla W)^2\right], \qquad (6)$$

and should be considered as a functional of the 4 potentials α, W, U, A. Here, we have additionally assumed symmetry with respect to the equatorial plane ($\theta = \pi/2$). The admissible functions must fulfill the boundary conditions $\alpha = 0$, $W = 0$, $A = 0$ on the rotation axis and $\alpha = 0$, $U = 0$, $A = 0$, $W - r\sin\theta = 0$ at infinity. The potential U must be regular on the rotation axis.

Note that it is an essential feature of our formulation that there is no distinction between the interior and the exterior of the star; both regions are treated simultaneously. The outside is just characterized by $p(V) = 0$, *i.e.*, the last term in (6) vanishes. The position and shape of the surface come out automatically from a self-consistent solution.

3 Numerical Procedure

In order to calculate the structure of rapidly rotating stars, a convenient procedure is to start from a non-rotating star and to increase the angular velocity gradually. The spherically symmetric solutions have been obtained by solving the Tolman-Oppenheimer-Volkoff equations, transformed to our coordinates as determined by the metric (3), specialized to the static case.

For rapidly rotating stars, we must solve the variational principle $\delta\mathcal{L} = 0$, *i.e.*, we have to determine those metric potentials α, W, U, A for which the variation of the integral (6) vanishes.

The domain $0 \leq r < \infty$, $0 \leq \theta \leq \pi/2$ of the coordinates (r, θ) in (6) is unbounded and thus not very suitable for the numerical treatment. Therefore, we transform the coordinate r to a new coordinate $\tilde{r} = r/(r + c_0)$, which has a finite domain, *e.g.*, $0 \leq \tilde{r} \leq 1$, where the constant c_0 is adapted approximately to the radius of the star. Additionally, instead of the angle θ we use $\mu = \cos\theta$ such that the domain of the variational integral (6) is the unit square in $(\tilde{r}, \mu)$. For details concerning the realization of the boundary conditions, we refer the reader to Herold and Neugebauer (1992).

For the discretization we use the *Finite Element* approach. We have applied rectangular 4-node bilinear finite elements with Gauß-Legendre integration formula. As the results with these ansatz were satisfactory, it was not necessary to turn to more complicated finite elements.

The discretized field equations are obtained by setting the derivatives of the discretized Lagrangian with respect to the node values of the potentials to zero. This non-linear system of algebraic equations has been solved by the Newton-Raphson method.

After convergence we have a solution represented by the node point values of the 4 potentials α, W, U, A. The actual procedure is to start from a non-rotating solution, to calculate the potentials on the finite element grid, to determine a converged solution for $\Omega = 0$ by the finite element iteration (the comparison with the solution from the modified TOV equations gives a good estimate of the discretization error), and then to change the input parameters Ω and/or V_0 slightly to obtain a new solution with the old solution as the starting point of the iteration. Actually, we fix the surface gravity parameter V_0 (this determines the mass, at least to a great extent) and increase the angular velocity Ω.

4 Results

An essential ingredient for the calculation of *realistic* neutron star models is the equation of state, *i.e.*, the relation $p = p(\varepsilon)$. Up to densities of roughly 10^{14} g/cm^3 the equation of state of cold catalyzed neutron star matter is well known, and the physics up to this region is well understood. But around the nuclear saturation density and especially above, the situation concerning the EOS is more complicated. Therefore, many different EOSs for neutron stars exist in the literature (for a review see Arnett and Bowers 1977; Glendenning 1988, *e.g.*). Since we concentrate here on the problem of rapid rotation, we do not want to discuss the problem of EOSs in more detail.

We have calculated rotating neutron star solutions for various EOSs for many combinations of the basic parameter V_0 and Ω. Especially we used EOS G (Canuto and Chitre 1974), the pion-condensed EOS π (Weise and Brown 1975) and the new EOS MPA (Wu *et al.* 1991).

For a discussion of the behavior of the metric potentials we refer the reader to Herold and Neugebauer (1992). In the following we only want to discuss some global properties.

4.1 Global Properties

First, there is the gravitational mass (total mass) M which characterizes a star. This quantity can be measured asymptotically through its gravitational action. Quantitatively, this leads to the mass formula $GM = \lim_{r\to\infty}(-rU)$. Alternatively, the total mass has been calculated by an integral over the matter distribution (see Neugebauer and Herold 1992). Similarly, the angular momentum J is an asymptotically measurable quantity (by Lense-Thirring precession, *e.g.*). Therefore, we have an analogous formula, $GJ = \lim_{r\to\infty}(r/2\sin^2\theta)\,A$. This relation can be evaluated for various values of θ, which should always give the same result and thus is a test of the numerical accuracy. Additionally, we have calculated the angular momentum of our solutions by the appropriate integral over the matter. All these different methods for the determination of M and J are in good agreement. The baryon mass (or equivalently the total number of baryons) cannot be determined from the asymptotic fields; rather one has to integrate over the interior of the star.

In Tables 1 and 2 we present a series of solutions which are typical for our results and have been calculated with the equation of state MPA. Beside the total mass M and the baryon mass M_0, the central density ε_c and the equatorial radius R (measured by the circumference) are shown. From the angular momentum J the moment of inertia I, defined by the (Newtonian) relation $I = J/\Omega$, has been calculated.

In each table the angular velocity increases from zero to the maximum possible angular velocity, which corresponds to that solution in which the star rotates at the Kepler frequency Ω_K, *i.e.*, the angular frequency of a particle in circular orbit at the equator. No rigidly rotating star can have $\Omega > \Omega_K$ and at this point mass shedding at the equator sets in. Certainly, this is the endpoint of such a series of increasing angular velocity. But that does not mean that the solutions up to this point are stable. There are instabilities due to non-axisymmetric modes which produce gravitational radiation, but those modes are damped by shear viscosity inside the star, which is believed to strongly depend on the temperature. These effects have been understood for non-relativistic stars, but a quantitative treatment in the relativistic regime is lacking (see Weber *et al.* and references therein).

As general trends in Tables 1 and 2 one recognizes that with increasing angular velocity the central density diminishes; the equatorial radius is going up as well as the moment of inertia. The last effects are clearly an indication of rotational flattening of the star.

TABLE 1
EOS MPA, $V_0 = -0.07$

Ω (s^{-1})	ε_c (10^{15} g/cm^3)	M ($M_\odot$)	M_0 ($M_\odot$)	R (km)	I (10^{45} g cm^2)
0	0.74	0.476	0.494	10.77	0.245
1255	0.73	0.474	0.492	10.90	0.248
2509	0.71	0.470	0.487	11.47	0.256
3137	0.70	0.467	0.484	12.04	0.263
3764	0.67	0.463	0.479	13.40	0.275
3952	0.67	0.462	0.477	14.41	0.279
3990	0.66	0.461	0.477	15.65	0.280

TABLE 2
EOS MPA, $V_0 = -0.4455$

Ω (s^{-1})	ε_c (10^{15} g/cm^3)	M ($M_\odot$)	M_0 ($M_\odot$)	R (km)	I (10^{45} g cm^2)
0	5.09	1.559	1.834	7.81	0.847
3137	4.90	1.565	1.839	7.90	0.863
6274	4.30	1.590	1.862	8.24	0.933
8783	3.65	1.624	1.890	8.82	1.048
10038	3.25	1.650	1.910	9.37	1.152
10872	2.92	1.677	1.932	10.06	1.271
11293	2.72	1.699	1.950	10.78	1.373
11456	2.63	1.711	1.961	11.74	1.434

4.2 Visualization by Embedding and 4D Ray-Tracing

There are certainly many possibilities to characterize the gravitational fields and the structure of the star in a coordinate-independent manner. We have chosen two methods. First, we have calculated embedding diagrams which visualize the intrinsic geometry of the surface of the neutron star and of internal surfaces of constant pressure (or density) by embedding them into the Euclidean 3-space. Typical embedding diagrams are presented in Figs. 1 and 2.

Figure 1 shows the surfaces of the solutions of Table 1, *i.e.*, relatively light neutron star models. Additionally, in Figure 2 the internal structure of the fastest star is depicted by surfaces of constant density. It can be recognized that the mass shedding at the equator begins with a bump there caused by the outermost layers only.

As a second visualization method of the space-time structure of rapidly rotating neutron stars, we show a typical picture of "what a rapidly rotating neutron star looks like". The idea is to assume that from the surface of the neutron star photons are emitted which are moving through the curved space-time and eventually reach the observer which is located near infinity. Obviously, this visualization method may be considered to be complementary to the embedding pictures of the interior of the

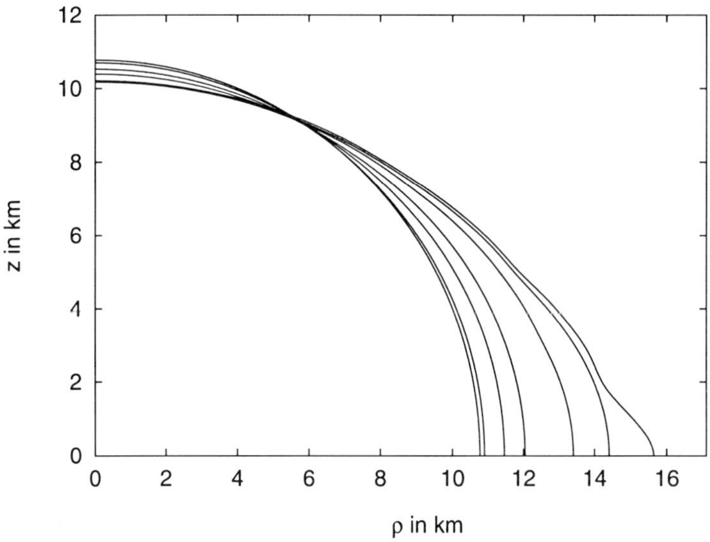

Fig. 1—Embedding diagrams of the surfaces of the stars of Table 1. The equatorial radius increases with angular frequency.

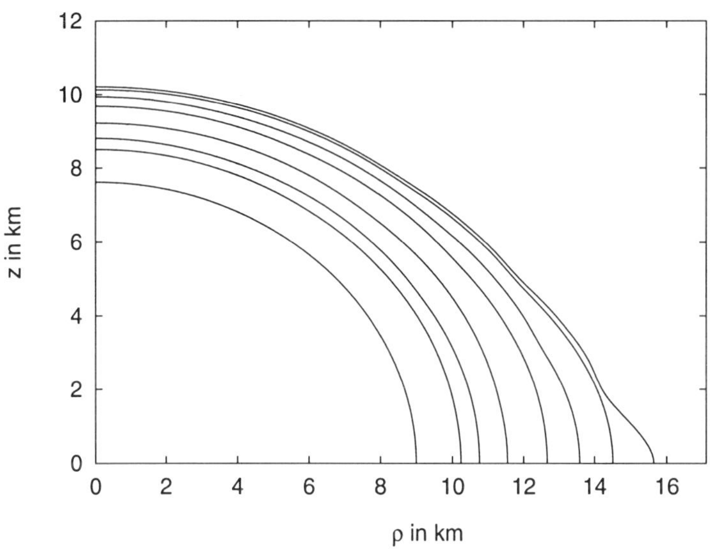

Fig. 2—Embedding diagrams of some internal constant-density surfaces of the fastest star of Fig. 1, which rotates just at the mass shedding limit. The outermost curve represents the surface of the star itself, while the other ones belong to the density values 10^7, 10^9, 10^{10}, 10^{11}, 10^{12}, 10^{13}, 10^{14} g/cm^3.

star, as the photons propagate in the region outside the star. Practically, we use a ray-tracing approach: from the observer's position the paths of photons are followed back in different directions by integrating the null geodesic equations, using the Christoffel symbols of the numerically determined metric (3), until each photon does (or does not) hit the surface of the rotating star. We call this procedure *4D ray-tracing*, since apart

Fig. 3—4D ray-tracing picture of a rapidly rotating neutron star near the mass shedding limit

from three-dimensional space the time also plays a role (the time when a photon hits the star determines the position on the surface).

In Figure 3 an example is presented. It is the fastest neutron star from Table 2 with $V_0 = -0.4455$ and $\Omega \approx 1.15\ 10^4\,\mathrm{s}^{-1}$ (near the mass shedding limit: $\Omega_K \approx 1.19\ 10^4\,\mathrm{s}^{-1}$). For the sake of visibility the surface of the star is painted in a checkerboard pattern (with 30° by 30° patches in the angles θ and φ).

Relativistic light deflection is responsible for the fact that one can see both poles simultaneously and more than the front hemisphere in equatorial regions (for non-rotating stars, see also Nollert *et al.* 1989). The effects of rotation are clearly visible as spectacular asymmetry. A closer inspection shows that in this picture time-of-flight effects and Lense-Thirring frame dragging of photon paths appear in combination.

References

Arnett, W. D., Bowers, R. L. 1977, *Ap. J. Suppl.*, **33**, 415.
Canuto, V., Chitre, S. M. 1974, *Phys. Rev. D*, **9**, 1587.
Datta, B. 1988, *Fund. Cosmic Phys.*, **12**, 151.
Friedman, J. L., Ipser, J. R., Parker, L. 1986, *Ap. J.*, **304**, 115.
Friedman, J.L., Ipser, J.R., Parker, L. (1989): *Phys. Rev. Lett.*, **62**, 3015.
Glendenning, N. K. 1988, *Phys. Rev. C*, **37**, 2733.
Hartle, J. B. 1967, *Ap. J.*, **150**, 1005 .
Hartle, J. B., Thorne, K. S. 1968, *Ap. J.*, **153**, 807.
Herold, H., Neugebauer, G. 1992, in *Proc. of the Bad Honnef Meeting on Relativity and Gravitational Physics*, ed. G. Schäfer, Lecture Notes in Physics (Berlin: Springer), in press.
Neugebauer, G. 1985, in *Proceedings of the Balatonszæplak Relativity Workshop*, ed. Z. Perjes (Budapest: KFKI Publ.), p. 103.
Neugebauer, G. 1988, in *Relativity Today, Proceedings of the 2nd Hungarian Relativity Workshop*, ed. Z. Perjes (Singapore: World Scientific), p. 134.
Neugebauer, G., Herlt, E. 1984, *Class. Quant. Grav.*, **1**, 695.
Neugebauer, G., Herold, H. 1992, in *Proc. of the Bad Honnef Meeting on Relativity and Gravitational Physics*, ed. G. Schäfer, Lecture Notes in Physics (Berlin: Springer), in press.
Nollert, H.-P., Ruder, H., Herold, H., Kraus, U. 1989, *Astr. Ap.*, **208**, 153.
Weber, F., Glendenning, N. K., Weigel, M. K. 1991, *Ap. J.*, **373**, 579.
Weise, W., Brown, G. E. 1975, *Phys. Lett. B*, **58**, 300.
Wu, X., Müther, H., Soffel, M., Herold, H., Ruder, H. 1991, *Astr. Ap.*, **246**, 411.

Neutron Star Crust Motion and Magnetic Field Evolution

Mal Ruderman

Columbia Astrophysics Laboratory, Columbia University

Abstract

The highly conducting crust of a spinning-down or spinning-up neutron star can be subject to stresses which exceed its yield strength. Consequent motions of the crust force changes in the stellar magnetic field which can lower the spin-down indices in young canonical radiopulsars and lead to very small orthogonal dipole moments located at the spin axis in accretion driven spun-up millisecond pulsars. Comparisons with observations of the radioemission properties of disk population millisecond pulsars indicate that most such pulsars had a spin-up genesis (through LMXB's) whereas the neutron stars which are now Globular Cluster millisecond pulsars were never greatly spun-up or spun-down since their formation (accretion induced collapse of White Dwarfs).

1 Introduction

The superconducting superfluid core of a neutron star is surrounded by a crustal lattice of nuclei. A great deal is now known quantitatively about the crust, *e.g.*, its composition, rigidity, and melting temperature, and about the properties of the superfluid neutrons which fill the internuclear space in the kilometer-thick lower crust. The deep crust is an extraordinarily good electrical conductor so that any magnetic field which passes through it from the core may be frozen in the crust. Such field would then change if, and only if, there is crust movement.

As rapidly spinning neutron star changes its spin rate because of pulsar spin-down or accretion induced spin-up (or spin-down), large stresses can build up in the solid crust which cause it to move ("plate tectonics"). Unfortunately a crucial input in determining crustal movement, the strength of the crust, is not yet known. The 10^{13} gcm^{-3} microcrystals that form the crustal lattice are expected to exceed their yield strengths when stressed so strongly that their dimensionless strains θ (change in length per unit length, or 1/3 the shear angle) exceeds $\theta_{\max} \sim 10^{-2}$ (Ruderman 1992a, hereafter RII, and references therein). When the lower crust temperature is about that within the Crab pulsar (several $\times 10^8$ K) or higher, overstressed microcrystals will yield by plastic flow (creep); at lower temperatures they become brittle and yield discontinuously with jumps which can be as large as $\Delta\theta \sim 10^{-1}\theta_{\max}$ (Ruderman 1992b, hereafter RIII). Large sections of crust will, of course, not be stronger than the microcrystals of which they are formed. They might, however, be very much weaker because of dislocation complexes (fault planes) and crumble (or crack, or creep) when subject to a stress much less than that typically needed to break their microcrystals. Therefore, the relevant effective maximum strain, θ_{max}, which designates the elastic limit beyond which large scale crust regions yield to a near fluid-like response, is not really known. It may well be 10^{-4} or less, typical of the strength (*i.e.*, the maximum strain before yielding) of bulk terrestrial examples prepared without special care.

2 Crust Breaking Stresses in a Spinning Neutron Star

When a neutron star rotates, its neutron superfluid can participate in the rotation only by establishing an array of quantized vortex lines. The instantaneous vortex line configuration determines the detailed superfluid flow. Within the deep crust at total densities $(2-8)\times10^{13}\mathrm{gcm}^{-3}$ these vortex lines are expected to be pinned to the crustal lattice nuclei (Anderson and Itoh 1975). As long as the crust and its superfluid corotate there is no important stress on the crust lattice because of the pinning. However, when the crust lattice spins-up (down), its internuclear neutron superfluid cannot spin-up (down) with it unless the pinned vortex lines move out (in) away from (toward) the stellar spin axis. Because of their pinning to lattice nuclei, the vortices are not free to do this and build up a shearing stress on the lattice. The pinning force per unit volume is

$$\mathcal{F} = 2\omega\Omega_n\rho_n\boldsymbol{r}_\perp. \tag{1}$$

Here $\boldsymbol{\Omega}_n$ is the angular velocity of the neutron superfluid a distance $\boldsymbol{r}_\perp$ from the axis, ρ_n is the superfluid density, and ω is the difference between Ω_n and the crustal lattice's spin rate (Ω). Pinning is ineffective at a crust density near $10^{13}\mathrm{gcm}^{-3}$ and at the crust core boundary ($\sim 2\times10^{14}\mathrm{gcm}^{-3}$). One consequence is that crustal lattice shear strength is needed to balance $\mathcal{F}$. It appears that in the strong pinning region, crust breaking would occur before vortex unpinning in all pulsars if $\theta_{max} \lesssim 10^{-4}$ and perhaps even if $\theta_{max} \sim 5\times10^{-3}$, its value if the entire crust was a single crystal (RII). However, other kinds of vortex motions might possibly preempt both vortex unpinning and lattice breaking (RII Appendix A).

A second possible strong shearing stress on the crust exists where the 10^{12}G magnetic field of canonical radiopulsars passes from the stellar core into the crust which it traverses before reaching the stellar surface. If, as expected, the stellar core protons form a Type II superconductor, the core magnetic field is organized into an array of quantized magnetic flux tubes which expand and merge when they enter the crust. Each quantized flux tube in the core contains an average magnetic field $B_c > 10^{15}$G. It is partly because the stellar core's superfluid neutron vortex lines covers are comparably magnetized when imbedded in a proton superconductor that single flux tubes and neutron vortex cores do not pass easily through each other. Core flux tubes may then possibly be moved by the expanding (contracting) array of core neutron superfluid vortex lines as a neutron star spins-down (up) (Srinivasan *et al.* 1990). The resulting stress on the bottom of a stationary conducting crust that anchors the core's flux tubes could then become as large as

$$S_B \sim \frac{BB_c}{8\pi}. \tag{2}$$

This stress could strain the crust beyond its yield strength if $B \gtrsim 3 \cdot 10^{10} (\theta_{\max}/10^{-4})$ G. Whereas a single core flux tube would move out with the expanding vortex array of a spinning-down neutron star, the response of a huge array of flux tubes may, however, be quite different. It is not at all clear that similar core flux tube motion would take place in a neutron star with a huge concentration of flux tubes ($\sim 10^{19} B_{12} \mathrm{cm}^{-2} \sim 10^{15}$ per neutron vortex line). Possible induced flux tube bunching and superconducting proton screening of electron currents may be important considerations if proton flux tube array motion is strongly coupled to that of neutron vortices. Much needs to be studied further. Here we shall simply assume that the $\mathcal{F}$ of equation (1), or the S_B of equation (2), or a combination of both, cause the deep crust to "flow" with those stresses when its yield strength is exceeded. The direction of such flow is indicated for spin-down in Figure 1. The magnetic field which traverses the crust is assumed to be frozen in it and move in that same direction. (This ignores crucial questions about backflow in the stratified crustal layers or just how the sinks and sources for the moving crust are formed.

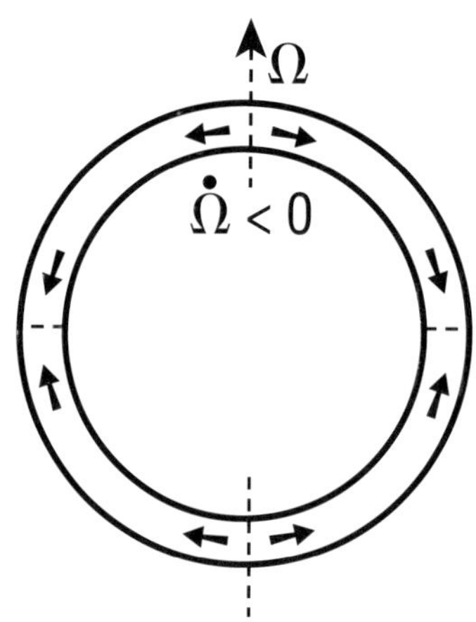

Fig. 1—Main deep crust stress-driven flow in a spinning-down neutron star.

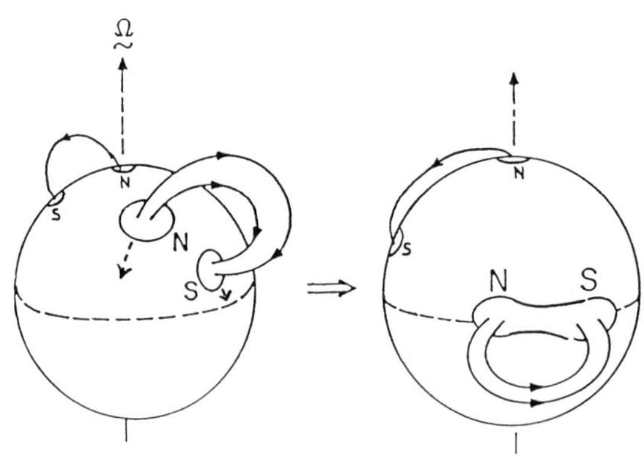

Fig. 2—Surface field evolution of a spinning-down neutron star when field line reconnection is ignored.

3 Magnetic Field Evolution During Spin-Down

The expected crust motion driven evolution of the surface magnetic field of a spinning-down neutron star is indicated in Figure 2. The motion toward the spin equator of the crust, which carries with it the strongest magnetic field regions, should have two observable consequences.

a) The spin-down index of a pulsar, $n \equiv -\Omega\ddot{\Omega}\dot{\Omega}^{-2}$, = 3 if a crust anchored dipole moment keeps a constant magnitude and angle with respect to $\mathbf{\Omega}$. A moving crust has $n \neq 3$. For the evolving geometry of Figure 2, $n < 2$ until part of one polar cap reaches the spin equator when n will increase. For the three young radiopulsars with measured indices, PSR 0531, PSR 1509, and PSR 0540, $n \sim 2.5, 2.8, 2.0$ respectively.
b) Within a few spin-down times $(-\Omega/\dot{\Omega})$ after the initial increase of dipole moment $(n < 3)$ most of a pulsar's magnetic polar caps reach the spin equator (Fig. 2). If core flux tubes do indeed move with the core superfluid neutron vortex lines which thread them, the flux tubes ultimately will be pushed into the crust-core interface. Because almost all of the stellar core's original flux ($\sim 10^{25}\mathrm{Gcm}^2$) is then squeezed into a very small area of highly conducting crust, crustal stress there will grow until the lower crust's yield strength is exceeded. The resulting crust breaking would allow North and South polar cap regions to move together, magnetic field recombination, and a reduction of the stellar magnetic moment. The dipole field would decrease roughly as the stellar spin (dipole $B \propto \Omega$) until B is so diminished that the crust strength is no longer exceeded by magnetic stresses (when $P \sim 10$ s according to RII). The initial evolution of dipole B vs. period P is indicated by the $\overline{01}$ segment of Figure 3. The subsequent spin-down evolution is the $\overline{12}$ segment; observations appear neither to support nor contradict it.

4 Spin-Up Driven Magnetic Field Evolution

When a spun-down neutron star is in a binary, it may be spun-up again by accretion from its companion. The evolution of the neutron star's magnetic field during such a spin-up phase depends upon the initial surface field geometry when spin-up begins. Three possible initial configurations and the evolution which results are shown in Figures 4 - 6 and Figure 3. In Figure 4, with N and S magnetic polar caps in opposite hemispheres, the magnetic dipole becomes aligned by spin-up and slightly increased. In Figure 5 (the "sunspot" configuration of Fig. 2), with both magnetic polar caps in the same hemisphere, magnetic N and S polar cap regions are pressed together at the same spin axis poles by spin-up driven crust motion. This results in a decreasing magnetic dipole moment proportional to $P^{1/2}$. The spin-up driven dipole magnetic field evolution for the above two initial configurations and an intermediate one are shown in Figure 3. During spin-down (perhaps controlled by interaction with its companion) it is assumed that crust strength prevents crust motion and the evolution of dipole B on the segment $\overline{33'}$. Similarly B is unchanged by any subsequent spin up on $\overline{3'3}$. Thereafter, further spin-up reduces only the sunspot-like contribution to the dipole moment. The largest reduction will occur when an accreting neutron star with

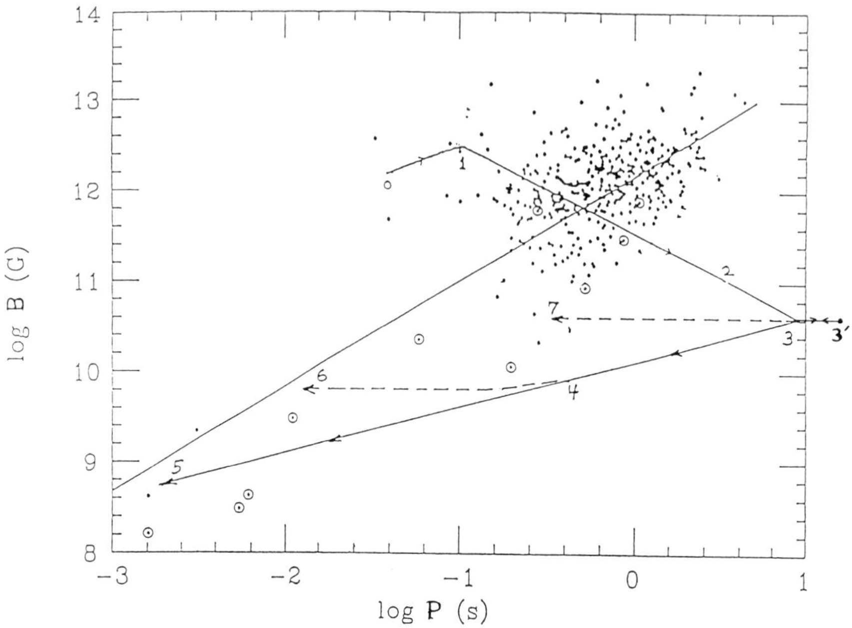

Fig. 3—Dipole P (from $P\dot{P}$) vs. period (P) for radiopulsars. The circled dots are pulsars in binaries. The diagonal is the "accretion spin-up line."

a pure sunspot-like field configuration is spun-up ($\overline{345}$) to the equilibrium "spin-up line" (the diagonal solid line of Fig. 3). Spin-up of the configurations of Figs. 5 and 6 gives the tracks $\overline{37}$ and $\overline{346}$ respectively.

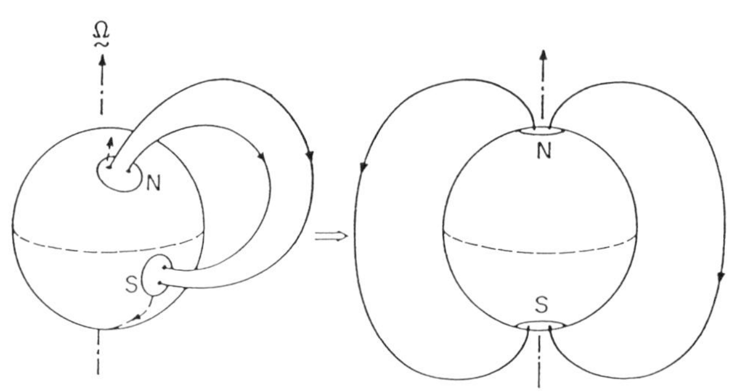

Fig. 4—Evolution of the spinning-up surface field of a spinning-up neutron star when surface flux goes between spin hemispheres.

After the spin-up line is approached X-ray emission from accretion-heated magnetic polar caps should no longer appear modulated to an observer: a sufficiently small polar cap at the spin axis is essentially always visible or always hidden. This feature may be related to the absence of observed spin-modulated X-ray emission in LMXB's.

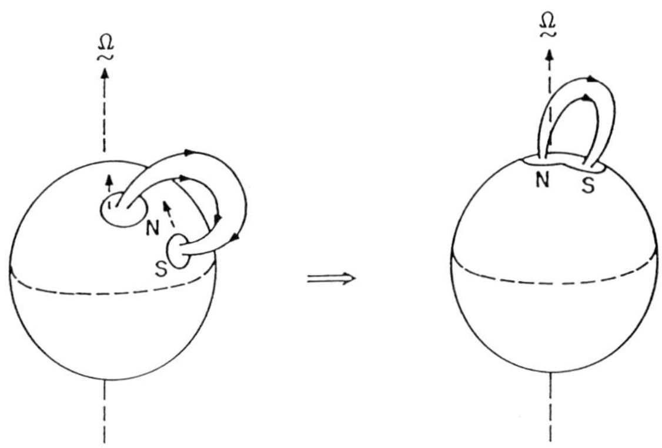

Fig. 5—Evolution of the surface field of a spinning up neutron star with an initial sunspot-like configuration.

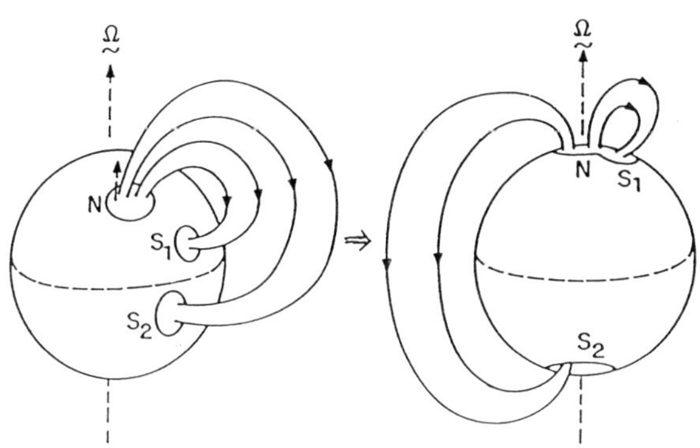

Fig. 6—Evolution of the surface field of a spinning-up neutron star whose initial surface field combines those of Figs. 4. and 5.

If an accretion spun-up neutron star reaches the lower left hand corner region of Figure 3, its magnetic field configuration would then be that of Figure 7 at either (or both) spin axis poles. This special configuration would have several unique consequences for the radio emission from such millisecond radiopulsars (Chen and Ruderman 1992, RII).

a) Because the magnetic dipole ($\boldsymbol{\mu}$) is perpendicular to $\boldsymbol{\Omega}$, the radiopulse profile should have two subpulses almost exactly 180° apart.
b) The pulse and interpulse widths (w) can be anomalously narrow. The narrowest pulse widths of canonical pulsars are those of "core components" with $w_0 \sim (\Omega R/c)^{1/2} \cdot |\hat{\boldsymbol{\Omega}} \times \hat{\boldsymbol{\mu}}|^{-1} \geq 2.5° P^{-1/2}$ (Rankin 1990). However, for the squeezed axial polar cap and $\boldsymbol{\mu} \perp \boldsymbol{\Omega}$ of Figures 5 and 6 , $w \sim 2.5° (P_3 P)^{-1/4} = 10°$ (FWHM) for PSR 1937 ($P = 1.6$ ms).
c) The usual open field line cone (vertical stripes in Fig. 7) is usually assumed to give rise to a similarly shaped cone of radio emission. However, in Figure 7 the open-field cone axis is tangential to the stellar surface. The General Relativistic bending of tangentially emitted rays (*e.g.*, beams a and b in Fig. 7) is very large ($\sim 40°$ for $GM/Rc^2 = 0.4$) and sensitive to small variations in emission distance

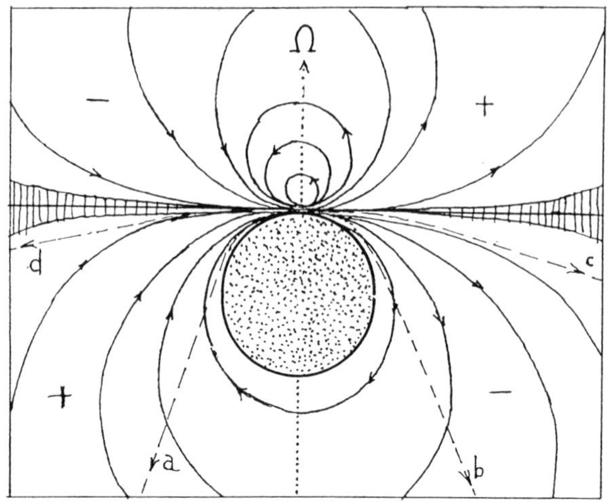

Fig. 7—Magnetosphere of a spun-up pulsar with the initial field geometry of Fig. 5 (or the North hemisphere of Fig. 6).

from the spin axis. As a result, an otherwise very narrow radiation emission cone is stretched into a fan beam very elongated in the $\boldsymbol{\mu} - \boldsymbol{\Omega}$ plane but with the same anomalously narrow azimuthal w discussed above. Thus the observed emission beam would be anomalously narrow but not easily missed.

d) The observed pulse and interpulse pass through different kinds of magnetospheric plasma (mostly on closed field lines). Where $\boldsymbol{\Omega} \cdot \boldsymbol{B} > 0$, the plasma consists mainly of (charge separated) positively charged ions which have no important effect on radio beams with frequencies in excess of several $\times 10^2$ MHz (*e.g.*, beams a and d of Fig. 7). However, the other beams (b and c) pass through mainly a pure electron plasma which, in the absence of local magnetic field, would be overdense for radio frequences $< 10^3$ MHz. Because of the large B, however, those waves which are polarized perpendicular to $\boldsymbol{B}$ can pass through. Thus one of the two pulsar radio fan beams should be almost 100% polarized at frequencies $< 10^3$ MHz, and the polarization angle should not vary significantly as that beam sweeps through a distant telescope.

Three of the four fastest spinning disk-population radiopulsars have the 180° double-pulsed profile predicted above in a). The best studied is the original millisecond pulsar PSR 1937. Its radio emission characteristics are shown in Figure 8. The very narrow observed width agrees with that described in b) and one of the pulses does indeed have the low-frequency polarization properties of d) (Thorsett and Stinebring 1991). Therefore the observed radio emission properties of PSR 1937 (and seemingly most of its siblings) are consistent with strong spin-up induced crust motion toward the spin axis. This movement could also account for their small asymptotic magnetic dipole moments. (The local polar cap field is not small.)

The radio-pulse shapes of the large number of millisecond pulsars ($P = 2 - 6$ ms) in the Globular Cluster 47 Tuc are shown in Figure 9. They show no evidence for large spin-up induced crust motion. Indeed, here the pulse shapes are quite broad, with

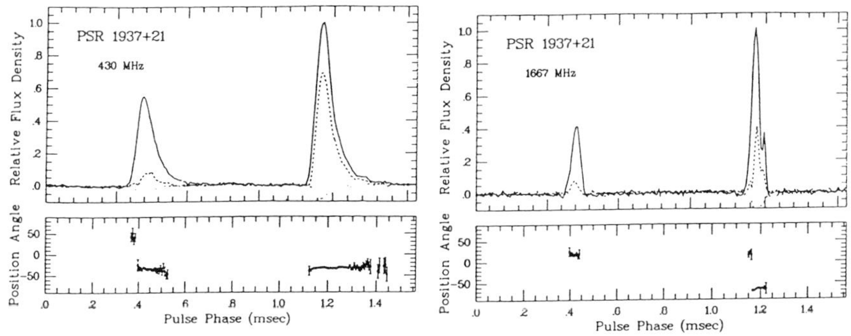

Fig. 8—Pulse shape and polarization for PSR 1937+21 at 430 and 1667 MHz. The dashed line is the linearly polarized intensity. Its position angle is given below (after Thorsett and Stinebring 1991).

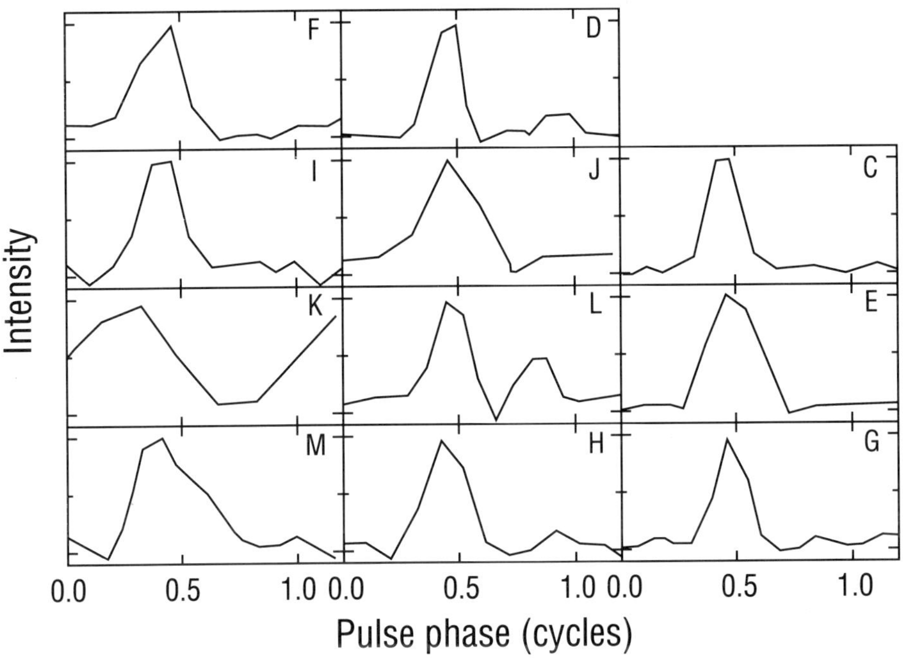

Fig. 9—Pulse profiles of millisecond pulsars in the globular cluster 47 Tuc (Manchester *et al.* 1991). The pulsar numbers are given by PSR0021-72 followed by the letter in each panel.

about the widths expected for canonical radiopulsars if these neutron stars had been born with millisecond periods. Therefore, if crusts and the magnetic fields anchored in them move as theory suggests, most globular cluster millisecond pulsars appear to have an origin different from that of their disk population cousins (*e.g.*, Accretion Induced Collapse of a White Dwarf).

5 Crust Cracking

If cool overstrained crusts relax by large scale cracking with $\Delta\theta \sim 10^{-4}$, the following properties should be associated with cracking events (RIII).

a) Spin period glitches with

$$\frac{\Delta\Omega}{\Omega} \sim \Delta\theta \frac{I_n}{I} \sim 10^{-6}, \tag{3}$$

where (I_n/I) is the ratio of crust superfluid moment of inertia (I_n) to that of the whole star (I).

b) An interval between glitches, independent of $\Delta\theta$,

$$\tau_g \sim \left(\frac{\Delta\Omega}{\Omega}\right) \frac{\Omega}{\dot{\Omega}} \frac{I}{I_n} \tag{4}$$

c) Very small "glitch activity," $(\Delta\Omega/\Omega)\tau_g^{-1}$, in warm crusted pulsars such as the Crab's.

d) A surface field structure in older spun-down pulsars consisting of patches of strong original field $(B \sim 3 \times 10^{12}\mathrm{G})$ separated by areas with very much weaker fields. A cyclotron resonance observation should then give the primeval field, whereas one which measures the dipole component only (*e.g.*, spin-down power; Alfven radius for accretion) would give a smaller value.

e) Around 10^{40} ergs of energetic X-ray/γ-ray emission accompanying a glitch in old neutron stars with almost empty magnetospheres.

Properties a), b), and c) are in agreement with observations. Property d) is not yet confirmed or contradicted by X-ray pulsar measurements. Verifying or contradicting property e) depends upon a resolution of the origin of Gamma-Ray Bursts.

This work was supported by the National Science Foundation under grant AST89-01681, and is contribution number 488 of the Columbia Astrophysics Laboratory.

References

Anderson, P. and Itoh, N. 1975, *Nature*, **256**, 25.

Chen, K. and Ruderman, M. 1992, "Radio Pulse Properties and the Origin of Millisecond Pulsars," to be published.

Manchester, R., Lyne, A., Robinson, C., D'Amico, N., Bailes, M., and Lim, J. 1991, *Nature*, **352**, 219.

Rankin, J. 1990, *Ap. J.*, **352**, 247.

Ruderman, M. 1991a, *Ap. J.*, **382**, 576 (RII).

Ruderman, M. 1991b, *Ap. J.*, **382**, 587 (RIII).

Srinivasan, G., Bhattacharya, D., Muslimov, A., and Tsygan, A. 1990, *Current Sci.*, **59**, 31.

Thorsett, S., and Stinebring, D. 1991, *Ap. J.*, **361**, 644.

Magnetic Field Evolution in Isolated and Recycled Pulsars

Roger W. Romani

Department of Physics, Stanford University

1 Introduction

Although the possibility of neutron stars (NSs) was discussed almost 60 years ago by Landau and by Baade and Zwicky, the central attribute that most controls their astrophysical appearance, strong large-scale magnetization, was not considered until shortly before the discovery of radio pulsars (*e.g.*, Pacini). However, in the 25 years that we have been studying pulsars, it has become clear that NS magnetic field histories are very important for understanding their origin and evolution. The observational situation has also been the key driver in our modeling of NS magnetic fields. Recent discoveries of new classes of pulsars, as well as improved surveys of well known populations have provided an expanded range of parameters for which we have measurements of magnetic properties.

This observational picture is not fully settled, but several basic trends are emerging — these should drive us towards a coherent picture of magnetic field evolution. There are three striking observational facts that seem robust: *i*) Young NS have (or acquire within $\sim$ 100y) a very characteristic initial field value $B_i \sim 10^{12.5}$G, with little dispersion (Narayan and Ostriker 1990). *ii*) During the $\sim 10^7$y over which these objects are visible as radio pulsars, there is little change in their magnetization. *iii*) Low field (10^{-2}-$10^{-4}B_i$) "recycled" pulsars exist and remain visible for over 10^9y. These lead to several questions. What determines the magic value of B_i? What is B in the pulsar "graveyard"? Why do low, but finite and persistent, fields appear in "recycled" pulsars? In this paper one scenario for interpreting this evolution in terms of initial field growth and accretion-induced decay is discussed.

2 Pulsar Field Observations

Here we wish only to set the stage for the subsequent discussions. For a clear treatment of recent observations see Battacharaya (1989) and the references therein. Pulsar magnetic field measurements are inferred from observed braking torques, where the surface dipole field is $B_s \approx 10^{12}\,(P/s)\,(\dot{P}/10^{-15}\mathrm{s\ s^{-1}})\,$G. Recent observations of pulsar inclinations (*e.g.*, Lyne and Manchester 1988) indicate that geometric effects do not greatly perturb these field estimates. Spindown and spinup torques on magnetic moments can also provide useful estimates for bright X-ray binaries and in some cases,

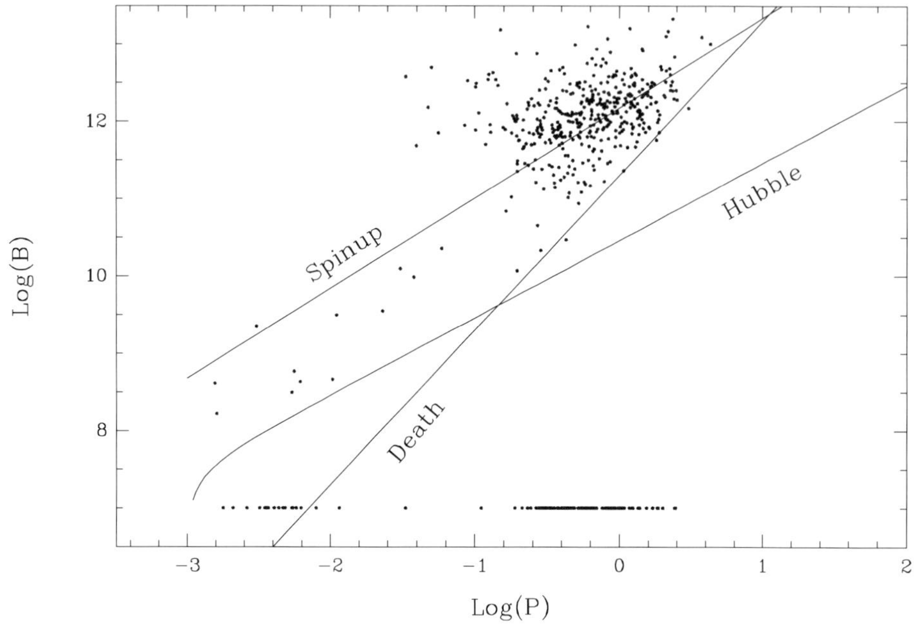

Fig. 1—Pulsars in the *P-B* diagram. Objects for which there is no published $\dot{P}$ are plotted at 10^7G.

direct measurement of surface fields are available from cyclotron resonances and related spectral features (Makishima, *et al.* 1992). These data support the radio pulsar inferences.

This is best displayed in the venerable $B-P$ digram (Figure 1). The pair-production 'death line' and the accretion pressure= magnetospheric pressure spinup line are also shown, with normalizations that should properly be viewed as phenomenological. The 'Hubble line' gives the final period after 10^{10}y evolution from the spinup line (or the $\sim$ 1ms dynamical limit) at constant magnetic field. If pulsars are born at the upper left and move through the diagram, then the objects at $P \sim 0.3$s, $B \sim 10^{11.5}$G were seen as evidence for decay of NS B fields. However, the continued presence of high field objects at these periods does then not allow the flow to be divergence free; this is the origin of the idea of "injection" (Vivekanand and Narayan 1981). The 'recycled pulsars' to the lower left give the key to long-term B evolution. Of course, to derive quantitative conclusions from this rather incomplete sample, detailed statistical analyses are required (*cf*. Wijers *et al.* 1992).

3 Isolated Pulsars Field Growth and Decay

In isolation we expect NS fields to suffer dissipative losses (ohmic decay), movement through the star (diffusive motion), and spatial re-arrangements (driven by stellar convective motions, Hall drift, *etc*.; *cf*. Goldreich 1992). The details of these processes depend on the locations of the fields, as well as the thermal history and structure of the star, so it is clear that the imprint left by the B field origin can have an effect on their late time evolution. In Romani (1990) one model for this initial field distribution is described, using a dynamo- or thermally generated field in the crust of the young NS (Blandford, Applegate and Hernquist 1983; Geppert 1992). The interest of such models in this context is that if the field reached is limited by the strength of the crust, then 10^{12-13}G fields naturally occur. The track of such field growth is shown in Figure 3.

Of the processes that can cause field loss, ohmic decay has been the the most extensively treated. Early fits to the data suggested a simple exponential decay with $\tau \sim 5$My (Figure 3). The value of ohmic processes for late time decay is that, for a given set of physical conditions, the decay remains (locally) a pure exponential independent of the field strength. Hence they are likely to be important in reaching low final field values. Of course, if this timescale remains so short, then all significant magnetization is rapidly lost. However, as the star cools, ohmic times in even the crust become quite long (Figure 2). It is clear, as shown by the late time track *iii*) in Figure 3, that ohmic processes alone will not give rise to low field NS. Note that the field configuration in this sum is most favorable to ohmic losses; field lines penetrating through the star will be carried by currents much less susceptible to this decay.

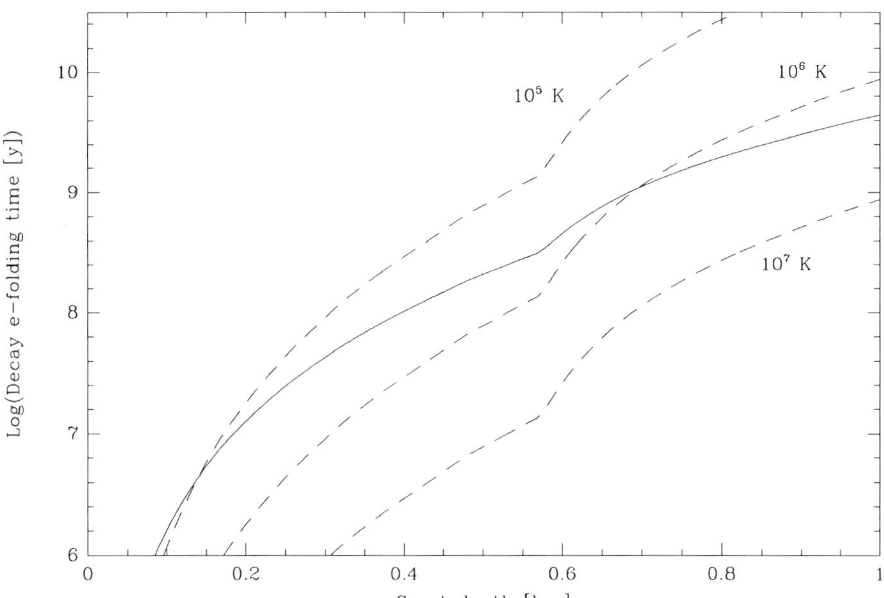

Fig. 2—Ohmic decay timescales as a function of depth. The solid line gives the decay due to impurity scattering, whereas dashed curves give the phonon scattering timescales at three temperatures. The shortest time will dominate, and pure ohmic decay to 10^8G requires $\approx 10\tau$, with τ that of the deepest significant current layer.

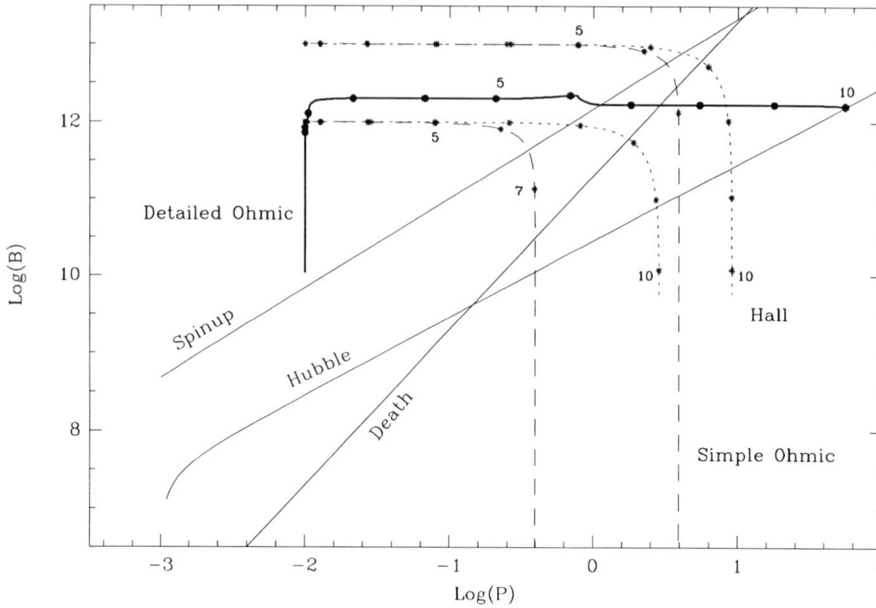

Fig. 3—NS field evolution in the $P - B$ diagram under *i)* dashed line - simple $\tau = 10^{6.5}$y ohmic decay, *ii)* dotted line - Hall cascade decay, *iii)* heavy line - Crustal growth and ohmic decay. Dots mark the time along the tracks in decade intervals; selected points are labeled with Log(t). Here pulsars are born at $P_i = 10^{-2}$s. Cases *i)* and *ii)* have pulsars injected at $B_i = 10^{12}$G and 10^{13}G.

With realistic thermal evolutions, ohmic losses in isolation will be small, leaving a NS with B^{11-12}G after a Hubble time. Other processes such as field re-arrangement *via* Hall currents may accelerate this decay at early times, if effective. Figure 3 includes tracks for Hall cascade-mediated decay with $\tau_H \sim 5 \times 10^7 \rho_{14} z_{km}^2 / B_{12}$y (Goldreich and Riesenegger 1992). However, since the cascade time gets quite long as the field decreases, decay beyond a few $\times 10^{10}$G may not occur, even as ohmic losses are proceeding on small scales.

4 Low Field Neutron Stars

All low field neutron stars at present are in binaries or can be inferred to have had an earlier binary phase. For radio pulsars this may be a tautology, since some addition of angular momentum above that represented by canonical ~ 1s spin periods is needed to make a low-field pulsar visible above the death line. Some independent evidence for low NS fields is inferred from the low mass X-ray binaries. The clear hypothesis that the binary interaction in some way causes the low NS fields has been adopted by a number of authors.

Observations of companions (Kulkarni 1990) and statistical arguments (*e.g.*, van den Heuvel *et al.* 1986, Narayan, Kulkarni, and Romani 1990) have proved important in outlining constraints on behavior during such binary decay. Most importantly, observations of cooling white dwarf companions (Kulkarni 1986) show that once fields reach values $< 10^{10}$G, the dipole fields do not spontaneously decay further (see also van den Heuvel, *et al.* 1986). Further, the value of this "asymptotic field" seems to span a wide range from 10^8G or less (*e.g.*, PSR1957+20) to perhaps $10^{11.5}$G (PSR0820+02, although this object may be too young to have experienced continued decay). Finally, statistical arguments suggest, particularly for the globular cluster sources, that the low B objects must descend from NS with high (or at least typical) fields.

It is apparent that neutron stars in binary systems are typically older than isolated pulsars, and indeed there is a general correlation of B with the binary's evolution time (Taam and van den Heuvel 1986). However, the first observation above shows that a simple age argument cannot fully describe the final field value. Instead, we must appeal to an interaction in the mass transfer itself, as opposed to the duration of the transfer phase. During the transfer, the binary acquires mass (ΔM), thermal energy ($E \sim G M_* \Delta M / R_*$), and angular momentum ($L \sim \Delta M v_{Kep}(r_{acc}) \cdot r_{acc}$). Scenarios that use each of these as the effective agent in B field reduction have been proposed (*e.g.*, E: Blondin and Freese 1986; L: Srinivasan *et al.* 1990). Although problems exist with each of these scenarios, no detailed comparison will be made here. Instead, since solutions that involve the mass directly (first apparently mentioned by Bisnovatyi-Kogan and Komberg 1975) are perhaps the simplest and most promising, one such picture will be described. Some observations that may help distinguish these models are mentioned at the end.

The basic idea is to use the overburden of the accreted matter to push magnetic field lines below the neutron star surface and thereby reduce the external dipole. In the magnetosphere, ionized plasma follows the field lines in an accretion column to the stellar surface, since flow times are very short, the conductivities are relatively high, and the magnetic pressure dominates. In the deep interior of the star, on the other hand, the flow speeds are very low (since the densities are large). Moreover, the conductivities are very high here and the magnetic pressures are negligible, so the field follows the matter motions. Accordingly, if material accretes down a magnetic pole in a high field neutron star, then it must flow towards the equator and into the stellar interior as the accretion continues. If the field lines can follow this motion, then the external dipole will be reduced.

Whether this advection of the magnetic field can be effected depends on the flow times of the matter and the field near the stellar surface. In particular, there will be a depth $z \sim B^2/(8\pi g\rho)$ in the accretion column, with g the NS surface gravity, at which matter pressure begins to dominate. At this point the matter will begin to flow freely. If the timescales for the magnetic field to slip past this flow are long enough, the field will follow. In Figure 4, the density ρ and depth z at which this free flow occurs are plotted as a function of the surface dipole field. The density structure of the neutron star atmosphere is based on a surface gravity of 10^{14}cm s^{-2}, a NS radius of 10km, and the accretion of a cosmic abundance plasma at $10^{-8} M_\odot$ y^{-1}. The local temperature is dominated by the heat released by the accretion, and the structure is followed as the degeneracy increases, the material solidifies, and the atomic Z of the accreted matter increases at high ρ (*cf*. Paczynski 1983).

In addition, several timescales of interest for the flow are plotted. At the accretion rate above, τ_{fl} is the time for the accretion column to be replenished to the depth z.

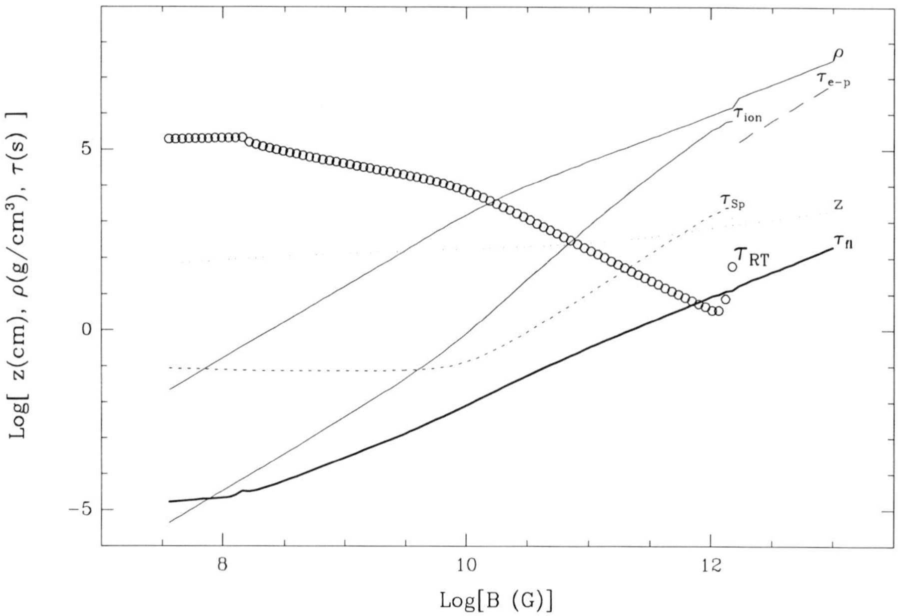

Fig. 4—Flow times and diffusion times on a NS polar cap under Eddington-rate accretion, as a function of the surface dipole field.

The other timescales describe effective diffusion times for the magnetic field to move through the matter. At very low densities, electron-ion scattering limits the conductivity, and the associated diffusion time through the accretion cap is plotted as τ_{ion}. At somewhat larger fields (higher densities) classical Spitzer diffusion proceeds more quickly, and the corresponding timescale is plotted as a short dash line. At high densities, the material solidifies, and at the relatively high temperatures maintained by the accretion energy, electron-phonon scattering limits the conductivity; τ_{e-p} shows the diffusion time through the accreting cap. Bohm diffusion and ambipolar diffusion are unimportant throughout the flow. The Raleigh-Taylor diffusion time in the fluid is also shown, where the length scale is limited to the local scale height of the atmosphere and if there are N such scales across the accretion cap, the time to percolate a fraction of order unity of the accreted material across the field is taken as $N^{1/2}$ times the dynamical time. Several effects may slow this overturn instability- for example, the divergence of field lines at the pole creates some shear stabilization, and the high temperature of the accreting matter helps decrease the density excess over the underlying field and crust. On the other hand, larger scale instabilities that have not been considered may shorten the interchange time.

Despite these uncertainties, some conclusions can be drawn from these timescales. When the flow time is less than the diffusion or interchange time, the field will be advected. For near-Eddington accretion rates it appears that advection can occur until the field is just below 10^8G. At this point, the accreted material flows freely in such a shallow layer of the star that resistive diffusion is adequate to allow it to slip past the magnetic field. Continued accretion will add mass to the upper crust, burying the underlying field generating currents deeper, but it will not decrease the external dipole further. This provides a natural "floor" for the field amplitude - preventing too many NS from being so strongly de-magnetized that they are lost as pulsars altogether. A second reason why fields $B_s \sim 10^8$G may provide an effective floor in accretion-induced decay can be noted by recalling that for such fields and $\dot{M} \sim \dot{M}_{Edd}$, the accretion ram pressure balances the magnetic pressure near the NS surface, and the accretion becomes nearly isotropic. Since anisotropic accretion is required to give the latitudinal field line motion and subduction near the magnetic equator, accretion should lose much of it's efficacy. For lower $\dot{M}$, although the decreased temperature increases the diffusion times, the flow times increase more rapidly. Below several percent of the Eddington accretion rate, $\tau_{fl} < \tau_{diff}$ through most of the B range; at low rates we can expect large amounts of mass accretion with little field reduction. At high $\dot{M}$, however, we need to accrete $\delta M \sim 10^{-4}$-$10^{-5} M_{\odot}$ to bury the screening currents deeply enough to change the external dipole by order unity. Accreted amounts of this order are expected in the common B star X-ray binaries and can provide the weakly-recycled pulsars that seemed to suggest field decay. This would resolve the question of injection. When accretion continues, we expect B to decrease as $\sim B_i(\delta M/M)$ until values $\sim 10^8$G are reached, carrying pulsars towards the lower left in Figure 3.

5 Conclusions

Scenarios allowing crustal field growth in young NS can provide an explanation for the "magic" pulsar field value. Ohmic decay alone cannot produce the recycled pulsars we see, although when mediated by other processes, such as Hall drift re-arrangements, it has some attractive features. Decay associated with mass accretion seems promising, but further tests of such models are needed.

Clearly, the most telling observation would be a measurement of the NS B in the pulsar graveyard. While it is possible that γ-Ray Burster cyclotron resonances provide precisely that, the recent BATSE results have made the situation less clear. To provide a clear test of accretion-induced decay models for recycled pulsars, we will need good estimates of the of mass transferred and accurate evolutionary models for a range of final B values. Unfortunately, this requires good measurements of the NS mass, which in general are available only for the NS-NS binaries, which have intermediate fields and therefore very small total mass accretion. The low-field objects are much more difficult to measure, generally involving studies of the faint companion, but some efforts are providing results (Aldcroft, Romani and Cordes 1992). It should be noted that the best hope for new data may lie in the globular clusters, where the exchange of a low-field strongly recycled pulsar into an existing relativistic binary can occur. AIC and direct collapse to fast pulsars may confuse the results to some degree.

I thank the organizers and sponsors of the Santa Barbara ITP for hospitality during this work and NASA grant NAGW-2963 for support.

References

Aldcroft, T., Romani, R. W. and Cordes, J. M. 1992, *Ap. J.*, submitted.
Bhattachraya, D. 1989, in *X-ray Binaries*, J. Hunt and B. Battrick, eds., (ESA: Noordwijk), 179.
Bisnovatyi-Kogan, G. S. and Komberg, B. V. 1975, *Sov. Astr.*, **18**, 217.
Blandford, R. D., Applegate, J. H and Hernquist, L. 1983, *M. N. R. A. S.*, **204**, 1025.
Blondin, J. M. and Frees, K. 1986, *Nature*, **323**, 786.
Geppert, U. 1992, this Workshop
Goldreich, P. 1992, this Workshop
Goldreich, P. and Riesenegger, A. 1992, Caltech preprint.
Kulkarni, S. R. 1986, *Ap. J.*, **306**, L85.
Kulkarni, S. R. 1990, in *Neutron Stars and their Birth Events*, W. Kundt, ed. (Kluwer: Dordrecht), 59.
Lynn, A. and Manchester, R. M. 1988, *M. N. R. A. S.*, **234**, 477.
Makishima, K, *et al.* 1992, Univ. Tokyo preprint.
Narayan, R. and Ostriker, J. 1990, *Ap. J.*, **352**, 222.
Paczynski, B. 1983, *Ap. J.*, **267**, 315.
Romani, R. W. 1990, *Nature*, **347**, 741.
Srinivasan, G, Bhattachraya, D., Muslimov, A. G. and Tsygan, A. I. 1990, *Curr. Sci*, **59**, 73.
Taam, R. E. and van den Heuvel, E. P. J. 1986, *Ap. J.*, **305**, 235.
van den Heuvel, E. P. J., van Paradijs, J. A. and Taam, R. E. 1986, *Nature*, **322**, 153.
Vivekanand, M. and Narayan, R. 1982, *J. Ap. Astr.*, **2**, 315.
Wijers, R. A. M. J., Verbunt, F., Bhattacharya, D., and Hartman, J. W. 1992 these Proceedings.

A Case for no Decay of the Magnetic Fields of Isolated Radio Pulsars

Ralph A. M. J. Wijers,[1] Frank Verbunt,[2] Dipankar Bhattacharya,[3] and Jan Willem Hartman[2]

[1]Princeton University Observatory
[2]Institute of Astronomy, University of Utrecht
[3]Raman Research Institute

Abstract

We have performed Monte Carlo simulations to investigate what the observed population of radio pulsars would be for various assumptions about the properties and evolution of the underlying total population. Comparison of simulated observed populations with the observed radio pulsars suggests that the magnetic field of a radio pulsar does not decay significantly during its active lifetime. This conclusion is not affected significantly by varying most model assumptions. A notable exception to this is the influence of the assumed dependence on pulse period of the pulsar beam width.

1 Introduction

The increase in pulse period observed in single radio pulsars corresponds to a loss in rotation energy of the neutron star. By equating this loss to the electromagnetic radiation emitted by a rotating magnetic dipole, one finds (Gunn and Ostriker 1970)

$$P\dot{P} = f(\alpha)\frac{8\pi R^6}{3Ic^3}B^2, \tag{1}$$

in which P and $\dot{P}$ are the pulse period and its derivative, R and I are the radius and moment of inertia of the neutron star, B the magnetic field strength (i.e BR^3 is the strength of the magnetic dipole), and α the angle between the rotation axis and the axis of the dipole field.

If the neutron star were radiating like a dipole in vacuo, $f(\alpha) = \sin^2\alpha$. It can be argued, though, that particle losses along the open field lines are a more efficient loss mechanism when α is small, so that $f(\alpha) \simeq 1$ for the combined energy loss via radiation and particles.

From a number of arguments, it appeared that the product $P\dot{P}$ decreases with pulsar age. This decrease has been mostly interpreted as evidence for the decay of the magnetic field strength B, although it could also be interpreted as evidence for alignment of the rotation and magnetic axes, if one keeps $f(\alpha) = \sin^2\alpha$ in equation (1). (We shall refer to the decrease of $P\dot{P}$ as 'field decay' in this paper.) With one exception, none of the specific arguments in favor of field decay has survived closer scrutiny combined with new observations (see, *e.g.*, the discussions by Bhattacharya 1989, 1992).

The exception is the statistical study of the radio pulsar population and its distribution in the (B, P)-plane (Narayan 1987, Narayan and Ostriker 1990). We decided, therefore, to investigate this remaining evidence for field decay more closely, also stimulated by the discovery that the extent of the electron density distribution perpendicular to the galactic plane is smaller than previously assumed, which affects pulsar distances determined from the dispersion measure (Bhattacharya and Verbunt 1991). Further incentive was provided by the fact that the notion of rapid field decay appears to conflict with our understanding of the evolution of certain X-ray binaries (*e.g.*, Verbunt *et al.* 1990).

Our results, reported in greater detail in Bhattacharya *et al.* 1992, indicate that $P\dot{P}$ does not decay significantly during the active lifetime of a pulsar. (This conclusion, however, is not a consequence of the smaller vertical scale height of the electron distribution.) In this paper we summarize our work and suggest some causes for the discrepancy between our conclusions and those of Narayan and Ostriker (1990).

2 A study of the local pulsar population

In our investigation, we simulate a population of radio pulsars and determine which pulsars of this simulated population would have been detected in four well-documented radio surveys. We then compare the simulated 'observed' pulsars with the real pulsars detected in the same four surveys and find the probability that the simulation correctly reproduces the properties of the observed pulsar population. By repeating the simulations for different parameters, we search for the parameter values that lead to the best simulation. To avoid an abundance of free parameters, we have limited our first study to pulsars whose projected position on the galactic plane is less than 3 kpc from the Sun, and assumed that the pulsar birth rate is uniform in this region, and constant in time.

2.1 The simulations

In more detail, we proceeded as follows. We assume that the properties with which a pulsar is born, viz. its pulse period P, magnetic field B, distance to the galactic plane z, and velocity in the z-direction v_z, are distributed according to the equations given in Table 1. We further assume that the fraction f_b of pulsars beamed towards the Earth is proportional to $P^{-\gamma}$, and that L, the luminosity at 400 MHz, is a function of P and $\dot{P}$ only, according to the relations given by either Narayan and Ostriker (1990, originally Proszyński and Przybycień 1984) or Stollman (1987). (The relations include a large spread of individual values of L around the average $L(P, \dot{P})$-relation, as well as the death line, $B/P^2 = 0.17 \times 10^{12}\,\mathrm{G\,s^{-2}}$, beyond which $L = 0$.)

Each single simulation is characterized by the parameters $\log B_0$, σ_B, h, σ_v, γ, τ, and the choice of luminosity function. In most simulations, we used standard values for h, σ_v, and γ, and the $L(P, \dot{P})$-relation given by Narayan and Ostriker (1990). We varied these parameters in a limited number of test simulations. For five values of τ, we performed simulations for a large number of $(\log B_0, \sigma_B)$ combinations.

TABLE 1

An overview of the key parameters and the values for which simulations were performed. For the first three parameters, a search of parameter space was conducted in the range indicated. For the others, we adopted the standard value listed and tried a more limited number of simulations for the values in brackets.

	Equation	Parameter Values
initial magnetic field strength ($y \equiv \log B_i$)	$p(y)\,dy = \frac{dy}{\sigma_B\sqrt{2\pi}} \exp\left(-\frac{1}{2}\left[\frac{y-\log B_0}{\sigma_B}\right]^2\right)$	$12.2 < \log B_0 < 12.6$ $0.18 < \sigma_B < 0.46$
field evolution	$B(t) = B_i \exp(-t/\tau)$	$5 < \tau < 300\,\mathrm{Myr}$
initial period	P_i	0.1 s
beaming law	$f_b \propto P^{-\gamma}$	$\gamma = \frac{1}{2}\ (\frac{1}{3},0)$
initial velocity ($v \equiv v_{zi}$)	$p(v)\,dv = \frac{1}{\sigma_v\sqrt{2\pi}} \exp\left(-\frac{1}{2}\left[\frac{v}{\sigma_v}\right]^2\right)\,dv$	$\sigma_v = 110(200)\,\mathrm{km\,s^{-1}}$
initial z ($z \equiv z_i$)	$p(z)\,dz = \frac{1}{2h} \exp\left(-\frac{\lvert z\rvert}{h}\right)\,dz$	$h = 60$ pc

Each individual simulation starts with a Monte Carlo drawing of the properties of the newborn pulsar, viz. P_i, B_i, z_i, and v_{zi}, from the distributions given in Table 1, as well as the age t that it lives before being 'observed'. P, B, z, and v_z are then evolved over time t to their current value, with use of equation 1 and of the galactic potential at the location of the Sun (from Kuijken and Gilmore 1989). We then calculate the current beaming factor and luminosity. The pulsar is given a projected (x, y)-position in a circle of 3 kpc radius centered on the Sun, and its dispersion measure DM is calculated from its position (x, y, z) and the electron distribution, for which we slightly modify the equation given by Lyne *et al.* (1985). We then check whether any of the four following surveys would have detected it: the Jodrell Bank survey (Davies *et al.* 1972), U Mass-Arecibo survey (Hulse & Taylor 1974), Second Molonglo survey (Manchester et al. 1978), and U Mass-NRAO survey (Damashek *et al.* 1978). (See Bhattacharya *et al.* 1992 for the exact detection criteria used.) This procedure is repeated until 2000 pulsars are 'detected'.

2.2 Measure of agreement between model and data

To measure the success of each simulation, we use four directly observable quantities, viz. P, B, L and $DM \sin b$, where b is the galactic latitude. $DM \sin b$ is a directly observable quantity that is closely related to the height z above the galactic plane. For each of these quantities i, we perform a Kolmogorov-Smirnov test to determine the probability $Q_{KS}(i)$ that the simulated 'observed' and the actually observed distributions are drawn from the same parent population. We then assign an overall measure of success to the simulation by taking the product $Q = Q_{KS}(P)\,Q_{KS}(B)\,Q_{KS}(L)\,Q_{KS}(DM \sin b)$. An important point to note is that the cutoff in projected distance at 3 kpc and the exact same detection criteria as in the simulations were also applied to select the 130 real observed pulsars that ended up in the comparison sample.

3 Results

In Figure 1, the positions and values of the maxima of Q are plotted in the ($\log B_0$, σ_B)-plane for 5 values for the field decay time scale τ, for the simulations where the other parameters were at their standard values (see Table 1). Also shown is one iso-Q contour for each τ at the level $1/e^2$ times the maximum value. (We note a subtlety in comparing the Q values of different τ: the $Q_{\rm KS}$ values are a priori probabilities that an observed distribution and a simulated distribution for a given choice of parameters are statistically identical. To get the likelihood of the best-fit model, a correction is required for the fact that we select the best value, *i.e.*, we have to calculate the probability that the highest Q for a given decay time has the value found, which is different from Q itself. The analogy of this in standard χ^2-minimization is reducing the number of degrees of freedom when fitting parameters.)

It is seen from Figure 1 that the best models are those with the longest decay times, and that the difference in quality between short and long decay times is substantial. A detailed discussion of the sensitivity of the position and Q value of the minimum to variation of our assumptions is presented in Bhattacharya *et al.* 1992. The main conclusions of the detailed discussion are:

(*i*) long decay times are superior also for other choices of the parameter values than the standard ones listed in Table 1. An example of this is shown in Figure 2, which is the analogue of Figure 1 with one change: we used the Stollman luminosity law. It is seen that the character of the contours and the trend of maximum Q with decay time is preserved.

(*ii*) the exception to (*i*) is the beaming law: if pulsar beaming does not depend on pulse period ($\gamma = 0$) short decay times are mildly favored, whereas in the intermediate case ($\gamma = 1/3$) the long decay times are favored, but the difference is small. Although the general opinion is that pulsar beaming does depend on pulse period, the exact form of this dependence and its relation to alignment are not agreed upon (Lyne and Manchester 1988, Rankin 1990). A final judgment on field decay requires better knowledge of the beaming law and the related problem of alignment (decrease of α with time).

(*iii*) The use of more quantities in the maximization procedure (notably the characteristic age) does not alter our main conclusion because other quantities are also well described by the good models. This point is illustrated in Figure 3: the cumulative distributions of observed and simulated populations are shown for decay times of 10 and 100 Myr. The quantities plotted are $\sin b$, DM, and $\dot{P}$. The main difficulty with the model for $\tau = 10$ Myr is the fact that it cannot simultaneously fit the distributions of B and P well (see Figs. 7 and 8 of Bhattacharya *et al.* 1992); it also has difficulties with the distributions of $\dot{P}$ and characteristic age. L and $DM \sin b$ are almost equally well described by simulations with short and long decay times.

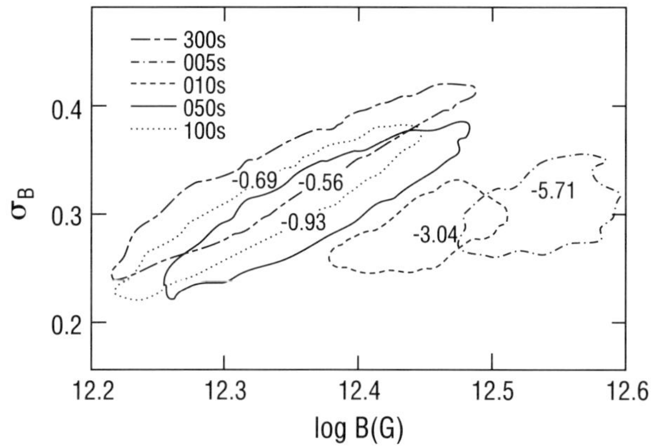

Fig. 1—The positions and values of the (logarithms of the) maxima of Q in the $(\log B_0, \sigma_B)$-plane for 5 values of the field decay time. The contours around the maxima are drawn at a value of Q equal to $1/e^2$ times the maximum.

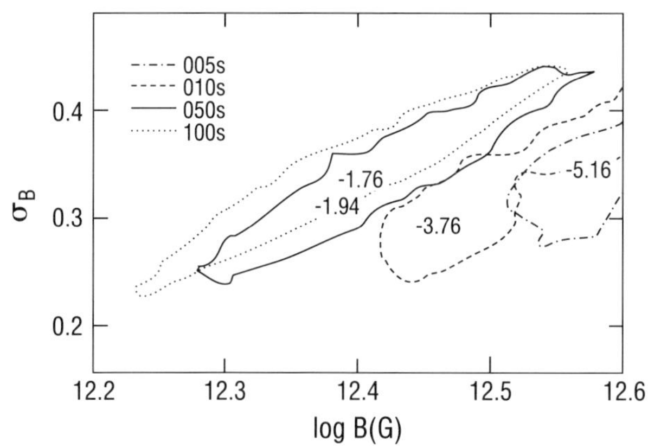

Fig. 2—As Fig. 1, for 4 values of the decay time, but with the Stollman luminosity law.

4 Discussion and Conclusion

We wish to briefly point out some differences between our work and that of Narayan and Ostriker (1990). First, they use 8 surveys in their comparison sample, including all the 4 we used. They applied no distance cutoff and simulated the entire galaxy, with the restriction that they also considered only vertical motion of the pulsars and fixed the dependence of the pulsar birth rate on distance from the center of the galaxy.

One of the surveys they included was the Jodrell Bank 1400 MHz survey of the northern galactic plane (Clifton *et al.* 1992). This survey is crucial, as the pulsars found by it are ten times younger on average than the typical pulsar found in previous surveys. However, a pulsar spectral index of 1.5 was assumed to compare this survey with the other ones, conducted at or near 400 MHz. The necessity of adopting a value for the poorly known spectral index introduces appreciable uncertainty. The highly non-uniform sky coverage of the Princeton-NRAO surveys (Dewey *et al.* 1985,

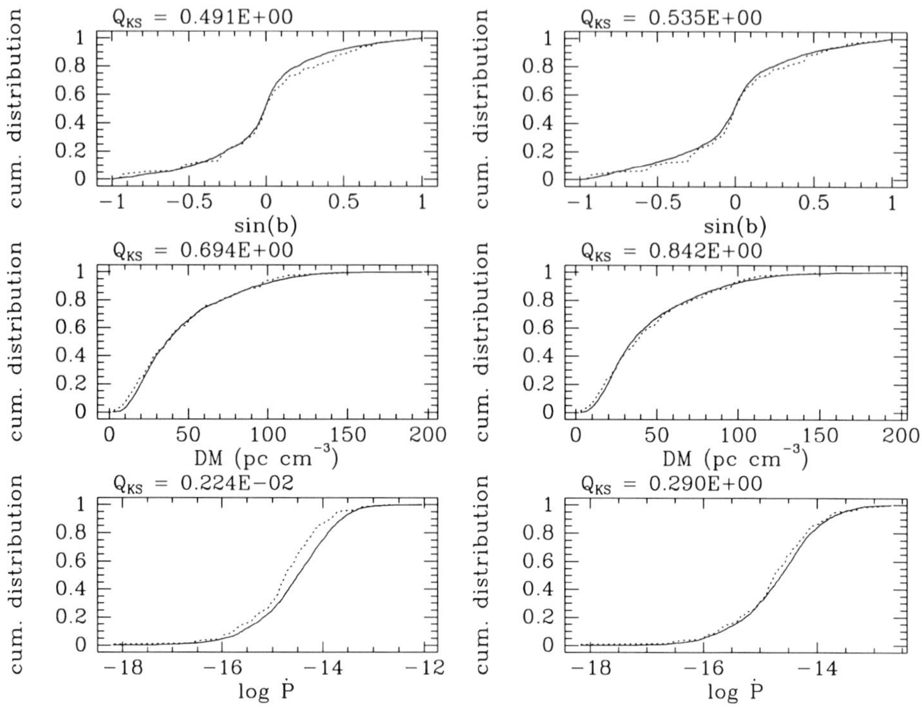

Fig. 3—The observed (dashed lines) and simulated (full lines) distributions of $\sin b$, DM, and $\dot{P}$. Note that the model with a short decay time produces too many observed pulsars with high values of $\dot{P}$. Results for the best simulation with standard parameters at 10 (100) Myr are shown in the left (right) panel.

Stokes *et al.* 1986) has not been taken into account, which also leads to appreciable uncertainties.

In addition, Narayan and Ostriker have fitted for more and different quantities than we have. They included the values of z rather than DM $\sin b$, and did not account for the finite scale height of the electron layer. These differences do not seem to contribute significantly to the discrepancy of our conclusions. They also included the measurements of pulsar velocities and the correlation of these with magnetic-field strength in their comparison, but they do not seem to have included any consideration of the selection effects on velocity observations. It is therefore not clear whether their inclusion of velocities is an advantage. The recent measurement of many more pulsar velocities has also weakened the velocity-field correlation (Harrison *et al.* 1992). A quantity they did not include in the fits was the pulsar luminosity.

Their model moreover is much more elaborate. It allows for two different populations whose birth parameters differ substantially. They find that two populations are required to fit the data. At short decay times, this agrees with our findings that a single population cannot fit the data well enough. Why they obtain a bad fit at long decay times and we find a good one is not understood at present and is the subject of further investigation by both groups.

RAMJW acknowledges support from NASA as a Compton Fellow under grant NAG 5-1816. FV acknowledges support from NWO under grant PGS 78-277.

References

Bhattacharya, D., 1989, in J. Hunt and B. Battrick (eds.) *X-ray Binaries,* Proceedings of the 23rd ESLAB Symposium, Bologna, Italy, ESA SP-296, p. 179

Bhattacharya, D., 1992, in J. Gil, T. H. Hankins, and J. Rankin (eds.) *Magnetospheric Structure and Emission Properties of Radio Pulsars*, Proceedings of IAU Colloquium 128, Pedagogical University of Zielena Gora press, Poland, p. 27

Bhattacharya, D., and Verbunt, F., 1991, *Astron Ap.* **242**, 128

Bhattacharya, D., Wijers, R. A. M. J., Hartman, J. W., and Verbunt, F. 1992, *Astron. Ap.* **254**, 198

Clifton, T. R., Lyne, A. G., Jones, A. W., McKenna, J., and Ashworth, M., 1992, *Mon. Not. R. Astron. Soc.* **254**, 177

Damashek, M., Taylor, J. H., and Hulse, R. A., 1978, *Ap. J.* **225**, L31

Davies, J. G., Lyne, A. G., and Seiradakis, J. H., 1972, *Nature* **240**, 229

Dewey, R. J., Taylor, J. H., Weisberg, J. M., and Stokes, G. H., 1985, *Ap. J.* **294**, L25

Gunn, J. E., and Ostriker, J. P., 1970, *Ap. J.* **160**, 979

Harrison, P. A., Lyne, A. G., and Anderson, B., 1992, in E. P. J. van den Heuvel and S. A. Rappaport (eds.) *X-ray binaries and the formation of Binary and Millisecond Pulsars*, Kluwer, Dordrecht, in press.

Hulse, R. A., and Taylor, J. H., 1974, *Ap. J.* **191**, L59

Kuijken, K., and Gilmore, G., 1989, *Mon. Not. R. Astron. Soc.* **239**, 605

Lyne, A. G., and Manchester, R. N., 1988, *Mon. Not. R. Astron. Soc.* **234**, 477

Lyne, A. G., Manchester, R. N., and Taylor, J. H., 1985, *Mon. Not. R. Astron. Soc.* **213**, 613

Manchester, R. N., Lyne, A. G., Taylor, J. H., Durdin, J. M., Large, M. I., and Little, A. G., 1978, *Mon. Not. R. Astron. Soc.* **185**, 409

Narayan, R., 1987, *Ap. J.* **319**, 162

Narayan, R., and Ostriker, J. P., 1990, *Ap. J.* **352**, 222

Proszyński, M. and Przybycień, D., 1984, in S. P. Reynolds and D. R. Stinebring (eds.) *Millisecond Pulsars*, NRAO, p. 185

Rankin, J., 1990, *Ap. J.* **352**, 247

Stokes, G. H., Segelstein, D. J., Taylor, J. H., and Dewey, R. J., 1986, *Ap. J.* **311**, 694

Stollman, G. M., 1987, *Astron. Ap.* **178**, 143

Verbunt, F., Wijers, R. A. M. J., and Burm, H. M. G., 1990, *Astron. Ap.* **234**, 195

Remarks on Triaxiality, Spin-Up and Magnetic Field Obliquity of Some Millisecond Pulsars

W. Y. Chau,[1,2] K. S. Cheng,[1] and J. L. Zhang[1,3]

[1]Department of Physics, University of Hong Kong, Hong Kong
[2]Department of Physics, Queen's University, Canada
[3]Department of Physics, University of Science and Technology, PRC

Abstract

We show how two "observed" properties of millisecond pulsars—a very small non-axisymmetry in the mass distribution and the near orthogonality between the rotational and the magnetic axis—can find a very natural explanation within the context of the collapse-formation and evolution of such objects modeled by uniform density ellipsoids. This could also provide alternate explanations to the observed spin-up of some pulsars in globular clusters.

1 Introduction

Interpretation of observational data on millisecond pulsars (MSPs) has led to some constraints on theoretical models for such objects. In particular, the observed period change together with the spin down via gravitational radiation sets a limit on the extent of non-axisymmetry in the mass distribution (Alpar and Pines 1985, Cheng and De Jager 1990). If the model is that of a uniform density ellipsoid, with axis a_1, a_2 and a_3 (parallel to the rotational axis), the limit can be expressed as one on the ellipticity, $\in$, defined as $\in = (a_1 - a_2)/a_1, \leq 10^{-7-8}$. In addition, observations of four weakest field fast MSPs show three to have a double pulse shape with frequency independent subpulse separation near 180° (Lyne and Manchester 1988, Fruchter, Stinebring and Taylor 1988). This indicates that the inclination angle between the rotational and magnetic axis, to be referred to as the obliquity χ in the rest of this paper, is near 90^0.

We will show in the following how these two "observed" features of MSPs can find very natural explanation within the context of the collapse - formation and evolution of young pulsars modeled by uniform density ellipsoids. Our scenario would not be applicable to those MSPs which are believed to be old neutron stars spun-up by accretion and which would have in any case nearly spin-aligned magnetic dipole moment (Ruderman 1991). Specifically, our ideas are relevant to those MSPs believed to be formed by accretion induced collapse of white dwarfs (Michel 1987, Narayan and Popham 1989), and would provide an alternate explanation for the nearly 90^0 obliquity observed to one that involves the breaking and movement of the crustal plates (see also Ruderman 1991). We will also remark on the question of spin-changes (notice, not necessarily spin-down) with the loss of angular momentum for these objects. This may be relevant to the observed spin-up of some pulsars in globular clusters which was explained by acceleration due to the mean gravitational potential of the cluster (Anderson *et al* 1990, Phinney 1991).

2 The Model

The model is that of a rotating, incompressible ellipsoid of uniform density ρ with axis a_1, a_2 and a_3 (being parallel to the axis of rotation Ω). Such objects have been extensively studied (see, *e.g.*, Chandrasekhar 1986), and have been invoked as the simplest, relevant model for a variety of pulsar phenomena. Examples include the evolution of pulsars with allowance for change of moment of inertia (Chau, Henriksen and Rayburn 1970), the possible relevance to bi-period, single x-ray pulsars (Henriksen, Feldman and Chau 1971), the emission of gravitational radiation from rapidly rotating, collapsing/exploding objects (Chau, Chia and Henriksen 1977.)

The sequence of equilibrium configuration of such objects can be characterized by a single parameter, the angular momentum J. For small values of J, the configuration is axisymmetric ($a_1 = a_2$), and these are classically known as the Maclaurin spheroids. The angular velocity Ω and the eccentricity e (defined to be $e^2 = 1 - \frac{a_3^2}{a_1^2}$) increase with increasing J to a value of $e_b = 0.81267$ and $\frac{\Omega_b^2}{\pi G \rho} = 0.37423$ at $J_b = 0.30375(GM^3 < a >)^{\frac{1}{2}}$, where M is the mass and $< a >= (a_1^2 a_3)^{\frac{1}{3}}$, when the sequence bifurcates into two, the original axisymmetric one **and** a triaxial one (classically known as the Jacobi ellipsoid). The Jacobi sequence is energetically more favorable than the Maclaurin one, and thus if the object can be spun-up to beyond the bifurcation value of J_b (or equivalently Ω_b, e_b) **BEFORE** breaking-up (via equatorial mass shedding, for example) at the critical value J_c, say, the object can then assume a triaxial shape. These Jacobi ellipsoids have one **markedly** different property as compared with the Maclaurin spheroids: they **SPIN-UP** with the loss of angular momentum.

This brief review of the properties of such objects would be adequate for the applications described in the following sections.

3 The Obliquity

We have shown (Chau and Zhang 1991) that the angle of inclination χ between the Ω-axis and the magnetic field $\vec{B}$ axis would increase to $\leq 90^0$ with the formation of the neutron star by collapse from a white dwarf, under quite general conditions. We assume frozen field condition, and also that the collapsing, rotating ellipsoidal object **always** maintains an ellipsoidal shape. The required velocity field (Hubble flow) then leads to the following relation between the final values of obliquity and eccentricity (χ_f, e_f) and the initial ones (χ_o, e_o), if $\vec{B}$ originally lies on the (a_1, a_3) plane:

$$\cot \chi_f = \left(\frac{1 - e_f^2}{1 - e_o^2} \right)^{1/2} \cot \chi_o. \tag{1}$$

In our calculation, the object collapses to become an extremely flattened, triaxial neutron star with therefore $e_f \to 1$. Thus, for nearly all values of e_o, χ_o, the final $\chi_f \leq 90^0$.

Of course, this final configuration could have an angular momentum J exceeding the critical one for the onset of equatorial mass loss, and we can then imagine mass shedding to occur, reducing J to the value for stable equilibrium. This would have no effect on the magnetic field axis, since it is virtually lying on the equatorial plane. Subsequent evolution of this triaxial object towards the bifurcation point with the loss of J via, say, gravitational radiation would not change χ far from 90^0, since $e_b \sim 0.81267$. Thus millisecond pulsars with $\vec{B}$ nearly perpendicular to $\vec{\Omega}$ seem to find a natural explanation in this scenario.

We realize, of course, that some time before the final configuration is attained, J could already have exceeded the critical value and mass shedding could have started. We have to assume that the mass loss during the short collapse time does not invalidate the use of equation (1) for the estimate of χ_f.

4 Triaxiality

As mentioned previously, the observed spin-down rate together with an analysis of spin down via gravitational radiation imposes a limit of $\in < 10^{-(7--8)}$ for millisecond pulsars. Consider a new-born pulsar with period P for which we can then immediately work out

$$\frac{\Omega^2}{\pi G\rho} = \frac{4\pi^2}{P^2\pi G\rho} \sim 0.3768 P^{*-2}\rho^{*-1}, \tag{2}$$

where the period P^* and density ρ^* are now in units of msec, and 5×10^{14} g/cc respectively.

The small value of $\in$ suggests that millisecond pulsars are very nearly axisymmetric. In fact, at the **exact** bifurcation point,

$$\left(\frac{\Omega^2}{\pi G\rho}\right)_b = 0.37423,$$

which would lead to:

$$P = 1.003\rho^{*-1/2}\text{ms}. \tag{3}$$

The above suggests that millisecond pulsars are located just in the neighborhood of the bifurcation point. This suggestion is intriguing, and indeed models with density ranging from 10^{14} g/cc to 10^{15} g/cc would yield a period range $2.243 \to 0.709$ ms **if** we require the new born (or millisecond) pulsar to be **at** the bifurcation point, which is, after all, a classical, hydrodynamical result for rotating uniform density configurations. We discuss the two possible situations separately:

(a) If $J < J_b$, then the object would be axisymmetric (Macluarin spheroids), and there should be **no more** gravitational radiation losses. The situation, of course, would be different if there is some mass ellipticity from non-hydrostatic origin (fossil ellipticity; see, *e.g.* Chau (1970), Finn and Shapiro (1990)). The effects on the spin down **rate** for such objects with the loss of angular momentum, via, for example, magnetic dipole radiation, allowing consistently for the change in moment of inertia of the object, have been studied for a long time (Chau, Henriksen and Rayburn 1971).

(b) **IF** $J > J_b$, **AND** the object can indeed assume a triaxial shape (we will return to this point later), then a series of interesting observational consequences may result.

In the first place, gravitational radiation would remove angular momentum, leading to a **decreasing** ellipticity **but** an **increasing** angular velocity. Thus, in this situation, we are talking **not** about a different spin-down rate as in (a), but a spin-**UP** instead! The time the object can spend in this phase can be estimated. Assuming constant $\in$, we have (Chau 1967)

$$\Omega^4(t) = \frac{\Omega_o^4}{1+kt}, \tag{4}$$

where $k \sim \Omega_o^4(\frac{256}{25})\frac{GM}{c^5} \in^2 < a >= 2.763\times10^{-7}(\in_{-8}^2/p_o^{*4})\text{year}^{-1}$ Here $\in_{-8}$ is in units of 10^{-8}, and we have take $M = 2\times10^{33}$g, $< a >= 10^6$cm.

Taking $\Omega_o \sim 6.8\times10^3$ rad s^{-1}, $\Omega(t) \sim 4.2\times10^3$ rad s^{-1}, corresponding to values at the critical point and bifurcation point respectively, for $\rho \sim 4\times10^{14}$ g/cc (stiff equation of state), we have $t \sim 5.53\times10^6\frac{1}{\in_8^2}$ years. To be more realistic, we should allow for $\in$ to change with Ω (Jacobi ellipsoids), but the above should give a plausible estimate for the time-scale that the object would spend on the Jacobi sequence. It would appear that even if the young pulsar assumes a triaxial shape, gravitational radiation could rapidly ($\sim 10^6$ years) reduce its ellipticity, drive it towards the Maclaurin sequence by removing its angular momentum, and **spin-up** the pulsar in the process. This could then explain very naturally the required small ellipticity for young pulsars. This can also provide an alternate explanation for the observed spin-up for some pulsars in globular clusters, as compared with the acceleration model (Anderson *et al.* 1990). We remark that spin-up with the loss of J for rapidly rotating **axisymmetric** polytropes had also been discussed by Shapiro *et al.* (1990).

It is interesting that Chandrasekhar (1970) has demonstrated the existence of secular instability at the bifurcation point induced by gravitational radiation reaction, and pointed out the possibility of fragmentation. The observational implications still remain to be explored.

We note, however, what actually happens at the bifurcation point depends critically on the presence of dissipation. With the inclusion of other dissipation than gravitational radiation, the situation becomes more complicated. Whether a Jacobi ellipsoid or a Dedekind ellipsoid, or even no triaxial configurations, will form depends on the relative strengths of viscosity and gravitational radiation reaction (Detweiler and Lindblom 1977; Shapiro and Teukolsky 1983). This whole topic of oscillations and stability of rapidly rotating objects receives increasing attention recently partly because of its relevance to MSPs (see, for example, Imamura, Friedman and Durisen 1985, Ipser and Lindblom 1990).

We now return to the question of whether the object can actually acquire a triaxial shape. As discussed in §2, this depends on a comparison of J_b and J_c. Without going into the necessary, fully dynamical calculation, an estimate can be made by assuming a polytropic equation of state for the object (Tsuruta snd Cameron 1967; Chau and Srulovicz 1971). For such an equation of state, $p = k\rho^{1+\frac{1}{n}}$, it has been shown that bifurcation before breaking-up is possible for $n \le 0.808$ (James 1964).

For complete polytropes, the ratio between the mean density $< \rho >$ and the central density ρ_c is a unique function of n. Thus

$$\frac{< \rho >}{3\rho_c} = -\left[\frac{3}{\xi}\frac{d\theta}{d\xi}\right]_1, \tag{5}$$

where θ and ξ are the dimensionless density and radius, and the bracketed quantity evaluated at the first zero ξ_1, *i.e.*, $\theta(\xi_1) = 0$. Neutron star models constructed with realistic equation of state (EOS) are of course **not** complete polytropes. We can nonetheless obtain an estimate by calculating an effective polytropic index n_{eff}, based on a computation of the ratio $< \rho > /3\rho_c$ for realistic EOS. The results are presented in Table 1. Thus, it appears that for a number of EOS, the models so constructed can indeed have an n_{eff} making it possible for them to assume a triaxial configuration.

5 Summary and Conclusion

Within the context of collapse-formation and evolution of pulsars modeled by uniform density ellipsoids, we have shown that the small non-axisymmetry in the mass distribution and the near orthogonality between the rotational and magnetic axis as suggested by observations find a very natural explanation. We also obtain the intriguing result that this model would place millisecond pulsars in just the neighborhood of the classical bifurcation point between the axisymmetric Maclaurin sequence and the triaxial Jacobi sequence. We show that a number of equations of state permits pulsar models to attain a triaxial, non-axisymmetric configuration. The object would evolve on a short time scale with the loss of angular momentum via gravitational radiation towards the bifurcation point, with decreasing ellipticity but increasing angular velocity. This evolution could provide a natural explanation for the observed spin-ups of some pulsars in globular clusters, as well as the rarity of such objects.

In the early section of this conference, Ruderman (this volume) has pointed out that there is another explanation for 180^0 phase separation of pulses of millisecond pulsars.

TABLE 1
Effective Polytropic Index of Various Models

Model$^{(*)}$	π	R	BJ	TI	MF
n_{eff}	0.922	0.756	0.808	0.752	0.741

The stellar mass is chosen to be $1.4 M_\odot$.
$^{(*)}$ Models used here are π (Maxwell and Weiss 1976), R (Pandharipande 1971), BJ (Bethe-Johnson 1974), TI (Pandharipande and Smith 1975a), MF (Pandharipande and Smith 1975b).

References

Alpar, M. A. and Pines, D. 1985, *Nature*, **314**, 334.
Anderson, S. B. *et al.* 1990, *Nature*, **346**, 42.
Bethe, H. A. and Johnson, M. B. 1974, *Nucl. Phys.*, **A230**, 1.
Chandrasekhar, S. 1970, *Phys. Rev. Letts.*, **24**, 611.
Chandrasekhar, S. 1986, *Ellipsoidal Figures of Equilibrium* (New York: Dover)
Chau, W. Y. 1967, *Ap. J.*, **147**, 664.
Chau, W. Y. 1970, *Nature*, **228**, 655.
Chau, W. Y., Chia, T. T. and Henriksen, R. N. 1977, *Ap. J.*, **214**, 576.
Chau, W. Y., Henriksen, R. N. and Rayburn 1971, *Ap. J.*, **168**, 179.
Chau, W. Y. and Srulovicz, P. 1971, *Phys. Rev.*, **D3**, 1999.
Chau, W. Y. and Zhang, J. L. 1991, *Ap. & Ap. Sci.*, in press.
Cheng K. S. and de Jager, O. C. 1990, *Astr. Ap.*, **234**, 273.
Detweiler, S. L. and Lindblom, L. 1977, *Ap. J.*, **213**, 193.
Henriksen, R. N., Feldman, P. A. and Chau, W. Y. 1972, *Nature*, **234**, 450.
Imamura, J. N., Friedman, J. L. and Durisen, R. N.1985, *Ap.J.*, **294**, 474.
Ipser, J. R. and Lindblom, L. 1990, *Ap. J.*, **355**, 226.
Finn, L. S. and Shapiro, S. L. 1990, *Ap. J.*, **359**, 444.
Fruchter, A., Stineberg, D. and Taylor, J. 1988, *Nature*, **333**, 237.
James, R. A. 1964, *Ap. J.*, **140**, 552.
Lyne, A. and Manchester, R. 1988, *M. N.R. A.S.*, **234**, 447.
Maxwell, O. and Weise, W. 1976, *Phys. Lett.*, **62B**, 159.
Michel, C. F. 1987, *Nature*, **329**, 310.
Narayan, R. and Popham, R. 1989, *Ap. J. Lett.*, **346**, L25.
Pandharipande, V. R. and Smith, R. A. 1975a, *Nucl. Phys.*, **A237**, 507.
Pandharipande, V. R. and Smith, R. A. 1975b, *Phys. Lett.*, **57B**, 205.
Pandharipande, V. R. 1971, *Nucl. Phys.*, **A178**, 123.
Phinney E. S. 1991, *M. N.R. A.S.*, in press.
Ruderman, M. A. 1991, *Ap. J.*, **366**, 261.
Shapiro, S. L. and Teukolsky, S. A. 1983, *Black Holes, White Dwarfs and Neutron Stars* (Wiley-Interscience).
Shapiro, S. L., Teukolsky, S. A. and Nakamura, T. 1990, *Ap. J.*, **357**, L17.
Tsuruta, S. and Cameron, A. G.W. 1966, *Nature*, **211**, 359.

X-Ray Emission from Single Pulsars

Hakkı Ögelman[1,2]

[1]Department of Physics, University of Wisconsin
[2]Max-Planck Institut für Extraterresrische Physik

1 Introduction

We classify the neutron stars in two main classes: *accretion powered* (binary) versus *rotation powered* (mostly single). The latter category is the subject of this workshop on isolated pulsars. The designation *pulsar* generally implies a radio-emitting single neutron star; however, I shall use it to mean a rotation-powered neutron star.

Of the presently known $\sim$ 500 pulsars there are only 10 confirmed and 6 unconfirmed X-ray detections. No doubt we shall discover more X-ray emitting pulsars as the state of technology and instrumentation advances. The point is, however, that the X-ray band is not the best wavelength range to detect the known radio pulsars which are mostly older than 10^6 years. On the other hand, pulsars younger than 10^5 years appear to be good candidates for X-ray emission. As a matter of fact, some of the very interesting pulsars like PSR 0540-69, PSR 1509-58, and Geminga have been first detected to pulse in X-rays.

The primary purpose of this review is to present an up-to-date check list of X-ray emitting pulsars (Table 1). In addition, I will also list briefly the different mechanisms envisaged for the X-ray emission of isolated pulsars and give the highlights known about these objects including some recent *ROSAT* findings.

2 Emission Mechanisms

The X-ray emission of pulsars comes either from the very close vicinity of the pulsar (inside the light cylinder or the surface) or perhaps from a compact synchrotron nebula associated with the relativistic electrons emitted by the pulsar. I am using the term "compact" nebula here in order to differentiate it from the larger supernova remnant (SNR) which is usually the thermal emission of hot shocked gas.

2.1 Compact Synchrotron Nebula

The X-rays associated with the compact nebulae of pulsars are believed to be the synchrotron emission of relativistic electrons accelerated by the low-frequency waves from the pulsar. There is ample energy in the rotational energy loss of the pulsar to create a powerful relativistic electron-positron wind as well as to transport magnetic fields in which they can synchrotron radiate.

2.2 Magnetospheric Emission

The non-thermal X-rays coming from within the light cylinder have been modeled by a number of authors and they can be classified into two main models (see the contributions by Ho, Ruderman, Harding, and Arons in these proceedings):

- outer gap model (Cheng, Ho and Ruderman 1986a,b)
- polar cap model (Cheng and Ruderman 1980; Harding and Daugherty 1991; Arons 1981)

The basic source of relativistic electrons in both models comes from the breakdown of the voltage gaps existing at the boundaries between the negative and positive magnetospheric charge distribution as determined by the open field lines extending to the light cylinder. The emerging X-ray spectrum is subsequently determined by the curvature radiation, synchrotron radiation and inverse Compton scattering of the electromagnetic cascade. In the polar caps, the pair production process in the strong magnetic fields also plays a role.

2.3 Thermal Emission from the Surface

Considering that neutron stars have no internal energy sources, one expects that they should be cooling from the moment they are created. Early calculations (Chiu 1964) indicated that the neutron stars are born very hot ($T \geq 10^{10}$ K) and cool rapidly by neutrino emission through the modified Urca process from the interior. The surface temperatures are expected to come down to $\sim 2 \times 10^6$ K in a few hundred years, slowly decrease to $\sim 5 \times 10^5$ K by about 10^5 years, and rapidly cool thereafter by means of photon emission from the surface. These temperatures imply that we should observe the bulk of the surface emission in the X-ray band. A variety of cooling models exists in the literature. The modified Urca process takes a long time to cool the neutron star (attractive for observers). The alternatives which cool the neutron star very fast (1 to 100 years) are the direct Urca process (Lattimer *et al.* 1991) and the cooling through exotic matter like pion condensates or quark matter (see Nomoto and Tsuruta 1987; Ögelman 1991 and references therein).

Reheating of the neutron star by internal friction (Shibazaki and Lamb 1989), accretion from the interstellar medium (Lamb *et al.* 1973), polar cap heating (Cheng and Ruderman 1980; Arons 1981) can prolong the cooling of the neutron star.

A more in-depth discussion of the topic of cooling and reheating can be found in these proceedings in the presentations by Tsuruta, Van Riper, Cheng, Miller, Ventura, and Shibanov.

3 Inventory of X-Ray-Emitting Isolated Pulsars

In Table 1 I list all the X-ray-emitting isolated pulsars (as of June 1992). Subsequently I outline the pulsars in the confirmed list individually. There are 10 confirmed and 6 possible detections. The list is in descending order of the energy loss rate $\dot{E} = I\Omega\dot{\Omega}$; for the confirmed pulsars it also turns out to be in increasing order of age as would be expected. Unless otherwise indicated, the confirmed pulsar X-ray data comes mostly from the *Einstein* results as summarized by Seward and Wang (1988). The X-ray data on unconfirmed pulsars is from Becker, Trümper, and Ögelman (1992). All X-ray luminosities are log of erg/s in the *Einstein* bandpass of .1-4 keV unless marked with (*) which indicates the *ROSAT* bandpass of 0.1-2.4 keV.

TABLE 1
X-ray-emitting single pulsars (as of June 1992)

pulsar	log $\dot{E}^{(1)}$	per(s)	dist(kpc)	log$L_X^{(2)}$	log$L_X^{(3)}$	log$L_X^{(4)}$	log t$^{(5)}$	log B(g)
0531+21	38.65	0.033	2.0	36.0	36.0	37.4	3.09	12.58
0540-69	38.17	0.050	55.0	36.4	36.4	37.0	3.22	12.70
1509-58	37.25	0.150	4.2	34.3	34.3	35.3	3.19	13.19
0833-45	36.84	0.089	0.5	33.0	31.6	33.0$^{(6)}$	4.05	12.53
1951+32	36.58	0.040	3.0	33.8	?	34.7	5.02	11.69
0656+14	34.58	0.385	0.4	32.5	31.7	?	5.04	12.67
0630+18$^{(7)}$	34.54	0.237	0.1	30.3*	29.6*	?	5.51	12.2
1055-52	34.48	0.197	0.9	32.5	?	?	5.73	12.04
1929+10	33.60	0.227	0.1	28.5	?	?	6.49	11.71
0950+08	32.75	0.229	0.1	29.4	?	?	7.24	11.39
			unconfirmed detections					
1813-36	33.14	0.387	3.2	33.0*	?	?	6.50	11.96
1642-03	33.08	0.388	1.3	31.9	?	?	6.54	11.93
1846-06	32.77	1.451	4.4	33.2*	?	?	5.70	12.92
0628-28	32.16	1.244	1.3	31.6*	?	?	6.40	12.48
1648-17	32.12	0.973	1.0	31.2*	?	?	6.70	12.24
1524-39	31.73	2.418	1.0	32.2*	?	?	6.30	12.84

(1) Log of $I\Omega\dot{\Omega}$ in erg/s
(2) The X-ray luminosity of the point-like object
(3) The pulsed X-ray luminosity of the point-like object
(4) The X-ray luminosity of the nebula plus point-like object
(5) Log of spin-down age $P/2\dot{P}$ in years
(6) the 1' compact nebula
(7) Geminga; Halpern and Holt 1992; Bertsch *et al.* 1992

3.1 PSR 0531+21 (Crab Pulsar)

- The Crab pulsar emits pulsed radiation in all wavelengths from radio to GeV gamma-rays. The polar cap or outer gap models appear to account for the pulsed emission which shows a power-law type spectrum.
- The luminosity of the nebular emission integrated over all wavelengths adds up to the rotational energy loss rate of the pulsar indicating that the relativistic pulsar wind is the origin of the energetic particles that synchrotron radiate in the nebula.
- It is the first pulsar for which the braking index has been measured: $n = \frac{\Omega\ddot{\Omega}}{\dot{\Omega}^2} = 2.51$ (Lyne, Pritchard, and Smith 1988)

- Hidden below the intense magnetospheric emission, it is difficult to see any cooling radiation from the Crab pulsar; an upper limit of T $\lesssim 3 \times 10^6$K exists (Toor and Seward 1977).
- In these proceedings there are reports on the EGRET observations of pulsed energetic gamma-rays (Thompson *et al.*), unpulsed gamma-rays (Mattox *et al.*) and Hubble Space Telescope observations (Biggs).

3.2 PSR 0540-69 (50 ms Pulsar in the LMC)

- PSR 0540-69 appears to be the Crab pulsar's twin in the LMC; it has a similar age, $\dot{E}$, and spectrum.
- It is one of the pulsars discovered first in X-rays (Seward *et al.* 1984), subsequently detected optically (Middleditch and Pennypacker 1985).
- It is also a pulsar for which a braking index measurement has been possible (Manchester and Peterson 1989 ; Ögelman and Hasinger 1990; Nagase *et al.* 1990, Gouiffes *et al.* 1992). The braking index measurement of Ögelman and Hasinger (1990), $n = 2.74 \pm 0.10$, utilizing *Einstein, EXOSAT* and CTIO data contemporaneous with the X-ray data was found to be incompatible with the *GINGA* and AAT measurements, $n = 2.04$ to 2.01 ± 0.02 respectively. The later optical measurements (Gouiffes *et al.* 1992) confirm the continuation of the *GINGA* braking index.
- Like the Crab pulsar, the intense magnetospheric emission makes it difficult to see any cooling radiation.
- In this Workshop there were further reports on this pulsar: *GINGA* observations by Kawai and *ROSAT* HRI observations by Seward.

3.3 PSR 1509-58 (Pulsar in the SNR MSH 15-58)

- This pulsar is another discovered in X-rays by the *Einstein* observatory (Harnden and Seward 1984). It has a power-law type spectrum indicating magnetospheric emission. It is surrounded by a synchrotron nebula.
- Compared with the other young pulsars like the Crab and PSR 0540-69 ($\sim 10^3$ yr), it has a longer period (150 ms) and higher $\dot{P}$. The implication is that pulsars can also be born with longer periods and higher magnetic fields.
- It is also a pulsar for which a braking index measurement has been possible: $n = 2.83$ (Manchester *et al.* 1985).
- Intense magnetospheric emission makes it difficult to see any cooling radiation.
- Recent *GINGA* observations have made it possible to phase relate the X and radio pulses (Kawai *et al.* this workshop).

3.4 PSR 0833-45 (Vela Pulsar)

- The Vela pulsar is the brightest gamma-ray source in the sky at around 100 MeV. Its spectrum turns down at energies below ~ 50 MeV (Kanbach *et al.* 1992) thus making it a weak source at X-ray wavelengths. The gamma-ray pulse shape is variable over time scale of months (Grenier 1992). It has been detected optically as a ~ 24 magnitude pulsed object (Wallace *et al.* 1977).
- Recent measurements with imaging X-ray detectors of *Einstein* and *EXOSAT* observatories have revealed a compact nebula of about $2'$ diameter surrounding the pulsar (Harnden *et al.* 1985, Ögelman and Zimmermann 1989). The spectral signatures suggest that while the point-like emission centered on the pulsar is the blackbody-like emission from the surface of the cooling neutron star, the $2'$-diameter compact nebula around the pulsar is a synchrotron emission feature.
- The *ROSAT* Survey data on the Vela SNR shows clearly (see the *ROSAT* calendar picture for May 1992) that the remnant is an X-ray emitting region of ~ 5° diameter with the pulsar lying near the center of this nebula.
- The *ROSAT* pointed observations on the Vela pulsar have been partially analyzed (Ögelman *et al.* 1991). Two interesting results are apparent in the preliminary analysis: i) The pulsar emits pulsed X-rays at the 4±2 % level (total including the point-like object and the compact nebula); almost all of the pulsed flux is in the soft channels (0.1-0.8 keV). ii) Although the ~ $30''$ resolution of the detector does not allow a clear separation, the point-like source is definitely softer than the surrounding compact nebula. This result supports the *EXOSAT* findings of Ögelman and Zimmermann (1989). We are presently modeling the radial intensity profiles to extract a spectrum for both of the components which should permit a temperature determination of the pulsar.

3.5 PSR 1951+32 (Pulsar in SNR CTB 80)

- This is a short period (40 ms) and low magnetic moment pulsar; anomalous in the opposite sense of PSR 1509-52 from the expected birth properties of pulsars.
- A point-like X-ray source and nebular emission has been seen in *Einstein* and *EXOSAT* data ((Wang and Seward 1984; Ögelman and Buccheri 1987). Preliminary analysis of our *ROSAT* observations with time error polluted data do not show any strong pulsations from the central object; final results await the reprocessing of the raw *ROSAT* data.

3.6 PSR 0656+14

- This pulsar is one of the best candidate pulsars showing evidence of blackbody-like radiation from its surface.
- The X-ray source E0656+14 was discovered in an *Einstein* satellite survey of ultrasoft sources and identified with the nearby radio pulsar PSR 0656+14. The

energy distribution of the *Einstein* satellite IPC X-ray data was found to be consistent with a blackbody at a temperature of 3-6$\times10^5$ K under the assumption of emission from the surface of a 16 km radius neutron star (Córdova *et al.* 1989).

- This pulsar was observed by *ROSAT* and pulsations at the radio period were detected at a pulsed fraction of 14$\pm$2 % (Finley *et al.* 1992, Finley and Ögelman 1992). The spectrum can be described by various model fits. A blackbody description yields a surface temperature of $\sim$ (9.0$\pm$0.4) $\times$ 10^5 K and a stellar radius of 10 km under the assumption of a 500 pc source distance. An alternative description is a blackbody modified by the presence of an He atmosphere with surface temperature $\sim$ (3.0$\pm$0.5) $\times$ 10^5 K. A positive correlation exists between the phase of the pulsed emission and the X-ray hardness, suggesting that the pulsations have their seat in a "hot spot" on the neutron star surface.

3.7 PSR 0630+18 (Geminga)

- First discovered by SAS-2 satellite in the 100 MeV gamma-ray band(Ögelman *et al.* 1975; Fichtel *et al.* 1975), Geminga was finally recognized to be an isolated pulsar from the 0.237 s pulsation period measured by *ROSAT* (Halpern and Holt 1992). Subsequently, the pulsations in gamma-rays were also confirmed (Bertsch *et al.* 1992).
- The X-ray emission appears very similar to the other three soft X-ray-emitting pulsars PSR 0656+14, PSR 0833-45 (Vela), and PSR 1055-52 and probably has its seat on the hot surface of the neutron star.
- Absence of radio emission from Geminga (Seiradakis 1992) raises the interesting possibility of other close-by pulsars of this nature missed in radio surveys.

3.8 PSR 1055-52

- PSR 1055-52 was one of the brightest sources detected in the first survey of radio pulsars by the *Einstein* observatory. Marginal HRI evidence for an extended source around the pulsar have led Cheng and Helfand (1983) to propose the idea of a synchrotron nebula around the pulsar. *EXOSAT* observations, on the other hand, show evidence of blackbody-like radiation from its surface (Brinkmann and Ögelman 1987).
- Our preliminary analysis of the *ROSAT* HRI and PSPC observations show that: i) the source has negligible extended component (at few arcsec scale); ii) has a very soft spectrum and emits pulsed X-rays like PSR 0656+14.
- This pulsar is among the best candidate pulsars showing evidence of blackbody-like radiation from its surface.

3.9 PSR 1929+10 and PSR 0950+08

- These are the oldest and nearest pulsars in the confirmed list. They were detected in the survey of radio pulsars by the *Einstein* observatory (Helfand 1983; Seward and Wang 1988).

- Results of a long *ROSAT* exposure of PSR 1929+10 have been presented in this meeting (Helfand).

3.10 The Unconfirmed Ones

- PSR 1642-03 is an *Einstein* candidate, the rest are *ROSAT* Survey candidates (see Becker, Trümper, and Ögelman 1992).
- Luminosities are around $31 \le \log L_X\,(\mathrm{erg/s}) \le 33$
- They are old and not so near-by pulsars. No cooling model predicts these pulsars as a priori candidates. Either new emission mechanisms must be considered, or these are accidental identifications.

Sensitive *ROSAT* observations are detecting the X-ray emission from more pulsars as this review is being written. We can expect to have a much better understanding of the thermal and magnetospheric processes on neutron stars in the near future.

References

Arons, J. 1981, *Ap. J.*, **248**, 1099.
Becker, W., Trümper, J., and Ögelman, H. 1992, these Proceedings.
Bertsch, D. L. *et al.* 1992, *Nature,* **357**, 306.
Brinkmann, W., and Ögelman, H.: 1987, *Astron. Astrophys.,* **182**, 71.
Cheng, A. F. and Ruderman, M. A. 1980, *Ap. J.*, **235**, 576.
Cheng, A. F., and Helfand, D. J.: 1983, *Ap. J.* **271**, 271.
Cheng, K. S., Ho, C., and Ruderman, M. 1986a, *Ap. J.*, **300**, 500.
Cheng, K. S., Ho, C., and Ruderman, M. 1986b, *Ap. J.*, **300**, 522.
Cheng, A. F., and Ruderman, M. A. 1980, *Ap. J.*, **235**, 576.
Chiu, H-Y.: 1964, *Ann. Phys.*, **26**, 364.
Córdova, F. A., Hjellming, R. M., Mason, K. O., and Middleditch, J. 1989, *Ap. J.*, **345**, 451.
Fichtel, C. E. *et al.* 1975, *Ap. J.* **198**, 163.
Finley, J. P. and Ögelman, H. 1992, these Proceedings.
Finley, J. P., Ögelman, H. and Kızıloğlu, Ü. 1992, *Ap. J.* , in print.
Gouiffes, C., Finley, J.P. and Ögelman, H., 1992 *Ap. J.* , in print.
Grenier, I. *et al.* 1992, this Workshop. Halpern, J. P. and Holt, S. S. 1992, *Nature,* **357**, 222.
Harding, A..K. and Daugherty, J. 1991 *Ap. J.* , **374**, 687.
Harnden, F.R.Jr. and Seward, F.D. 1984, *Ap. J.* **283**, 274.
Harnden, F.R.Jr., Grant, P.D., Seward, F.D. and Kahn, S.M. 1985, *Ap. J.* **299**, 828.
Helfand, D.J.: 1983, in *Supernova Remnants and their X-ray Emission, IAU Symposium No. 101,* eds. J. Danziger and P. Gorenstein, (Reidel, Dordrecht), p. 471.
Kanbach, G. *et al.* 1992, this Workshop. Lamb, F.K., Pethick, C.D., Pines, D., 1973 *Ap. J.* **184**, 271.
Lattimer, J. M., Pethick, C. J., Prakash, M., and Haensel, P. 1991, *PRL,* **66**, 2701.
Lyne, A.G., Pritchard, R.S. and Smith, F.G. 1988, *M. N. R. A. S.,* **233**, 667.
Manchester, R.N., Durdin, J.M. and Newton, L.M. 1985, *Nature,* **313**, 374.
Manchester, R.N. and Peterson, B.A. 1989, *Ap. J. Lett.* **342**, L23.
Middleditch, J. and Pennypacker, C. 1985, *Nature,* **313**, 659.
Nagase, F., Deeter, J., Lewis, W., Dotani, T., Makino, F., and Mitsuda, K. 1990, *Ap. J. (Letters),* **351**, L13.
Nomoto, K., and Tsuruta, S. 1987, *Ap. J.*, **312**, 711.
Ögelman, H. *et al.* 1975, *Bull. Am. Phys. Soc.*, **20**, 656.
Ögelman, H., and Buccheri, R.: 1987, *Astron. Astrophys.,* **186**, L17.
Ögelman, H. and Zimmermann, H.-U. 1989, *Astron. Astrophys.,* , **214**, 179.
Ögelman, H., and Hasinger, G. 1990, *Ap. J. (Letters),* **353**, L21.
Ögelman, H. 1991, in *Neutron Stars: Theory and Observation,* ed. J. Ventura and D. Pines, (Kluwer, Dordrecht), p. 87.
Ögelman, H., Finley, J. P., Aschenbach, B., Trümper, J., and Zimmermann, U. 1991, *BAAS,* **23**, 1349.
Seiradakis, J.H. 1992, *IAU Cir.* # 5532
Seward, F.D., Harnden, F.R. Jr. and Helfand, D.J. 1984, *Ap. J. Lett.* **287**, L19.
Seward, F.D. and Wang, Z-R.: 1988, *Ap. J.* **332**, 199.
Shibazaki, N., and Lamb, F. K. 1989, *Ap. J.*, **346**, 808.
Toor, A. and Seward, F.D.: 1977, *Ap. J.* **216**, 560.
Wallace, P.T. *et al.* : 1977, *Nature,* **266**, 692.
Wang, Z.R., and Seward, F.D.: 1984, *Ap. J.* **285**, 607.

Search for Cooling Neutron Stars in the ROSAT Survey

Werner Becker, Joachim Trümper, and Hakki Ögelman

Max-Planck-Institut für Extraterrestrische Physik

1 Introduction

With the Einstein observatory (1978-1981) about 40 radio pulsars (including those in supernova remnants) were observed, leading to the detection of about 10 of them (Seward and Wang 1988). The ROSAT *all sky survey* now provides the unique opportunity to observe all known radio pulsars in the soft X-ray band (0.1-2.4 keV) with a typical survey sensitivity of $F_X \approx 5 \cdot 10^{-13}\,\mathrm{erg\,cm^{-2}\,s^{-1}}$. This sensitivity corresponds to a neutron star temperature of $T_S \approx 5.3 \cdot 10^5\,(d\,/\,1\,\mathrm{kpc})^{\frac{1}{2}}\,(10\,\mathrm{km}\,/\,R_0)^{\frac{1}{2}}$ °K for low interstellar column densities. The cooling history of neutron stars has been investigated by various theory groups (Tsuruta and Nomoto 1987, Shibazaki and Lamb 1989, Van Riper 1990). If cooling were the only relevant process, only 2% of the 520 radio pulsars having a characteristic age $\leq 10^5$ years would be detectable in the ROSAT survey. But the surface temperature might remain high due to reheating processes:

- accretion from interstellar medium (Lamb *et al.* 1973)
- superfluid-crust interaction (Shibazaki and Lamb 1989)
- particles streaming back onto the polar caps (Cheng and Ruderman 1983).

In the following we present the results of a search for cooling neutron stars in the ROSAT *all sky survey* data. The results are preliminary since the algorithms of the standard analysis system (SASS) have been used for the detection or determination of upper limits. They are designed to search for point sources with unknown position. A somewhat higher sensitivity for the search can be achieved by inspecting the pixels at known radio pulsar positions in a more detailed investigation. This investigation is already on the way.

2 The ROSAT All Sky Survey

The ROSAT *all sky survey*, which started on July 30, 1990, was completed within 6 months, using the PSPC in the focus of the X-ray telescope. With a field of view of 2° and a nominal survey progression of 1° per day in ecliptic longitude, a celestial position was visible for at least two days. The total integration time was ≈ 100 - 700 seconds for sources near the ecliptic equator and up to ≈ 10000 seconds near the ecliptic poles. The average exposure time for the sample of radio pulsars was ≈ 300 seconds. The location accuracy for point sources obtained from a correlation between the ROSAT positions computed via the standard analysis software and the positions taken from the SIMBAD Star Catalog indicates that $r_{90\%} = 30.5$ arcsec, *i.e.*, for 90%

of the tested samples the difference between the ROSAT positions and the SIMBAD positions is ≤ 30.5 arcsec (Voges 1991). In the present preliminary search we have excluded all detections with deviations ≥ 60 arcsec from the known radio position.

3 Data Analysis

We have used the results from the standard analysis SASS to search for significant detections at pulsar radio positions. A source was detected, if its detection likelihood was ≥ 10, where the likelihood is defined as $-\ln(P)$, and P is the probability that the detection is just background. If no significant signal was found, a 2σ count rate upper limit was computed using all photons collected within a radius of 300 arcsec around the pulsar's radio position. No background correction or detailed source analysis has been applied during the analysis, so that the results are still contaminated with photons from the background or nearby X-ray sources. The results given in this work should be regarded as preliminary for that reason.
With the knowledge of the counting rates and rate upper limits, it is possible now to compute upper limits for the neutron star blackbody temperature, using the following equation:

$$\frac{2\,\pi\,(1+z)R_0^2}{h^3\,c^2\,d^2}\int_{0.07\,keV}^{2.4\,keV}\frac{E^2\,A_{eff}(E,\theta)\cdot e^{-\sigma(E)\int_0^d n_H(s)\,ds}}{e^{[(1+z)\,E/K\,T_0]}-1}\,dE - N = 0$$

The meaning of the parameters is as follows:

- N = measured source count rate or 2σ upper limit, computed using an extraction radius of 300 arcsec. If the source was found in more than one survey strip, the different results have been merged together weighted with respect to their likelihood.
- Neutron star radius: $R_0 = (1+z)^{-1} R_\infty$ where z is the gravitational red shift. In order to compare our results with cooling curves from Shibazaki and Lamb (1989), we have used the parameters of their *standard* cooling model: $M = 1.4\,M_\odot$ and $R_0 = 15.8$ km resp. $R_\infty = 18.39$ km.
- Characteristic pulsar age $t_c = P/2\,\dot{P}$, column density N_H, and distance d: These basic pulsar parameters, as well as the associated errors, were taken from the pulsar list of Lyne, Manchester, and Taylor 1990. N_H was estimated from the dispersion measure assuming 10 hydrogen atoms for each free electron. From the error of the dispersion measure we have computed the error for the column density N_H. The distances are uncertain within a factor of two.
- $\sigma(E)$ is the photoelectric absorption cross section per hydrogen atom (Morrison and McCammon 1983) and $A_{eff}(E,\theta)$ is the energy and off-axis angle dependent effective area of the ROSAT PSPC detector (ROSAT technical appendix 1991).

A pure blackbody spectrum was assumed for the neutron star. No deviations due to the influence of the magnetic field or the neutron star atmosphere were taken into account.

4 Results

4.1 Neutron Stars in SNR

The well known synchrotron sources in the remnants of Crab, Vela, MSH 15-52, CTB 80 and 0540-69.3 were found with high likelihoods. In order to search for thermal neutron stars, a detailed structural and time analysis has to be performed. The results will be published elsewhere. The pulsars PSR 0656+14 (lik=1328) and PSR 1055-52 (lik=58) were well detected during the ROSAT survey. For PSR 0656+14 (possibly related to the Gemini-Monoceros enhancement [Thompson *et al.* 1991]) we have computed a surface temperature of $\log T_S = (6.0034 \pm 0.1387)$ °K which is in good agreement with the value received from ROSAT pointed observations (Finley and Ögelman 1992). For PSR 1055-52 we have found a surface temperature of $\log T_S = (6.069 \pm 0.0932)$ °K. Since the pulsar is claimed to be surrounded by a 10 arcsec X-ray emitting synchrotron nebula (Cheng and Helfand 1983), the pointed data has to be used for a more detailed analysis. Finally we note that the putative neutron star 1E 161348-5055.1 in RCW 103 could not be detected (Becker *et al.*1992).

4.2 New Candidates

- **PSR 1524-39**
  A source was found near the radio pulsar PSR 1524-39 with a count rate of 0.25 ± 0.03 cts/s and a likelihood of 163.6. Examining the surrounding of PSR 1524-39 for other possible X-ray sources, we have found only an F5V star (SAO 206666) with $m_v = 9.2$ and $m_b = 9.7$ resp. $\log(f_X/f_V) = -3.24$, located ΔRA=2.24 arcsec and ΔDEC=18.8 arcsec away from the known pulsar position listed in Lyne, Manchester, and Taylor (1990). From an investigation of bright and/or nearby stars detected in the ROSAT *all sky survey*, one finds that this f_X/f_V ratio is well above the value expected for F5V stars (Schmitt 1991). However, we cannot exclude a binary nature of the F5V star which might enhance its X-ray emission.
- **PSR 1648-17**
  Soft X-ray emission of 0.026 cts/s was detected from a position nearby the radio pulsar PSR 1648-17 with a likelihood of 11.8 (99.9992 % confidence). The difference between the radio pulsar position and the detected X-ray source is only ΔRA=25 arcsec and ΔDEC=1.6 arcsec. An examination of the surroundings of this pulsar for other possible X-ray sources was negative within a range of 30 arcmin search radius and $m_V \leq 17$. This source is the best candidate for a new neutron star detection in the soft X-ray band since the EINSTEIN mission. One possible explanation for the origin of the detected X-ray photons is that PSR 1648-17 is reheated by superfluid-crust interaction. However, a theoretical count rate estimate using the model from Shibazaki and Lamb (1989) with $J = 31 \cdot 10^{44}\,\mathrm{g\,cm^2\,rad\,s^{-1}}$ and $N_H = 9.56 \cdot 10^{20}$ atoms/cm^2 predicts a count rate of $\approx 2.68 \cdot 10^{-5}$ cts/s, which is still a factor 10^3 smaller in comparison to the measured ones.

- **PSR 1846-06, 0628-28, 1813-36**

Three further radio pulsars have a detection likelihood just below the detection threshold (lik=10) and a position which differs only few arcsec from the known radio positions taken from the pulsar list of Lyne, Manchester and Taylor 1990 (see Table 1). The measured count rates are much higher than those expected from the cooling curves.

TABLE 1

PSR	age year	rate cts/s	exposure sec	conf. of source in chan.(7-240)	ΔRA arcsec	ΔDEC arcsec
1846-06	$5.0\ 10^5$	0.0365	354	99.94 %	3.22	15.84
0628-28	$2.7\ 10^6$	0.0274	587	99.84 %	16.52	39.24
1813-36	$2.8\ 10^6$	0.0685	140	99.77 %	10.48	11.16

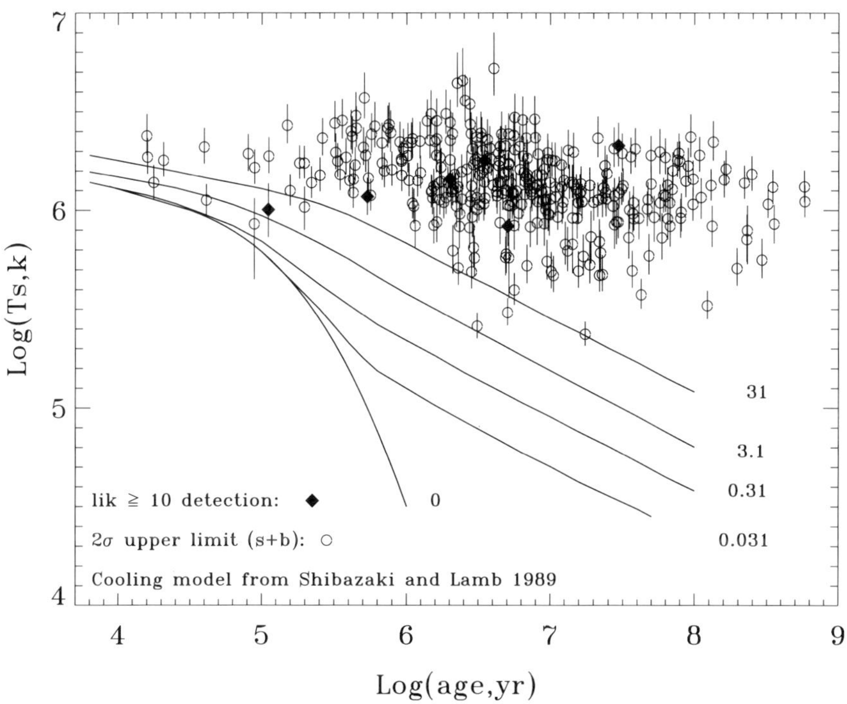

Fig. 1—Preliminary 2σ temperature upper limits for radio pulsars with known characteristic age. The vertical lines display the uncertainty range due to the lack of exact values for distance and column density.

4.3 Preliminary Upper Limits

For the remaining radio pulsars, upper limits were found which are summarized in Figure 1. This graph displays the resulting 2σ temperature upper limits for radio pulsars with known characteristic age $t_C = P/2\dot{P}$ together with the cooling curves from the *standard* cooling model of Shibazaki and Lamb (1989). The curves show the thermal evolution of neutron stars for various rates of internal heating caused by friction between the neutron superfluid and the crust. They are labeled with the values of the differential angular momentum J (in units of $10^{44}\,\mathrm{g\,cm^2\,rad\,s^{-1}}$) of the neutron superfluid in the inner crust.

Some of the temperature upper limits come close to the cooling curves for $J = (3.1 - 31)\cdot 10^{44}\,\mathrm{g\,cm^2\,rad\,s^{-1}}$. Their temperature values can be found in Table 2. The columns T_S and T_∞ give the neutron star surface temperature and the temperature as observed at infinity. The values are listed together with their likelihood of detection. *ul* indicates a 2σ *upper limit* which was calculated from an extraction radius of 300 arcsec around the given radio pulsar position using a maximum likelihood algorithm. ΔRA and ΔDEC label the difference between the known radio position and the centroid of the collected X-ray photons inside the extraction circle.

References

Arons, J. 1981, *Ap. J.*, **248**, 1099

Becker, W., Trümper, J., Hasinger, G., and Aschenbach, B. 1992, these Proceedings.

Cheng, A. F., Ruderman, M. A. 1980, *Ap. J.* **235**, 576

Finley, J. P. and Ögelman, H. 1992, these Proceedings.

Lyne, A. G., Manchester, R. N., Taylor, J. H. 1990, *Observed and Derived Parameters for Pulsars*, from *"anonymous ftp"* 128.112.128.165

Lamb, F. K., Pethick, C. D., Pines, D., 1973 *Ap. J.* **184**, 271

Nomoto, K., Tsuruta, S. 1986, *Ap. J. Lett.*, **305**, L19

Morison, R, Mc.Cammon, D. 1983, *Ap. J.*, **270**, 119

ROSAT AO-2 technical appendix 1991, Call for Proposals, MPE

Schmitt, J. 1992, 21 IAU General Assembly, High Lights Vol. 9

Seward, D. F., Wang, Z. R. 1988, *Ap. J*, **332**, 199

Shibazaki, N., Lamb, F. K. 1989, *Ap. J.*, **346**, 808

Thompson, R. J. Jr., *et al.* 1991, *Ap. J. Let.*, **366**, L83

Voges, W. 1991, *Proc. 2th Conf. on Digitised Optical Sky Surveys*

TABLE 2

Temperature upper limits of radio pulsars which come close to the range predicted by theoretical cooling models

PSR	$P/2\dot{P}$ log(yr)	T_S Log(°K)	T_∞ Log(°K)	likelihood	ΔRA arcsec	ΔDEC arcsec
1800-21	4.1980	6.3777	6.3118	4.4 ul	73.75	91.44
1259-63	4.2030	6.2705	6.2045	1.1 ul	137.10	24.84
1705-44	4.2456	6.1409	6.0750	3.3 ul	17.09	41.76
1737-30	4.3143	6.2541	6.1882	0.0 ul	12.12	38.52
1930+22	4.5977	6.3216	6.2557	0.0 ul	131.66	82.80
2334+61	4.6116	6.0515	6.9855	4.9 ul	34.92	15.84
1727-47	4.9048	6.2873	5.2214	0.0 ul	13.46	15.84
1916+14	4.9470	5.9311	5.8651	1.3 ul	24.10	8.64
0611+22	4.9493	6.2162	6.1503	0.0 ul	92.71	32.04
0906-49	5.0478	6.2749	6.2090	0.1 ul	47.23	2.88
0740-28	5.1958	6.0994	6.0335	0.4 ul	82.11	63.36
0154+61	5.2948	6.0160	5.9501	0.4 ul	53.93	13.32
1001-47	5.3432	6.1396	6.0737	0.1 ul	100.17	4.32
1702-19	6.0585	5.9224	5.8565	3.0 ul	0.36	43.56
0053+47	6.3215	5.7963	5.7304	0.7 ul	78.07	9.72
0450+55	6.3597	5.7088	5.6429	0.3 ul	23.95	80.28
2045-16	6.4526	5.6894	5.6235	0.0 ul	3.11	97.20
0834+06	6.4725	5.7598	5.6939	0.3 ul	9.36	0.00
1929+10	6.4917	5.4145	5.3485	1.8 ul	58.60	114.12
0823+26	6.6882	5.7623	5.6964	1.4 ul	35.64	43.20
0943+10	6.6927	5.7791	5.7132	0.0 ul	117.36	25.92
1133+16	6.7026	5.4831	5.4172	0.4 ul	24.12	36.00
2151-56	6.7114	5.7606	5.6946	0.1 ul	102.89	43.20
2327-20	6.7497	5.5975	5.5316	0.2 ul	116.52	138.96
0149-16	7.0064	5.6906	5.6247	0.6 ul	66.60	0.00
1112+50	7.0223	5.6715	5.6056	0.1 ul	42.84	108.36
2152-31	7.1210	5.7957	5.7298	0.0 ul	18.00	76.32
1919+21	7.1964	5.6935	5.6276	4.3 ul	42.84	32.76
0950+08	7.2430	5.3733	5.3073	0.3 ul	33.48	86.76
1604-00	7.3392	5.6723	5.6064	0.2 ul	27.36	39.24
1237+25	7.3584	5.6754	5.6095	0.0 ul	52.92	74.52

ROSAT Observations of PSR 0656+14: a Cooling Neutron Star

John P. Finley[1] and Hakkı Ögelman[1,2]

[1]Department of Physics, University of Wisconsin-Madison
[2]Max-Planck Institut für Extraterresrische Physik

Abstract

The Position Sensitive Proportional Counter (PSPC) onboard *ROSAT* observed PSR 0656+14 for ~ 3200 seconds in 1991 March. PSR 0656+14 is found to be pulsating at the radio period of 0.384 s with a pulsed fraction of 14±2 %. These observations firmly establish PSR 0656+14 as one of only a handful of known isolated X-ray emitting radio pulsars. The spectral distribution of counts can be described by various models. A blackbody spectrum yields a surface temperature of $T_S = (1.1\text{-}1.2)\times10^6$ K and a neutral hydrogen column density of $(0.8\text{-}1.2)\times10^{20}\text{cm}^{-2}$. For an assumed distance to the source of 400-600 pc, this model yields a stellar radius of 8-12 km for an observer at infinity. Alternative models incorporating the effects of a stellar atmosphere on the emergent blackbody spectrum also produced acceptable fits to the data. The He atmosphere model with B=0 of Romani (1987) yields a surface temperature of $T_S = (2.7\text{-}3.1)\times10^5$ K and neutral hydrogen column density of $(1.6\text{-}2.3)\times10^{20}\text{cm}^{-2}$ for a neutron star radius of 8-16 km. The implied source distance for this model is 100-250 pc. Alternatively, the He atmosphere model of Miller (1991) with $B=9.4\times10^{11}$ G yields a surface temperature in the range $(0.95\text{-}1.2)\times10^6$ K and a neutral hydrogen column density of $(0.45\text{-}0.85)\times10^{20}\text{cm}^{-2}$ for a stellar radius of 8-16 km. The implied source distance in the case of this model is 450-1100 pc. A positive correlation exists between the pulsed emission and the X-ray hardness, suggesting that the pulsations have their seat in a "hot spot" on the neutron star surface. Implications of the models with regard to the cooling curve of neutron stars is discussed.

1 Introduction

E0656+14 was discovered in an *Einstein* satellite survey of ultrasoft sources (Córdova *et al.* 1991) and subsequently identified with the radio pulsar PSR 0656+14 (Córdova *et al.* 1989). The *Einstein* IPC energy distribution was found to be consistent with the distribution expected from a blackbody at a temperature of $3\text{–}6 \times 10^5$ K at a distance of ≤ 550 pc and an equivalent hydrogen column density of $\leq 4 \times 10^{20}$ cm^{-2} (Córdova *et al.* 1989). At the relatively young spin-down age ($\tau = P/2\dot{P}$) of ~ 10^5 years PSR 0656+14 is an important object to study with regard to the thermal evolution of neutron stars. We report on a 3200 second observation of PSR 0656+14 acquired in 1991 March with the Position Sensitive Proportional Counter (PSPC) onboard *ROSAT*. A detailed account of most of the work presented here can be found in

Finley *et al.* (1992), and this paper is intended to capsulize that work as well as present some complementary results not found there. In Section 2 we will briefly describe the observations. Section 3 will contain the results of a search for pulsations and the spectral modeling whereas a discussion of the results can be found in Section 4.

2 Observations

PSR 0656+14 was observed with the Position Sensitive Proportional Counter (PSPC) at the focus of the X-ray telescope onboard *ROSAT*. Detailed descriptions of the satellite, X-ray mirrors, and detectors can be found in Trümper (1983) and Pfeffermann *et al.* (1986). Briefly, the PSPC is sensitive over the energy range 0.1-2.4 keV with an energy resolution $\Delta E/E \sim 0.43$ at 0.93 keV. The spatial resolution is $\sim 25''$ in the center of the focal plane. The observations reported here were obtained on 25 March 1991 (JD 244 8341.5) with a total effective exposure time of 3190 seconds. Because of earth occultation, radiation belt passage, and observation of other targets the data consisted of two uninterrupted intervals of $\sim$ 1330 and $\sim$ 1860 seconds separated by $\sim$ 15000 seconds.

The net count rate from PSR 0656+14 was 1.96 ± 0.03 cts/sec in the 0.1-2.4 keV range, excluding a deadtime correction of $\sim$ 3%. The position of PSR 0656+14 was determined to be $\alpha = 6^h\ 59^m\ 48^s.0$, $\delta = 14^\circ\ 14'\ \ 12''$ which is in good agreement with the radio position of $\alpha = 6^h\ 59^m\ 48^s.11$, $\delta = +14^\circ\ 14'\ \ 21''.12$ (Taylor 1991) considering the $6''$ accuracy of the satellite attitude. The radial profile of the events was consistent with the expectation for a soft point source. Data preparation and handling was carried out with the MIDAS/EXSAS software as provided by the European Southern Observatory (ESO) and the Max Planck Institut für Extraterrestrische Physik (MPE), respectively.

3 Results

3.1 Pulsations

To test for the presence of pulsations at the radio period of P=0.384883 seconds referenced to JD 244 8063.5 (Taylor 1991), the PSR 0656+14 photon arrival times were propagated to the Solar System Barycenter using the JPL DE200 ephemeris and the adopted coordinates $\alpha = 6^h\ 59^m\ 48^s.11$, $\delta = +14^\circ\ 14'\ 21''.12$ (Taylor 1991). The $Z^2{}_1$ statistic (Buccheri *et al.* 1986) was then calculated over a narrow range of periods bracketing the expected radio period. At the radio period of 0.384884 s (referenced to JD 244 8341.0), the $Z^2{}_1$ statistic maximized at a value of 40.38 which has a probability of arising by chance of $\sim 2\times10^{-9}$. This detection firmly establishes PSR 0656+14 as an X-ray pulsar for the first time. The pulse profile is a broad enhancement over half a cycle in pulsar phase and is displayed in Figure 1.

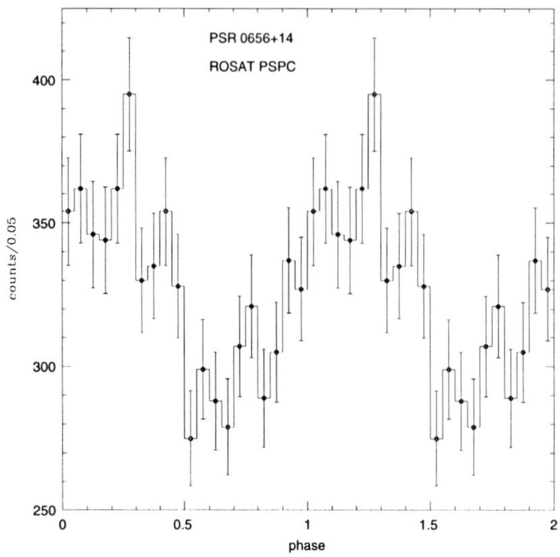

Fig. 1—The pulse profile of PSR 0656+14. Phase 0 corresponds to JD 244 8341.0 and two cycles are plotted for clarity.

3.2 Spectrum

The high counting rate of PSR 0656+14 allowed statistically meaningful spectral fits to be performed to the pulse height distribution of counts. A blackbody spectrum gave a marginally acceptable reduced χ^2 of 1.43 for the 161 degrees of freedom. The 90% confidence level parameters were T_∞ = (8.6-9.4)$\times10^5$ K and a neutral hydrogen column density of N_H = (0.8-1.2)$\times10^{20}$ cm^{-2}. For a source distance of 400-600 pc, the implied neutron star radius is 8-12 km. The unabsorbed 0.08-2.4 keV flux of F = (5-15) $\times$ 10^{-12} ergs cm^{-2} sec^{-1} corresponds to a luminosity of 3 $\times$ 10^{32} ergs sec^{-1} (D/500 pc)^2(F/10^{-11} ergs cm^{-2} sec^{-1}). A steep power law of photon index −4.8 also produced an acceptable fit to the data as would be expected for the Wien end of a blackbody curve.

We investigated the possibility that a neutron star atmosphere modifies the emergent spectrum (Romani 1987, Miller 1991). Spectra of pure Fe and pure He atmospheres for zero surface field (B=0 G) (Romani 1987) and pure He with a surface field of B=9.4$\times10^{11}$ G (Miller 1991) were available for fitting. The zero field pure Fe atmosphere model failed to produce an acceptable fit whereas both He atmosphere models produced acceptable reduced χ^2 of ~ 1.2 per degree of freedom. The 90% confidence level parameter values for these models as well as the blackbody model are presented in Table 1.

Since PSR 0656+14 pulsates at the radio period of 0.384 s, we have investigated the possibility of spectral variation as a function of pulsar phase. The statistics did not allow meaningful phase-resolved spectroscopy, so we employed a hardness parameter defined as the ratio of counts in channels 26-50 (~ 0.26-0.50 keV) to the counts in channels 7-25 (~ 0.07-0.25 keV) for this purpose. When this ratio was calculated as a function of pulsar phase, a positive correlation was found to exist with < a 1% chance of arising by chance from an uncorrelated population. In addition, the average energy as a function of pulsar phase showed a similar correlation and is displayed in Figure 2. The simplest interpretation of these results is that the pulsed emission has its

TABLE 1
PSR 0656+14 Spectral Fitting Results

Parameter	Blackbody	He B=0 G	He B=9.4×10^{11} G
reduced χ^2	1.43	1.20	1.31
$N_H(10^{20}cm^{-2})$	0.8-1.2	1.6-2.3	0.45-0.85
$T_\infty(10^5K)$	8.6-9.4	2.2-2.6	8.6-9.8
R_∞(km)	8-12	11.5-18.6	11.5-18.6
Distance(pc)	500	100-250	450-1100

seat in a stellar "hot spot" which sweeps through the field of view once each rotation. Under the assumption of blackbody emission, the 14% pulsed fraction corresponds to a temperature change of ~ 4%.

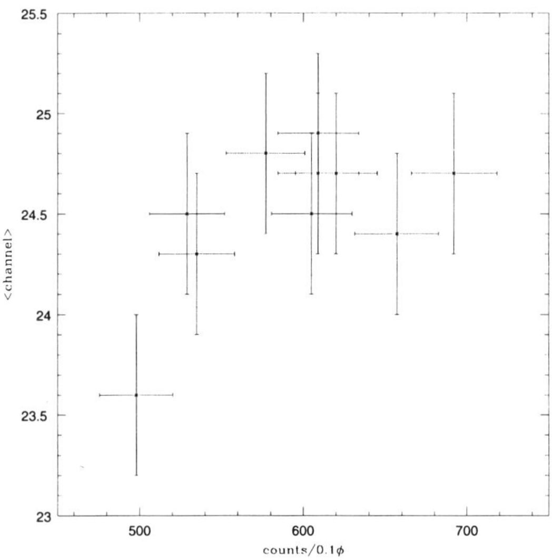

Fig. 2—The average channel number as a function of the phase density of the 0.384 s rotation period. The average channel is proportional to the average energy (<E> ≈ <channel>/100). The correlation coefficient is 0.72.

4 Discussion

The detection of pulsed X-rays at the radio period from PSR 0656+14 makes it only the fifth pulsed X-ray emitting isolated pulsar. At a spin-down age, $\tau_{SD} = P/2\dot{P}$, of 1.1× 10^5 years PSR 0656+14 is a "middle-aged" pulsar, about 10 times younger than the bulk of radio pulsars ($\tau_{SD} \sim 10^6$ yrs). The pulsed X-ray emisssion of PSR 0656+14 is ~1% of the spin-down luminosity of ~ 4× 10^{34} ergs s^{-1} not unlike younger X-ray pulsars such as the Crab and Vela which are believed to emit pulsed X-rays via magnetospheric processes. However, the predominantly soft spectrum of PSR 0656+14 argues against this origin for the pulsed emission according to conventional interpretation (Cheng, Ho, and Ruderman 1986a,b). In fact, a neutron star with the spin-down age of PSR 0656+14 is expected to emit blackbody radiation in accord with various cooling scenarios (see Ögelman 1991 for a review and further references). These facts lend strong support to the cooling neutron star hypothesis.

In accord with the cooling neutron star hypothesis, the most straightforward interpretation of the positive correlation of the X-ray hardness with phase of the pulsed emission is that the pulsed emission orginates in a region of higher temperature on the neutron star surface. Various models of the temperature distribution on the surface of a neutron star incorporating geometric effects (Greenstein and Hartke 1983), the heat conducting properties of the outer crust (Gudmundsson, Pethick, and Epstein 1982), and the effects of the large magnetic field on the heat conduction (Van Riper 1988) have appeared in the literature. In all cases the deviations from a black-body spectrum are expected to be small, $\leq 10\%$. The possibility that the hot spots may be due to polar cap heating via electrostatic accretion of particles in the magnetosphere (Cheng and Ruderman 1980, Arons 1981) seems untenable as the area of the polar cap necessary to explain the observations is commensurate with the total stellar surface even if PSR 0656+14 is only at a distance of 200 pc.

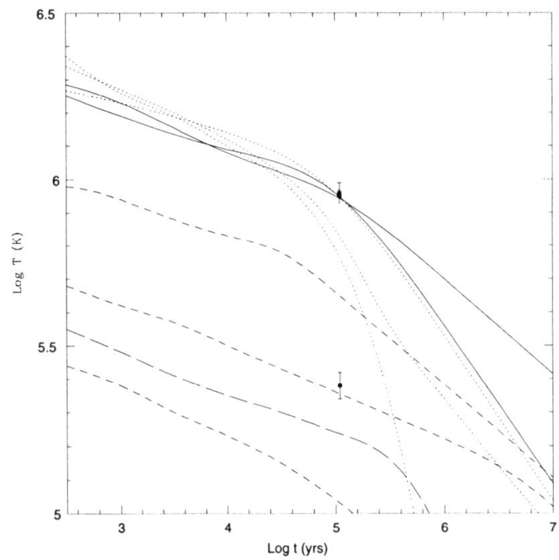

Fig. 3—Various neutron star cooling curves as published in the literature: standard cooling without superfluid (solid), standard cooling with superfluid (dot), standard cooling with superfluid and exotic matter (short dash), and direct URCA enhanced standard cooling (long dash). The data points are the measured *ROSAT* parameters for the models presented here.

Various authors have discussed the thermal evolution of neutron stars in the recent literature (Nomoto and Tsuruta 1987, Shibazaki and Lamb 1989, Ögelman 1991 and references therein). The results for the spectral model fits presented here lead to different conclusions in terms of the appropriate cooling scenario. In Figure 3 we display various cooling curves as reported in the literature as well as the results for PSR 0656+14 presented here. The surface temperature of $\sim 10^6$ K from the blackbody fit and the model of Miller (1991) for pure He atmosphere and surface field $B = 9.4\times10^{11}$ G can be accommodated by cooling scenarios with and without a superfluid component (dotted curves and solid curves in Fig. 3 respectively) but lie well above what would be expected for the presence of exotic condensates in the inner core such as pions or kaons (short dash curves in Fig. 3). On the other hand, the surface temperature of $\sim 300,000$ K from the He atmosphere model with surface field $B = 0$ G of Romani (1987) would favor the presence of exotic condensates in that the majority of the standard cooling scenarios do not yield such a low temperature after 10^5 years. However, recent work on the direct URCA process (Lattimer *et al.* 1991) has shown that if the proton concentration, which is determined by the poorly known symmetry energy of matter above nuclear density, is above a critical value, the cooling rate of a neutron star by neutrino emission is enhanced by almost an order of magnitude after ~ 100 years (the long dash curve in Fig. 3). Lattimer *et al.* (1991) estimate that this condition on the proton concentration is satisfied in most of the current calculations.

In summary, we find PSR 0656+14 to be emitting pulsed X-rays at the radio period. The pulsed emission has its seat in a region of higher temperature on the stellar surface as indicated by the hardness-pulse phase correlation. The data are described by various thermal models but lead to different conclusions with respect to PSR 0656+14's thermal evolution and point out the uncertainty in the current understanding of neutron star atmospheres. For a given model, the parameter ranges are found to be small indicating the high quality of the data. Independent observations of PSR 0656+14, such as may be provided by *EUVE* or *ALEXIS*, which can further constrain the parameters of the source will provide valuable insight into the thermal evolution of neutron stars.

This work was supported in part by NASA grant NAGW-2643.

References

Arons, J., 1981, *Ap. J.*, **284**, 1099.
Buccheri, R., *et al.* 1983, *Astr. Ap.*, **128**, 245.
Cheng, K. S., Ho, C., and Ruderman, M. 1986a, *Ap. J.*, **300**, 500.
Cheng, K. S., Ho, C., and Ruderman, M. 1986b, *Ap. J.*, **300**, 522.
Cheng, A. F., and Ruderman, M. A. 1980, *Ap. J.*, **235**, 576.
Córdova, F. A., Hjellming, R. M., Mason, K. O., and Middleditch, J. 1989, *Ap. J.*, **345**, 451.
Córdova, F. A., Kartje, J., Rodriguez-Bell, T., Mason, K. O., and Harnden, F., R., Jr. 1991, in *Extreme Ultraviolet Astronomy*, ed. R. F. Malina and S. Bowyer, (Pergamon Press), p. 30.
Finley, J. P., Ögelman, H., and Kızıloğlu, Ü. 1992, *Ap. J.*, **394**, *to be published.*
Greenstein, G., and Hartke, G. J. 1983, *Ap. J.*, **271**, 283.
Gudmundsson, E. H., Pethick, C. J., and Epstein, R. I. 1982, *Ap. J. (Letters)*, **259**, L19.
Lattimer, J. M., Pethick, C. J., Prakash, M., and Haensel, P. 1991, *PRL*, **66**, 2701.
Miller, M. C. 1991, Ph. D. thesis, California Institute of Technology.
Nomoto, K., and Tsuruta, S. 1987, *Ap. J.*, **312**, 711.
Ögelman, H. 1991, in *Neutron Stars: Theory and Observation*, ed. J. Ventura and D. Pines, (Kluwer, Dordrecht), p. 87.
Pfeffermann, E. *et al.* 1986, *SPIE*, **733**, 519.
Romani, R. W. 1987, *Ap. J.*, **313**, 718.
Shibazaki, N., and Lamb, F. K. 1989, *Ap. J.*, **346**, 808.
Taylor, J. H. 1991, private communication.
Trümper, J. 1983, *Adv. Spa. Res.*, **2**, 241.
Van Riper, K. A. 1988, *Ap. J.*, **329**, 339.

ROSAT Observations of RCW 103

Werner Becker, Joachim Trümper, Günther Hasinger, and Bernd Aschenbach

Max-Planck-Institut für Extraterrestrische Physik

1 Introduction

The X-ray point source 1E 161348-5055.1 detected with the EINSTEIN HRI detector close to the center of the supernova remnant RCW 103 has been considered as one of the best candidates for a cooling neutron star. The absence of an optical or radio counterpart for the central X-ray source supported the view that this object is a hot neutron star with a temperature $T_\infty = 2.2 \times 10^6$ °K, 3.3 kpc away from the sun. A total of 50 counts made up of 30 source counts and $\approx$ 20 background counts were recorded with the EINSTEIN HRI during a 7972 seconds exposure, corresponding to a count rate of $(3.8 \pm 0.7) \times 10^{-3}$ cts/s in the energy range (0.6 – 2.0) keV (Tuohy and Garmire 1980). However, the small number of counts and the lack of spectral information (the point source 1E 161348-5055.1 is not seen in the EINSTEIN IPC data) did not allow any detailed analysis with the EINSTEIN data.

In the present work, we discuss ROSAT pointed observations of RCW 103 and report about the search for the X-ray point source 1E 161348-5055.1 in the ROSAT-PSPC and HRI data. Since no point source has been detected, we give only a new temperature upper limit for 1E 161348-5055.1 which is below the value published by Tuohy *et al.* (1980, 1983)

2 Observations

The supernova remnant RCW 103 was one of the first targets observed in the pointing program of the ROSAT X-ray observatory between February and March 1991. A 40 ksec PSPC (Position Sensitive Proportional Counter) pointed observation was performed to search for the point source, to possibly determine its spectral parameters and to search for a periodic modulation of the X-ray flux. Another goal was to obtain an improved map of RCW 103 with spectrophotometric resolution. In addition, a 20 ksec pointing observation with the ROSAT High Resolution Imager (HRI) was proposed to support these investigations and to establish the existence of the point source beyond any doubt. The PSPC observations and about 40% of the HRI observations have been done with the putative point source 13 arcmin off-axis. The corresponding angular resolution for this observation was $\approx$ 35" FWHM for the PSPC and $\approx$ 25" FWHM for the HRI detector. The remaining 12 ksec HRI observations will be performed on-axis with improved angular resolution during a later observational period.

3 Data Analysis

Analyzing the ROSAT data available up to now, neither the PSPC nor the HRI data shows any evidence for a point source at the position of 1E 161348-5055.1. For comparison, Figures 1 and 2 show the brightness pictures of RCW 103 obtained from an analysis of the EINSTEIN HRI - and the ROSAT HRI data.

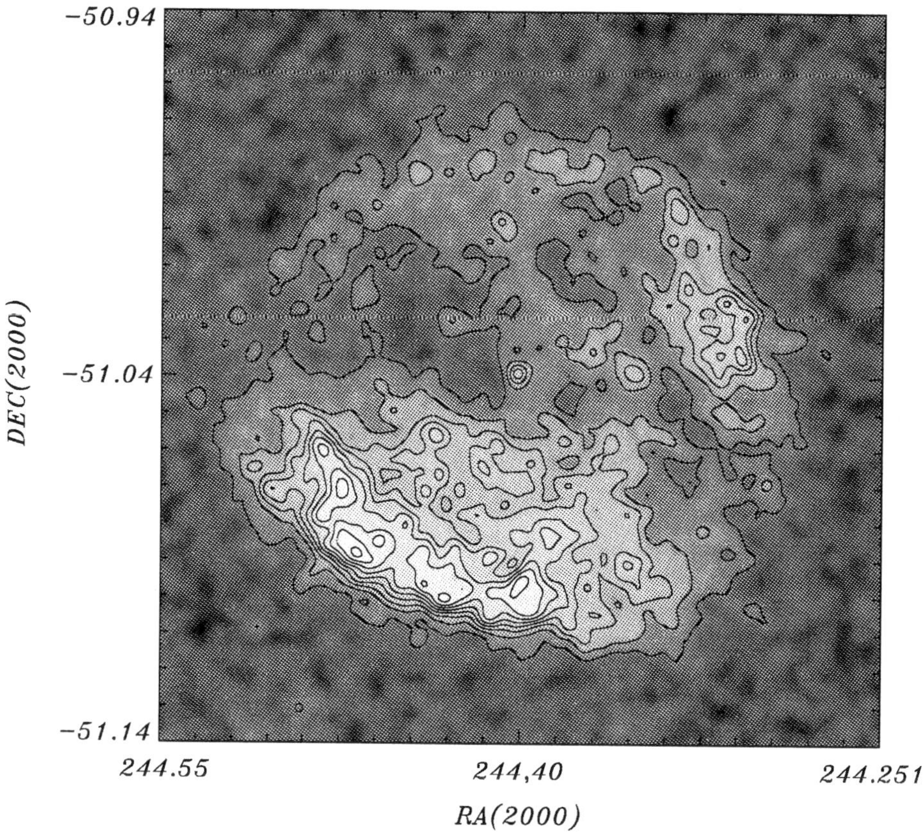

Fig. 1—The SNR RCW 103 as seen with the EINSTEIN HRI detector. The point source 1E 161348-5055.1 shows up just at the center of the remnant.

The EINSTEIN HRI data were collected during a 7972 second pointed observation of RCW 103 (Tuohy *et al.* 1980). Figure 2 shows the remnant as it was seen with the ROSAT HRI detector during a 7910 second pointed observation.

There is no evidence for a central point source in the ROSAT HRI image. For the other parts of the remnant the structures found in each picture are in good agreement.

To give an impression of the main differences between the ROSAT and the EINSTEIN HRI detectors the effective areas of the two detectors are plotted in Figure 3. For the ROSAT HRI we have used the effective area for a 13 arcmin off-axis observation whereas the effective area of the EINSTEIN HRI is that for an on-axis observation. The

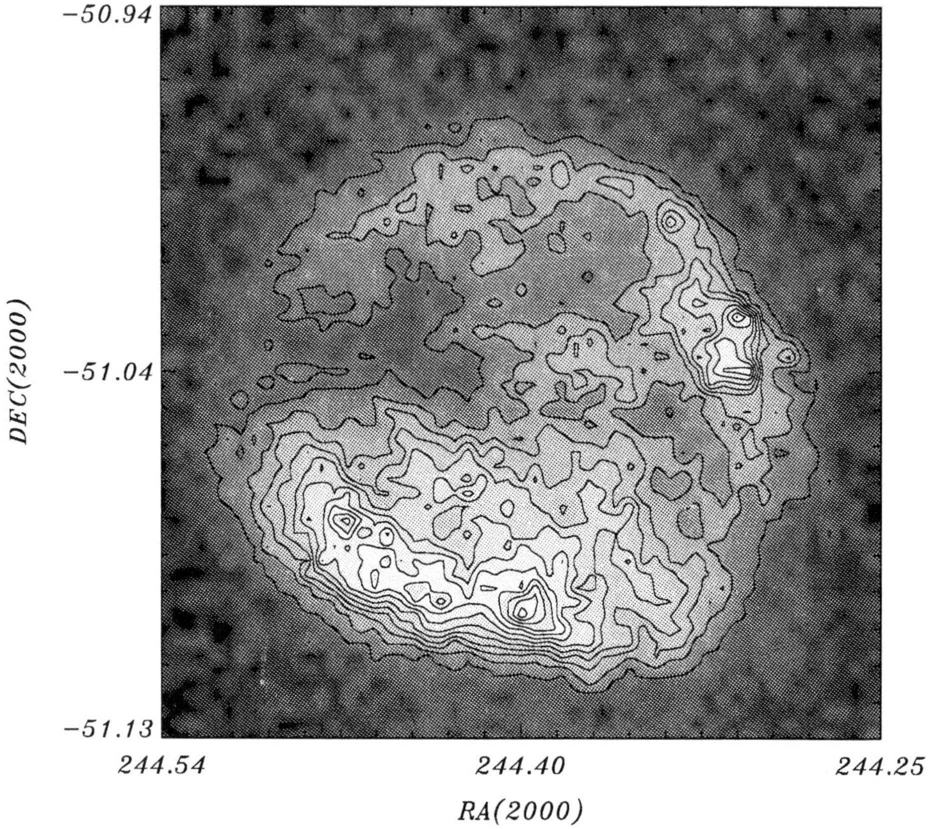

Fig. 2—The SNR RCW 103 as seen with the ROSAT HRI detector.

ROSAT HRI effective area exceeds that of the EINSTEIN detector for $0.1 \leq E \leq 2.4$ keV by a factor 3-4, but it has a cut-off for energies ≥ 2.4 keV. The EINSTEIN HRI energy band exceeds that of the ROSAT detector, and it is still sensitive for energies above 2.4 keV.

However, if the putative point source 1E 161348-5055.1 in RCW 103 is a cooling neutron star with a temperature $T_\infty = 2.2 \times 10^6$ °K, as it was claimed by Tuohy *et al.* (1983), it will emit primarily in the energy range (0.4-2.0) keV (c.f. Fig. 4).

For a quantitative analysis of the ROSAT data we have determined a 3σ count rate upper limit by summing all photons within a radius $r_{2\sigma} = 2 \times \mathrm{FWHM}/2.354$ centered on the position of 1E 161348-5055.1 and using the following formula:

$$ul(3\sigma) \approx 3\sqrt{\frac{2\,(s+b)\,100}{95.45}}$$

This approach is a good approximation for background-limited observations, and it will result in a conservative count rate upper limit. The results from the ROSAT PSPC and HRI observations are summarized in Table 1. Assuming a cooling neutron star with the parameters given by Tuohy *et al.* except the surface temperature, the count rate upper limit corresponds to an upper limit for the temperature. We note, that our

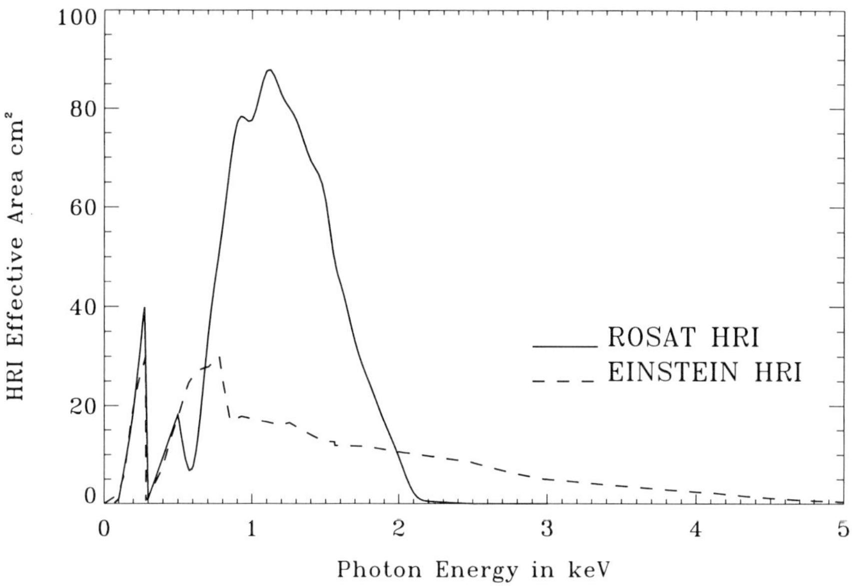

Fig. 3—Effective areas of the EINSTEIN - and the ROSAT HRI detectors.

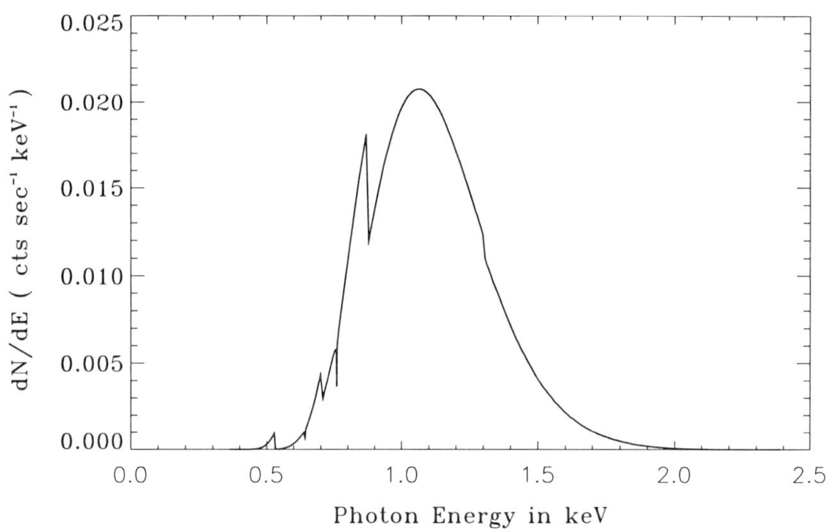

Fig. 4—Absorbed blackbody spectrum expected for the point source 1E 161348-5055.1 in RCW 103 using $T_\infty = 2.2\times10^6$ °K, $N_H = 10^{22}\,cm^{-2}$, $d = 3.3$ kpc, $R_\infty = 11.15$ km and $M = 1.3\,M_\odot$ (Tuohy *et al.* 1980).

3σ upper limits are well below the temperature of 2.2×10^6 °K found by Tuohy *et al.* (1980, 1983). However, the temperature limits derived are just compatible with the predictions by the neutron star cooling models (c.f. Nomoto and Tsuruta 1987 or Van Riper 1991).

TABLE 1

Temperature Upper Limits for 1E 161348-5055.1

ROSAT Detector	Energy keV	$r_{2\sigma}$ arcsec	3σ ul cts/s	$T_\infty^a \times 10^6$ °K	$T_\infty^b \times 10^6$ °K	$T_\infty^c \times 10^6$ °K
PSPC	0.10-0.40	44.7	0.0016	—	—	—
PSPC	0.40-2.40	30.0	0.0082	≤ 1.60	≤ 1.30	≤ 1.20
PSPC	0.40-1.00	31.5	0.0060	—	—	—
PSPC	1.00-2.40	31.0	0.0062	—	—	—
PSPC	0.10-2.40	29.6	0.0082	≤ 1.60	≤ 1.30	≤ 1.20
HRI	0.10-2.40	21.4	0.0075	≤ 1.75	≤ 1.45	≤ 1.35
EINSTEIN Detector	**Energy keV**	**extraction radius**	**source cts/s**	**$T_\infty^d \times 10^6$ °K**		
HRI	0.60-2.00	6.0"	0.0038	2.20		

[a] Parameter set from Tuohy (1980)
[b] Standard cooling model from from Shibazaki and Lamb (1989) and column density from Nugent *et al.* (1984)
[c] FP model from Nomoto and Tsuruta (1987)
[d] Tuohy *et al.* (1980, 1983)

Finally, we report some preliminary results of a spectroscopic study of RCW 103 obtained from an analysis of the PSPC data which show new and interesting features. Figure 5 depicts a grey scale representation of the hardness ratio HR2=(D-C)/(D+C) where C(0.4-1.0 keV) and D(1.0-2.4 keV) are the counts in the energy range given in parentheses. The spectrum is hardest along the low-intensity belt which runs diagonally from the south-west to the north-east across the remnant. We note that the hardness distribution peaks close to the position of 1E 161348-5055.1. Although it is tempting to speculate about a separate component in RCW 103 responsible for the hard spectrum, we conclude that this emission is not associated with a cooling neutron star because of its size and the absence of a point-like source.

We would like to thank Fred Seward for making the Einstein IPC and HRI images of galactic supernova remnants available to us, including the image of RCW 103 shown in this paper.

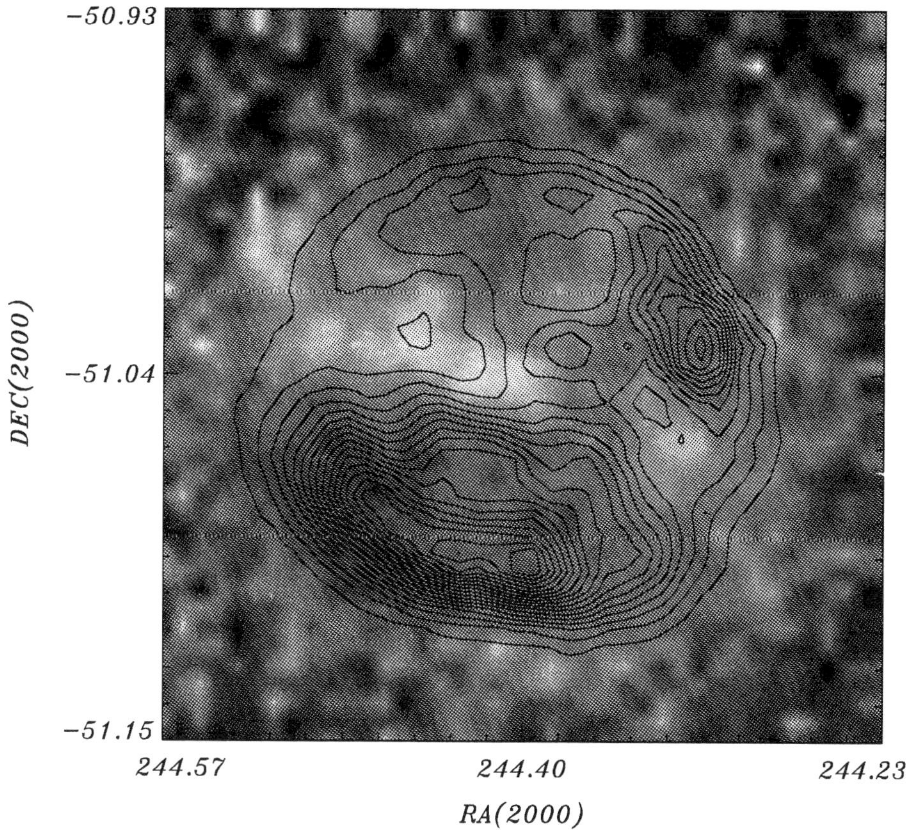

Fig. 5—Hardness ratio HR2=(D-C)/(D+C) of RCW 103 (dark = small HR2, white = large HR2). The contours show the brightness distribution.

References

Nomoto, K., Tsuruta, S. 1987, *Ap. J.*, **312**, 711
Nugent, J. J., *et al.* 1984, *Ap. J.*, **284**, 612
Van Riper, K. A., 1991, *Ap. J. Sup..*, **75**, 449
Shibazaki, N., Lamb, F. K. 1989, *Ap. J.*, **346**, 808
Tuohy, I. R., *et al.* 1980, *Ap. J. Lett.*, **239**, L107
Tuohy, I. R., *et al.* 1983, *Ap. J. Lett.*, **268**, 778
Seward, F. D., Wang, Z. R. 1988, *Ap. J.*,**332**, 199
Seward, F .D. 1990 *Ap. J. Suppl.*, **73**, 781

Rapid Cooling and Structure of Neutron Stars

Kenneth A. Van Riper[1] and James M. Lattimer[2]

[1]Los Alamos National Laboratory
[2]SUNY-Stony Brook

1 Direct Urca Neutrino Emission

Neutrino emission in the core of a neutron star by the modified URCA process

$$\begin{aligned}(n,p)+p+e^- &\rightarrow (n,p)+n+\nu_e\\ (n,p)+n &\rightarrow (n,p)+p+e^-+\bar{\nu}_e\end{aligned}$$

requires the existence of a spectator nucleon to simultaneously conserve energy and momentum. Each degenerate nucleon involved in the reaction contributes a factor $T/T_F \ll 1$ to the emission rate, where T is the temperature and T_F is the Fermi temperature. Lattimer *et al.* (1991) have pointed out that the direct Urca process

$$\begin{aligned}n &\rightarrow p+e^-+\bar{\nu}_e\\ p+e^- &\rightarrow n+\nu_e\end{aligned}$$

can occur in the degenerate matter of the neutron star core if the proton/nucleon ratio is in excess of some critical value lying in the range 0.11–0.15. We have constructed a number of high density equations of state (EOSs 1–6) which satisfy this criterion; our numerical calculations make use of these for the neutron star model. Prakash *et al.* (1992) showed that matter with *any* proton/nucleon ratio can rapidly cool by the direct Urca process if Λ hyperons exist. Nearly all studies of supranuclear matter to date indicate that one or the other, or both, of these situations will take place.

The direct Urca emissivity is

$$\begin{aligned}\epsilon_{\mathrm{Urca}} &= \frac{457\pi}{10080}\frac{G_F^2(1+3g_A^2)}{\hbar^{10}c^5}m_n m_p \mu_e (kT)^6\Theta_t\\ &= 4.24\times10^{27}(Y_e n/n_s)^{1/3} m_n^* m_p^* T_9^6\Theta_t \text{ erg cm}^{-3}\text{ s}^{-1}\end{aligned}$$

where $T_9 = T/10^9$ deg, $n_s = 0.16\text{ fm}^{-3}$ is nuclear density, m_n and m_p are the effective nuclear masses, m_n^* and m_p^* are these masses divided by the nucleon mass, and the threshold factor $\Theta_t(x) = 1$ if $x > 0$, 0 otherwise. For the dense matter equations of state that we assume, the threshold limits direct Urca emission to densities greater than 6.9×10^{14} g cm^{-3}. The overall coefficient is 5 orders of magnitude larger than the rate for the modified Urca process $\approx 10^{22}(Y_e n/n_s)^{1/3}T_9^8$ erg cm^{-3} s^{-1}, and the direct Urca falls more slowly with T. When either the neutrons or protons, or both, are superfluid, the direct Urca rate is suppressed by the factor $\exp\left(-\frac{T}{T_c^n}\right)\exp\left(-\frac{T}{T_c^p}\right)$, where T_c are the critical temperatures corresponding to the superfluid gaps.

2 Thermal Evolution Models

We made numerical calculations of the evolution of neutron stars cooling by the direct Urca and other accelerated cooling mechanisms using the computer code developed by Van Riper (1988, 1991). This code uses a diffusion algorithm to follow both the conduction of heat and energy losses by neutrino emission inside the star; Van Riper (1991) describes the thermal conductivities, neutrino emissivities, and heat capacities used. We compute the temperature distribution interior to the density 10^{10} g cm^{-3} and treat the atmosphere external to this as a boundary condition using the models of Van Riper (1988). For most models, a surface magnetic field strength $B = 10^{12}$ G was assumed. Magnetic field effects are important only at densities $< 10^{10}$ g cm^{-3}. At higher densities a field $B < 10^{13}$ G makes little impact on the conductivity, heat capacity, and equation of state, and thus will not effect the cooling wave.

Accelerated core cooling, such as by the direct Urca process, results initially in a temperature inversion. After a few months, the core of matter with densities greater than nuclear has temperatures about a few $\times 10^8$ deg, while the overlying crust is an order of magnitude hotter. Energy in the crust conducts to the core, resulting in a cooling wave which moves outward on a time given by the thermal conduction time in the crust. The surface temperature drops precipitously, as shown in Figure 1, when the cooling wave reaches the surface.

We define a cooling time t_w, indicated by the filled circle in Figure 1, as the time at which the slope of the cooling curve is the steepest. Before this time, the surface temperature T_s is the same as for the standard cooling case (and depends only on the properties of the crust). After t_w, the interior of the star has become isothermal, and T_s is determined by the global balance of emission and residual thermal energy.

A number of models with direct Urca and other accelerated cooling mechanisms, such as the the Urca process on percolating quarks (Kiguchi and Sato, 1981), give a range of cooling times $1 \le t_w \le 700$ years. The cooling time increases with the radius R of the star, as is shown in Figure 2. A power law fit to $t_w(R)$ ($\propto R^6$ is best) is rather poor. A much better power law results if we consider t_w as a function of the crust thickness $R_{\rm Shell} = R - r\left(\rho = \rho_{\rm nuclear}\right)$. The exponent of the fit, 1.8-2.0, is the value expected from our analytic considerations below. The solid curves in Figure 2 are a sequence of models of $M = 1.4M_\odot$ stars with different high-density equations of state. The dashed curve is from varying the mass with one equation of state—our EOS 2 which has compressibility of 180 MeV at nuclear density and gives a maximum neutron star mass of $1.45M_\odot$. When superfluid effects are not included, the cooling time increases by a factor of 3. Because of the lack of superfluid suppression of the neutron heat capacity C_V in the crust and the dependence $t_w \propto C_V$. The factor of 3 depends on an average in both space and time of the fraction of superfluid

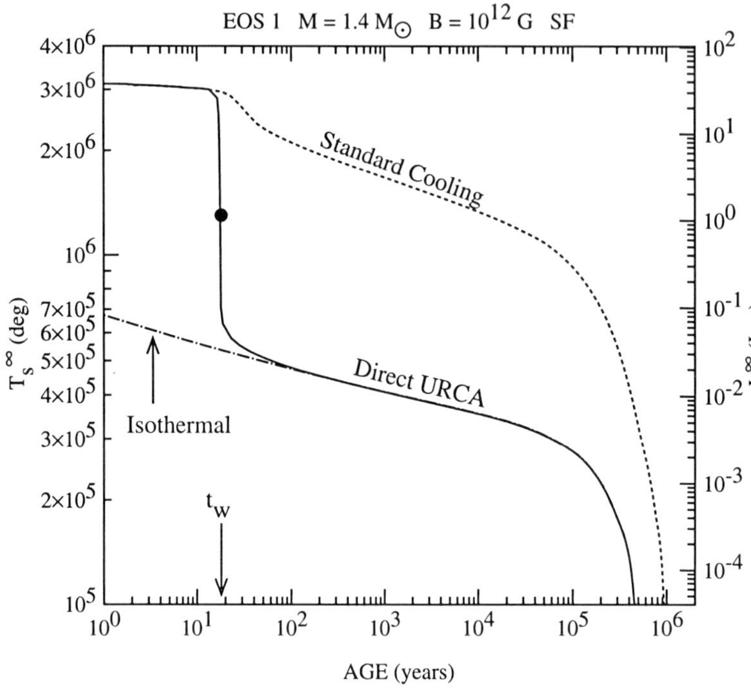

Fig. 1.— Cooling curves for a 1.4 $M_\odot$ neutron star with a surface magnetic field strength 10^{12} G constructed with EOS 1 and including standard superfluid effects. The dashed line assumes "standard" neutrino emissivities. The solid curve results when direct Urca emission occurs in the core. The filled circle shows our definition of the cooling time t_w. The dot-dashed curve assumes an isothermal interior and the direct Urca process; the interior of the realistic star has become isothermal when this and the solid curve coincide.

neutrons in the crust and the contribution of the neutrons to C_V relative to the electron contribution, which is unaffected by superfluidity.

3 Analytic Model for Diffusion Through the Crust

We have developed analytic and semi-analytic models for the cooling wave in the crust to find the dependence of the cooling time on stellar properties such as the crust thickness. We consider only conductive transport of energy in the crust; neutrino emission there takes place a time scale longer than t_w at an age of t_w. General relativistic terms are also neglected for the sake of simple equations. (These approximations are not made in the numerical calculations.) The equations of radiative transport and energy balance are

$$L = -4\pi r^2 K \frac{\partial T}{\partial r} \quad \text{and} \quad \frac{\partial L}{\partial r} = -4\pi r^2 C_V \frac{\partial T}{\partial r},$$

where L is the luminosity, r is the radial coordinate, and K is the thermal conductivity. A good fit to the latter for a large region of (T,ρ) space is $K \simeq A(\rho/\rho_o)/T$ with $A = 4\times 10^{29}$ erg cm^{-1} s^{-1} and $\rho_o = 2.7\times 10^{14}$ g cm^{-3}. The specific heat can be approximated

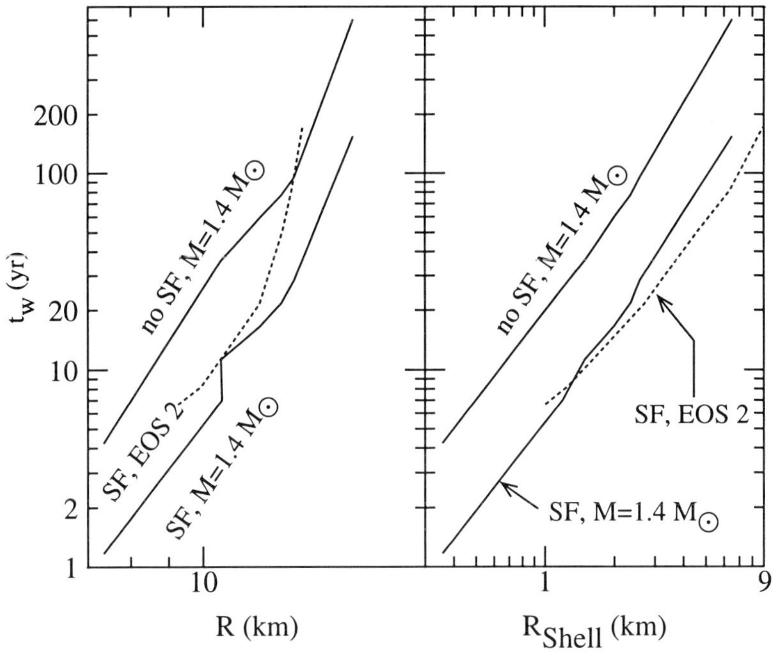

Fig. 2.— The cooling time t_w as a function of stellar radius (left panel) and the crust thickness (right panel)

by $C_V \simeq B(\rho/\rho_o)^{1/3}T$ with $B = 1.6 \times 10^{11}$ erg K^{-2} cm^{-3}. Combining all this gives a single equation

$$\frac{1}{r^2}\frac{\partial}{\partial r}\left(\frac{A\rho r^2}{\rho_o T}\frac{\partial T}{\partial r}\right) = B\left(\frac{\rho}{\rho_o}\right)^{1/3} T\frac{\partial T}{\partial t}$$

for conduction in the crust. Here we consider the simple case ρ = constant; more general solutions can be found in Lattimer *et al.* (1992). We assume that the solution for T is separable in time and space, $T = T_o\psi(r)\phi(t)$, where T_o is the initial temperature at the outer edge of the crust (*i.e.*, at $t = 0$, $r = R$). The boundary conditions are taken to be $\psi(R) = 1$, $\psi'(R) = 0$, $\phi(0) = 1$, and $\psi(R_{core}) = T_{core}/T_{surface} \sim 1/30$. Defining a dimensionless depth

$$x = \frac{R - r}{R - R_{core}}$$

and shell thickness

$$q = \frac{R}{R - R_{core}},$$

we have

$$\frac{1}{\psi^2}\frac{d}{dx}\frac{1}{\psi}\frac{d\psi}{dx} + \frac{2}{\psi^3(x+q)}\frac{d\psi}{dx} = \frac{B}{A}\left(\frac{\rho}{\rho_o}\right)^{-2/3}(R - R_{core})^2 T_o^2 \phi\frac{d\phi}{dt} = -\alpha$$

where α is a separation constant. The solution for the time function is

$$\phi = \sqrt{1 - \frac{t}{\tau}}$$

where

$$\tau = \frac{BT_o^2 (R - R_{\rm core})^2}{2A\alpha(\rho/\rho_o)^{2/3}} \simeq 67 \left(\frac{T_o}{10^9\ \rm K}\right)^2 \left(\frac{R - R_{\rm core}}{\rm km}\right)^2 \frac{1}{\alpha(\rho/\rho_o)^{2/3}}\ \rm yrs.$$

Note the proportionality to $B \propto C_V$ and the $2^{\rm nd}$ power of the shell thickness. If we assume a thin shell, $q \ll 1$, the radial equation becomes

$$\frac{1}{\psi^2}\frac{d}{dx}\frac{1}{\psi}\frac{d\psi}{dx} = -\alpha,$$

which is solved by

$$\psi = \frac{1}{\cosh[x\sqrt{\alpha}]}$$

with

$$\alpha = [\cosh^{-1}\psi(1)^{-1}]^2 \simeq 17,$$

which has a weak dependence on the core temperature.

4 Core Superfluidity

The direct Urca process is suppressed by a factor $\exp\left(-\frac{T}{T_c}\right)$ for neutrons and/or protons when those nucleons are superfluid. The reduction in the cooling rate and the consequent effect on the cooling curves can be large if a large portion of the direct Urca emitting core is superfluid. The effect is much more than in the standard cooling case where crust bremsstrahlung emission, which persists in superfluid regions, is comparable in total emissivity to the modified Urca process.

Figure 3 shows cooling curves for three variations of superfluid treatment. Superfluid suppression of the emissivity and heat capacity does not occur in the "no SF" case. The "SF" model treats suppression with gaps based on realistic calculations (Takatsuka 1972, Chao, Clark, and Yang 1972). The "extreme SF" model exaggerates the neutron $^3P_2$ and proton $^1S_0$ gaps above nuclear density by increasing both their height and density range. The standard and extreme gaps are shown in Figure 4. The neutron $^1S_0$ gap in the crust is the same in both models. In addition to exploring the effects of uncertainties in these two gaps, the extreme model is useful in estimating the consequences of gaps at higher densities due to higher order pairings.

Direct Urca cooling was calculated for three equations of state. The stiff EOS 3, which has a maximum neutron star mass of $2.06M_\odot$, results in a $1.4M_\odot$ neutron star having a central density ρ_c just above the direct Urca threshold. A soft case and a high ρ_c is represented by EOS 4, for which the maximum mass is $1.44M_\odot$. Figure 4 shows the ρ_c of these three neutron stars. The direct Urca emitting region extends from the threshold of 6.9×10^{14} g cm^{-3} to the central density; the magnitude of the superfluid effect depends on the overlap of the gaps with this density range. The relative overlap decreases with increasing central density, and Figure 3 shows a large

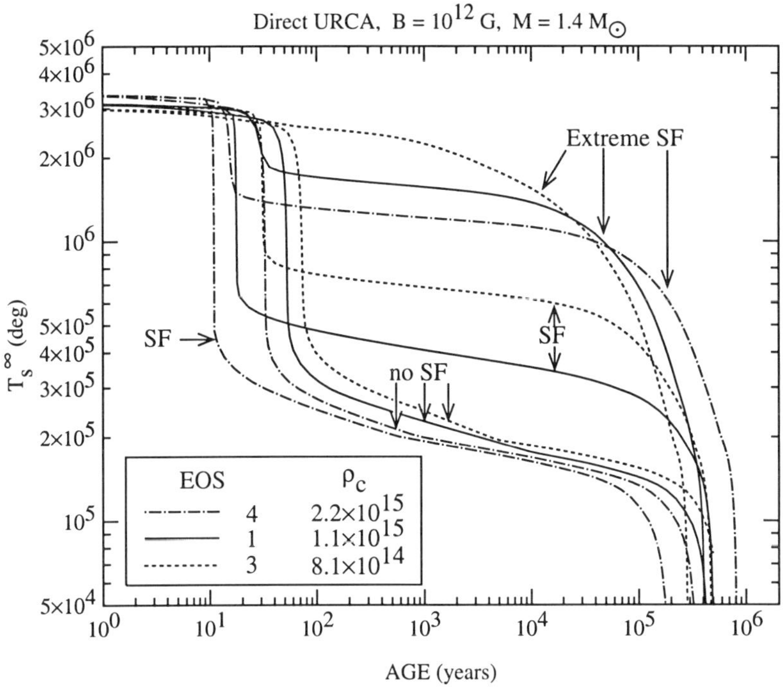

Fig. 3.— Cooling curves for $1.4M_{\odot}$ neutron stars cooling by the direct Urca process. The different line styles correspond to 3 different EOSs which result in the central densities shown. For each EOS, superfluid suppression of emissivities and heat capacities was neglected (no SF) or treated with standard superfluid parameters (SF) or with enhanced gaps above nuclear density (Extreme SF).

change in the cooling curve for the SF models for equations of state 3 and 1 which have a large amount of superfluid matter in the core. The central density for equation of state 4 is well above the reach of the standard neutron $^3P_2$ gap; the major difference between the SF and no SF models for equation of state 4 is the change in t_w due to the superfluid suppression of the heat capacity in the crust. The extreme SF gaps nearly obviate the role of the direct Urca process as an accelerated cooling mechanism.

5 Conclusions

A neutron star cooling very rapidly in the center will undergo a sharp decrease in surface temperature at the time t_w for thermal diffusion through the crust. This time depends on the square of the thickness of the crust, and is also influenced by the details of the neutron superfluidity in the crust. An observational determination of t_w will limit the radius of the star, and hence the mass and high density equation of state.

For stars cooling by the direct Urca process, the surface temperature in the epoch between the drop at t_w until an age of about 10^5 years depends strongly on the details of the superfluid gaps above nuclear density and on the central density of the star.

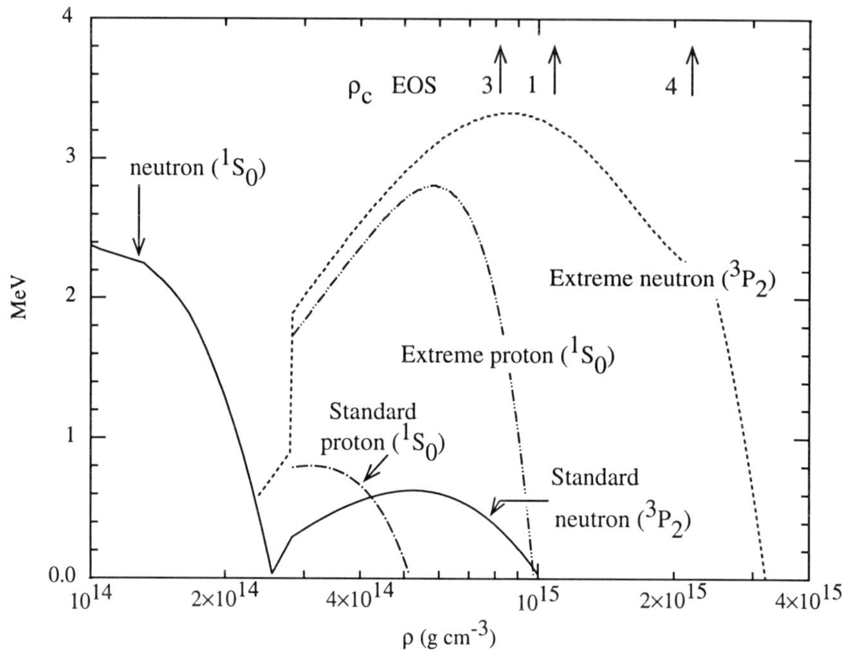

Fig. 4.—The superfluid gaps used the calculations. The same neutron $^1S_0$ gap is used for both the standard and extreme models. The arrows along the top show the central densities of the stars whose cooling is shown in Figure 3.

The parameters of the high density gaps are uncertain, as is the central density. It is also possible that higher order gaps exist for both the neutron and protons; these gaps will also influence the surface temperature for $t > t_w$.

References

Chao, N.-C., Clark, J. W., and Yang, C.-H. 1972, Nucl. Phys. A., 179, 320.
Kiguchi, M., and Sato, K. 1981, Progr. Theor. Phys. Lett., 66, 725.
Lattimer, J.M., Pethick, C.J., Prakash, M. and Haensel, P., *Phys.Rev. Lett.*, **66**, 2701 (1991)
Lattimer, J.M., Van Riper K. A., Prakash, M., and Prakash, M. (1992) in preparation
Prakash, M., Prakash, M., Lattimer, J.M. and Pethick, C.J., *Ap. J. (Lett.)*, in press (1992)
Takatsuka, T. 1972, Progr. Theor. Phys., 48, 1517.
Van Riper, K. A. 1988, Ap. J., 329, 339.
Van Riper, K. A. 1991, Ap. J. Suppl., 75, 449.

Testing Pulsar Thermal Evolution Theories by Observations

Sachiko Tsuruta

Department of Physics, Montana State University

Abstract

We briefly summarize the results from the most recent calculations of neutron star cooling involving the pion, quark, kaon, and direct URCA process, and frictional heating. Our latest cooling and heating models are carefully compared with the currently most up-dated pulsar observational data. Some implications are discussed.

1 Introduction

A pulsar can cool by the 'standard' cooling scenario where the conventional neutrino processes such as the modified URCA and bremsstrahlung neutrino processes cool the stellar interior (*e.g.*, Tsuruta 1986 for the general review), and the 'non-standard' scenario involving drastically fast cooling mechanisms such as the pion cooling and the direct URCA process (*e.g.*, Tsuruta 1992 for the general review). In the absence of additional heat input, in either case a pulsar will become cold relatively quickly, within less than $\sim 10^5$ years. Several heating processes have been suggested, such as heating due to friction, crust cracking and accretion. In this paper we shall review some of the recent developments in this area. Our major emphasis, however, will be given to the problems which are not covered by other contributors to this volume.

We present in Section 2 the results from the most recent calculations on cooling, both standard and various non-standard, and frictional heating. These theoretical models are compared in Section 3 with the currently most up-dated observational data. Concluding remarks are given in the last section, 4.

2 Theoretical Models

Thermal evolution of pulsars is calculated by solving the standard hydrostatic equations and the energy balance and energy transport equations. The internal energy is governed by heating and cooling. The total cooling rate includes the photon luminosity from the surface and various neutrino luminosities, both standard and non-standard. In the 'exact' evolutionary calculations these equations in the fully general relativistic form are solved simultaneously and locally as a function of time and the radial distance from the stellar center (see *e.g.*, Nomoto and Tsuruta 1987, hereafter NT87). The major neutrino mechanisms in the standard cooling are the modified URCA, plasmon, and bremsstrahlung processes (*e.g.*, NT87). The non-standard processes which have been suggested are the neutrino processes involving the pion and kaon condensates and quarks (*e.g.*, Umeda *et al.* 1992a; Page and Baron 1990; Iwamoto 1980), and the direct URCA process (Lattimer *et al.* 1991; Baym, this volume).

The 'exact' evolutionary calculations are carried out, both for standard and non-standard cooling, using the currently most up-dated input (Umeda *et al.* 1992a). The results for the standard cooling are essentially the same as in NT87, except at very low temperatures. Here the low temperature region is treated more carefully. Among the non-standard scenarios the physics has been more carefully investigated for the pion cooling than for the others (Umeda *et al.* 1992a). According to the recent, careful work by Muto and Tstsumi (1988) it is reasonable to assume that the pion phase appears at $\geq \sim 2\ \rho_N$ (where ρ_N is the nuclear density). Therefore, we have conducted the 'exact' cooling calculations for stars with pion cores within this range (Umeda *et al.* 1992a), using the pion emissivity recently derived by Muto and Tatsumi (1988). The pion condensation is characterized by the Landau-Migdal parameter $\tilde{g}'$ which describes the baryon-baryon short-range correlations. Some of the interesting results are that: the star cools much faster in the presence of the pion condensate; a sudden drop in temperature starts at ~ 30 - 100 years after birth; and that the use of the 'exact' evolutionary method is critical in order to conduct realistic pion cooling calculations. This is because the earlier calculations with the similar microphysical input but conducted with the isothermal method (Tsuruta 1987) indicate that in the isothermal case cooling is far less drastic, and furthermore there should be no sudden drop of luminosity during ~ 1 - 100 years. The effects of stellar mass, the equation of state, and superfluidity are very important. For instance, with a reasonably stiff equation of state, *e.g.*, the PS model (Pandharipande, Pines and Smith 1976), the central density of a star with the baryon mass $M_A\ < \sim (1.5 - 2)\ M_\odot$ is lower than the critical density above which the pion condensate appears, and hence such a less massive star will possess no pion core. For the medium FP model (Friedman and Pandharipande 1981) this critical mass is about 0.6 $M_\odot$. With very soft equations of state, such as the BPS model (Baym, Pethick and Sutherland 1971), stars within a reasonable mass range all possess pion cores, and the sudden drop of temperature takes place earlier. If the effect of superfluidity is properly taken into account, the pion cooling can be substantially slowed down and the sudden drop in temperature become much milder.

We have carried out similar calculations for quark and kaon stars using the same input physics and method as for pion stars except that the quark and kaon emissivities are used. The results are compared with the pion case in Figure 1 of Tsuruta 1992. Qualitatively, the general behavior is similar to the pion case. For kaon stars qualitatively similar results were obtained by Page and Baron (1990) also; although their equations of state are somewhat different from ours. Recently it was pointed out that a neutron star would cool much faster even without the fast-cooling exotic particles, if the direct Urca process is in operation (Lattimer *et al.* 1991). Therefore, with the BPS and FP models we explored the direct URCA cooling. The results are shown in Figure 1 of Tsuruta 1992. With the FP model the direct URCA cooling is very similar to the pion case. However, a significant difference is noted when the soft BPS model is used, in the sense that for the direct URCA cooling, the sudden drop of temperature takes place earlier. Very recently, the direct URCA cooling was calculated by Page and Applegate (1992) and Van Riper (this volume). Their results are qualitatively similar to ours.

The frictional interaction of neutron superfluids with normal matter in the interior of pulsars will dissipate rotational energy of superfluids and generate heat. Tsuruta (1987) and Shibazaki and Lamb (1988) carried out frictional heating calculations using the isothermal approach, and predicted that heating may play a crucial role for neutron stars older than about ten thousand years. Recently Van Riper (1991) conducted phenomenological studies of the effect of an additional constant heat source on thermal evolution of neutron stars, although the specific heating mechanisms were not considered. Very recently the 'exact' evolutionary calculations were carried out for the frictional heating (Umeda *et al.* 1992b). The results may be summarized as follows. If a stiff equation of state is adopted, the frictional heating can greatly increase the temperature in the photon cooling era, significantly changing the thermal evolution of relatively old neutron stars. This effect, however, is less important for stars with a medium to soft equation of state. The details are found in Umeda *et al.* 1992b.

In addition to frictional heating, some other authors considered other possible heating mechanisms. The work on heat generated by crust cracking is reported by Cheng *et al.* (this volume), and heating due to accretion by Nelson, Salpeter and Wasserman (this volume). We are currently conducting systematic studies of various potentially significant heating mechanisms.

window6pc

3 Comparison with Observation

The *Einstein* survey of thermal emission from radio pulsars, followed by the EXOSAT mission, yielded ten positive detections out of nearly two dozens of observed candidates (Helfand 1983; Seward and Wang 1988; see Ögelman 1991 for general review and other references). The ROSAT mission, currently in operation, is expected to double the list of positive detections. Some ROSAT results already started appearing (see *e.g.*, Ögelman; Helfand; Seward; Becker *et al.*.; Becker, Trümper and Ögelman; in this volume). Since there exist problems of distinguishing between the pulsar surface emission and others such as the magnetospheric or nebula emissions, the observed count rates should be generally considered to be the upper limits to the surface thermal emission unless there exists some spectral evidence. As of April 1992, there are three detections for which there is such spectral evidence. These sources are: (i) PSR 0833-45, the Vela pulsar, (from EXOSAT by Ögelman and Zimmermann 1989), (ii) PSR 0656+14 (from ROSAT by Finley, Ögelman and Kiziloğlu 1992), and (iii) PSR 1055-52 (from EXOSAT by Brinkmann and Ögelman 1987). We carefully compared our thermal evolution models with these detection points. The results are summarized below.

In Figure 1 our cooling models are compared with these detection points, which are shown with the vertical error bars. The solid curve refers to the standard cooling of the PS model star with the baryon mass $M_A = 1.4\ M_\odot$. The dashed and dot-dashed curves represent the pion cooling of the 1.65 $M_\odot$ PS model and 1.4 $M_\odot$ FP model stars, respectively. These equations of state are chosen to illustrate the measure of uncertainty arising from uncertainty in the nuclear theories. The standard curve is not shown for the FP model because all stars with this equation of state and $M_A \; > \sim 0.5$ $M_\odot$ possess pion cores, whereas the evolutionary curves for the FP model stars with

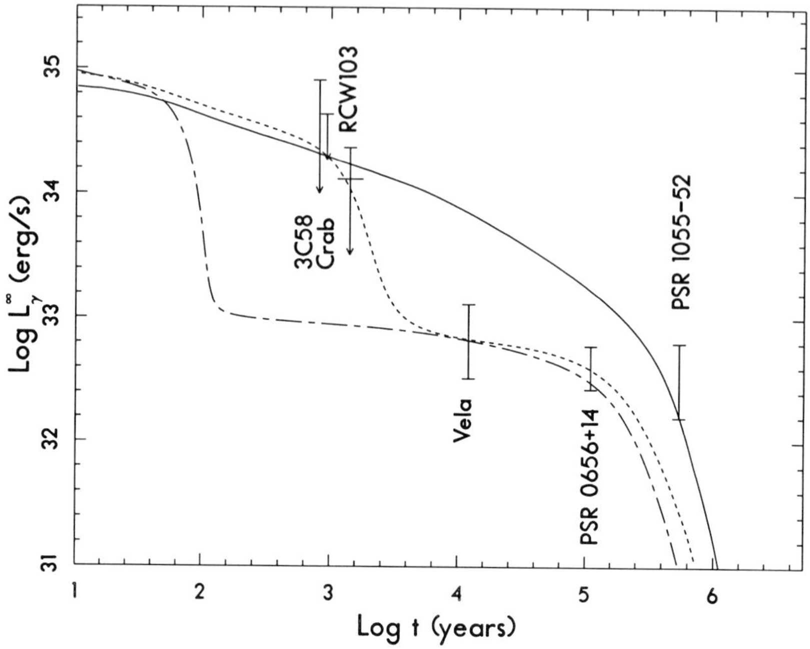

Fig. 1—The photon luminosity L_γ^∞ to be observed at infinity as a function of age t. The three detection points are indicated by the vertical bars, and the upper limits for some younger pulsars are shown by the downward arrows.

$M_A \leq \sim 0.5\ M_\odot$ are not consistent with these detection points (Umeda *et al.* 1992a). The BPS model is excluded from our consideration because the maximum mass of pion stars with such a soft equation of state is less than 1.4 $M_\odot$. The effect of superfluidity is taken into account by adopting one of the currently most realistic estimates of the energy gap as given by Takatsuka and Tamagaki 1982 (referred to as the TT82 model) for the standard cooling, whereas we adopted for pion cooling a variation of the TT82 model properly modified for the pion condensate (see Umeda *et al.* 1992a for details). As a representative case, $\tilde{g}' = 0.5$ is adopted. The upper limits for some younger pulsars (the Crab, 3C 58 and RCW 103) are shown by the downward arrows. The Crab and 3C58 data are taken from Nomoto and Tsuruta 1986, while the RCW 103 data comes from the latest ROSAT observation reported by Becker *et al.* (this volume). Our results show that with medium to stiff equations of state, the observed surface temperatures of the Vela pulsar and PSR 0656+14 are too low to be consistent with the standard cooling scenario, whereas they are consistent with pion cooling when a realistic superfluid model is adopted. On the other hand, without superfluidity, the pion curves for all stars with a reasonable range of mass and equation of state are too steep to pass through the error bars of both Vela and PSR 0656+14 data (Umeda *et al.* 1992a). The detection point of PSR 1055-52 is consistent with the standard cooling of the PS model, but the pion curves are too low.

Let us now compare our heating models with observation. Figure 2 shows our results for frictional heating of the standard PS model with $M_A = 1.4\ M_\odot$. Note that the

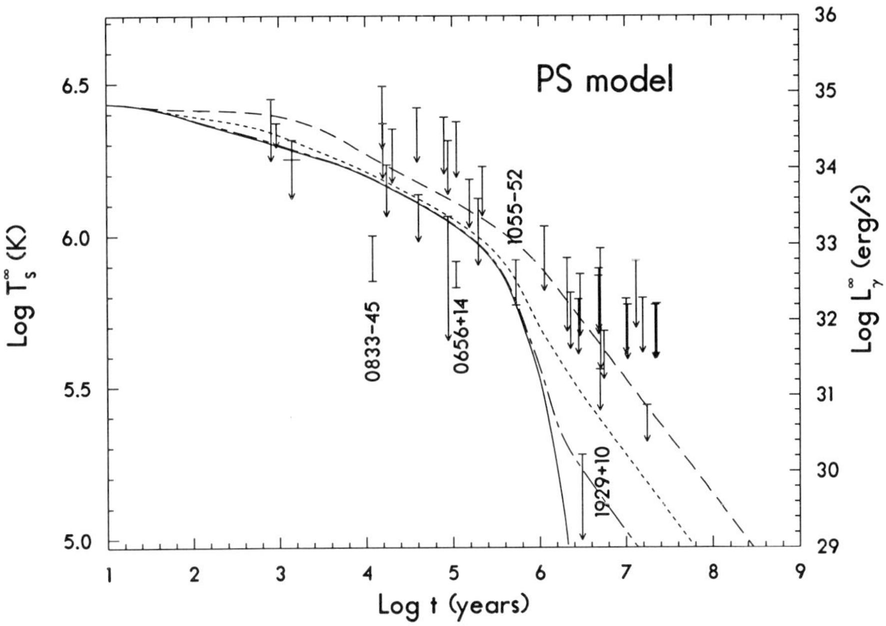

Fig. 2—The effect of frictional heating. The surface temperature T_s^∞ and photon luminosity L_γ^∞ to be observed at infinity as a function of age t. In addition to the observation points in Fig. 1 many ROSAT upper limits are shown.

frictional heating effect is most pronounced for the stiffest model since it possesses the thickest crustal layer. The dot-dashed, dashed and long-dashed curves refer to the differential angular momentum $J = J_1$, J_2 and J_3, respectively, which correspond to the critical angular frequency $\omega_{cr} = 0.1$, 1 and 10 in the vortex creep theory (Pines, this volume; Alpar, Cheng and Pines 1989). ω_{cr} refers to a measure of the pinning strength, with the regimes $\omega_{cr} \leq 0.1, 0.1 \leq \omega_{cr} \leq 1$, and $1 \leq \omega_{cr}$ representing the superweak, weak, and strong pinning. For comparison, cooling without heating is shown by the solid curve. In addition to the data points given in Figure 1 the upper limits to many other pulsars are shown by the downward arrows. The value for PSR 1929+10 is taken from the the EXOSAT observation reported by Alpar *et al.* 1987, while the other points are taken from the recent ROSAT all sky survey data as reported by Becker, Trümper and Ögelman (this volume).

The upper limit to PSR 1929+10 may be used to set some constraint on the strength of the internal frictional heating. This upper limit is significantly below the heating curves with J_2 and J_3 (dashed and long-dashed curves in Fig. 2) corresponding, respectively, to the weak - strong boundary and the strongest pinning in the vortex creep theory. The conclusion is that for a stiff model with large crustal layers such as the PS model, this constraint supports the superweak regime with $J \leq\sim J_1$, $\omega_{cr} \leq\sim 0.1$ in the vortex creep theory. Other observational and theoretical considerations related to this theory also support the superweak pinning (Alpar, Cheng and Pines 1989). The surface temperature of PSR 1055-52 is consistent with the standard evolutionary scenario for the PS model, both with and without the additional frictional heating. The heating curve even with relatively strong frictional heating with J_2 (dashed curve) is within the range of this source, although the strongest case with $J \sim J_3$ (long-dashed curve) should be excluded. We are currently investigating the combined effects of both superfluidity and heating on the pion case. The preliminary results indicate that when the superfluid effect is properly taken into account, for some pion stars within a reasonable range of mass and equation of state the same evolutionary curves may simultaneously pass through the error bars of all of these three detection points (the Vela, PSR 0656+14 and PSR 1055-52) if the maximum frictional heating is imposed. However, such strong heating contradicts with the constraint imposed by the upper limit to PSR 1929+10.

We shall now summarize possible implications of the comparisons reported above. Since the stellar mass is most likely $> \sim 0.5\ M_\odot$, all stars with a medium to soft equation of state probably possess pion cores, and pion cores, even with the effect of superfluidity, will cool too fast to be consistent with the PSR 1055-52 data point. This may imply that the equation of state is stiffer than the medium FP model. With the stiff PS model the PSR 1055-52 data is consistent with the standard scenario, both with cooling alone and with additional heating, whereas the Vela and PSR 0656+14 data are consistent with both cooling and heating of a superfluid pion star. The implication may be, as already proposed earlier (*e.g.*, Tsuruta 1986), that there are two classes of neutron stars, less massive stars evolving with the standard scenario and more massive stars which evolve with the pion scenario. For the PS model this critical mass dividing these two classes of stars is $\sim (1.5 - 2)\ M_\odot$ (for $\tilde{g}' = 0.5 - 0.6$), whereas it can become $\sim (1 - 1.3)\ M_\odot$ for an equation of state softer than the PS model but stiffer than the FP model. Alternatively, some of the superfluid pion curves, if the maximum heating is imposed, may pass through the error bars of all of the three detection points, and the implication could be that all pulsars possess pion cores and cool relatively fast. However, if such an option is valid, we would have to disregard the strong constraint given by the upper limt to PSR 1919+10.

Most of other upper limits shown in Figure 2 are sufficiently high enough to be consistent with all theoretical curves, but at least three of these points are below the maximum heating (long-dashed curve). Combined with the PSR 1929+10 constraint, these data may be used to exclude the very strong heating near the maximum. We may emphasize here that relatively short time intervals were spent for each source during the ROSAT survey period. By conducting longer period observations in the near future, the upper limits can be significantly lowered, and then further useful constraints are expected to be given.

Here we chose the pion cooling as a representative example of the fast, non-standard cooling scenarios. Since other non-standard scenarios also exhibit qualitatively similar behavior, the above conclusion may be extended to other suggested non-standard scenarios, such as the kaon, quark, or direct URCA cooling. However, before similar detailed comparison with observation can be conducted, more careful theoretical work would be required on these alternative scenarios, *e.g.*, the realistic estimate of the superfluid energy gap in these cases.

Before closing this section, some words of caution are in order. All the evolutionary curves considered above were obtained without including the effects of magnetic fields. Although in the early cooling calculations the effects of magnetic fields were considered to be important, Yakovlev (1982) and Hernquist (1985) showed that these effects on cooling itself are negligible. Among the various magnetic-field-related pulsar problems yet to be solved properly is the effect of possible modulation of temperatures over the surface. Due to the anisotropic electron conduction under strong magnetic fields a significant difference arises between the polar and equatorial temperatures if heat flow from the hotter polar region to the cooler equatorial region is neglected (Umeda and Tsuruta 1992; see also Shibanov *et al.* in this volume). It is important to solve this problem properly, because if the effect of this temperature modulation is indeed significant, the observed temperature will depend on the inclination of the polar axis to the line of sight. Therefore, we are currently exploring this problem carefully, adopting a two-dimensional approach (Umeda and Tsuruta 1992). Our preliminary results indicate that this effect is very important. Also, in obtaining the observation points, a black-body spectrum is assumed. The possible implication of this assumption is discussed by some other contributors to this volume (*e.g.*, Miller; Lai, Salpeter and Shapiro).

4 Concluding Remarks

We have summarized the recent major developments in the area of pulsar thermal evolution. This area will remain important for various reasons. For instance, the heating theories could be better tested by long observations of older neutron stars through the space missions already in operation, such as ROSAT, and those planned for the immediate future, such as ASTRO-D and AXAF. Moreover, the satellite programs such as ROSAT, AXAF and XMM could potentially offer the actual temperature measurements (not just upper limits), significantly reduce the error bars, and detect more new sources. The current upper limits definitely would be lowered by a substantial degree. These observational improvements are expected to offer more definitive and exciting conclusions when carefully compared with theories.

We acknowledge with thanks useful discussions with many of our colleagues, especially H. Umeda, T. Ögelman, R. Epstein, K.S. Cheng, W. Becker, N. Shibazaki, K. Nomoto, T. Tatsumi, T. Muto, and R. Tamagaki. This work was supported, in part, by the NASA grant NAGW-2208 and the NSF grants AST-9013290 and RII-8921978.

References

Alpar, M. A., Brinkmann, W., Kiziloğlu, Ü., and Ögelman, H. 1987, *Astron. Astrophys.*, **177**, 101.
Alpar, M. A., Cheng, K. S., and Pines, D. 1989, *Ap. J.*, **346**, 823.
Baym, G., Pethick, D. J., and Sutherland, P. 1971, *Ap. J.*, **170**, 299.
Brinkmann, W., and Ögelman, H. 1987, *Astron. Astrophys.*, **346**, 823.
Finley, J. P., Ögelman, H., and Kiziloğlu, Ü. 1992, *Ap. J.*, submitted.
Friedman, B., and Pandharipande, V. R. 1981, *Nucl. Phys. A.*, **361**, 502.
Helfand, D. J. 1983, in *Supernova Remnants and Their X-Ray Emission*, ed. J. Danziger and P. Gorenstein (D. Reidel Publ. Co.), p. 471.
Hernquist, L. 1985, *M.N.R.A.S.*, **213**, 313.
Iwamoto, N. 1980, *Phys. Rev. Lett.*, **44**, 1637.
Lattimer, J. M., Pethick, C. J., Prakash, M., and Haensel, P. 1991, *Phys. Rev. Lett.*, **66**, 2701.
Muto, T., and Tatsumi, T. 1988, *Prog. Theor. Phys.*, **79**, 461.
Nomoto, K., and Tsuruta, S. 1986, *Ap. J. Lett.*, **305**, L19.
Nomoto, K., and Tsuruta, S. 1987, *Ap. J.*, **312**, 711.
Ögelman, H. 1991, in *Neutron Stars: Theory and Observation*, ed. J. Ventura and D. Pines (Kluwer Academic Publ., Dordrecht), p. 87.
Ögelman, H., and Zimmermann, H.-U. 1989, *Astron. Astrophys.*, **214**, 179.
Page, D., and Applegate, J. H. 1992, preprint.
Page, D., and Baron, E. 1990, *Ap. J. Lett.*, **354**, L17.
Pandharipande, V. R., Pines, D., and Smith, R. A. 1976, *Ap. J.*, **208**, 550.
Seward, F. D., and Wang, Z.-R. 1988, *Ap. J.*, **332**, 199.
Shibazaki, N., and Lamb, F. K. 1988, *Ap. J.*, **346**, 808.
Takatsuka, T., and Tamagaki, R. 1982, *Prog. Theor. Phys.*, **67**, 1649.
Tsuruta, S. 1986, *Comments on Astrophys.*, **11**, 151.
Tsuruta, S. 1987, in *13th Texas Symp. on Rel. Ap.*, ed. M. P. Ulmer (World Sci.), p. 499.
Tsuruta, S. 1992, in *The Structure and Evolution of Neutron Stars*, ed. D. Pines, R. Tamagaki, and S. Tsuruta (Addison-Wesley Publ. Co.).
Umeda, H., Nomoto, K., Tsuruta, S., Muto, T., and Tatsumi, T. 1992a, *Ap. J.*, submitted.
Umeda, H., Shibazaki, N., Nomoto, K., and Tsuruta, S. 1992b, *Ap. J.*, submitted.
Umeda, H., and Tsuruta, S. 1992, in preparation.
Van Riper, K. 1991, *Ap. J.*, **372**, 251.
Yakovlev, D.G. 1982, *Astron. Zh.*, **59**, 683.

Effects of Evolving Rotating Equilibrium Configurations on the Cooling and Spin-Down of Pulsars

Kwong-Sang Cheng,[1] W. Y. Chau,[1,2] J. L. Zhang,[1,3] and H. F. Chau[1]

[1]Department of Physics, University of Hong Kong
[2]Department of Physics, Queen's University, Canada
[3]Department of Physics, University of Science and Technology, PRC

Abstract

We model a pulsar by an uniform density rotating spheroid covered by a solid crust. During the spin-down (up) epoch, pulsars can adjust their shapes to their equilibrium configurations, which depend on the stellar rotation velocity Ω, caused by plastic flow and continuous crust breaking that results from various mechanisms even though the crust is supposedly rigid. We show that this behavior results in a coupling among the stellar temperature (T), the rotation velocity (Ω) and the eccentricity (e); cooling and spin-down therefore mutually affect each other. Theoretically, crust cracking could be the dominant heating mechanism for pulsars instead of the vortex creeping mechanism if the crust is an ideal coulomb lattice. But the present observed X-ray data can only provide constraints on the parameters of these two heating mechanisms and cannot determine which is the dominant one. For weak field millisecond pulsars, either one can lead to heating that balances the blackbody cooling, thus making these pulsars persistent soft X-ray emitters.

1 Introduction

The thermal and rotational evolution of neutron stars has been intensively studied since the discovery of a pulsar (Hewish *et al.* 1968) in view of their importance in providing information about the internal structure of neutron stars and their electrical, thermal and dynamical properties. However, these two topics are usually treated as two weakly correlated subjects, and in fact, the internal temperature of the star (T) and the stellar rotation velocity Ω do not couple to each other explicitly. The slow-down of a pulsar is usually considered to result mainly from the electromagnetic dipole radiation (Manchester and Taylor 1977), and the cooling of the star is caused mainly by various types of neutrino emission and blackbody radiation (for a general review, c.f. Tsuruta 1986).

Recently, a number of authors (Baym 1981; Alpar and Ögelman 1990; Takatsuka 1991) have studied the direct relation between the thermal evolution and the spin-up (down) of the star. It has been shown that the thermal contraction resulting from rapid neutrino emission could lead to changes by up to 20%–25% in Ω and 50%–90% in the rotation rate ($\dot{\Omega}$) over several years after the star was born. On the other hand, the spin-down of the star can also affect stellar thermal evolution because the differential rotation between the crust and the neutron superfluid can generate internal heating

$\propto \dot{\Omega}$ (Alpar *et al.* 1984*a*,*b*; Shibazaki and Lamb 1989), which significantly affects cooling of neutron stars after $\sim 10^6$ years since they were born. Such effect is even more important for weak field pulsars with $B \sim 10^8 - 10^9$ G. Although their $\dot{\Omega}$ is an order of magnitude smaller than that of strong field pulsars, their typical life time is 3 orders of magnitude longer than that of strong field pulsars, and it is likely that they concentrate in the globular cluster (Kulkarni 1990; Manchester *et al.* 1991). The observational implication will be discussed in Section 5.

2 Oblateness of Pulsars

For a rapidly rotating pulsar, the stellar oblateness changes appreciably as a result of pulsar spin-down. It is interesting to ask if the stellar oblateness can still maintain its equilibrium value because the star is actually covered by a thick crust.

Baym and Pines (1971) study a self-gravitating elastic incompressible spheroid and show that a mean stress σ in the crust will arise if the stellar shape deviates from its equilibrium value, and it is given by $\sigma = \tilde{\mu}(\epsilon_o - \epsilon)$ and $\epsilon \equiv (I - I_o)/I_o$, where $\tilde{\mu}$ is the mean shear modulus of the crust, I is the moment of inertia of the star, I_o is its non-rotating value and ϵ_o the equilibrium value of ϵ. For a Macluarin spheroid (Chandrasekhar 1986), $\epsilon_o = (1 - e^2(\Omega))^{-1/3} - 1$ where $\Omega^2 = 2\pi G\rho f(e) = 2\pi G\rho[\frac{(1-e^2)^{1/2}}{e^3}(3 - 2e^2)\sin^{-1} e - \frac{3(1-e^2)}{e^2}]$ for $0 \le e \le e_{sec} = 0.812670$ with e_{sec} being the maximum eccentricity for a secularly stable Maclaurin spheroid and ρ the mean density of the star. If ϵ is close to ϵ_o, then

$$\sigma = \tilde{\mu} g(e) \frac{\delta e}{e}, \tag{1}$$

where $g(e) = \frac{2}{3}\frac{e^2}{(1-e^2)^{4/3}}$ is a monotonic increasing function. There are several ways to reduce this stress.

2.1 Plastic Flow

According to Smolukowski and Welch (1970), the crust lattice can relieve the stress by means of creeping or plastic flow for $T > 10^{8\,o}$K, therefore, the star can maintain its equilibrium configuration through plastic flow.

2.2 Crust-Breaking By Gravitational Force

For $T < 10^{8\,o}$K, the growth rate of elastic strain becomes dominant, ϵ continuously depart from its equilibrium value and hence the stress increases until it reaches some critical value σ_c which is given by

$$\sigma_c = \tilde{\mu}\theta_c, \tag{2}$$

where θ_c is the maximum (critical shear) angle before fracture. For an ideal coulomb lattice $\theta_c \simeq 10^{-1} - 10^{-2}$ (Smolukowaki 1970), but for real crystals in which fracture is strongly determined by lattice imperfections, *e.g.* dislocations, Smolukowski estimates that for neutron stars, $10^{-5} \leq \theta_c \leq 10^{-4}$. Substituting equation (2) into equation (1), we obtain

$$\left(\frac{\delta e}{e}\right)_{\max} \approx \theta_c / g(e) \tag{3}$$

Equation (3) gives the maximum deviation of eccentricity from its equilibrium value due to the gravitational force.

2.3 Crust-Breaking By Crustal Superfluid

According to the canonical neutron star models (Shapiro and Teukolsky 1983), the inner crust ($4\times10^{11}\,\mathrm{g\,cm^{-3}} \leq \rho \leq 2\times10^{14}\,\mathrm{g\,cm^{-3}}$) consists of a lattice of neutron-rich nuclei together with a superfluid neutron and electron gas. The velocity of the neutron superfluid is completely determined by the distribution of its quantized vortex lines which are believed to pin in the nuclei (Alpar *et al.* 1984). As a result of pulsar spin-down, the neutron vortices, which corotate with the crust, will rotate differently from the superfluid angular velocity (Ω_n). This difference causes a large strain in the crust (Ruderman 1991). Again, this strain is limited by the critical shear angle θ_c which gives the maximum deviation of the stellar eccentricity from its equilibrium value due to the crustal superfluid.

$$\left(\frac{\delta e}{e}\right)_{max} = 6 \times 10^{-5} h(e) \, (\theta_c/10^{-3}) \tilde{\mu}_{30} R_6^{-2} \Omega_3^{-2} (\rho_n)_{13}^{-1}, \tag{4}$$

where $h(e) \equiv 2f(e)/e\frac{df}{de}$ and $0.966 \leq h(e) \leq 1$ for $0 \leq e \leq 0.812670$.

3 A Simplified Model of Neutron Star Evolution

In this section, we use a two-component neutron star model consisting of a solid crust and a uniform density superfluid core with the gravitational energy, rotational energy, and equilibrium configuration approximated by those of a Maclaurin spheroid (Chandrasekhar 1986). Although such a star is covered by a solid crust, it can still maintain its equilibrium configuration for reasons discussed in the previous section. We also assume that the star can only lose energy and angular momentum via neutrino emission, blackbody radiation from the stellar surface, and magnetic dipole radiation. Furthermore, T and T_s are the core and surface temperatures related by (Gudmundsson, Pethick and Epstein 1983)

$$T_8 = 1.288 (T_{s6}{}^4 / g_{s14})^{0.455}, \tag{5}$$

where g_{s14} is the surface gravity in the units of 10^{14} cm s^{-2}, and T_8 and T_{s6} in the units of $10^{8\,o}$K and $10^{6\,o}$K, respectively. The angle between the rotation axis and the magnetic axis is denoted by χ. Since the shape of the star is expected to change substantially and the magnetic field is likely to be frozen in the crust for at least millions of years because the crustal conductivity is extremely high (Baym, Pethick, and Pines 1969), then as long as the motion of the crust follows the form of Hubble flow, Chau and Zhang (1991) show that χ can be simply related to e by

$$\tan\chi(t) = \left[\frac{1 - e^2(o)}{1 - e^2(t)}\right]^{1/2} \tan\chi(0). \tag{6}$$

The basic equations determining the evolution of the system and the energy and angular momentum equations can then be written as

$$\frac{d}{dt}J_{star} = -\frac{d}{dt}J_\nu - \frac{d}{dt}J_d - \frac{d}{dt}J_{bb} \tag{7}$$

and

$$\frac{d}{dt}(E_{grav} + E_{rot} + E_{int}) = -L_d - L_{bb} - L_\nu. \tag{8}$$

In the above equations, the gravitational energy $E_{grav} = 2GM^2 \sin^{-1} e/5a_1(e)e$, the rotational energy $E_{rot} = I\Omega^2/2$, with the moment of inertia I given by $2Ma_1^2(e)/5$, the luminosity of dipole radiation $L_d = 2\mu^2 \sin^2\chi(e)\Omega^4/3c^3$ where $\mu \equiv B_s R^3$ is the stellar dipole moment, B_s its surface magnetic field, R its radius of a spherical star, and the luminosity of blackbody radiation $L_{bb} \approx 4\pi a_1^2(0)\sigma T_s^4$ (Strictly speaking, $4\pi a_1^2(0)$ should be replaced by the surface area of the spheroid). The stellar angular momentum $J_{star} = I\Omega$. The angular momentum carried away by neutrinos $dJ_\nu/dt = 2a_1^2(e)\Omega(t)L_\nu/5c^2$, that by blackbody photons $dJ_{bb}/dt \approx 2a_1^2(e)\Omega(t)L_{bb}/3c^2$, and that by the dipole radiation $dJ_d/dt = L_d/\Omega$. The rate of change of internal energy can be expressed as

$$\frac{dE_{int}}{dt} = C\frac{dT}{dt} - \frac{dE_{strain}}{dt} - \frac{dE_f}{dt}. \tag{9}$$

The first term represents the change of the thermal energy of the star and C (Maxwell 1979) is heat capacity. The second term represents the average rate of strain energy change of the crust. In Section II, we point out that we need only consider $T < 10^{8o}$K when the deformation caused by the stress reaches a critical angle θ_c, and the crust will break for various reasons. We (Cheng *et al.* 1992) have estimated that the mean rate of strain energy relased is given by

$$\frac{dE_{strain}}{dt} = 2B\theta_c^2\frac{\dot{e}}{e}S(10^{8o}\mathrm{K} - T). \tag{10}$$

The third term of equation (10) represents the energy loss of the crustal neutron superfluid resulting from the frictional interaction with the crust which dissipates the free energy of the crustal neutron superfluid differential rotation. Such heating term can be expressed as (Alpar *et al.* 1984)

$$\frac{dE_f}{dt} = -I_p\omega_{cr}\dot{\Omega} \equiv -\beta\dot{\Omega} \tag{11}$$

where I_p is the moment of inertia of the crustal neutron superfluid and ω_{cr} is the critical angular velocity lag between the superfluid and the crust. The detailed effects of this particular heating term on cooling of rigid neutron stars have already been discussed (Alpar *et al.* 1984, Shibazaki and Lamb 1989, Tsuruta 1987).

4 Model Results and Discussion

4.1 Spin-down Curves

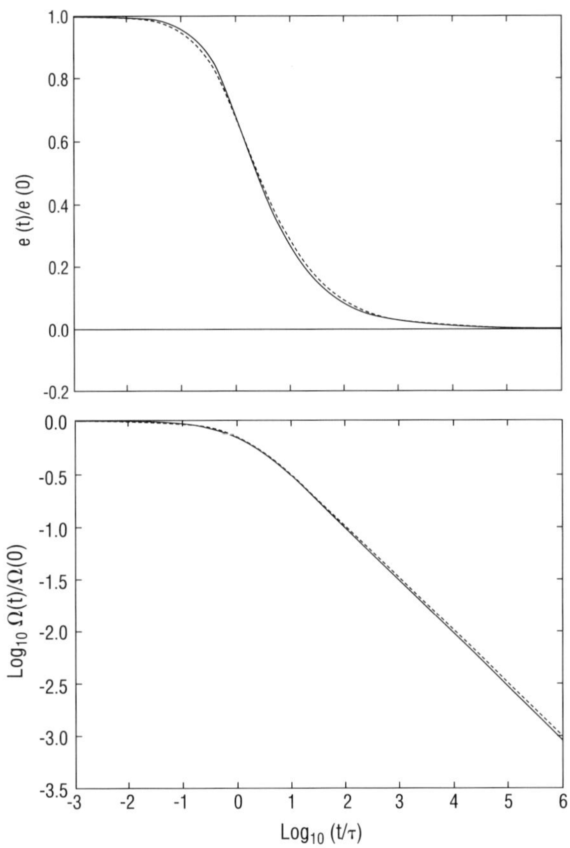

Fig. 1—Master curves describing the evolution of (a) the eccentricity $e(t)/e(0)$, and (b) angular velocity $\Omega(t)/\Omega(0)$ as function of time in the unit of year. In Figs. 1a–1b, the solid lines are the model curves of the evolving configuration stars and the dashed lines are for the spherical rigid stars of the same mass and density with electromagnetic dipole radiation as the only dissipative mechanism. In Fig. 1a, the additional dotted line is a reference line which is given by equation (12) with $\tau = \tau_{eff}$.

Figure 1 shows two master curves. The solid curves are our model results and the dashed curves are for spherical rigid stars of the same mass and density with electromagnetic dipole radiation as the only dissipative mechanism. For the latter, the spin-down curves can be described by

$$\Omega(t) = \frac{\Omega_o}{(1 + t/\tau)^{1/2}} \tag{12}$$

with

$$\tau = 6.5 \times 10^2 \mu_{30}^{-2} \sin^{-2} \chi_o \Omega_{o3}^{-2} I_{45} \text{ year}, \tag{13}$$

where μ_{30} is the magnetic moment in the units of 10^{30}G cm^3, I_{45} is the stellar moment of inertia in the units of 10^{45} g cm^2, and Ω_o and χ_o are chosen to have the same initial values as in our corresponding calculations. Also, in Figure 1 we have made all quantities dimensionless by scaling with their initial values, and time is expressed in the units of τ.

We would like to identify two interesting points for the master curves of these two types of stars. (a) The eccentricity (Fig. 1a) changes rapidly around $t/\tau \sim 1$. In fact, $e(t)/e(0)$ is very close to the functional form of $(1 + t/\tau_{eff})^{-1/2}$, where $\tau_{eff} = 1.2\tau$, the dotted line shown in Fig. 1a. This closeness can be understood by the fact that $\dot{e}/e = h(e)\dot{\Omega}/\Omega \simeq \dot{\Omega}/\Omega$.

For the rigid star model, the solution of Ω is given by equation (12). For an evolving configuration star $\dot{\Omega} = \dot{\Omega}[I(t), \chi(t), T(t), \Omega(t)]$, and $\tau(t)$ is defined by the relation $\dot{\Omega} = -\Omega^3/\Omega_o^2\tau(t)$, very weakly dependent on $T(t)$, and

$$\frac{\tau(t)}{\tau} = \frac{I(t)\sin^{-2}\chi(t)}{I_o\sin^{-2}\chi_o}\begin{cases}1.0 & \text{for } t = 0,\ e(0) = e_{sec},\ \chi(0) = \chi_o\ , \\ 1.3 & \text{for } t = \infty\ .\end{cases} \tag{14}$$

where χ_o and I_o are the inclination angle and moment of inertia for a rigid neutron star. Hence, the qualitative behavior of $e(t)$ can be approximated by

$$\frac{\dot{e}}{e} \simeq \frac{\dot{\Omega}}{\Omega(t)}(\tau_{eff}, \Omega(t)) \tag{15}$$

with $\tau_{eff} \sim 1.2\tau$. (b) Again, Ω of the evolving star (solid curve of Fig. 1b) can be described by equation (12)with τ_{eff} replacing τ.

4.2 Cooling Curves

TABLE 1
Constraints on the Heating Parameters

PSR	$\log t$/year	$\log B_s$/Gauss	$\log T/^oK$	β_{43}^a	θ_{c-3}^b
0531 − 21	3.09	12.58	< 8.49	< 1.5	—
0833 − 45	4.05	12.51	7.22/8.11	0/0.22	0/—
0656 + 14	5.04	11.90	6.14/7.76	0/0.27	0/17
1055 − 52	5.73	12.03	7.15/8.07	0.2/25	8.5/—
1929 − 52	6.49	11.71	< 6.93	< 0.45	< 12

[a] Assuming $\theta_c = 0$.
[b] Assuming $\beta_{43} = 0$. The upper bound of θ_c for those pulsars with $T > 10^8 K$ cannot be determined because crust-breaking has not occurred yet.

In Table I, we use the observed data (summarized by Ögelman 1991) to put constraints on the value of heating parameters, namely, β and θ_c. From the fitting results, we see that (a) the possible values of β indicate that the strong pinning regions cannot be the dominant regions in the outer crust and (b) the values of θ_c tend to favor an

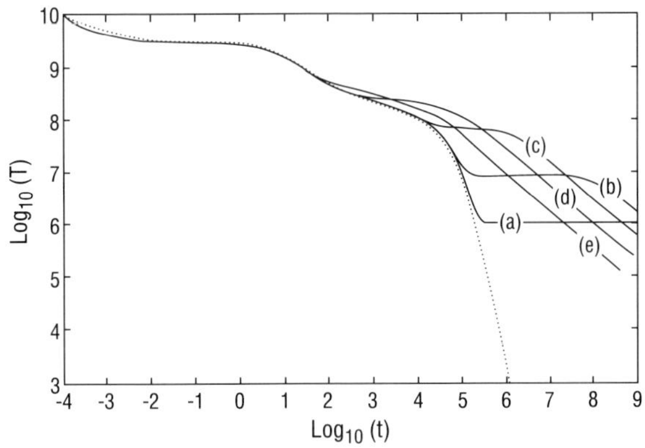

Fig. 2—Cooling curves for pulsars with various surface magnetic fields ($B_s = 10^8$ G, 10^9 G, 10^{10} G, 10^{11} G and 10^{12} G for curve a, b, c, d and e respectively), $\beta_{43} = 0.22$ which is the maximum value allowed by the observed data and assuming no crust quake heating ($\theta_c = 0$).

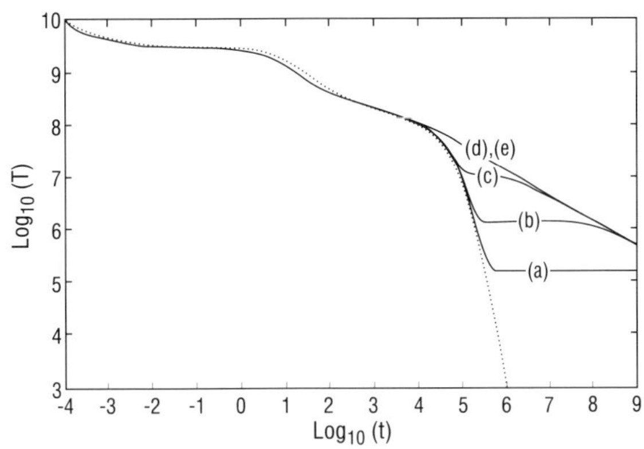

Fig. 3—Cooling curves for pulsars with various surface magnetic fields ($B_s = 10^8$ G, 10^9 G, 10^{10} G, 10^{11} G and 10^{12} G for curve a, b, c, d and e respectively), $\theta_c = 0.012$ which is the maximum value allowed by the observed data and assuming no frictional heating of crustal superfluid ($\beta_{43} = 0$).

ideal coulomb lattice structure in the neutron star crust over the model dominated by lattice imperfections.

Figure 2 shows cooling curves for different values of surface magnetic field and a fixed $\beta_{43} = 0.22$, which corresponds to the maximum value allowed by the observed data, but no crust quake heating ($\theta_c = 0.$).

Figure 3 is similar to Figure 2, but the frictional heating is replaced by the crust quake heating.

So far, the present observed X-ray data can only provide constraints on the parameters of the heating mechanisms if they are assumed to operate one at a time. Which is the dominant one cannot yet be determined. Future detection of thermal X-ray from non-accreting millisecond pulsars, however, can discriminate between the two mechanisms because they predict different soft X-ray luminosity (c.f. Fig. 2 and Fig. 3) for millisecond pulsars with the same B_s and P.

It is interesting to point out that the weak field pulsars ($B_s = 10^8 - 10^9$ G.) can maintain constant temperatures for a time scale of $10^8 - 10^9$ years. Physically, it is very easy to understand why strong field pulsars cannot maintain constant temperature but weak field pulsars can. The frictional heating term is proportional to $\dot{\Omega}$ which is a constant for weak field but decreases as $t^{-3/2}$ for strong field pulsars. This argument also applies the crust quake heating term.

Recent observations (Kulkarni 1990; Manchester *et al.* 1991) indicate that some globular clusters may contain a large number of millisecond pulsars with magnetic fields around $10^8 - 10^9$ G. We suggest that these globular clusters should also be soft X-ray sources with $kT \sim 100$ eV. In particular, 47 Tuc could be a very good example (Grindlay 1981). On the contrary, if a globular cluster is a soft X-ray emitter with no known X-ray sources (*i.e.*, accreting X-ray sources), it may contain a large number of millisecond pulsars. Furthermore, some galactic X-ray sources with unknown origins may also be millisecond pulsars for which we have missed their conical radio beams but from which we receive the isotopic X-ray emission.

References

Alpar M. A. , 1984, *Ap. J.*, **276**, 325.
Alpar M. A. and Ögelman, 1990, *Ap. J. Lett*, **349**, L55.
Baym G., 1981, *Ap. J.*, **248**, 767.
Baym G. and Pines D., 1971, *Annals of Physics*, **66**, 816.
Chandrasekhar S., 1986, *Ellipsoidal Figures of Equilibrium* (New York: Dover).
Chau W. Y. and Zhang J. L., 1991, *Ap. Spa. Sci. Rev.*, in press.
Grindlay J. E., 1981, *Galatic X-ray Sources*, ed. P. W. Sanford, P. Laskarides and J. Salton (John Wiley & Sons).
Gudmundsson E. H., Pethick C. J. and Epstein R. I., 1983, *Ap. J.*, **272**, 286.
Hewish *et al.*, 1968, *Nature*, **217**, 709.
Kulkarni S. R., 1990, *Bull. Am. Astr. Soc.*, **22**, 1308.
Manchester R. N. *et al.*, 1991, *Nature*, **352**, 219.
Maxwell O. V., 1979, *Ap. J.*, **231**, 201.
Ögelman H., 1991, *Neutron Stars: Theory and Observation*, ed. Pines D. and Ventura J. (Kluwer Acad. Pub.).
Ruderman M. A., 1991, *Ap. J.*, **366**, 261.
Shapiro, S. L. and Teukolsky, S. A., 1983, *Black Holes, White Dwarfs, and Neutron Stars*, (John Wiley and Sons:New York).
Shibazaki N. and Lamb F. K., 1989, *Ap. J.*, **346**, 808.
Smolukowski R. and Welch D., 1970, *Phys. Rev. Lett.*, **24**, 1191.
Takatsuka T., 1991, *Prog. Theor. Phys.*, **85**, 427.
Tsuruta S., 1987, *13th Texas Sympsium on Relativistic Astrophysics*, ed. M. P. Ulmer (World Scientific), **499**.

Cyclotron Emission From Neutron Stars Accreting From the Interstellar Medium

Robert W. Nelson, Edwin E. Salpeter, and Ira Wasserman

Center for Radiophysics and Space Research, Cornell University

Abstract

We predict that isolated *magnetic* neutron stars accreting from the interstellar medium should emit a significant fraction of their total luminosity in X-rays, just below the electron cyclotron energy $E_B = 11.6B_{12}$ keV, but well above the effective temperature expected from thermal emission, $kT_e \sim 50$ eV. If neutron star magnetic fields do not decay, a large number of old neutron stars may be detectable by ROSAT as isolated, low-luminosity X-ray pulsars accreting from the dense cores of molecular clouds.

1 Introduction

Neutron stars are usually discussed in the context of the approximately 500 known radio pulsars and 100 accreting X-ray binaries. Yet estimates of the neutron star birth rate, typically $(100\text{ yr})^{-1}$, suggest there are 10^8-10^9 old neutron stars in the galaxy. Clearly, the vast majority of neutron stars are not bright radio pulsars or members of accreting binary systems. Detection of thermal emission from isolated neutron stars has been a long term goal (for a review see Ögelman 1991). More recently, the controversy over the origin of gamma-ray bursts has heightened the need to determine the spatial distribution of neutron stars in the galaxy (Paczyński 1990; Hartmann, Epstein, and Woosley 1990; Blaes and Rajagopal 1991). Here we argue that if neutron star magnetic fields do not decay, a large number of old neutron stars may be detectable as isolated, low-luminosity X-ray pulsars accreting from the dense cores of molecular clouds.

An isolated neutron star of mass M, moving at speed v through a molecular cloud of mass density $\rho = m_p n_H$, accretes material with impact parameters less than the gravitational capture radius, $r_g = 2GM/v^2$ (Bondi and Hoyle 1944; Salpeter 1964; Alcock and Illarionov 1980). The mass accretion rate is then

$$\dot{M} \simeq \pi r_g^2 v\rho \simeq 10^{-15}\left(\frac{100\text{ km s}^{-1}}{v}\right)^3\left(\frac{n_H}{10^2\text{ cm}^{-3}}\right)M_{1.4}^2\, M_\odot\text{ yr}^{-1}, \qquad (1)$$

so that the accretion luminosity and effective temperature are

$$L = \frac{GM\dot{M}}{R} \simeq 10^{31}\left(\frac{100\text{ km s}^{-1}}{v}\right)^3\left(\frac{n_H}{100\text{ cm}^{-3}}\right)M_{1.4}^3 r_6^{-1}\text{ erg s}^{-1}, \qquad (2)$$

$$T_e \simeq 10^6\ {}^\circ\mathrm{K} \left(\frac{100\ \mathrm{km\ s^{-1}}}{v}\right)^{3/4} \left(\frac{n_H}{100\ \mathrm{cm^{-3}}}\right)^{1/4} \left(\frac{10^{11}\mathrm{cm^2}}{A}\right)^{1/4} M_{1.4}^{3/4} r_6^{-1/4} \tag{3}$$

where $M_{1.4} \equiv M/(1.4M_\odot)$, $r_6 \equiv R/10^6$ cm, and A is the surface area of the accretion hot spot. Newly formed neutron stars cool below this temperature after about 10^5 years, so that heating by accretion may be the best way to keep old neutron stars hot (Ögelman 1991).

Unfortunately, *thermal* emission from isolated neutron stars accreting from the ISM is difficult to detect except from very nearby stars. The hydrogen column density to photoelectric absorption for a photon of energy $h\nu$ is

$$N_H \sim 5 \times 10^{19} \left(\frac{h\nu}{100\ \mathrm{eV}}\right)^3 \mathrm{cm^{-2}}, \tag{4}$$

so that thermal photons may only travel $\sim$ 16 pc $(100\ \mathrm{eV}/h\nu)^3 n_H^{-1}(\mathrm{cm^{-3}})$ without being absorbed. Estimates of the number of neutron stars emitting thermal radiation detectable by ROSAT have been given by Treves and Colpi (1991).

We show below, however, that isolated neutron stars with moderately strong magnetic fields ($10^{11} \lesssim B < 10^{13}$ G) accreting from the ISM should emit a significant fraction of their total luminosity in X-rays just below the magnetic cyclotron energy $E_B = 11.6B_{12}$ keV ($B_{12} \equiv B/10^{12}$ G), where photoelectric absorption is not important. In this regime, each accreting proton loses much of its energy in magnetic Coulomb collisions with atmospheric electrons, producing large numbers of $0 \to n$ electron Landau transitions. These excited electrons then decay back to the ground state, producing large numbers of cyclotron photons with energies near E_B. Under the right circumstances, a significant fraction of these photons will escape the atmosphere without being absorbed and will be emitted as a nonthermal X-ray component, possibly a (broad) line feature, superposed on the Wien tail of a lower energy thermal spectrum. Consequently, a neutron star accreting from the core of a molecular cloud with a column density $N_H \lesssim 8 \times 10^{25} B_{12}^3$ cm^{-2} should be observable as a low luminosity X-ray pulsar. X-rays resulting from accreting neutron stars in clouds may be readily detectable for $B_{12} \gtrsim 0.05 M_{c,5}^{1/3} R_{c,10}^{-2/3}$, where the cloud mass and radius are $10^5 M_{c,5} M_\odot$ and $10R_{c,10}$ pc, respectively.

We ignore mechanisms that may prevent or reduce the accretion rate from the Bondi-Hoyle result, such as the "propeller" effect, magnetic mirroring, or collisionless flow (*cf.* Alcock and Illarionov 1980). There may, in fact, be a minimum allowed mass accretion rate for a given neutron star rotation period and magnetic moment, so that only a fraction, $f_{acc} < 1$, of old neutron stars are able to accrete from the ISM. As yet we have not attempted to estimate f_{acc}.

2 Proton Stopping in Neutron Star Atmospheres

2.1 Nonmagnetic Atmospheres

Early calculations of the physics of accretion onto neutron stars ignored complications due to the presence of ultrastrong magnetic fields (Zel'dovich and Shakura 1969; Alme and Wilson 1973). In these models, the accreting gas deposits its energy into the atmosphere mainly by multiple Coulomb scattering of accreting protons with atmospheric electrons. Models which assume the existence of a collisionless, adiabatic shock have also been considered (Shapiro and Salpeter 1975; Langer and Rappaport 1982). However, since collisionless shocks are poorly understood and their stability in accreting neutron stars has been questioned (Alme and Wilson 1973; Braun and Yahel 1984), we neglect this possibility and assume that accreting ions reach the surface at free-fall velocities. Accreting protons then enter the atmospheric plasma with energies

$$E_p = \frac{GMm_p}{R} = 194M_{1.4}r_6^{-1}\ \text{MeV}. \tag{5}$$

A fast proton moving through an ionized plasma loses energy to binary Coulomb collisions with atmospheric electrons as well as to generation of collective plasma oscillations. The resulting plasma stopping power for a proton moving with velocity $\mathbf{v} = v\hat{z}$ is

$$\frac{dE_p}{dz} = -\frac{4\pi n_e Z^2 e^4}{m_e v^2}\ln\Lambda_c, \tag{6}$$

where the Coulomb logarithm is $\ln\Lambda_c = \ln m_e v^2/\hbar\omega_p$ and $\omega_p^2 = (4\pi e^2 n_e/m_e)^{1/2}$ is the plasma frequency. For typical neutron star atmospheres $\ln\Lambda_c \simeq 10 \approx$ constant. Moreover, when $B \equiv 0$, most of the stopping power is due to impulsive collisions in which electron recoil velocities are directed perpendicular to the incident proton momentum. The fraction of energy that goes into transverse electron motion is

$$\frac{E_\perp}{E_0} = 1 - \frac{1}{\ln\Lambda_c}. \tag{7}$$

In terms of the Thomson optical depth, $d\tau = -n_e\sigma_T dz$, the energy profile of a decelerating proton is then

$$E_p = E_0\sqrt{1-\frac{\tau}{\tau_s}}, \tag{8}$$

where the nonmagnetic stopping depth is

$$\tau_s = \frac{1}{6}\frac{m_p}{m_e}\left(\frac{v_{ff}}{c}\right)^4\frac{1}{\ln\Lambda_c} = 5.1M_{1.4}^2 r_6^{-2}\left(\frac{10}{\ln\Lambda_c}\right), \tag{9}$$

where n_e is the electron number density, and $\sigma_T = 8\pi e^4/3m_e^2c^4$ is the Thomson scattering cross section.

When $B \equiv 0$, the energy lost by the accreting gas, both to individual proton-electron collisions and to collective plasma oscillations, is thermalized in atmospheric Coulomb collisions and eventually emitted as thermal bremsstrahlung radiation.

2.2 Magnetic Atmospheres

In the strong magnetic fields of magnetic neutron stars, atmospheric electrons are free to move along the direction of the field lines, but are quantized in discrete Landau orbitals with energies

$$E_n \approx \frac{p_z^2}{2m_e} + nE_B, \tag{10}$$

where

$$E_B \equiv \frac{\hbar e B}{m_e c} = 11.6 B_{12}\ \mathrm{keV}. \tag{11}$$

Moreover, for $k_B T << E_B$ radiative de-excitation greatly dominates collisional excitation, so that atmospheric electrons are almost entirely in the ground state. Effectively, the plasma has a longitudinal temperature $T_{\parallel} \sim T_e$ but a perpendicular excitation temperature $T_{\perp} = 0$. An atmospheric plasma in a strong magnetic field is thus semi-one-dimensional in character, with electrons moving like "beads on a wire" along the magnetic field lines.

The anisotropy of a magnetic plasma strongly affects the stopping physics of the accreting protons. In particular, a proton moving along a magnetic field line and scattering with an atmospheric electron has center of mass energy $m_e v^2/2$. The highest possible $0 \to n$ electron Landau excitation that can result from a magnetic Coulomb collision is

$$n_{max} \equiv \frac{m_e v^2}{2E_B} = 9.0 \left(\frac{v}{v_{ff}}\right)^2 B_{12}^{-1} M_{1.4} r_6^{-1}. \tag{12}$$

Protons with $n_{max} < 1$ cannot induce any Landau excitations and only lose energy to electron recoil *along* the magnetic field lines, as well as to magnetic plasma oscillations. As in the $B \equiv 0$ case, the accretion energy is eventually radiated as thermal bremsstrahlung. However, when

$$n_{max} >> 1 \tag{13}$$

Coulomb collisions can excite large $0 \to n$ electron Landau transitions, and much of the accretion energy goes into producing these transitions. Moreover, for fields $B \gtrsim 10^{11}$ G, dipole radiative de-excitation of Landau electrons dominates collisional de-excitation. Consequently, an excited electron in the nth Landau level decays back to the ground state primarily by emitting n cyclotron photons via the cascade $n \to n-1$, $n-1 \to n-2, \ldots\ 1 \to 0$.

We have simulated proton stopping in a homogeneous, strongly magnetized plasma, allowing for proton energy loss to magnetic Coulomb scattering and to collective (magnetic) plasma oscillations (Nelson, Salpeter and Wasserman 1992). Our calculations are similar to those done by Miller, Salpeter and Wasserman (1987, 1989), who considered magnetic fields strengths and incident velocities where $n_{max} \lesssim 1$. In that case only $0 \to 1$ excitations were possible, and little of the incident accretion energy went into cyclotron photons. Our simulations allow for smaller magnetic field strengths where $n_{max} >> 1$ and large $0 \to n$ Landau excitations are possible. In this case much of the incident accretion energy goes into Landau excitations which then decay to produce cyclotron photons. The apportionment of the initial energy into

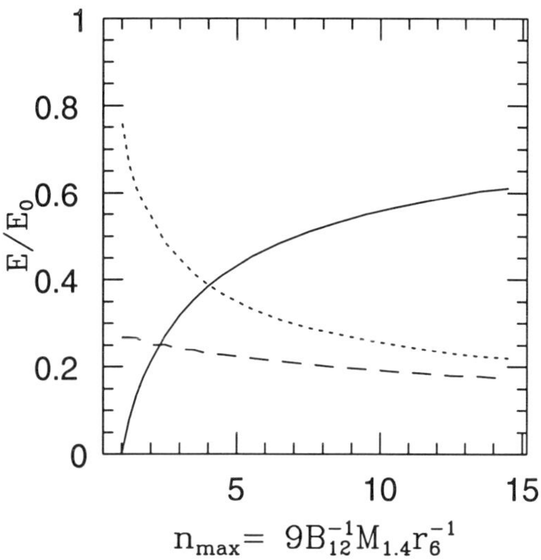

Fig. 1—The fraction of the initial accretion energy that goes into cyclotron photons (solid line), plasma oscillations (dotted line), and electron recoil along the magnetic field (dashed line). For $n_{max} >> 1$ (but $B \gtrsim 10^{11}$ G) most of the accretion energy goes into cyclotron photons distributed along the path of the decelerating proton.

cyclotron photons, plasma oscillations, and electron recoil along the magnetic field line as a function of n_{max} is summarized in Figure 1.

When $n_{max} >> 1$, the stopping power of a magnetic plasma is similar in form to the nonmagnetic case in equation (6) but with the approximate substitution

$$\ln \Lambda_c \to \ln 2n_{max}. \tag{14}$$

Consequently, the proton energy profile is nearly the same as equation (8) but with a larger magnetic stopping stopping length, $\tau_B \simeq \tau_s(\ln \Lambda_c / \ln 2n_{max})$. Physically, a strong magnetic field prevents energy transfer to Landau states with $n > n_{max}$ and thus suppresses the proton energy loss rate, resulting in larger stopping depths. Moreover, analogous to equation (7), the fraction of the total energy that goes into transverse Landau excitations is

$$\frac{E_{0 \to n}}{E_0} \approx 1 - \frac{1}{\ln 2n_{max}}. \tag{15}$$

These excitations subsequently decay to produce large numbers of cyclotron photons. *Thus, for field strengths in the range* $10^{11} \lesssim B << 10^{13}$ G, *most of the accretion energy is converted directly into cyclotron photons, at least initially.*

3 Spectral Signatures of Low-Luminosity Magnetic Accretion

We do not attempt realistic magnetic radiative transfer calculations here, but appeal to "naive" random walk arguments to give a semiquantitative account of the cyclotron radiative transport and resulting emission spectrum (Nelson, Salpeter, and Wasserman 1992).

Three characteristic optical depths, which describe much of the cyclotron photon transfer physics, are plotted in Figure 2 as a function of the magnetic field strength. Cyclotron photons are born with energies near E_B, and are distributed in space along a path from $\tau = 0$ to the proton *stopping depth*, τ_B (solid line). Photons born above the

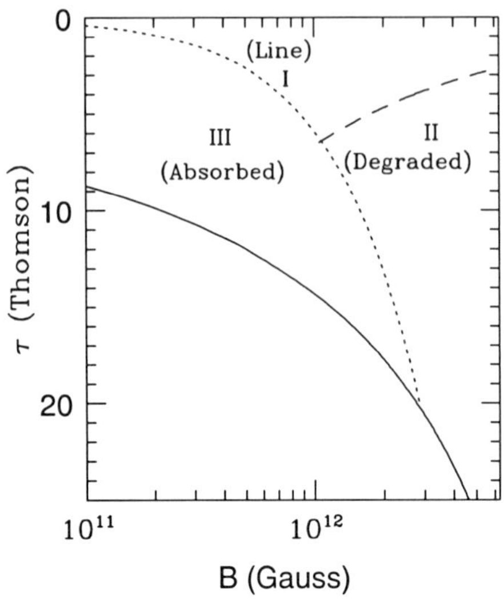

Fig. 2—The proton stopping depth (solid line), cyclotron thermalization depth (dotted line), and cyclotron "Compton" depth (dashed line), as a function of magnetic field strength. Cyclotron photons born in region I escape as a line feature in the emission spectrum. Photons from region II escape but suffer substantial energy degradation from Compton scattering. Photons born in region III are absorbed before escaping.

thermalization depth (dotted line), escape the atmosphere without being absorbed by inverse bremsstrahlung, while those born below this depth are absorbed. At higher energies (or field strengths), photons born below the *Compton depth* (dashed line) suffer significant energy loss ($E/\Delta E \gtrsim 1$) due to large numbers of electron scatters with Compton recoil.

Two examples summarize much of our results: Protons accreting onto a $1.4M_\odot$ neutron star with $R = 10^6$ cm and $B = 3 \times 10^{12}$G can excite Landau transitions to $n_{max} = 3$ and penetrate to a Thomson depth $\tau_B \simeq 23$, releasing about 30% of their incident energy into cyclotron photons with energies $E_B = 35$ keV. Nearly all of these will escape the atmosphere, but with significant Compton degradation, so that the initial line feature is smeared out by the time the photons reach the surface. On the other hand, if $B = 3 \times 10^{11}$ G, $n_{max} = 30$, protons stop at a depth $\tau_B \simeq 12.5$ releasing $\sim$ 75% of their incident energy in cyclotron photons of energies 3.5 keV. Although most of these photons will be absorbed, those born above the thermalization depth escape with little Compton energy loss resulting in a (broad) emission line feature.

The fraction of cyclotron photons that escapes without being absorbed is (roughly) given by the ratio of the thermalization depth to the proton stopping depth. The fraction of the energy flux emitted in X-rays , $f_X \equiv F_X/F$, is of order

$$f_X \simeq \frac{E_{0\to n}}{E_0}\left(\frac{\text{thermalization depth}}{\text{stopping depth}}\right) \simeq 0.1 B_{12}^{7/6} F_8^{1/12} r_6^{5/2} M_{1.4}^{-9/4} \left(\ln 2n_{max} - 1\right) \quad (16)$$

where $F_8 \equiv F/10^{-8}F_E$ and $F_E = 1.4 \times 10^{25} M_{1.4} r_6^{-2}$ erg cm^{-2} s^{-1} is the Eddington flux (Nelson, Salpeter, and Wasserman 1992). This result depends only very weakly on the effective temperature of the accreting star since a fraction, f_X, of the energy of *each accreting proton* is emitted as Comptonized cyclotron photons with energies $\lesssim 11.6 B_{12}$ *keV*.

We conclude that an isolated magnetic neutron star accreting from the ISM should have a telltale signature: a nonthermal X-ray component, possibly a cyclotron line feature, superposed on the Wien tail of a much lower energy thermal spectrum.

4 Observational Strategies

Good places to search for isolated accreting neutron stars are in the dense cores of giant molecular clouds (GMCs). Typically, GMCs have radii of $R_c \sim 20$ pc, column densities $N_H \sim 10^{22}\text{cm}^{-2}$, with mean hydrogen densities of $n_H \sim 200\ \text{cm}^{-3}$, whereas dense subclumps within these clouds have radii $r_s \sim 5$ pc and densities $n_H \sim 10^3 - 10^4\ \text{cm}^{-3}$ (Blitz 1987). For a density of old neutron stars in the solar neighborhood of $N_{ns} \sim 10^{-3}\ \text{pc}^{-3}$ (Paczyński 1990, although Hartmann, Epstein, and Woosley 1991 give a somewhat smaller value), we expect there to be ~ 30 neutron stars within each GMC, and ~ 1 per subclump.

The total X-ray flux of radiation at a distance D from the source is

$$F_X \simeq 10^{-14}\left(\frac{f_X}{0.1}\right)\left(\frac{D}{1\ \text{kpc}}\right)^{-2}\left(\frac{100\ \text{km s}^{-1}}{v}\right)^3\left(\frac{n_H}{10^2\ \text{cm}^{-3}}\right)M_{1.4}^3 r_6^{-1}\ \text{erg cm}^{-2}\ \text{s}^{-1}. \tag{17}$$

The limiting flux sensitivity for the ROSAT observatory in the $0.1-2$ keV band depends on the shape of the emission spectrum, but is typically $F_{min} \sim 10^{-14}$ erg cm^{-2} s^{-1} for deep exposures of $\gtrsim 10^4 s$ (Shanks *et al.* 1991). An observation of a GMC at a distance D to a limiting flux, F_{min}, should thus detect all accreting neutron stars with velocities $v \lesssim 200(f_X/0.1)^{1/3}(D/1\ \text{kpc})^{-2/3}(n_H/10^2\ \text{cm}^{-3})^{1/3}M_{1.4}r_6^{-1/3}$ km s^{-1} Thus ROSAT should be able to detect most accreting neutron stars in the dense cores of GMCs to a distance of 1 kpc.

One isolated X-ray pulsar accreting from a molecular cloud may already have been detected. Bhatt (1990) has argued that the Einstein source H 0253+193, coincident with the closest known molecular cloud, MBM 12 (d = 65 pc), is an isolated neutron star accreting from the ISM. Recent GINGA observations found X-ray pulsations in the 2 – 20 keV band with a period $P_{puls} = 206s$, indicating a rotating magnetic neutron star or white dwarf (Takano *et al.* 1989). However, there is no detection of a visual companion or any modulation (other than the 206s pulsation period) that can be identified with an orbital period. The average flux, $F_X = 6 \times 10^{-11}$ erg cm^{-2}s^{-1}, could be due to accretion in a particularly dense region of the cloud. Others have cautioned, however, that the observations are also consistent with a magnetic white dwarf, which, by chance, is coincident with the molecular cloud (Halpern and Patterson 1990; Koyama 1991).

This work was supported by NASA grant NAGW-666 and NSF grant AST 89-13112. RWN thanks the Los Alamos National Laboratory for travel support.

References

Alcock, C., and Illarionov, A. 1980, *Ap. J.*, **235**, 541.
Alme M. and Wilson, J. 1973, *Ap. J.*, **186**, 1015.
Bhatt, H. C. 1990, *Astron. and Astrophys.*, **232**, 367.
Blaes, O., and Rajagopal, M. 1991, *Ap. J.*, **381**, 210.
Blitz, L. 1987, in *Physical Processes in Interstellar Clouds*, Morfill and Scholer, eds., (Reidel: Dordrecht), p. 35.
Bondi, H. and Hoyle, F. 1944, *M.N.R.A. S.*, **104**, 273.
Braun, A. and Yahel, R. Z. 1984, *Ap. J.*, **278**, 349.
Halpern, J.P., and Patterson, J. 1987, *Ap. J. Lett.*, **312**, L31.
Hartmann, D., Epstein R.I., and Woosley, S. 1990, *Ap. J.*, **348**, 625.
Koyama, K. *et al.* 1991 *Ap. J.*, **377**, 240.
Langer, S. and Rappaport, S. 1982 *Ap. J.*, **257**, 731.
Miller, G.S., Salpeter, E. E. and Wasserman, I.1987, *Ap. J.*, **314**, 215.
Miller, G.S., Salpeter, E. E., and Wasserman, I. 1990 *Ap. J.*, **346**, 405.
Nelson, R.W., Salpeter, E. E., and Wasserman, I. 1992, *in preparation.*
Ögelman, H. 1991, in *Neutron Stars: Theory and Observation*, ed. J. Ventura and D. Pines (Dordrecht: Kluwer Academic Publishers), p. 87.
Paczyński, B. 1990, *Ap. J.*, **348**, 485.
Patterson, J. and Halpern, J.P. 1990, *Ap. J.*, **361**, 173.
Salpeter, E., E. 1964, *Ap. J.*, **140**, 796.
Shanks, T., Georgantopoulos, I., Stewart, G., Pounds, K., Boyle, B. and Griffiths, R. 1991, *Nature*, **335**, 315.
Shapiro, S., and Salpeter E. 1975, *Ap. J.*, **198**, 671.
Takano *et al.* 1989, *IAU Circular No. 4745.*
Treves, A. and Colpi, M. 1991, *Astron. and Astrophys.*, **241**, 107.
Zel'dovich, Ya. and Shakura, N. 1969, *Sov. Astron. -A. J.*, **13**, 173.

Thermal Emission from Isolated Neutron Stars

M. Coleman Miller

Department of Physics, University of Illinois at Urbana/Champaign

1 Introduction

Thermal emission from neutron stars may provide us with information about these objects that we could not obtain in any other way. Correct interpretation of thermal spectra could yield data about neutron star magnetic fields, surface composition, or surface temperature. In addition, the accurate determination of the surface temperatures of neutron stars with varying ages would provide vital clues about neutron star thermal histories and cooling processes and would allow us to evaluate current theories of the dense matter in neutron star interiors and neutrino emission processes operating there.

To carry out such a program, we need fairly accurate model atmospheres. Since detectors cannot observe all wavelengths, we must base our inferences on the emission in the detector bandpass, which for observations of neutron star thermal emission is in the soft X-rays or extreme UV (*e.g.*, *ROSAT* is sensitive from 100 eV to 2 keV). Since stellar opacities are frequency-dependent, thermal emission may be channeled into or out of the detector bandpass. In fact, calculations using opacities computed for zero magnetic field (Romani 1987) have indicated that, especially for temperatures $T < 3 \times 10^5$ K and low-Z atoms such as helium, the color temperature in the *Einstein* bandpass could be nearly three times the effective temperature of the surface. If applicable, this result would strengthen greatly the upper limits on surface temperature found by Einstein, possibly requiring accelerated cooling of neutron stars.

Most radio pulsars are believed to have magnetic fields greater than 10^{11} G. In fields of this strength, atomic physics is much different than it is at $B = 0$. Thus, although pulsars with low magnetic fields may be treated in the non-magnetic approximation, accurate modeling of the atmospheres of most solitary pulsars requires opacity tables and ionization equations computed in the strong-field limit. To this end, many authors have computed energy levels and transition strengths for hydrogen in strong fields (Simola and Virtamo 1978; O'Connell 1979; Kara and McDowell 1980; Rösner *et al.* 1983; Rösner *et al.* 1984; Forster *et al.* 1984; Ruder *et al.* 1985; Wunner 1986; Wunner and Ruder 1987; Wunner *et al.* 1987). However, neutron star photospheres are not necessarily composed mostly of hydrogen. Neutron stars have very complicated life histories, and some of the processes that might influence the composition of the atmosphere are fallback from the supernova, accretion from the interstellar medium (which may or may not be channeled by the magnetic field), spallation, and fusion processes on the surface. Thus, the dominant source of opacity in the photosphere could be any of a number of elements, including hydrogen, helium, carbon, nitrogen, oxygen, neon, silicon, iron, or others. It is probable that the photosphere is composed of a pure element rather than, say, a cosmic composition.

This composition results primarily because, as many authors have pointed out, the surface gravity of neutron stars is so large that gravitational separation should take place on a time scale 1–100 s. Thus, the lightest element present will rise to the top and dominate the opacity.

It is therefore of interest to calculate wavefunctions, energy levels, and transition strengths in strong magnetic fields not only for hydrogen, but also for elements with atomic number $Z > 1$. This calculation has been done by Miller and Neuhauser (1991). For the present report, hydrogen, helium, carbon, and nitrogen were chosen as representative compositions. It is important to stress that, since the only objects believed to have magnetic fields in the $10^{10} - 10^{13}$ G range are neutron stars, and the thermal X-ray flux from these objects is extremely low, only low-resolution spectroscopy is likely to be possible in the near future. Combined with the uncertain value of the surface gravitational redshift and distribution of magnetic field strength on the surface, this means that energy calculations more accurate than a few percent are unwarranted. Similarly, for a first-pass calculation, transition strengths need not be computed to better than $\sim$ 20% accuracy. It is, therefore, acceptable to make certain simplifications that, while greatly reducing the computational burden, retain the requisite precision.

This talk is a report on the first attempt to produce magnetic model atmospheres for neutron stars and is a summary of the results presented in Miller (1992). In §2 the fundamentals of atomic physics in strong magnetic fields are summarized and the applicability of the strong field assumption is discussed. In §3 the modeling procedure is described. Of particular importance is the treatment of polarization, which may have significant quantitative effects (see Ventura *et al.* and Shibanov *et al.*, this volume). In §4 the results of the modeling are discussed, and the conclusions are presented in §5.

2 Atomic Physics in Strong Magnetic Fields

In the following sections we scale lengths in units of the Landau length

$$\hat{\rho} = \sqrt{\frac{\hbar c}{eB}} = 2.5 \times 10^{-10} B_{12}^{-1/2} \text{ cm} , \tag{1}$$

where $B_{12} = B/10^{12}$ G, and scale magnetic fields in units of the reference field

$$B_0 = \frac{2\alpha^2 m_e^2 c^2}{e\hbar} = 4.7 \times 10^9 \text{ G} , \tag{2}$$

which is the field at which the Coulomb and magnetic energies are equal for hydrogen.

For $B \gtrsim B_0$, atoms are cylindrical to a first approximation. In this approximation, the quantum numbers are ν in the direction of the magnetic field, m in the azimuthal direction, and n in the radial direction. In strong fields, wavefunctions may be expanded in a cylindrical basis set

$$\Psi_{m\nu}(z, \rho, \phi) = \sum_n f_{nm\nu}(z) \Phi_{nm}(\rho, \phi) ,$$

where z is along the magnetic field, ρ is the radial coordinate, and ϕ is the azimuthal coordinate. The functions Φ_{nm} are the Landau wave functions

$$\Phi_{nm}(\rho, \phi) = \frac{\sqrt{n!}}{\sqrt{2\pi(n+|m|)!\hat{\rho}^2}} x^{|m|} e^{-x^2/2} L_n^{|m|}\left(x^2\right) e^{-im\phi} , \tag{5}$$

where $x = \rho/(\sqrt{2}\hat{\rho})$ and $L_n^{|m|}$ are the associated Laguerre polynomials. In this work, only $n = 0$ states are considered; that is, electrons are constrained to be in their lowest Landau state. This is a good approximation if

$$B \gg \frac{Z^2}{(2m+1)^3} B_0 ,$$

where Z is the atomic number of the nucleus. The $n = 0$ approximation is good for the valence electrons of any atom in a 10^{12} G field, and for all of the electrons of, say, carbon in 10^{12} G, but becomes inaccurate for the inner electrons of, *e.g.*, iron in lower fields.

Atoms in strong fields are stretched along B. For instance, hydrogen has a characteristic radius across the field of $\hat{\rho}$ and a length along the field of

$$l \approx \left[\frac{a_0/\hat{\rho}}{\log(a_0/\hat{\rho})}\right] \hat{\rho} ,$$

where $a_0 \approx 5 \times 10^{-9}$ cm is the Bohr radius (Ruderman 1971). For example, hydrogen in a 10^{12} G field may be represented as a cylinder with a radius$\approx a_0/20$ and a length$\approx a_0/3$. Though the atom is cylindrical, it is smaller in every dimension than in zero field. As a result, the binding energy is much greater than in zero field. For example, the ground state binding energy of hydrogen in a 10^{12} G field is $\approx$ 160 eV, which is much greater than the zero-field binding energy of 13.6 eV. The ionization edge is, consequently, moved closer to the detector energies, which are a few hundred eV, leading to higher opacities in the bandpass and lower color temperatures than in $B = 0$.

3 Model Assumptions

Our calculation of model atmospheres follows the standard treatment given by, *e.g.*, Mihalas (1978) for plane-parallel model atmospheres in local thermodynamic equilibrium (LTE). Both the plane-parallel assumption and the assumption of LTE are likely to be better for neutron star atmospheres than for those of main sequence stars. For soft X-rays at optical depth unity, the atmosphere has a thickness < 10 cm, small compared to the 10^6 cm radius of the star, so the atmosphere is plane parallel to a high degree of accuracy. Moreover, since the characteristic densities in the atmosphere are $\sim 1\ \mathrm{g/cm^3}$, the collision frequency is much higher than in main sequence stars, so that atoms are more apt to thermalize and establish LTE.

The question of what opacity source is most important for given temperature and pressure is unresolved. Bound-bound opacity is most important when the temperature is low ($< 3 \times 10^5$ K) and there are few other opacity sources below the ionization energy. At these low temperatures, line broadening may be so substantial that line opacities dominate the continuum; however, above 3×10^5 K these opacities are negligible for most frequencies. Inverse bremsstrahlung opacity becomes important when the ionization fraction is high, which occurs in high temperature and high-pressure environments. Thus, for example, neutron stars only a few years old may have surface temperatures of several million Kelvin, so that atoms would have a very high degree of ionization and inverse bremsstrahlung opacity could be dominant in the photosphere. At $\approx 10^6$ K, which is probably representative of neutron stars $\approx 10^3 - 10^5$ years old, bound-free opacity may be the most important. Even for higher temperatures, if atoms with $Z > 1$ (such as carbon) are important in the photosphere, the inner electrons of these atoms may not be completely ionized, so that bound-free absorption could produce identifiable features in the spectrum. We included only bound-free opacity. As a result, our models are restricted to effective temperatures $T_{\mathrm{eff}} = 10^6$ K, where our inclusion only of the bound-free opacity is most likely to be acceptable.

Perhaps the most important approximation in the work reported here is our treatment of polarization. In very strong magnetic fields, the cross section $\sigma_{\parallel}$ for absorption of light polarized along the field is substantially greater than the cross section $\sigma_{\perp}$ for absorption of light polarized transverse to the field. This difference occurs essentially because light polarized along the field is interacting with the long axis of the atom. The ratio $\sigma_{\parallel}/\sigma_{\perp}$ can be as large as several thousand for hydrogen in a 10^{12} G field. Thus, for an exact treatment one should consider the transfer of polarized radiation. However, as a first approximation the cross sections from the modes were simply added, rather than weighted as in a Rosseland mean.

Since the cross section for transverse light is so small, the light would tend to come from much deeper levels than the field-aligned light. It would, therefore, come from regions where pressure ionization is sufficient to completely ionize at least hydrogen, resulting in a relatively featureless spectrum (Pavlov *et al.*, in preparation). Since the transverse light comes from deeper and thus hotter regions than the longitudinal light, it might have greater total flux, and could tend to wash out ionization edges. There are two effects that could alter this conclusion.

First, there may be substantial mode mixing in the atmosphere, converting transverse light into light polarized along the field. This conversion would reduce the flux in the transverse mode and make it relatively less important. Second, the relative fluxes in the polarization modes depend on the temperature profile in addition to the relative absorption cross sections. The dependence of temperature on depth is weaker if transverse modes are included than if only longitudinal modes are considered, and the resultant difference between transverse flux and longitudinal flux is less than the difference between their cross sections. It is, therefore, possible that spectral features characteristic of the longitudinal opacity could be observed, even when polarization is correctly included. Some polarization effects are addressed by Ventura (this volume) and Shibanov (this volume).

4 Results

The main results of this work are summarized in Table 1 and Figures 1 and 2. Table 1 shows the ratio of the color temperature that would be measured by *Einstein* (T_E)

TABLE 1
Color Temperatures for $T_{\text{eff}} = 10^6$K

B (G)	Element	T_E/T_{eff}	T_R/T_{eff}
9.4×10^{11}	H	1.3	0.93
9.4×10^{11}	He	0.87	1.1
9.4×10^{11}	C	0.78	1.1
9.4×10^{11}	N	0.78	1.1
2.35×10^{12}	H	1.3	1.0
2.35×10^{12}	He	0.87	1.1
2.35×10^{12}	C	0.71	1.1
2.35×10^{12}	N	0.71	1.1
4.7×10^{12}	H	1.0	1.1
4.7×10^{12}	He	0.83	1.1
4.7×10^{12}	C	0.74	1.1
4.7×10^{12}	N	0.76	1.1

and *ROSAT* (T_R) satellites to the effective temperature (fixed at 10^6 K) for model atmospheres with several compositions and magnetic fields. Here the color temperature is defined as the temperature of a blackbody that would produce the same number of counts by the detector as did the model atmosphere. In all the cases considered, the color and effective temperatures were fairly close. This agreement is in contrast to the ratio of nearly three found for low-temperature helium atmospheres in the absence of a magnetic field. Physically, this is because in $\sim 10^{12}$ G fields the characteristic binding energies are > 100 eV, so the peak of the photoabsorption opacity is close to the energy of the X-ray detector bandpass (*i.e.*, several hundred eV). Therefore, photons at these energies come from layers that are shallower, and thus cooler, than the layers which contribute most in zero field.

Figures 1 and 2 show model spectra for atmospheres composed of pure helium (Figure 1) and pure carbon (Figure 2) relative to a blackbody spectrum in different magnetic fields. The effective temperature is again 10^6 K. The surface gravity is fixed at $g = 3 \times 10^{14}\,\text{cm}\,\text{s}^{-2}$, though the surface gravity has little effect on the shape of the unredshifted spectrum (Miller 1992). For hydrogen (not shown) and helium, the features in the spectrum produced by the photoabsorption edges are quite distinct. The position of these edges depends only on the surface magnetic field and redshift, so if these features were identified, the magnetic field could be estimated to within a factor of ten. Since such an estimate would be independent of other magnetic field estimates (such as from dipole braking), this would provide a valuable check on current magnetic field estimates. The carbon spectrum has many features, corresponding to the various electrons, and hence it is more difficult to identify outstanding features that would allow estimates of the magnetic field strength. This fact is also true of nitrogen (not shown). Identification of spectral features from any atmosphere with atomic number $Z > 2$ is difficult at the current level of modeling precision.

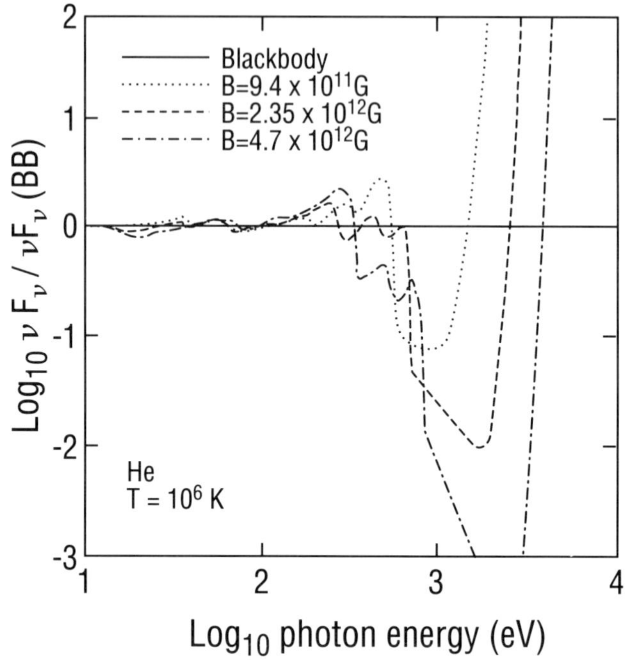

Fig. 1—Flux relative to a blackbody for Helium, $T = 10^6$K.

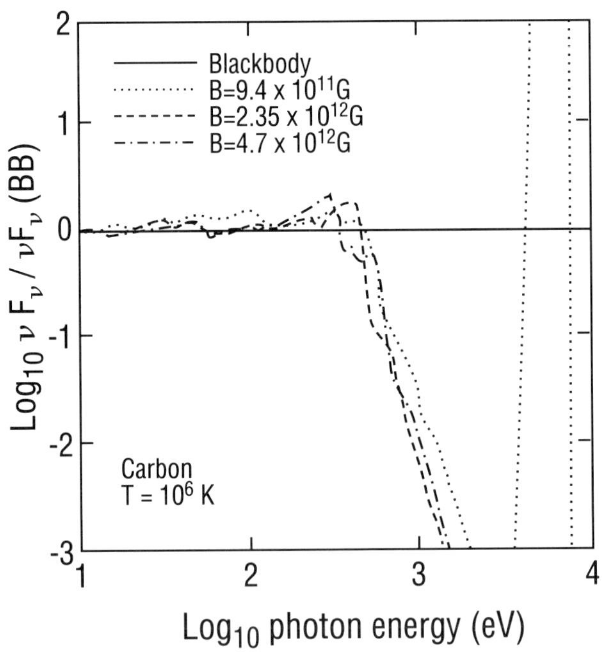

Fig. 2—Flux relative to a blackbody for Carbon, $T = 10^6$K.

5 Summary

According to non-magnetic calculations at low temperatures ($< 10^6$ K), the opacity in the bandpasses of existing detectors (100 eV–5 keV) is small, so photons observed at these frequencies come from deep, hot layers of the atmosphere. As a result, the color temperatures of these models can be much greater than the effective temperature—as much as a factor of three greater for pure helium atmospheres at $T_{eff} = 3 \times 10^5$ K. However, when the effects of 10^{12} G magnetic fields are included, the opacities at detector energies are much greater, and the color and effective temperatures are about equal.

If the bound-free opacity is important in neutron star atmospheres, it is possible that the strength of the surface magnetic field can be estimated directly from the energy of the ionization edge, which is very prominent for low-Z elements such as hydrogen and helium. This would provide an estimate of the field which is independent of estimates from, *e.g.*, the braking torque. Spectra are currently not precise enough to distinguish between models computed with exact absorption opacities and opacities accurate to within a few percent, and it is unlikely that in the near future we will be able to estimate more than the approximate value of the magnetic field and the surface temperature from neutron star thermal spectra. For current applications, opacities accurate to $\sim$ 10% are sufficient. However, in the longer term it is possible that a combination of detailed modeling and precise spectra will allow the determination of surface compositions, and could eventually become one of the most productive sources of information about neutron stars.

This work was done at the suggestion of Sterl Phinney, and discussions with Sterl and Fred Lamb have helped greatly to clarify its content. I am also indebted to Joseph Ventura and Yura Shibanov for pointing out the potential importance of polarization. This research has been supported by NSF grant PHY 91-00283 and NASA grant NAGW 1583.

References

Forster, H., *et al.*, 1984, *J. Phys. B: At. Mol. Phys.*, **17**, 1301.
Kara, S. M. and McDowell, M. R. C., 1980, *J. Phys. B: At. Mol. Phys.*, **13**, 1337.
Mihalas, D., 1978, *Stellar Atmospheres*, (San Francisco: Freeman).
Miller, M. C. and Neuhauser, D., 1991, *M. N. R. A. S.*, **253**, 107.
Miller, M. C., 1992, *M. N. R. A. S.*, **255**, 129.
O'Connell, R. F., 1979. *Phys. Lett.*, **70A**, 389.
Pavlov, G., *et al.* in preparation.
Romani, R. W., 1987, *Ap. J.*, **313**, 718.
Rösner, W., Herold, H., Ruder, H., and Wunner, G., 1983, *Phys. Rev.*, A **28**, 2071.
Rösner, W., Wunner, G., Herold, H., and Ruder, H., 1984, *J. Phys. B: At. Mol. Phys.*, **17**, 29.
Ruder, H., Herold, H., Rösner, W., and Wunner, G., 1985, *Physica B and C*, **127**, 11.
Ruderman, M. A., 1971, *Phys. Rev. Lett.*, **27**, 1306.
Simola, J., and Virtamo, J., 1978, *J. Phys. B: At. Mol. Phys.*, **11**, 3309.
Wunner, G., 1986, *J. Phys. B: At. Mol. Phys.*, **19**, 1623.
Wunner, G., Geyer, F., and Ruder, H., 1987, *Astro. Sp. Sc.*, **131**, 595.
Wunner, G., and Ruder, H., 1987, *Phys. Scr.*, **36**, 291.

Hydrogen Molecules and Chains in a Magnetic Neutron Star Atmosphere

Dong Lai, Edwin E. Salpeter, and Stuart L. Shapiro

Center for Radiophysics and Space Research, Cornell University

Abstract

We study the electronic structures of different forms of hydrogen in a typical superstrong magnetic field ($B \sim 10^{12}$ Gauss) found on the surface of a neutron star, including atoms, poly-molecules H_n ($n = 2, 3, 4, \cdots$), molecular ions(H_2^+) and negative ions(H^-). The binding energies are greatly increased over their zero-field values, typically by a factor of $\sim (\ln b)^2$, where $b = B/B_0$, $B_0 = m_e^2 e^3 c/\hbar^3 = 2.35 \times 10^9$ G. We also consider the excitations of H_2 molecules. We estimate the equilibrium abundances of different forms of H in a neutron star atmosphere and find that H atoms and molecules will be present if the atmosphere is cool ($T_{\rm eff} \sim 10^5 - 10^6$ K).

1 Introduction

The intense magnetic fields ($B \sim 10^{12}$ G) believed to exist on the surfaces of some neutron stars motivate the study of atoms, molecular chains, and condensed matter in fields of extreme magnitude (see Ruderman 1974 for general review). Understanding of the chemical composition and other physical properties of the neutron star atmosphere is important for confronting theory with observation and is also a crucial ingredient for possible future spectroscopy of neutron stars.

In superstrong magnetic fields the structure of atoms and condensed matter is dramatically changed by the fact that the magnetic force on an electron is stronger than the Coulomb force it experiences; *i.e.*, the cyclotron energy $\hbar\omega_e = \hbar eB/m_e c = 11.57 B_{12}$ keV is much bigger than the atomic binding energies. In the direction perpendicular to the field, the electrons are confined to move on cylindrical Landau orbitals around a nucleus. The orbitals have radii

$$\rho_m = (2m+1)^{1/2}\hat{\rho}, \quad m = 0, 1, 2, \cdots, \tag{1}$$

where $\hat{\rho}$ is the cyclotron radius

$$\hat{\rho} = (\hbar c/eB)^{1/2} = 2.57 \times 10^{-10} B_{12}^{-1/2} \text{ cm}, \tag{2}$$

and B_{12} is the magnetic field strength in units of 10^{12} G. If the magnetic field is sufficiently strong so that

$$\hat{\rho} \ll \frac{a_0}{Z}, \quad \text{or } b \equiv \frac{B}{B_0} \gg Z^2, \tag{3}$$

(Z is the nuclear charge, a_0 is the Bohr radius, $B_0 = m_e^2 e^3 c/\hbar^3 = 2.35 \times 10^9$ G is the atomic unit for the magnetic field strength), then the electrons are confined to the ground Landau level, and the atom will have a cigar-like structure. The sizes (in atomic units) of the atomic wavefunctions perpendicular and parallel to the field are, respectively,

$$R \sim \rho_m = \sqrt{\frac{2m+1}{b}}, \quad L_z \sim \frac{1}{Zl_m}, \quad \text{with } l_m = \ln\left(\frac{1}{Z\rho_m}\right). \tag{4}$$

Because of this extreme confinement of electrons in the transverse direction by the magnetic field, the Coulomb force becomes much more effective for binding electrons in the parallel direction, therefore giving greatly increased binding energy. For a hydrogenic atom, the energies of the 'deep-bound' states are

$$E_m \sim -Z^2 l_m^2 \text{ (au)}. \tag{5}$$

Moreover, it is possible for these elongated atoms to form molecular chains by covalent bonding along the field direction.

There have been several recent studies on the binding energies of the infinite chains in a strong magnetic field (*e.g.*, Neuhauser, Koonin and Langanke 1987, Jones 1985, 1996). However, as we show below, strictly one-dimensional infinite chains rarely exist in typical neutron star surfaces. For sufficiently cool atmospheres, the matter will predominantly exist in the form of individual atoms and small molecules. We focus on hydrogen, since we expect that the lightest elements, like H and He, may be the most important chemical species in a neutron star atmosphere due to their predominance in the accreting gas and also to strong gravitational separation of light and heavy elements.

We have recently studied the ground state binding energies of different forms of hydrogen in a strong magnetic field (Lai, Salpeter and Shapiro 1992, hereafter LSS). In §2 we shall give a brief review of the basic physics of molecules in a strong magnetic field. In §3 we discuss preliminary results for the molecular excitations of H_2. In §4 we discuss the relative abundances of different forms of H in a typical neutron star atmosphere. In §5 we summarize our findings.

2 Binding energies of H_n molecules

The process of forming molecules in a high magnetic field is very different from the zero-field situation. In the zero-field case, two H atoms in their ground states with spins opposite to each other form a H_2 molecule by covalent bonding. In a high magnetic field, the spins of the electrons in the atoms are all aligned antiparallel to the magnetic field, and two atoms in their ground states cannot bind together. Instead, one of the atoms has to be excited to a higher state first before the two atoms can form a molecule. However, this activation process costs very little energy, as can be seen from equation (5), and hence the resulting molecule is stable. In this way, more atoms can be added to form a bigger molecule, in contrast to the field-free case.

For a given B field, as the number of H atoms n increases, the electrons occupy more and more Landau orbitals (up to $m = n - 1$); beyond some critical number, it becomes energetically more favorable for electrons to settle into the inner Landau orbitals with nodes in their longitudinal wavefunctions. We have estimated this critical number to be

$$n_s \sim \left[\frac{b}{(\ln b)^2}\right]^{1/5}, \tag{6}$$

beyond which the energy per atom asymptotes to a value independent of n. For a typical magnetic field strength of interest ($B \sim 10^{12} - 10^{13}$ G), the saturation point is $n_s \sim 3 - 5$. Thus we are able to understand binding energies of all H_n molecules without calculating all of the big molecules.

We have recently performed Hartree-Fock calculations to determine the ground state structures of molecules with $n = 2, 3, 4$; for $n = \infty$, we have used a variational method. For details of the calculations and complete numerical tables, see LSS, where we also obtained the energies of H_2^+ and H^-, which are also of potential importance to the atmosphere opacities. We note that the dissociation energies of the small molecules (before saturation) scale as $(\ln b)^2$, but they are numerically smaller than the ionization energy of the H atom for $B \sim 10^{12}$ G.

3 Excitation Levels of H_2

In the ground state, the directions of the molecular axis and magnetic field coincide, and the two electrons occupy the $m = 0$ and $m = 1$ orbitals, respectively. We therefore specify the electronic ground state by $(m_1, m_2) = (0, 1)$. The molecules can have many excitation levels.

3.1 Electronic Excitations

The electrons can occupy different orbitals other than $(0, 1)$, giving rise to the electronic excitations. For example, the first excited level is $(0, 2)$, the second excited level is $(0, 3)$, etc. The energy excitation $\Delta\varepsilon(m_1, m_2)$ of the (m_1, m_2) state relative to the ground state $(0, 1)$ is approximately proportional to $\ln b$, as in the case for atoms.

3.2 Vibrational Excitations

The ions can vibrate along the magnetic field axis, giving the vibrational excitations. In the Born-Oppenheimer approximation, the motion of the ions is governed by the interatomic potential $U(Z, R_\perp)$, which is simply the electronic energy when the relative positions of the ions are Z along the field direction and $R_\perp$ perpendicular to the field. For vibrational motion along the field direction $\hat{z}$, we can approximate U by $U(Z, 0)$, which can be calculated using the Hartree-Fock method described in LSS.

For small-amplitude vibrations around the equilibrium separation, we get a harmonic vibrational spectra with excitation quanta $\hbar\omega_{\rm vib}$. We estimate that $\hbar\omega_{\rm vib}$ scales with the magnetic field strength according to

$$\hbar\omega_{\rm vib} \sim (\ln b)^{3/2} M^{-1/2} \ \text{(au)}, \tag{7}$$

where M is the reduced mass of the ions in units of electron mass.

3.3 Rotational (Oscillational) Excitations

The molecular axis can deviate from the magnetic field direction, and it precesses around the field, giving the constrained rotational (oscillational) excitations. The interatomic potential for this rotational motion is much harder to calculate accurately, because the symmetry of the electron wave functions is lost when the molecular axis does not coincide with the field direction. Nevertheless, we can obtain upper limits to the interatomic potential curve (and hence to the excitation energy $\hbar\omega_{\rm rot}$) if we assume the electron transverse wavefunctions are still cylindrical Landau wavefunctions. We estimate the errors associated with this approximation will not exceed $\sim 30\%$ in the computed rotational excitation energies.

For small-amplitude rotations (the angle between the molecular axis and field axis is sufficiently small), we obtain the approximate scaling law

$$\hbar\omega_{\rm rot} \sim (b\ln b)^{1/2}/M^{1/2}\ (\mathrm{au}). \tag{8}$$

We see that $\hbar\omega_{\rm rot}$ increases as B increases even faster than $\hbar\omega_{\rm vib}$ and the dissociation energy increase.

In Table 1, we present a sample of our numerical data for the H_2 molecule. From the table, we notice that in a high magnetic field, the electronic, vibrational and rotational excitations are all comparable, unlike the zero-field case where one usually has $\Delta\varepsilon_{\rm elec} \gg \hbar\omega_{\rm vib} \gg \hbar\omega_{\rm rot}$.

TABLE 1

Properties of H_2 in High Magnetic Fields[a]

B_{12}	I(H)	$D(H_2)$	$\Delta\varepsilon(0,2)$	$\Delta\varepsilon(0,3)$	$\hbar\omega_{\rm vib}$	$\hbar\omega_{\rm rot}$
0	13.6	4.75			0.52	
1	161.5	45.6	33	47	10	14
5	257.1	109	51	73	24	42
10	309.6	149	60	88	30	65

[a]Energies are in units of eV. I(H) is the ionization energy of the H atom, $D(H_2)$ is the dissociation energy of H_2, $\Delta\varepsilon(0,2)$ and $\Delta\varepsilon(0,3)$ are the first two electronic excitations in H_2, $\hbar\omega_{\rm vib}$ and $\hbar\omega_{\rm rot}$ are the small-amplitude vibrational and rotational excitation quanta of H_2 in the ground electronic state.

4 Abundance of H Molecules in A Neutron Star Atmosphere

Here we shall examine a neutron star atmosphere consisting of pure hydrogen. Recall that the scale height of the atmosphere of temperature T is $h \sim kT/m_p g \simeq 0.08\, T_5 g_{14}^{-1}$ (cm), where m_p is the proton mass, T_5 is the temperature in units of 10^5 K, g is the gravitational acceleration on the surface, and $g_{14} = g/(10^{14}\ \mathrm{cm/s^2})$. Therefore the atmosphere is the very thin layer on the top of the surface.

Consider the physical conditions characterizing a neutron star atmosphere. In a high magnetic field, the electron Fermi temperature $T_F = E_F/k$ is given by

$$T_F = 2.67 B_{12}^{-2} \left(\frac{\rho}{\mu_e}\right)^2 \ (\mathrm{K}), \tag{9}$$

where μ_e is the mean molecular weight per electron, and ρ is in units of 1 g/cm^3 (see, *e.g.*, Fushiki, Gudmundsson and Pethick 1989). Notice that equation (9) is valid only when density is sufficiently small, *i.e.*,

$$\rho \le 7.08 \times 10^3 \mu_e B_{12}^{3/2} \ (\mathrm{g/cm^3}), \tag{10}$$

so that only the ground Landau level is occupied. Equation (9) is to be compared with the Fermi temperature in field-free case,

$$T_F = 3.0 \times 10^5 (\rho/\mu_e)^{2/3} \ (\mathrm{K}). \tag{11}$$

Clearly, the high magnetic field lifts the degeneracy of the electrons even at relatively high density. The typical neutron star surface temperature is $\gtrsim 10^5$ K, hence for densities $\lesssim 1$ g/cm^3 typical of the photosphere (see below), the electrons are nondegenerate. Therefore the equation of state of the atmosphere is simply the ideal gas law, *i.e.*, $P = 2\rho kT/m_p$.

In the zero-field case, assuming free-free opacity $\kappa = \kappa_0 \rho T^{-3.5}$, $\kappa_0 \simeq 7.4 \times 10^{22}$ (cgs), the photosphere (optical depth $\tau = 2/3$) density-temperature relation is simply

$$\rho(0) = \left(\frac{m_p g T^{2.5}}{3k\kappa_0}\right)^{1/2} = 4.2 \times 10^{-3} g_{14}^{1/2} T_5^{5/4} \ (\mathrm{g/cm^3}). \tag{12}$$

In a strong magnetic field, the radiative opacity becomes anisotropic. The free-free absorption and electron scattering opacities are reduced below their zero field values by a factor of $(\omega/\omega_e)^2$ for photons with polarization vector perpendicular to the magnetic field. Silant'ev and Yakovlev (1980) have calculated the appropriate average Rosseland mean free-free opacity, which we have fitted to the approximate relation

$$\kappa(B) \simeq 400 \left(\frac{kT}{\hbar\omega_e}\right)^2 \kappa(0). \tag{13}$$

The resulting photosphere density-temperature relation is then

$$\rho(B) \simeq 0.3 g_{14}^{1/2} T_5^{1/4} B_{12} \ (\mathrm{g/cm^3}). \tag{14}$$

Other sources of opacity such as bound-free and bound-bound absorption will increase the opacity and reduce the photosphere density, but our above estimates define the general range of the physical parameters in the atmosphere.

We now consider the equilibrium abundances in the atmosphere. The Saha equation for the ionization-recombination process e+p=H+Q_1 in a high magnetic field is given by (Khersonskii 1987):

$$\frac{n[H]}{n_p n_e} = \frac{1}{2} \frac{\tanh \eta_e}{\eta_e} \frac{\sinh \eta_p}{\eta_p} \left(\frac{2\pi\hbar^2}{m_e kT}\right)^{3/2} z_H(B,T) e^{Q_1/kT}, \tag{15}$$

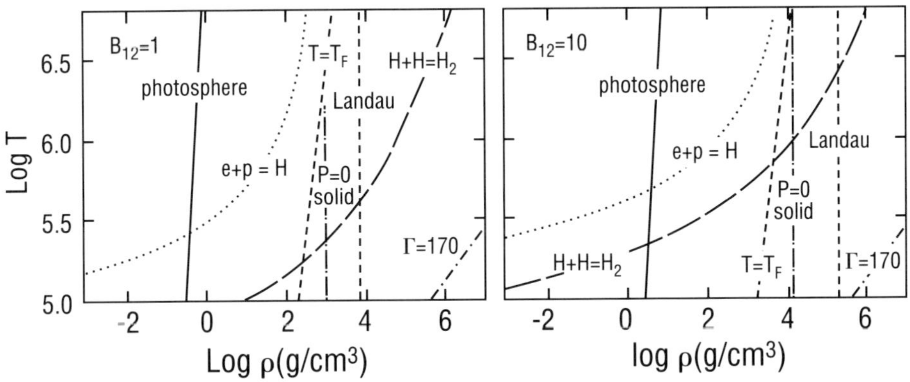

Fig. 1—Physical conditions in a neutron star atmosphere for $B_{12} = 1$ and $B_{12} = 10$.

where Q_1 is the ionization potential of the atom in ground state, z_H is the atom internal partition function, $\eta_e = \hbar\omega_e/2kT$, $\eta_p = \hbar\omega_p/2kT$, and $\hbar\omega_p$ is the cyclotron energy of proton. However, this equation is only approximate, because of the uncertainties in the coupling of the motion of the atom with its internal structures (Herold, Ruder, and Wunner 1981). For the temperature range of interest $\hbar\omega_e \gg kT$, equation (15) reduces to

$$\frac{1-x}{x^2} \simeq 2\pi\hat{\rho}^2\lambda_D n z_H e^{Q_1/kT} \simeq 6\times 10^{-3} B_{12}^{-1}\rho T_5^{-1/2} z_H e^{Q_1/kT}, \tag{16}$$

where $n = n_p + n_H$ is the baryon density, $x = n_p/n$, and $\lambda_D = (2\pi\hbar^2/m_e kT)^{1/2}$.

Ignoring the magnetic field effect on the motion of the atom, the Saha equation for the equilibrium process H_n+H=H_{n+1} + Q_{n+1} is

$$\frac{n[H_{n+1}]}{n[H_n]n[H]} = \left(\frac{n+1}{n}\right)^{3/2}\left(\frac{2\pi\hbar^2}{m_p kT}\right)^{3/2}\frac{z[H_{n+1}]}{z[H_n]z[H]}e^{Q_{n+1}/kT}, \tag{17}$$

where $n[H_{n+1}], n[H_n], n[H]$ are the number densities of H_{n+1}, H_n, H respectively, and $z[H_{n+1}]$, $z[H_n]$, $z[H]$ are their internal partition functions.

To get a qualitative overview of the relative abundances of different species of H, we have determined the H atom half-ionization and H_2 half-dissociation curves in a $\rho-T$ diagram. In Figure 1, we show the results for $B_{12} = 1$ and $B_{12} = 10$; we also plot the electron degeneracy line and photosphere density-temperature relation. To the left of the half-ionization line (labeled 'e+p=H') the atmosphere is mostly ionized, while to the right of the half-dissociation line (labeled 'H+H=H_2'), most of the material is in the molecular states, H_2, H_3, etc. Between the ionization line and the H_2 dissociation line, the dominant species is H atom. We see that for $B_{12} = 1$, the photosphere is dominated by atoms when $T_5 \lesssim 3$. We thus expect to see atomic line features in this regime; molecular states cannot be seen unless the temperature drops below 10^5 K. For $B_{12} = 10$, the photosphere is dominated by H_2 when $T_5 \lesssim 3$. Even at relatively high temperatures, however, it is still possible that the existence of even a small abundance of molecules will significantly affect the atmosphere opacities and thermal structure, although molecular line features may not be directly observed.

As the matter density increases, larger molecules are expected to form in equilibrium. However, when the matter density becomes comparable to the internal density of the molecule, which is roughly equal to the zero-pressure solid density

$$\rho_0 \sim 1000\, B_{12}^{6/5}\ \mathrm{g/cm^3}, \tag{18}$$

(see LSS), the molecules lose their identities. In Figure 1, we plot the zero-pressure line (labeled '$P = 0$ solid') beyond which the matter should be considered as an electron-proton condensed gas. As the matter density increases further, additional Landau levels are filled by the electrons (labeled by 'Landau'), the material approaches the properties of condensed matter at zero magnetic field. Finally, at the well-known melting line $\Gamma = e^2/kTr_s \simeq 170$, *i.e.*, $T_5 = 0.013\rho^{1/3}$, the plasma condenses into solid. Of course, in the high density region, hydrogen cannot exist, so the high density portion of Figure 1 is only a schematic illustration of the transition from the gaseous to the solid phase. We notice that the outmost solid crust as discussed by some authors (*e.g.*, Fushiki et al. 1989) is unlikely to be relevant to the observable region of a neutron star due to the finite temperature at the surface.

5 Summary

Recent data from ROSAT have shown promising signs for the ultimate detection of thermal radiation from cooling neutron stars (Ögelman 1992). Such detection will provide important clues which will contribute to our understanding of the cooling mechanisms and may help to constrain the equation of state of nuclear matter (Baym 1992). However, before any such confrontation of theory and observation can bear fruit, detailed studies of the neutron star atmosphere are needed.

On the surfaces of those neutron stars in which the magnetic field B is strong, bound atoms, molecules, and chains may be important constituents, and their existence in the neutron star atmosphere may have some observable consequences for the X-ray spectra from cooling and accreting neutron stars. The ionization-recombination equilibrium of atomic hydrogen depends mainly on the ratio $kT/(\ln b)^2$, where T is the temperature and b is the magnetic field strength in units of 2.35×10^9 G. If the ratio is sufficiently small to give an appreciable abundance of neutral atomic hydrogen, then at least diatomic molecular hydrogen can have a non-negligible abundance. For a fixed value of $kT/(\ln b)^2$, the long chains and solid hydrogen grains become more important in principle as b increases. However, if the field strengths are confined to $B_{12} \lesssim 10$, these larger structures are likely to have low abundance for the temperature regime of interest $T \sim 10^5 - 10^6$ K. Since H_2 has many excited state levels, some spectral lines may be observable or, if the collisional broadening should turn out to be strong, the neutron star opacities may be enhanced by the presence of H_2. For instance, for $B_{12} \sim 10$ and $kT \sim 30$ eV, the abundance of H_2 is appreciable (although H is more abundant) and some of the excitation energies are comparable to kT, therefore the excitation spectra of H_2 may be important for the opacities. We plan to study these excitations in the future.

D.L. thanks the Los Alamos National Laboratory for financial support to attend the 'Physics of Isolated Pulsars' Workshop. This work was supported by NSF Grants AST 89-13112, AST 90-15451 and by NASA Grant NAGW-2364 to Cornell University.

References

Baym, G. 1992, this proceeding.
Fushiki, I., Gudmundsson, E. H., and Pethick, C. J. 1989, *Ap. J.*, **342**, 958.
Herold, H., Ruder, H., and Wunner, J. 1981, *J. Phys. B*, **14**, 751.
Neuhauser, D., Koonin, S. E., and Langanke, K. 1987, *Phys. Rev. A*, **36**, 4163.
Jones, P. B. 1985, *Mon. Not. Roy. Astro. Soc.*, **216**, 503.
Jones, P. B. 1986, *Mon. Not. Roy. Astro. Soc.*, **218**, 477.
Khersonskii, V. K. 1987, *Sov. Astron.*, **31**, 225.
Lai, D., Salpeter, E. E., and Shapiro, S. L. 1992, *Phys. Rev. A*, in press (April 1).
Ögelman, H. 1992, this proceeding.
Ruderman, M. 1974, in *Physics of Dense Matter* (I.A.U. Symp. No. 53), ed. C. Hansen (Dordrecht-Holland/Boston), p.117.
Silant'ev, N. A., and Yakovlev, D. G. 1980, *Ap. Space Sci.*, **71**, 45.

Modeling Anisotropic Thermal Emission from Pulsars

Joseph Ventura,[1] Yurii A. Shibanov,[1,2,3] Vyacheslav E. Zavlin,[2] and George G. Pavlov[2,3]

[1]Department of Physics, University of Crete, and Research Center of Crete, Heraklion, Crete, Greece
[2]A. F. Ioffe Physico-Technical Institute, St. Petersburg, Russia
[3]Department of Astronomy and Astrophysics, Pennsylvania State University, USA

1 Introduction

The deployment of imaging X-ray telescopes into orbit has revived interest in detecting thermal radiation from the surfaces of cooling neutron stars, and has motivated detailed theoretical studies of spectral deviations from the Planck law, expected to arise due to the interplay of the photospheric temperature profile with the frequency dependence of realistic opacities at the neutron star's surface layers. Along this line, Romani (1987) explored the influence of a non-degenerate outer atmosphere on the emitted radiation, showing that the spectrum would depend on the outer layer's chemical composition. Of particular interest is the spectral hardening he found for a light-element composition of the surface layers.

In a recent generalization of the above reported elsewhere (Shibanov *et al.* 1992), allowing for the anisotropic and polarization dependent photon transport in a strong magnetic field we have found that the hard spectral tail, though somewhat reduced relative to the nonmagnetic case, is still prominent (see also Shibanov *et al.* 1992a). In this paper we focus on an anisotropy with respect to the pulsar magnetic field direction exhibited by the above hard spectral tail which may help explain the low amplitude periodic modulation and spectral behavior reported in recent *ROSAT* observations of PSR 0656+14 and 1929+10 (Finley *et al.* 1992, 1992a, Helfand 1992).

2 Anisotropy in computed spectra

In Figure 1 we show the photon mean free path (the photon diffusion length) for two normal (polarization) modes as a function of photon frequency. The mean free path is modified by the presence of a $\hbar\omega_c = 54.5$ keV magnetic field in a model neutron star (pure hydrogen) photosphere at an assumed effective surface temperature $T_e = 10^6$ K. Here, ω_c is the electron cyclotron frequency, τ represents the Thomson optical depth, and complete ionization has been assumed (Shibanov *et al.* 1992). At low frequencies, $\omega \ll \omega_c$, the mean free path of the ordinary mode (j=2) displays the typical Kramers' law spectral behavior,

$$D_2 \propto \frac{u^3}{1-e^{-u}}\ , \qquad u = \frac{\hbar\omega}{kT}\ ,$$

up to frequencies where the Thomson scattering becomes important. At the same time, the extra-ordinary mode (denoted as j=1) follows a flatter spectral curve taking higher values,

$$D_1 \sim (\omega_c/\omega)^2 D_2.$$

Note that magnitudes of both D_1 and D_2 depend on orientation with respect to the magnetic field. The symbols $\parallel$ and $\perp$ denote weighted averages employed to describe photon diffusion parallel and perpendicular to the external field.

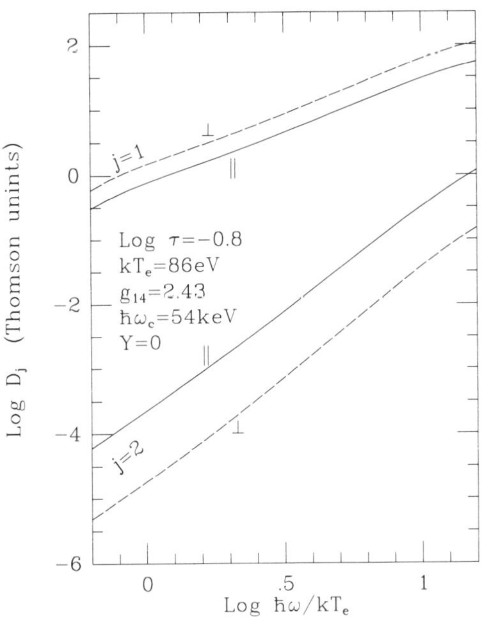

Fig. 1—Spectral dependence of the photon diffusion length (mean free path) for diffusion along and across the magnetic field. The values of the plasma temperature and density (kT=57 eV, ρ=0.89 g cm^{-3}, and kT=63 eV, ρ=0.79 g cm^{-3} for $\parallel$ and $\perp$, respectively) are those at Thomson optical depth τ = −0.8 in model atmospheres with magnetic field parallel and perpendicular to the normal to the atmosphere.

Photon transport is described in terms of two coupled diffusion equations for the transport of the ordinary (j=2 with **e**-vector almost $\parallel$ **Bk**-plane), and extraordinary (j=1, **e**-vector $\perp$ **Bk**) polarization modes in the magnetized plasma (Pavlov and Gnedin 1984, Ventura 1979, Mészáros and Ventura 1979, Kaminker, Pavlov and Shibanov 1982,1983). The two equations are coupled via the magnetic Thomson scattering important at high frequencies. Typically, photon diffusion is dominated by the normal mode with the longer mean free path. Thus, the Rosseland mean of the extinction coefficient,

$$K_R^{-1} = \frac{3}{2}\langle(D_1 + D_2)\rangle \propto \int (D_1 + D_2)\frac{\partial B_\nu}{\partial T}\,d\nu\,,$$

which adequately describes photon diffusion at large optical depths, is dominated by the extraordinary photons in our case. At high frequencies, the radiation emitted at the neutron star surface shown in Figures 2 and 3 is thus almost totally composed of the extraordinary photons, as these escape from deeper and hotter layers of the photosphere. In these graphs we compare the emitted spectra for the two extreme

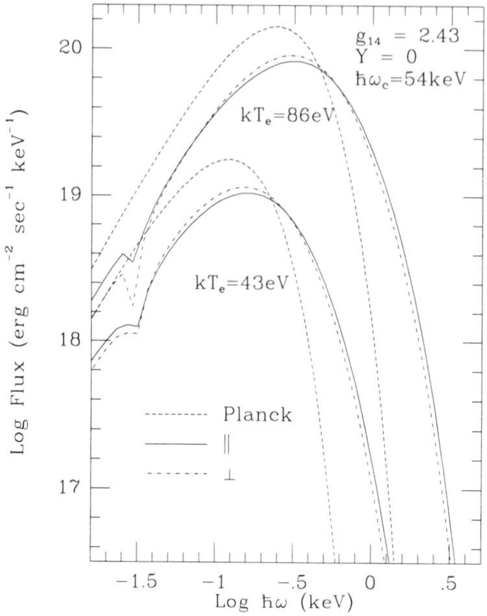

Fig. 2—The emitted spectral flux for atmosphere models with $T_e = 5 \cdot 10^5$ and 10^6 K, $B = 4.7 \cdot 10^{12}$ G. Solid and dash-dotted curves correspond to the magnetic field oriented parallel and perpendicular to the normal to the atmosphere surface. Dashed curves show the corresponding Planck spectra. Absorption features at $\hbar\omega \simeq 0.03$ keV are the ion cyclotron lines.

geometries considered, *i.e.*, photon diffusion $\|$ and $\perp$ to **B**. A marked change is seen in the degree of polarization characterizing these two geometries.
Indeed, the more extreme divergence of the mode diffusion lengths in the $\perp$ case is reflected in the lower emitted flux of ordinary photons in this geometry. The total flux for these two geometries is also shown. Concentrating on the behavior of the hard spectral tail of the $5 \cdot 10^5$ K model, we find that the two curves intersect at about 0.25 keV, and diverge at higher photon energies with the parallel model yielding a harder spectrum. Table 1 shows the size of this anisotropic behavior, which amounts to more than 45% at 1keV.

It is thus clearly seen that diffusion parallel to the magnetic field gives a harder spectral tail in excess of the transverse diffusion case. However, this behavior is reversed at intermediate photon energies between 35 and 260 eV, where emission in the parallel case is below the transverse case. At low (optical-UV) energies, the situation is again reversed, with parallel emission exceeding transverse emission. Similar behavior is seen in the 10^6 K model, but with higher values of the flux, and the intersections of the two curves shifted to higher frequencies.

3 Discussion

We thus conclude that emission at the pulsar surface is expected to be anisotropic with respect to the direction of the magnetic field, even when a constant effective surface temperature is assumed along the neutron star surface. Assuming that this is the sole factor contributing to the observed anisotropy, we would thus also expect to find a modulation in the optical-UV emission in phase with that found in the hard X-ray tail, while a phase reversal is expected at intermediate energies. It would thus be interesting to pursue observations in the optical-UV band to confirm whether there

TABLE 1
Hard Spectral Tail Anisotropy

$\hbar\omega$ (keV)	$F_{\parallel}$ (10^{16} erg/s cm^2)	$F_{\perp}$ (10^{16} erg/s cm^2)	%
0.27	813.0	807.0	0.7
0.50	255.5	220.1	16.1
0.79	50.30	37.63	33.7
1.08	10.79	7.314	47.5
1.25	4.26	2.737	55.6
1.46	1.48	0.903	64.0

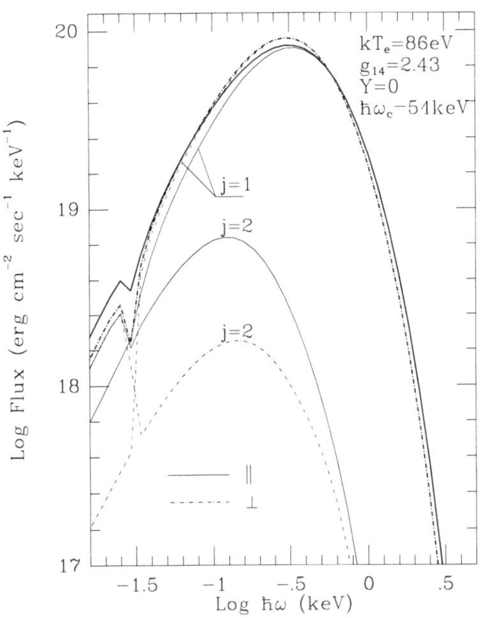

Fig. 3—The emitted spectral flux in the extraordinary (j=1) and ordinary (j=2) polarizations. Heavy curves show the flux spectra summed over the polarizations.

is a periodic modulation in the emission simultaneous to that observed in the X-rays, as this would also confirm the thermal nature of the emission. As the existing cooling calculations for magnetic neutron stars (Hernquist 1984, Van Riper 1988) are unable to predict with certainty any degree of effective temperature variation along the stellar surface, it is difficult to exclude this possibility. It is thus desirable to use the observational data to confirm or disprove the above tendency for a reversal at intermediate energies. Yet, interstellar absorption may render the source unobservable at intermediate energies. Confirmation of a spectral hardening in phase with maximum light in the X-rays would then also serve as a test for the mechanism we have proposed. This

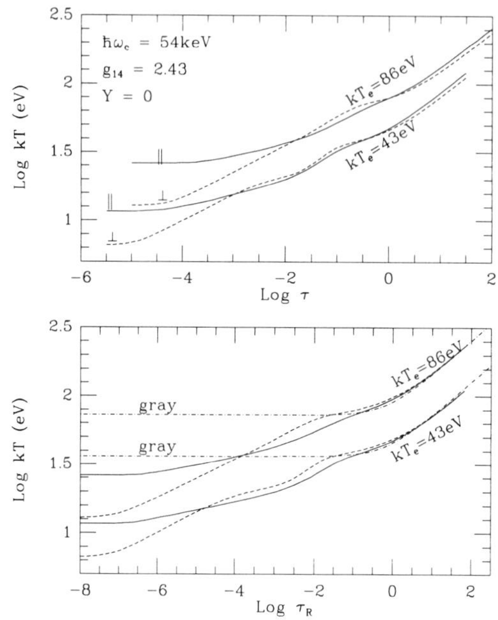

Fig. 4—Thermal structure of the neutron star atmosphere for pure hydrogen model atmospheres at $T_e = 5 \cdot 10^5$ and 10^6 K, magnetic field $B = 4.7 \cdot 10^{12}$ G, and the two orientations considered. For comparison we also give the temperature profile for the usual gray atmosphere.

last effect should be accompanied by a simultaneous brightening in the optical-UV band.

Having discussed the connection of our results to the existing or future observations, let us now try to interpret them in terms of the temperature distribution in our model atmospheres. In Figure 4 we compare the temperature profiles of the $\parallel$ and $\perp$ models to the gray profiles. A general property of the non-gray models—opacities decreasing with frequency—is their lower surface temperature due to the relative transparency of their surface layers to harder photons. This explains the drop in low frequency emission as compared to the gray model (which emits close to the Planck spectrum). Furthermore, in the non-gray models, the drop in the opacity at high frequency results in the hard photons coming from deeper and hotter photospheric layers, and leads to the appearence of the hard spectral tails. Let us now note that the transverse model shows a turnover of its temperature profile from the asymptotic $\tau_R^{1/4}$ behavior (τ_R is the Rosseland optical depth) at a higher depth than in the parallel case (see Shibanov *et al.* 1992). This results in a higher temperature "bump" at intermediate optical depth in the case of transverse diffusion, and gives rise to stronger emission at intermediate energies. On the other hand, spectral hardening at high energies can be elucidated as follows. The outgoing flux is proportional to the source (Planck) function at a thermalization depth for the extraordinary mode, $\tau_{th} \sim (D_1/\kappa_1)^{1/2}$, where κ_1 is the angle-averaged absorption coefficient (see, *e.g.*, Kaminker *et al.* 1983). The high-energy photons escape from deep layers where $T^4 \propto \tau_R$. Hence $T^4(\tau = \tau_{th}) \sim \tau_{th} K_R \sim (D_1/\kappa_1)^{1/2} \langle D_1 \rangle^{-1}$. Since κ_1 is independent of the field direction, and D_1 is smaller for diffusion along the field (see Fig. 1), the source function at the thermalization depth is larger when the field is parallel to the normal to the atmosphere. It should be emphasized that the difference between $D_1^{\parallel}$

and $D_1^{\perp}$, and thereby the anisotropy of the hard spectral tail, is rather sensitive to polarization characteristics of the normal modes. For instance, if one were to make the simplifying assumption that the polarizations of the normal modes are *exactly* linear (as, *e.g.*, in the electron-positron vacuum polarized by superstrong magnetic field), then $D_1^{\parallel}$ would be equal to $D_1^{\perp}$ (Pavlov and Yakovlev 1977) leading to an escaping high-energy flux independent of the magnetic field orientation.

This research has been partly supported through NASA grant NAGW-1522.

References

Finley, J. P., Ögelman, H., and Kiziloğlu, Ü. 1992, *Ap. J. Lett.*, submitted.

——— 1992a, this Conference.

Helfand, D. J. 1992, this Conference.

Hernquist, L. 1984, *Mon. Not. R. Astr. Soc.*, **213**, 313.

Kaminker, A. D., Pavlov, G. G., and Shibanov, Yu. A. 1982, *Ap. Sp. Sci.*, **86**, 249.

Kaminker, A. D., Pavlov, G. G., and Shibanov, Yu. A. 1983, *Ap. Sp. Sci.*, **91**, 167.

Mészáros, P. and Ventura, J. 1978, *Phys. Rev.*, **D19**, 3565.

Pavlov, G. G. and Gnedin Yu. N. 1984, in *Astrophys. Space Phys. Rev.*, R.A. Sunyaev ed., Vol. 3, Harwood Academic Publishers.

Pavlov, G. G., and Yakovlev, D. G. 1977, *Astrofizika (Sov. Astroph.)*, **13**, 173.

Romani, R. W. 1987, *Ap. J.*, **313**, 718.

Shibanov, Yu. A., Zavlin, V. E., Pavlov, G. G., and Ventura, J. 1992, *Astr .Ap.*, submitted.

Shibanov, Yu. A., Zavlin V. E., Pavlov, G. G., Ventura, J., and Potekhin, A. Yu. 1992a, this Conference.

Van Riper, K. A. 1988, *Ap. J.*, **329**, 339.

Ventura, J. 1979, *Phys. Rev.*, **D19**, 1684.

Model Atmospheres of Magnetic Neutron Stars

Yurii A. Shibanov,[1,2,3] Vyacheslv E. Zavlin,[1] George G. Pavlov,[1,3]
Joseph Ventura,[2] and Alexander Yu. Potekhin[1]

[1]A. F. Ioffe Physico-Technical Institute, St. Petersburg, Russia
[2]Department of Physics, University of Crete, Heraclion, Greece
[3]Department of Astronomy and Astrophysics, Pennsylvania State University, USA

1 Introduction

Recent detections of thermal-like X-ray radiation from radio pulsars (Ögelman 1991, and references therein) allow one, in principle, to determine surface temperatures of neutron stars (NSs) of different ages, *i.e*, investigate thermal evolution of NSs. This investigation is especially important because the NS cooling rate essentially depends upon still poorly known properties of very high density matter in NS interiors (equation of state, existence of pion condensate and free quarks, superfluidity of nucleons, *etc.*).

Accurate interpretation of the observations in terms of an effective surface temperature requires taking into account that the spectrum emitted by the NS surface may deviate substantially from blackbody due to the presence of a non-grey plasma atmosphere. Since the temperature in the atmosphere is essentially non-uniform and the photon mean free path depends on frequency, the emitted spectra do not correspond to any definite temperature. Generally, high-frequency radiation escapes from deeper and hotter layers giving rise to a hard spectral tail in excess of the Wien dependence for the same effective temperature T_e. This excess obviously may lead to substantial overestimation of the effective surface temperature when we interpret the X-ray spectra in terms of blackbody radiation. In particular, it has been demonstrated by Finley *et al.* (1992) that the blackbody fit of the ROSAT spectrum of PSR 0656+14 yields $T_e \sim 9 \times 10^5$ K, whereas fitting with a model spectrum of the (non-magnetic) helium atmosphere (Romani 1987) results in $T_e \sim 3 \times 10^5$ K, leading thereby to quite different conclusions on the NS parameters.

The shape of the spectrum can be found from stellar atmosphere modeling which is strongly complicated in the case of NSs by *the high densities* of the atmospheric matter (up to ~ 10 g cm^{-3}) and, especially, *the huge magnetic fields* (up to 10^{12} - 10^{13} G), which drastically change opacities and make the atmosphere anisotropic. Due to this anisotropy, radiation propagates in the atmosphere as two so-called normal modes with different (almost orthogonal) polarizations and opacities (see, *e.g.*, Gnedin and Pavlov 1974). Moreover, properties of the outgoing radiation are mainly determined (Pavlov and Shibanov 1978) by the "extraordinary" mode; its opacity is by a factor of

$\sim (\omega_c/\omega)^k$ lower (*i.e.*, photon mean free path longer) than that of the "ordinary" mode or the field-free opacity (ω_c is the electron cyclotron frequency; $k \sim 1$ - 2, depending on the sourse of opacity).

A first attempt to model a magnetic NS atmosphere has been undertaken by Miller (1992). Following the numerical scheme by Romani (1987) developed for modeling non-magnetic NS atmospheres, he applies polarization-averaged bound-free opacities in strong magnetic field. In such an approach, the outgoing radiation is largely dominated by the "ordinary" mode with larger opacity, in contradiction with what we argued in the previous paragraph. Miller's results should, therefore, be revised with accurate allowance for polarization and anisotropy in the radiative transfer.

In this presentation we describe briefly the general approach to modeling magnetic NS atmospheres and report on first results for hydrogen/helium atmospheres of typical cooling NSs with strong magnetic fields $B \sim 10^{11} - 10^{13}$ G, and effective temperatures $T_e \sim 10^5 - 10^6$ K.

2 Basic Equations

To construct our model atmosphere, we describe the radiative transfer in terms of two coupled diffusion equations for the two normal modes of radiation in the magnetized plasma (*e.g.*, Nagel 1980, Kaminker *et al.* 1982a):

$$\frac{d}{d\tau} D_j \frac{dJ_j}{d\tau} - s(J_j - J_{3-j}) = \kappa_j \, (J_j - B_\nu/2). \tag{1}$$

Here $d\tau = n_e \sigma_T dz$ is the Thomson optical depth, $(4\pi/c) J_j$ denotes the radiative spectral energy density in the jth normal mode ($j = 1, 2$), whereas $B_\nu = B_\nu(T)$ is the Planck function. D_j and κ_j are frequency dependent diffusion and absorption coefficients (in Thomson units $(n_e\sigma_T)^{-1}$ and $n_e\sigma_T$, respectively) defined below.

Due to the medium anisotropy, the diffusion coefficient, D_j, is a function of the angle, Θ_B, between the magnetic field **B** and the normal **N** to the surface of our plane-parallel atmosphere:

$$D_j = D_j^{\parallel} \cos^2 \Theta_B + D_j^{\perp} \sin^2 \Theta_B \,, \tag{2}$$

$$D_j^{\parallel} = \int_0^1 \frac{\mu^2 d\mu}{K_j(\mu)}, \quad D_j^{\perp} = \frac{1}{2} \int_0^1 \frac{(1-\mu^2) d\mu}{K_j(\mu)}, \tag{3}$$

where μ is the cosine of the angle between the wave vector **n** and the magnetic field. The extinction coefficient, K_j, is the sum of the absorption and scattering coefficients,

$$K_j(\mathbf{n}) = K_j^a(\mathbf{n}) + K_j^s(\mathbf{n}) \,, \tag{4}$$

and the integrated scattering coefficient, K_j^s, is

$$K_j^s(\mathbf{n}) - \sum_{i=1}^{2} \int d\mathbf{n}' K_{ji}^s(\mathbf{n}, \mathbf{n}') \,. \tag{5}$$

We define in addition the angle-averaged coefficients, κ_j (true absorption of mode j), and s (mode conversion), as

$$\kappa_j = \frac{1}{4\pi}\int d\mathbf{n}\, K_j^a(\mathbf{n}), \qquad s = \frac{1}{4\pi}\int d\mathbf{n}\, d\mathbf{n}'\, K_{12}^s(\mathbf{n},\mathbf{n}'). \tag{6}$$

To ensure the continuity of the normal flux component and the absence of externally incident radiation, the radiation field must satisfy the following boundary conditions:

$$2D_j\frac{dJ_j}{d\tau} = J_j, \quad (\text{for } \tau = 0), \qquad J_j \to \frac{1}{2}B_\nu(T), \quad (\text{for } \tau \to \infty). \tag{7}$$

The transfer coefficients D_j, κ_j and s are determined, generally, by free-free, bound-free, bound-bound transitions, and Thomson scattering. The free-free transitions of electrons and Thomson scattering on electrons in a magnetized plasma have been investigated in detail by Kaminker, Pavlov and Shibanov (1982b, 1983). For photon energies below or near the ion cyclotron energy, $\hbar\omega_{ic} = (Zm_e/Am_p)\hbar\omega_c = 6.3(Z/A)B_{12}$ eV, the free-free transitions of ions may also be important, and we include these in the absorption coefficients in a relaxation-time approximation. Existing results on the bound-free transitions (*e.g.*, Wunner *et al.* 1983) cannot be directly applied to calculation of the polarization-dependent opacities, so we re-derived the cross-sections for arbitrarily polarized photons (Potekhin and Pavlov 1992) and calculated the opacities with allowance for pressure effects significant in the dense NS atmospheres. According to our preliminary estimates, the bound-bound transitions only slightly affect the overall shape of the emitted spectrum; therefore, we do not include these in our models.

To solve the radiative diffusion equations, one should know how the temperature and densities of the atoms, ions, and electrons depend on the depth. These dependences can be found from equations of thermal, hydrostatic, and ionization equilibrium supplemented by the equation of state of the atmospheric matter. We consider atmospheres in radiative thermal equilibrium which requires the outward flux to remain constant throughout the atmosphere, *i.e.*,

$$H = 4\pi\int d\nu\left(D_1\frac{dJ_1}{d\tau} + D_2\frac{dJ_2}{d\tau}\right) = \sigma T_e^4, \tag{8}$$

$$\int d\nu \sum_j \kappa_j\left(J_j - \frac{1}{2}B_\nu(T)\right) = 0. \tag{9}$$

Although hydrostatic equilibrium is not affected by the force-free magnetic field, ionization equilibrium is substantially modified (Gnedin *et al.* 1974, Khersonskii 1987). For instance, for a hydrogen plasma it gives

$$n_a = n_e n_i \frac{1}{2}\left(\frac{2\pi\hbar^2}{m_e T}\right)^{3/2} \frac{\tanh\eta_e}{\eta_e}\frac{\sinh\eta_i}{\eta_i}\, U, \tag{10}$$

where n_a, n_e, and n_i are the number densities of atoms, electrons, and ions, $\eta_{e,i} = \hbar\omega_{c,ic}/kT$; ω_c (ω_{ic}) is the electron (ion) cyclotron frequency, and U is the partition function of the hydrogen atom in a strong magnetic field.

Assuming the ideal gas equation of state as a lowest order approximation, we can finally close the system to obtain the temperature profile and source function in the star's atmosphere.

To solve the system of the above equations, we start with the usual grey profile

$$T(\tau_R) = T_e \left[\frac{3}{4} \left(\tau_R + \frac{2}{3} \right) \right]^{1/4} , \tag{11}$$

which ensures the proper behavior at large optical depth, solve numerically the coupled diffusion equations (1), and apply corrections to the temperature until the condition for equilibrium, Equation (9), is satisfied. τ_R above denotes the Rosseland depth, $d\tau_R = K_R \, d\tau$, defined in terms of the Rosseland mean of the absorption coefficient,

$$K_R^{-1} = \frac{3\pi}{4\sigma T^3} \int d\nu \frac{D_1 + D_2}{2} \frac{\partial B_\nu}{\partial T} . \tag{12}$$

After the new $T(\tau)$ and depth dependencies of different species are obtained, we solve the diffusion equations anew, and iterate until convergence is reached (see Zavlin and Shibanov 1991, and Shibanov *et al.* 1992, for details).

3 Results

Some representative results of our modeling are shown in Figures 1 through 5. Figure 1 shows how the emitted spectra depend on the value of the magnetic field. The spectra in Figures 1 and 2 were computed neglecting the bound-free opacities. Although the bound-free transitions are by no means negligible at $T_e \lesssim 10^6$ K (see below), the assumption on full ionization allows one to analyze qualitatively dependence of the results on the model parameters applying much faster codes (Shibanov *et al.* 1992).

We see that all the "magnetic" spectra in Figure 1 are somewhat *softer at high frequencies than the corresponding non-magnetic spectrum* (obtained by the same method we have described) but, nevertheless, significantly *harder than the Planck spectrum* corresponding to the same T_e. This difference can be explained by the fact that the opacity κ_1 of the normal mode with longer mean free path (the so called "extraordinary" mode) depends on frequency more gradually than the field-free opacity: $\kappa_1 \sim (\omega/\omega_c)^2 \, \kappa(B = 0) \propto \omega^{-1}$ at $\omega \ll \omega_c$, whereas $\kappa(B = 0) \sim \kappa_2 \propto \omega^{-3}$. Since the shape of the outgoing spectrum is determined by the slope of the extraordinary opacity spectrum more than by its absolute value, the overall spectra in Figure 1 *only slightly depend on B* if $\hbar\omega_c \gg kT_e$. However, the field dependence is clearly displayed in the electron and ion *cyclotron absorption lines* centered at frequencies ω_c and ω_{ic} (see also Figure 2).

Since the diffusion coefficients depend on the *orientation of the magnetic field* with respect to the normal to the atmosphere, there is a noticeable difference in spectra escaping from the magnetic pole and equator of the neutron star (Figure 3). The anisotropy of the spectra and its observational implications are discussed in more detail by Ventura *et al.* (1992). Figure 3 also demonstrates another major feature of the cooling NS radiation, its high degree of *polarization*. It arises due to the fact that

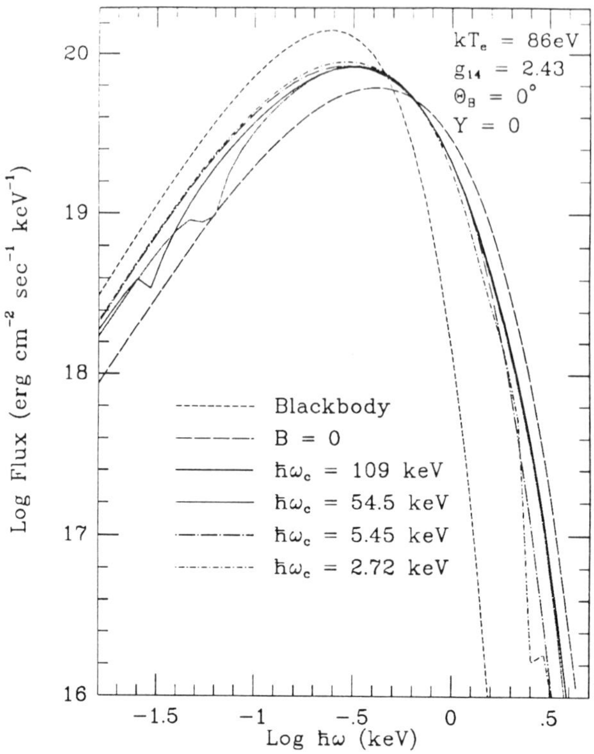

Fig. 1—Spectra of the emitted flux at the NS surface at the effective temperature of 10^6 K and surface gravity $g = 2.43\times 10^{14}$ cm s^{-2} for different values of the magnetic field oriented along the normal to the pure hydrogen atmosphere are compared to the non-magnetic and Plank spectra shown as dashed curves. Spectral features in the Rayleigh-Jeans and Wien parts of the spectra are ion and electron cyclotron lines, respectively.

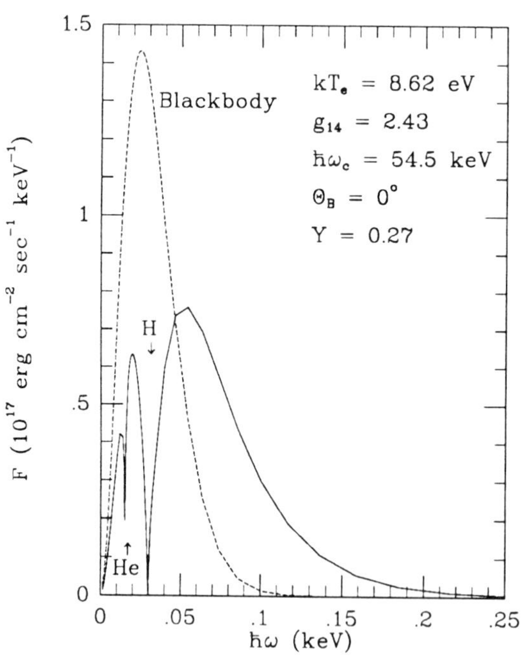

Fig. 2—Ion-cyclotron lines in the spectrum emitted by a H-He atmosphere with $T_e = 10^5$ K, $B = 4.7\times 10^{12}$ G. Since the positions of the lines depend not only on the magnetic field but also on the ratio Z/A, detection of these spectral features may provide information on both the field and H/He abundance in NS atmospheres.

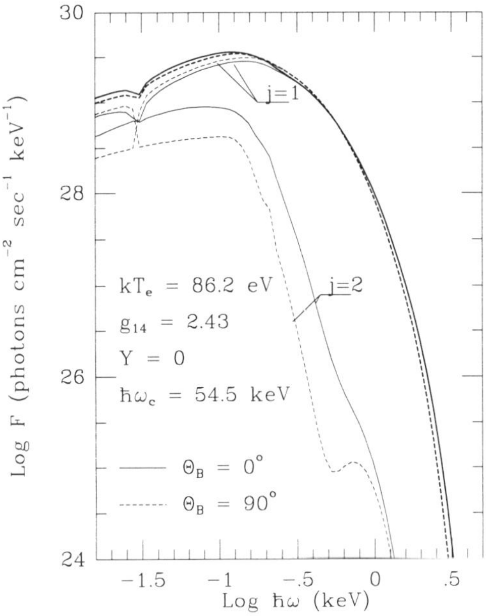

Fig. 3—Spectral fluxes for magnetic field oriented parallel and perpendicular to the normal to the atmosphere (solid and dashed lines, respectively). Light curves show the fluxes in the ordinary (j=2) and extraordinary (j=1) polarization modes. The spectra were computed with allowance for bound-free opacities caused by a small number of non-ionized hydrogen atoms (*cf.* Fig. 5). The radiation in the ordinary mode is more sensitive to the orientation of the magnetic field since the orientation effect on the temperature profile is more pronounced in outer layers (Fig. 5).

the extraordinary mode (almost linearly polarized perpendicular to B) escapes from deeper and hotter layers than the ordinary one.

The effect of small admixture of non-ionized hydrogen atoms on the spectra and temperature run is shown in Figures 4 and 5. We see that the effect is noticeable

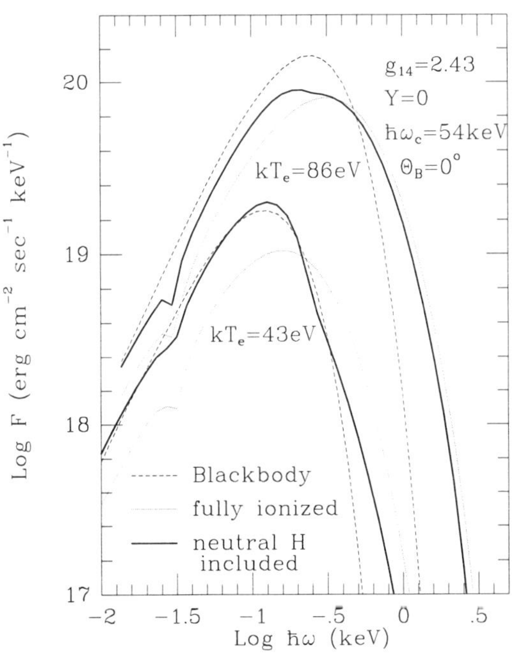

Fig. 4—Spectral fluxes for $T_e = 10^6$ and 5×10^5 K with and without allowance for the bound-free transitions in neutral hydrogen atoms (heavy and light solid lines, respectively). Dashed lines show corresponding blackbody spectra. At $B = 4.7 \times 10^{12}$G, the main photoionization threshold for an isolated atom lies at $\hbar\omega \simeq 250$ eV. The threshold is redshifted and smoothed by pressure effects in dense layers of the atmosphere.

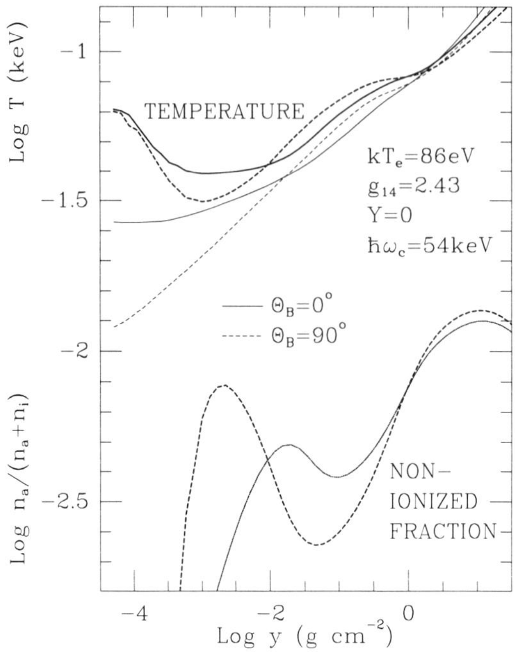

Fig. 5—Dependences of the temperature and relative density of non-ionized hydrogen atoms on the column density in a hydrogen atmosphere for two orientations of the magnetic field. Light curves show the temperature run when existence of the neutral atoms is neglected. The photoionization absorption heats the outer layers thereby increasing the Rayleigh-Jeans parts of the emitted spectra (*cf.* Fig. 4).

even if the effective temperature is as high as 10^6 K. Note that due to pressure effects there are no sharp "jumps" in the spectra associated with the photoionization thresholds. On the other hand, the spectral features are seen more clearly in the ordinary mode (Figure 3) as it escapes from less dense layers; they are better displayed in the transverse geometry due to a larger temperature gradient in the radiating layers. Additional heating of the outer layers enhances low-frequency tails of the spectra which is important for detecting cooling NSs in optical/UV range (Pavlov 1991).

Concluding, we note that the presence of the magnetic field in the atmospheres of cooling NSs substantially affects spectra, angular distribution, and polarization of their radiation. In particular, the emitted spectra deviate significantly from both the Planck law and nonmagnetic spectra. This deviation should certainly be taken into account to draw realistic conclusions on the NS physics from observations of the thermal-like radiation of these objects.

The authors are grateful to Peter Mészáros for helpful comments. This research has been partly supported through NASA grant NAGW-1522.

References

Finley, J. P., Ögelman, H., and Kiziloğlu, Ü 1992, this volume.

Gnedin, Yu. N. and Pavlov, G. G. 1974, *Sov. Phys. JETP*, **38**, 903.

Gnedin, Yu. N., Pavlov, G. G., and Tsygan, A. I. 1974, *Sov. Phys. JETP*, **39**, 301.

Kaminker, A. D., Pavlov, G. G., Silant'ev, N. A., and Shibanov, Yu. A. 1982a, *Astrofizika (Sov. Astrophysics)*,**18**, 283.

Kaminker, A. D., Pavlov, G. G., and Shibanov, Yu. A. 1982b, *Ap. Sp. Sci.*,**86**, 249.

Kaminker, A. D., Pavlov, G. G., and Shibanov, Yu. A. 1983, *Ap. Sp. Sci.* **91**, 167.

Khersonskii, V. K. 1987, *Astron. Zh. (Sov. Astronomy)*, **64**, 433.

Miller, M. C. 1992, *M. N. R. A. S.* **255**, 109.

Nagel, W. 1980, *Ap. J.*, **236**,904.

Ögelman, H. 1991, in *Neutron Stars: Theory and Observations*, ed. J. Ventura and D. Pines, (Kluwer, Dordrecht), p. 87.

Pavlov, G. G. 1991, in Proc. IAU Colloquium #128, *The Structure and Emission Mechanisms of Radio Pulsars*, ed. T. Hankins.

Pavlov, G. G. and Shibanov, Yu. A. 1978, *Astron. Zh. (Sov. Astronomy)* **49**, 741.

Potekhin, A. Yu. and Pavlov, G. G. 1992, *Ap. J.*, submitted.

Romani, R. W. 1987, *Ap. J.*, **313**, 718.

Shibanov, Yu. A., Zavlin, V. E., Pavlov, G. G., and Ventura, J. 1992, *Astr .Ap.*, submitted.

Ventura, J., Shibanov, Yu. A., Zavlin, V. E., and Pavlov, G. G. 1992, this volume.

Wunner, G., Ruder, H., Herold, H., and Schmitt, W. 1983, *Astr. Ap.*, **117**, 156.

Zavlin, V. E. and Shibanov, Yu. A. 1991, *Astron. Zh.(Sov. Astronomy)* , **68**, 999.

Recent Observations of Rotation-Driven Pulsars

James M. Cordes[1]

[1] Astronomy Department and NAIC, Cornell University

1 Introduction

Recent work on radio pulsars has led to a better understanding of beaming and altitudes of emission for radio emission. Magnetic field structure is remarkably dipolar for many long period pulsars, but is as yet poorly constrained in millisecond pulsars. Radio timing programs are summarized, as is recent work on measurements of pulsar space velocities and precessional effects. Fluctuations in radio emission from the Crab pulsar, particularly those that may have a counterpart in gamma rays, are outlined. The Crab's pulse component separation varies with energy so as to suggest a covariation in emission altitude.

2 Beaming

Radio beams from long-period pulsars comprise (1) hollow-cone and (2) core components. The relative strength of the core and cone components is strongly dependent on pulse period, observation frequency, and, most likely, on other properties of the neutron star. Core components typically occur later than the midpoint between the twin conal components, suggesting nonaxiality of beam components or that the core and conal components are emitted at different altitudes, as discussed below.

Observed pulse widths depend on the intrinsic beam width and on geometrical factors, including the angle between the spin vector and magnetic moment, α, and the 'impact' angle, σ, the minimum angle between the line of sight, and the magnetic moment. Pulsar beams are thought to be coaxial, or nearly so, with the magnetic moment. For orthogonal rotators ($\alpha = \pi/2$) the observed pulse width (in angular units) is identical to the beam width for $\sigma = 0$. Nearly aligned rotators show pulse widths $\gg$ beam widths. The possibility remains that perfectly aligned rotators would be 'DC' pulsars, should such objects be capable of radiating. Recent work shows that the scaling of pulse width with period for core beams is remarkably consistent with the beam size implied by the Goldreich-Julian (1969) size of the magnetic polar cap

$$\theta_{beam} \approx 2^\circ\!.5 \left(\frac{r}{R_{NS}}\right)^{1/2} P^{-1/2}, \tag{1}$$

so long as the radius of emission is nearly equal to the neutron star's radius, $r \approx R_{NS}$ (Rankin 1990). Lyne and Manchester (1988) derive a weaker scaling ($\propto P^{-1/3}$) as a consequence of not recognizing distinct classes of conal components. Pulsars that (evidently) are orthogonal rotators with $P \gtrsim 0.04$ s show core widths having a lower envelope that conforms to equation (1); objects with larger pulse widths correspond to non-orthogonal rotators. The fastest millisecond pulsar, PSR 1937+21, does not conform to this relation, as it has a pulse width about 8 times smaller.

3 Radio Polarization and Magnetic Field Structure

The radio emission mechanism is understood only to the extent that it involves coherent radiation from a relativistic outflow along magnetic field lines, combined with corotational motion. With the additional assumptions that (1) emission at a particular frequency comes from a narrow range of radii and (2) no propagation effects in the magnetosphere alter the state of polarization after emission, one may calculate the expected plane of polarization as a function of pulse phase, as did Radhakrishnan and Cooke (1969). There are many examples of pulsars where the measured and predicted polarization angles (based on dipolar fields) agree remarkably well (*e.g.*, Blaskiewicz, Cordes, and Wasserman 1991). Other pulsars show departures from predictions, but these are understood in terms of superposition effects due to (1) the tendency for the emission to occur in one of two (approximately) orthogonal modes, the relative strength varying with pulse phase, and (2) coexistence of core and conal emission. In summary, apart from superposition effects, radio polarization gives strong support to the view that the magnetic field is close to dipolar in form *in the radio emission regions.* This conclusion holds for longer period pulsars. The polarization of millisecond pulsars does not conform to the Radhakrishnan and Cooke model (*e.g.*, Thorsett 1991). Departures of the magnetic field from a dipolar form are possible (*e.g.*, Ruderman 1992).

4 Altitudes of Radio Emission

Altitudes of emission have been determined or limited by the following methods:

(1) *The period-pulse width relation:* As described above, this relation suggests that core components are radiated very near the neutron star surface, $r_{core} \approx R_{NS}$. Conal components, being much broader typically, require correspondingly larger radii of emission, in the range of $r_{cone} \approx 5 - 100 \times R_{NS}$.

(2) *Aberration/retardation/field-sweepback limits:* A difference in altitude of emission Δr (*e.g.*, between core and cone pulse components or between the same components at different frequencies) implies an arrival time difference $(1+\sin\alpha)\Delta r/c$ from aberration and retardation; field-line sweepback will lessen this delay by model-dependent amounts. If a radius-to-frequency mapping is assumed, the absolute altitudes of emission may be solved for in terms of Δr. Upper bounds $\Delta r/c \lesssim 1 - 10$ ms imply upper bounds on emission radii of a few per cent of the light cylinder radius, $R_{LC} \equiv c/\Omega = 10^{4.7} P$ km (Cordes 1978; Phillips and Wolszczan 1990).

(3) *Conformance of polarization with magnetic dipole topology:* The aforementioned conformance of polarization data with magnetic dipole configurations, which holds for *conal* components, suggests that $R_{NS} \ll r_{cone} \ll R_{LC}$. Higher order

multipoles near the surface can alter the magnetic structure there while sweep-back of field lines modifies the structure near the light cylinder.

(4) *v/c effects in polarization waveforms:* Particle trajectories include motion along frozen-in field lines as well as corotation motion, $\Omega \times \mathbf{r}$. The corotational motion therefore breaks the symmetry of the dipolar field, yielding an offset between the centroids of the intensity pulse shape and the polarization angle waveform, the latter arriving later by an amount $\Delta t \approx 4r_{core}/c$ (Blaskiewicz, Cordes, and Wasserman 1991). The implied altitudes of emission are in the range of a few to $100 \times R_{NS}$, consistent with other methods.

(5) *Interstellar scintillation limits:* Diffraction from electron-density variations in the interstellar medium has an implied isoplanatic angle $\lesssim 0.1$ μarc sec, comparable to the size of the light cylinder of a $P = 1$ s pulsar at 1 kpc distance. Comparisons of scintillations in different pulse components can determine or limit the transverse separations of emission regions; with adoption of outflow models (*e.g.*, in a dipolar field), transverse separations yield altitudes of emission (Cordes, Weisberg, and Boriakoff 1983). Scintillation limits are typically poorer than those from other methods, although in some cases they imply upper bounds of 5% of R_{LC}. In one case, the transverse offset requires an altitude $r \approx R_{LC}$ (Wolszczan and Cordes 1987), but the observations involved an anomalous episode of scintillation whose interpretation may be problematic.

5 Radio Timing Programs

Radio timing programs at Arecibo, Jodrell Bank, Green Bank, and Parkes are currently observing nearly all the known ($\sim$ 500) pulsars. Radio data are time-tagged to an accuracy better than 1 μs, although the ultimate precision of times-of-arrival (TOAs) is approximately 1 milliperiod, due to the phase and amplitude variations of single pulses and the finite number of pulses summed before calculating a TOA. Spin period derivatives, $\dot{P}$, are available for about 90% of the known pulsars (J. Taylor, R. Manchester, and A. Lyne, personal communication).

Rotation-driven pulsars show a wide range of departure from the smooth spin-down, including (1) *glitches*, discontinuities in Ω and/or $\dot{\Omega}$ seen in $\sim$ 12 objects; and (2) *'timing noise'* seen in most objects that behaves statistically like a random walk in phase, Ω, or $\dot{\Omega}$, or some mixture of these processes. For some objects, individual events that comprise the timing noise have been identified (Cordes and Downs 1985). Timing noise events differ from the form of glitches and there seems to be no obvious relationship between glitches and timing noise.

In addition to studying neutron star spins, radio timing data are used to study (1) binary companionship; radio pulsars' companions include other neutron stars, white dwarfs, substellar dwarfs, and planets; (2) binary evolution through determination of orbital elements of present day binaries; (3) General Relativity through the form and time dependence of NS-NS binaries (Taylor, Wolszczan, and Weisberg 1991); (4) turbulence in the interstellar medium through wave-propagation delays (Cordes *et al.* 1990; Foster, Backer, and Wolszczan 1990; Phillips and Wolszczan 1991); (5) gravitational wave backgrounds that perturb arrival times at the submicrosecond level (Stinebring *et al.* 1991).

6 Space Velocities of Neutron Stars

Radio pulsars were recognized as a runaway population soon after their discovery because the population showed a large galactic scale height and a few individual objects showed large proper motions. Pulsar space motions (above that due to galactic rotation) are measured with three techniques: (1) direct interferometric measurement of the proper motion; (2) pulse timing measurements; and (3) interstellar scintillation measurements. The last method is insensitive to the direction of motion, unlike the first two methods. However, the scintillation estimate of the velocity depends on distance as $D^{1/2}$ whereas the other methods yield velocities $\propto D$. Because the distance scale is good to only a factor of two (Lyne, Manchester, and Taylor 1985), one must be cautious in estimating the velocity distribution. In any case, the distribution of transverse speeds shows a mean in the range of 150 km s^{-1} and a tail extending to several times the mean. The largest speed is nearly 10^3 km s^{-1} (measured through interferometry; Harrison, Lyne, and Anderson 1992).

Recent discoveries give further evidence for large space velocities. A young, large $\dot{P}$ pulsar (1853+01) in the supernova remnant W44 (Wolszczan, Cordes, and Dewey 1991) may have been in a binary with the small $\dot{P}$ pulsar 1854+00 that is now about 20 arc min away and has a 10% smaller dispersion measure (DM). To account for the separation and difference in DM, a space velocity of about 500 km s^{-1} is needed for both objects, and the electron density in or near the remnant must be $\sim$ 1 cm^{-3}. The pulsar 1757-24, another young, large $\dot{P}$ object (Frail and Kulkarni 1991; Manchester *et al.* 1991), is surrounded by a synchrotron nebula at the head of a peculiar supernova remnant identified by Becker and Helfand (1985). To explain the location of the pulsar outside the remnant, a velocity $V_\perp \approx 1 - 2 \times 10^3$ km s^{-1} is needed.

The eclipsing millisecond pulsar 1957+20 moves at $V_\perp \approx 120$ km s^{-1} (Ryba and Taylor 1991) oriented along the axis of a bow shock in Hα whose structure depends in part on spindown losses ($\sim$ few $\times 10^{35}$ erg s^{-1}; Kulkarni and Hester 1988).

The large space velocities of pulsars derive from at least two contributions: (1) disruption of binary motion and (2) kicks imparted from asymmetry of supernova explosions. A third contribution may arise from the magnetic-dipole rocket effect (Harrison and Tademaru 1975). A Monte Carlo simulation of neutron-star formation in binaries (Dewey and Cordes 1987) concluded that there were insufficient compact binaries to account for the large number of high-velocity pulsars. A kick asymmetry yielding $\sim$ 100 km s^{-1} accounts for both the velocity distribution and the small percentage of binary membership among radio pulsars. Work by Bailes (1989) supports the conclusion of asymmetric explosions for pulsars in general.

7 Precession Effects

Another consequence of asymmetric explosions is that spin and orbital angular momenta in surviving binaries are generally misaligned in post-supernova systems, though they were probably aligned by tidal effects in pre-explosion systems. The Hulse-Taylor binary pulsar 1913+16 appears to have undergone a misaligning kick, implying that the pulsar's spin vector will gyrate around the total angular momentum by the geodetic precession process with a period of 300 yr. The precession period is 600 yr for a similar NS-NS binary, 1534+12 (Wolszczan 1991). Spin-axis precession produces pulse-shape changes that depend on the form of the pulsar beam. Secular pulse-shape changes at the level of 1% yr^{-1} are seen from 1913+16 (Weisberg, Romani, and Taylor 1989; Cordes, Wasserman, Lundgren, and Blaskiewicz, in preparation) that have no known counterpart in other pulsars, therefore suggesting that the changes are due to geodetic precession with a spin-orbit misalignment angle $\approx 1 - 10°$. The spin-orbit angle is a direct measure of supernova asymmetry. Measurements of geodetic precession are proceeding and will ultimately reveal (1) the cross-sectional structure of the pulsar beam; and (2) information on the orbital elements, the helium star's mass, and the kick velocity. Burrows and Woosley's (1986) work on explosions of helium stars also supports the conclusion of an asymmetric supernova for 1913+16. Wijers, van Paradijs, and van den Heuvel (1992) have constrained helium star masses and velocity kicks through a statistical study of high-mass binary radio pulsars.

8 Radio Fluctuations

Radio emissions show variability on a wide range of time scales, both $\ll P$ and $\gg P$. It is not known if radio variability arises mostly from variations in densities or energies of relativistic particles or from mere changes in the coherence mechanism that must underlie the large brightness temperatures (*e.g.*, 10^{25} to 10^{30} K). The radio luminosity is a small fraction of the total spindown energy loss rate (1% for slow rotators, 10^{-10} for the Crab pulsar). However, regular radio variations, such as drifting subpulses with quantized drift rates (Wright and Fowler 1981), and fast switching between two or more preferred pulse shapes (Bartel *et al.* 1982) suggest that radio phenomena reflect the global state of the magnetosphere. If so, the $\sim$ 100% variations in the radio signals would imply proportionate variations in plasma flow and, perhaps, torque on time scales of microseconds to minutes or hours. Such time scales are insufficient to allow buildup of spin phase variations to measurable values. A direct method for testing the nature of radio variability and whether it reflects the overall energetics of the pulsar magnetosphere is to look for similar variations at wavelengths where the emission mechanism is incoherent. More robustly, one can search for *correlations* between radio and high-energy emissions. The absence of any correlation would signify that either (1) the particles responsible for coherent and incoherent emission are different and that the hypothesis is untestable; or (2) the particles are the same and that radio variations are due solely to variations in the efficiency of the coherence mechanism caused by subtle changes in the particle distribution function.

9 Multiple-Wavelength Studies of the Crab Pulsar

In an attempt to explore the nature of radio emissions in particular and the overall magnetosphere in general, the Crab pulsar has been studied in the radio, infrared, and in γ-rays. The Crab pulsar differs from all well-studied radio pulsars by displaying a long tail in its pulse amplitude distribution. Pulse amplitudes > 100 times the mean are seen. Figure 1 shows the differential distribution of the pulse area ('pulse energy') at 0.8 GHz, obtained simultaneously with the first pointing by the Compton Gamma Ray Observatory (1991 May). The distribution is power law with slope ~ -3.5; the roll-off at low energies is due to incompleteness. The time interval *between* 'giant pulses' (defined as those with pulse energy > 40× the mean) is close to being exponential with a mean of 10 s, implying that giant pulses occur with Poisson statistics (Lundgren *et al.* 1992).

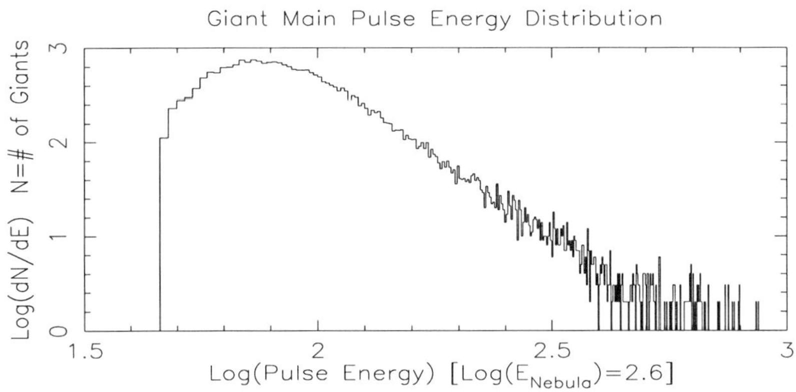

Fig. 1—Differential distribution of pulse areas ('energies') of single giant pulses from the Crab pulsar at 0.8 GHz. Some of the pulses are stronger than the contribution from the entire Crab Nebula.

Figure 2 shows pulse shapes for the Crab pulsar from 0.8 GHz radio data, 2.2μ data from Palomar (Lundgren, Cordes, and Beckwith, in preparation), and OSSE data from the Compton Gamma Ray Observatory between 60 and 250 keV (Ulmer *et al.* 1992). The radio and IR data were aligned through a proper timing analysis where a radio average pulse was used as a template. The OSSE data have been aligned by eye only. The figure shows that the separation between pulse components evolves with wavelength, the IR separation being largest, followed by the radio, then the OSSE data. Perusal of the literature (*e.g.*, Hasinger *et al.* 1984) shows a general trend where the separation between components reaches a maximum in the IR and decreases monotonically up to γ−ray energies. The radio community interprets pulse shape evolution with wavelength in terms of varying altitudes of emission and the changing opening angle of the open field line region of the magnetosphere (Cordes 1978; Phillips and

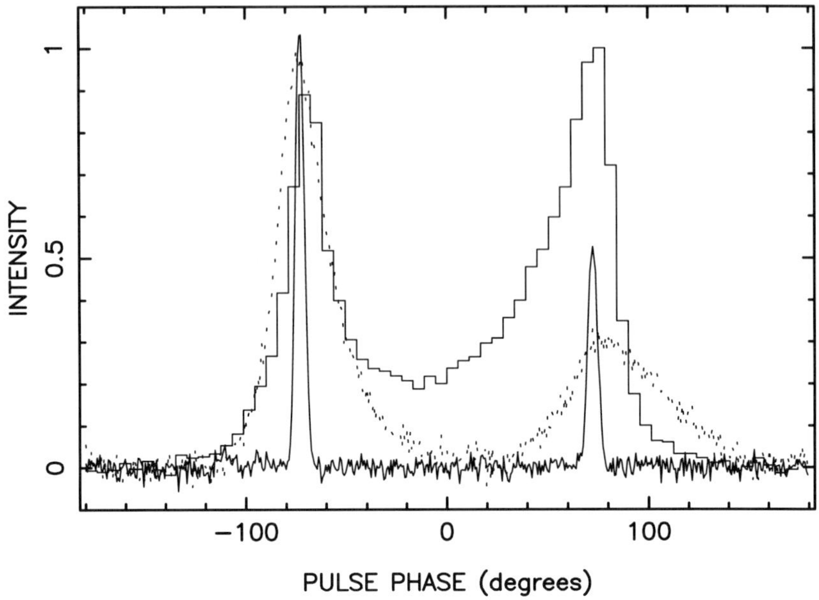

Fig. 2—Average pulses from the Crab pulsar at 0.8 GHz (solid line, continuous), 2.2μ (dotted), and at 60-250 keV from the Compton GRO OSSE detector (solid line, staircase). Amplitudes are in arbitrary units.

Wolszczan 1990). Whether a similar view may be adopted for the high-energy, incoherent emission of the Crab pulsar is, as yet, uncertain.

Crab radio data were obtained at the National Radio Astronomy Observatory, Green Bank, WVa and the National Astronomy and Ionosphere Center, Arecibo Observatory which are operated under cooperative agreements with the NSF. IR data on the Crab were obtained with the 5m Hale Telescope under a cooperative agreement between Cornell University and Caltech. The author thanks K. Mathews for use of his IR photometer. OSSE data were provided by Mel Ulmer.

References

Bailes, M. 1989, *Ap. J.*, **342**, 917.
Becker, R. and Helfand, D. J. 1985 , *Nature* , **313**, 115.
Blaskiewicz, M., Cordes, J. M., and Wasserman, I. 1991, *Ap. J.*, **370**, 643.
Burrows, A. and Woosley, S. E. 1986, *Ap. J.*, **308**, 680.
Cordes, J. M. 1978, *Ap. J.*, **222**, 1006.
Cordes, J. M. and Downs, G. S. 1985, *Ap. J. Suppl.*, **59**, 343.
Cordes, J. M., Weisberg, J. M., and Boriakoff, V. 1983, *Ap. J.*, **268**, 370.
Cordes, J. M., Wolszczan, A., Dewey, R. J., Blaskiewicz, M., and Stinebring, D. R. 1990, *Ap. J.*, **349**, 245.
Dewey, R. J. and Cordes, J. M. 1987, *Ap. J.*, **321**, 780.
Foster, R. S., Backer, D. C., and Wolszczan, A. 1990, *Ap. J.*, **356**, 243.
Frail, D. and Kulkarni, S. R. 1991, *Nature*, **352**, 785.
Goldreich, P. and Julian, W. H. 1969, *Ap. J.*, **160**, 971.
Harrison, E. and Tademaru, E. 1975, *Ap. J.*, **201**, 447. .
Harrison, P. A., Lyne, A. G., Anderson, B. 1992, in *X-ray Binaries and the Formation of Binary and Millisecond Pulsars*, NATO ASI, in press.
Hasinger, G. *et al.* 1984, *Adv. Sp. Res.*, **3**, 63.
Kulkarni, S. R. and Hester, J. J. 1988, *Ap. J.*, **335**, 801.
Lundgren, S., Cordes, J. M., Foster, R. S., Hankins, T. H., Ulmer, M., and Garasi, C. 1992, in *The Compton Observatory Science Workshop*, ed. C. Shrader, N. Gehrels, and B. Dennis, NASA Conference Pub. 3137, 260.
Lyne, A. G., Manchester, R. N., and Taylor, J. H. 1985, *M.N.R.A.S.*, **213**, 613.
Lyne, A. G. and Manchester, R. N. 1988, *M.N.R.A.S.*, **234**, 477.
Manchester, R. N., Kaspi, V. M., Johnston, S., Lyne, A. G., and D'Amico, N. 1992, *M.N.R.A.S.*, in press.
Phillips, J. A. and Wolszczan, A. 1990 in *Low Frequency Radio Astronomy from Space*, ed. K. Weiler and N. Kassim, (New York: Springer-Verlag Lecture Notes in Physics Series, 175.
Phillips, J. A. and Wolszczan, A. 1991, *Ap. J. Letts.*, **382**, L27.
Radhakrishnan, V. and Cooke, D. J. 1969, *Ap. L.*, **3**, 225.
Rankin, J. M. 1990, *Ap. J.*, **352**, 247.
Ruderman, M. 1992, these Proceedings
Ryba, M. and Taylor, J. H. 1991, *Ap. J.*, **380**, 557.
Stinebring, D. R., Ryba, M. F., Taylor, J. H., and Romani, R. W. 1990, *Phys. Rev. Lett.*, **65**, 285.
Taylor, J. H., Wolszczan, A., and Weisberg, J. M. 1991, *Nature* , **355**, 132.
Thorsett, S. 1991, Ph. D. Thesis, Princeton University.
Ulmer, M. *et al.* 1992, in *The Compton Observatory Science Workshop*, ed. C. Shrader, N. Gehrels, and B. Dennis, NASA Conference Pub. 3137, 260.
Weisberg, J. M., Romani, R., and Taylor, J. H. 1989, *Ap. J.*, **347**, 1030.
Wijers, R., van Paradijs, J., and van den Heuvel, E. 1992, *Ast. Ap.*, in press.
Wolszczan, A. 1991, *Nature*, **350**, 688.
Wolszczan, A. and Cordes, J. M. 1987, *Ap. J.*, **320**, L35.
Wolszczan, A., Cordes, J. M., and Dewey, R. J. 1990, *Ap. J. Letts.*, **372**, L99.
Wolszczan, A. and Frail, D. 1992, *Nature*, **355**, 145.
Wright, G. A. and Fowler, L. A. 1981, *Ast. Ap.*, **101**, 356.

Profile Classification Scheme and Physics of Pulsar Emission

Janusz A. Gil[1,2]

[1]Max-Planck Institute for Radioastronomy
[2]The Astronomical Center of Zielona Góra, Poland

Abstract

It is argued that the beam of mean pulsar emission is organized into a central core and two conal components. It is suggested that the conal structure is a consequence of a subpulse drifting phenomenon caused by the subpulse beams rotating around the magnetic axis. A period dependence of the altitude of emission regions is also discussed.

1 Profile Classification Scheme

Recently Rankin (1990, 1992a,b) has empirically recognized the existence of two conal beams around a pencil-like core beam in the mean pulsar emission. They seem to be a common property of all pulsars, not only the ones with four of five components. The original hollow-cone model (Radhakrishnan and Cooke, 1969; Komesaroff, 1970) was able to explain only double profiles (when the observer's sight line cuts the beam centrally) and single profiles (when the beam is grazed) along with polarization position angle behavior. The position angle curve in single profiles is usually shallow whereas in double profiles it is usually S-shaped with the large value of the central gradient being explained as a consequence of a central cut through the beam. To explain triple profiles, Backer (1976) has introduced a central-pencil beam responsible for a "core" pulsar emission. To explain 5-component profiles, which are also observed, Rankin (1983, 1992a,b) has postulated that the cone splits into an inner and an outer sub-cone.

The inner and outer conal emission, obviously present in five component pulsars, has also been recognized in some triple and even double component pulsars (Rankin, 1992a,b). In some pulsars the central component is not associated with core emission, but instead with the inner cone.

The way by which Rankin obtained her important results is interesting and worthy of short discussion. First, she noticed that in pulsars with two-pole interpulse emission (interpulsars), for which the inclination angle θ between the rotation and magnetic axes is nearly 90°, the FWHM width of core components at frequencies near 1 GHz scales with pulsar period P as $\Delta\varphi_{1/2}^{\rm core} = 2.45° \cdot P^{-1/2}$ (Rankin, 1990, Fig. 1) with surprising accuracy. Using this observational fact she developed a method of determining the inclination angle θ. This method is based on the fact that $\Delta\varphi_{1/2}^{\rm core} = 2.45° \cdot P^{-1/2}/sin\theta$ independently of the model of pulsar emission. But what is equally important is that Rankin (1990) has obtained values of the opening angle α of the core emission which are also model independent. In fact, for pulsars with core components, the intercept angle σ of the closest approach of the line-of-sight to the magnetic axis should be small and for $\sigma \approx 0$, $\Delta\varphi \approx 2\alpha/sin\theta$. Since the opening angle α does not depend on θ, one can use the interpulse scaling law found by Rankin (1990) to obtain

$$\alpha_{\rm core} = 1.2^\circ\, P^{-1/2}, \tag{1}$$

near 1 GHz for all pulsars.

For pulsars with triple or multiple profiles one can determine the inclination angle θ by applying Rankin's method to their core components. Rankin (1992b) has also calculated the intercept angle σ for a large number of pulsars from the maximum gradient of the polarization position angle $d\psi/d\varphi_{max} = \sin\theta/\sin\sigma$. Having values of θ and σ, she has used an exact description of the geometry of pulsar emission proposed by Gil (1981, 1984a) and calculated the opening angles

$$\alpha = 2\sin^{-1}\left\{\sin^2(\Delta\varphi/4)\sin\theta\sin(\theta+\sigma) + \sin^2(\frac{\sigma}{2})\right\}^{1/2} \tag{2}$$

for a large number of pulsars, where $\Delta\varphi$ is an appropriate half power width of pulse profile near 1 GHz. Surprisingly, they follow exclusively two well defined parallel lines

$$\alpha_{cone}^{in} = 4.3^\circ P^{-1/2} \tag{3}$$

for the inner conal components and

$$\alpha_{core}^{out} = 5.7^\circ P^{-1/2} \tag{4}$$

for the outer conal components in multiple profile pulsars. When Rankin (1992a,b) attempted to fit the outer components of triple profiles into this geometry, they all appeared to satisfy either equation (3) or equation (4).

I would like to emphasize that calculations of the opening angles (eqs. [1], [3], and [4]) from equation (2) involve two implicit assumptions: (i) the radiation is emitted tangentially to the magnetic field lines and (ii) the emission is strongly linearly polarized in the planes of field lines. Therefore, the fact that Rankin (1992a,b) has found the opening angles for all pulsars to follow three distinct, parallel lines instead of scattering all over the (α, P) plane, strongly argues for curvature radiation as the pulsar radio emission mechanism. A straightforward interpretation of the observational fact expressed by equations (1), (3), and (4) is the organization of the mean pulsar beam into core and two conal components surrounding it.

2 Drifting Subpulses and Profile Components

In my opinion, the multi-conal structure of mean pulsar beams is closely associated with the phenomenon of drifting subpulses, observed in many pulsars either as a systematic drift across the pulse window or a longitudinal stationary intensity modulation at the outer profile components (Rankin 1986). Drifting subpulses can be described as a complex phenomenon of phase-intensity-position angle modulation

(Taylor *et al.* 1971). The intensity of subpulse peak increases systematically toward the profile centre and at the same time the position angle at the subpulse peak varies according to the Radhakrishnan and Cooke (1969) model, reflecting the orientations of the planes of magnetic field lines in the emission region.

A good example of the phase-intensity-position angle modulation is demonstrated by PSR 2303+30. A typical sequence of individual pulses from this pulsar is presented in Figure 1. The apparent drift bands visible in this figure correspond to a phenomenon which I call a secondary drift. These bands are formed from even and odd pulses separately. Individual pulses alternate between a single-subpulse (odd) and double-subpulse (even) form with astonishing precision. Odd pulses usually have a strong central subpulse while even pulses have two weaker subpulses. The primary subpulse drift rate is about $2.5°/P$, which is very fast. Its subpulses drift across the pulse window from the edge of the profile to its centre during one period.

One can distinguish two different types of phase-versus-intensity modulation in the pulse sequence presented in Figure 1. Type A, visible in panel a, demonstrates constant phase and intensity in both single and double pulses. Type A modulation is preceded and followed by Type *B* in which both phase and intensity of subpulses are modulated along secondary driftbands. The modulation is very precise with subpulse intensity strongly depending on its phase. The subpulses at the trailing edge are weak, the intensity gradually increases until it reaches a maximum near the center of the pulse window, and then it gradually decreases again toward the leading edge of the profile.

This extremely regular modulation visible in Figure 1 is consistent with rotation of angular beams associated with individual subpulses around the pulsar beam axis. Type *A* corresponds to a velocity of rotation $v = 360°/(2nP)$, where n is the total number of subpulse beams within the instant pulsar beam and type *B* corresponds to a velocity slightly different from that of type *A*. In principle, the drift velocity v depends on a number n of subpulse beams. From visual inspection of Figure 1, the actual number of observed beams is probably 4. This means that $v = 45°/P$, and a subpulse beam completes one revolution around the magnetic axis in about $360°/45° = 8$ periods of pulsar rotation. If this is exactly 8 periods, both phase and intensity of subpulse are constant (type *A*). In type *B* modulation, of which a long sequence is presented in Figure 1a, the velocity of rotation slightly differs from $45°/P$. This explains the strong dependence of the subpulse intensity on its phase within the pulse window. The position angle also shows approximately 8-periods of modulation (see below). The strong phase-intensity-position angle correlation discussed here is very similar to that observed in PSR 0809+74 (Taylor *et al.* 1971).

The single-pulse polarization characteristics of PSR 2303+30 are also very interesting. The position angle variation is presented in lower panels of Figure 1. Panel *b* shows the position angle at the peak of several subpulses (open circles whose size is proportional to the peak intensity). Filled symbols represent an intensity weighted average which can be interpreted as the Radhakrishnan and Cooke (1969) curve. The mean position angle is plotted as a solid line. Panel *c* shows swings of the position angle of the individual subpulses presented in panel *a* (above 50% of peak intensity

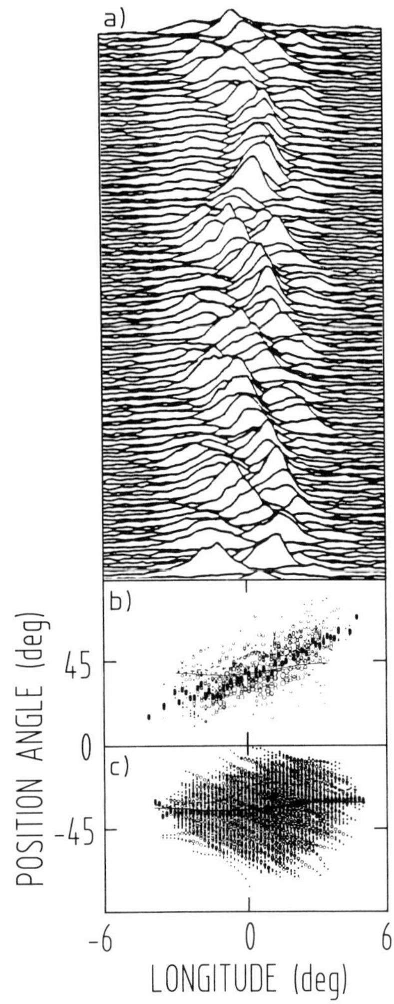

Fig. 1—Single pulses from PSR 2303+30 at 400 MHz; a. sequence of 200 pulses with pulse number increasing upwards, b. position angle values at subpulse peaks, c. position angle swings across subpulses.

threshold). The size of the open circles is proportional to the subpulse peak intensity. Filled symbols represent the intensity weighted average, which coincides with the mean position angle curve, also shown as a solid line in panel *b*. The position angle swings of individual subpulses (panel *c*) seem to be symmetrical about the Radhakrishnan and Cooke (1969) slope in panel *b* rather than about the mean position angle curve.

The subpulse-associated beams should be separated from each other by distances comparable with their sizes, so their rotation around the magnetic axis automatically implies a multi-conal structure of the mean pulsar beam. An association of drifting subpulses with the profile components was first suggested by Hankins and Wolszczan (1987). From the observed ratio of subpulse-to-profile width one can anticipate an existence of two cones surrounding the core emission. Figure 2 shows simulation of pulsar emission within the above model. I have considered one subpulse beam near the magnetic axis and several approximately equi-distant beams surrounding it. All

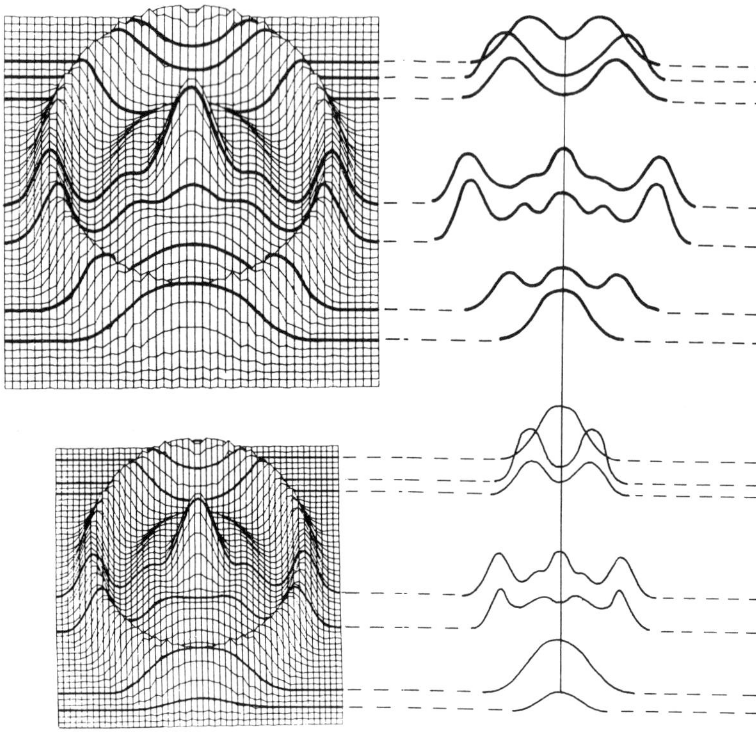

Fig. 2—A "double sombrero" structure of the average beam of pulsar emission at two different frequencies. The relative intensities of the core, inner cone, and outer cone were arbitrarily chosen.

beams were uniformly rotating around the magnetic axis, completing one revolution in several pulsar periods (for details of calculations see Gil 1984b). The mean intensity of the pulsar beam for two frequencies corresponding to two different emission radii is illustrated in Figure 2. Different cuts of the observer's sight line through the emission cones correspond to different types of the observed mean pulsar profiles, along with their frequency evolution.

3 Location of Emission Regions

It is widely accepted now that pulsar radio-emission originates close to the star rather than near the light-cylinder, but how close still remains an open question. Cordes (1978) was the first to set constraints on the emission radii for three pulsars, PSRs 0525+21, 0950+08, and 1133+16. Assuming that retardation and aberration delays are below the uncertainities of determination of a cold plasma dispersion fiducial point (Craft 1970), he obtained the emission radii for frequencies between 430 MHz and 1400 MHz to be less than 100 to 1000 km. It is worth mentioning that Cordes (1978) assumed that some sort of radius-to-frequency mapping operated in the emission region.

Assuming dipolar structure of the magnetic field lines in the emission region, the opening angle in equations (1), (3), and (4) can be expressed in the form

$$\alpha = 0.00124° s r^{1/2} P^{-1/2}, \tag{5}$$

where $r(\nu)$ is the altitude of the emission region at a given frequency ν and $0 < s = d/r_p < 1$ is the parameter describing locus of field lines on the polar cap corresponding to a given component (d is the distance from the magnetic axis to the base of field lines, and r_p is the Goldreich-Julian (1969) polar cap radius). The parameter s can be determined observationally from the ratio of the full pulse width measured at the profile wings to the pulse width measured for appropriate profile components. Thus, knowing the opening angle computed from equations (1), (3), or (4), one can estimate the radius of emission region from equation (5). Table 1 presents result of calculations performed for a number of pulsars for which $\Delta\varphi, \theta, \sigma$ and s can be well determined. In the case of the millisecond pulsar PSR 1937+21 I have used $\theta = 50°$ and $\sigma = 30°$ which explains narrow 10° pulse profile width at 430 MHz.

TABLE 1
Values of the Magnetic Field in the Emission Region

PSR	P [s]	$\dot{P} \cdot 10^{15}$ [s/s]	$B_0/10^{12}$ [Gauss]	$r_{430}(P)$ [km]	$B(r)/10^7$ [Gauss]
0525+21	3.74	40.06	12.4	825	2.2
2303+30	1.57	2.89	2.15	515	1.57
1237+25	1.38	0.95	1.16	520	0.8
0329+54	0.7	2.05	1.2	400	1.8
0950+08	0.25	0.23	0.24	240	1.7
1937+21	0.00156	0.00011	0.0004	23	3.3

Let us note that the radius of emission region seems to depend on the pulsar period as

$$r(P) = 23 \text{ km } (P/0.00156\text{s})^{0.46}, \tag{6}$$

suggesting that the emission region is close to the polar cap in milliseconds pulsars and at about 1000 km in long period pulsars. Note that the range of P is 2500, $\dot{P}$ is 400000, $B_\circ = 3.2 \cdot 10^{19}(P\dot{P})^{1/2}$ is 30000, $r_{430}(P)$ is 40, whereas the range of $B(r) = B_0(R/r(P))^3$ is only 4 around 10^7 Gauss. It seems that the actual mechanism of pulsar coherent radiation prefers magnetic fields around 10^7 Gauss in the emission region. It should also be mentioned that in order to satisfy equations (1), (3), and (4) together with equation (6), the parameter s in equation (5) should be slightly period dependent $s \propto P^{-0.23}$.

Finally, let us explore the consequences of period dependence of the emission radii expressed in equation (6). It is believed that the region of pulsar radio emission lies well within the light-cylinder radius $R_{LC} = cP/2\pi$, so the magnetic field can fully control the relativistic motion of charged particles. This can be expressed approximately by the condition $r/R_{LC} < 0.1$, where according to equation (6)

$$\frac{r}{R_{LC}} \approx 10^{-2}P^{-1/2}. \tag{7}$$

In order to have the magnetic field in the emission region around 10^7 Gauss (Table 1), one has to express the pulsar period P in equation (7) through the spin-down magnetic field $B_\circ = 3.2(P \cdot \dot{P})^{1/2}$ Gauss. Hence, the condition for observing radio pulsars has the form

$$3.2 \cdot 10^{19}\dot{P}^{1/2}/B_\circ < 10. \tag{8}$$

Since in the ordinary pulsars $\dot{P} > 10^{-17}s/s, B_\circ > 10^{10}$ G. On the other hand, in the millisecond pulsars $\dot{P} < 10^{-18}$ s/s and thus $B_\circ < 4 \cdot 10^9$ G. This means, that the magnetic field between 10^9 and 10^{10} Gauss should not be generally observed. This apparent gap in the spin down magnetic fields seems really to exist.

I would like to acknowledge receipt of an Alexander von Humboldt Fellowship. I thank E. Gil for technical assistance and editorial work. This paper is supported in part by the Polish KBN Project 2 0385 9101.

References

Backer, D. C. 1976, *Ap. J.*, **209**, 895
Cordes, J. M. 1978, *Ap. J.*, **222**, 1006
Craft, H. D. 1970, Ph. D. thesis, Cornell University, Ithaca, New York
Gil, J. A. 1981, *Acta Phys. Pol.*, **B12**, 1081
Gil, J. A. 1984a, *Astr. Ap.*, **132**, 312
Gil, J. A. 1984b, *Astr. Ap.*, **131**, 67
Goldreich, P. and Julian, W. H. 1969, *Ap. J.*, **157**, 869
Hankins, T. H., and Wolszczan, A. 1987, *Ap. J.*, **318**, 410
Komesaroff, M. N. 1970, *Nature*, **225**, 612
Radhakrishnan, V., and Cooke, D. J. 1969, *Ap. Lett.*, **3**, 225
Rankin, J. M. 1983, *Ap. J.*, **274**, 333
Rankin, J. M. 1986, *Ap. J.*, **301**, 901
Rankin, J. M. 1990, *Ap. J.*, **352**, 314
Rankin, J. M. 1992a, *Proceedings of IAU Colloquium #128 "The Magnetospheric Structure and Emission Mechanisms of Radio Pulsars , Łagów, Poland, June 1991* (Hankins, T. H., Rankin, J. M., and Gil, J. A., eds.; Univ. Zielona Góra Press), p. 133
Rankin, J. M. 1992b, *Ap. J.*, submitted
Taylor, J. H., Huguenin, G. R., and Hirsch, R. M. 1971, *Ap. Lett.*, **9**, 205

Main Pulse-Interpulse Correlation in PSR 1702-19

James D. Biggs

Universities Space Research Association, NASA/Goddard Space Flight Center

Abstract

At radio frequencies, the average profile for PSR 1702–19 has a very large proportion of circular polarization in both its main pulse and interpulse. This, together with the approximately 180° separation between main pulse and interpulse and nearly 90° separation between magnetic and rotation axes, suggests that we are observing emission from near the two magnetic poles of this pulsar. In a study of its individual radio pulses, we find evidence for a correlation between the emission from the main pulse and interpulse. We suggest that this is evidence for 'communication' between its two magnetic poles - an effect that has also been observed for PSR 1055-52 and PSR 1822-09.

1 Introduction

The spin parameters of PSR 1702–19 are within the normal range for the population of observed pulsars. Its 299 ms pulse period is slightly shorter than the median for the population of known pulsars, and the values for its characteristic age and surface magnetic field, 1.1×10^6 yr and 1.1×10^{12} G, respectively, are not far from the median value for these quantities.

The pulse structure of PSR 1702–19 has been studied in detail by Biggs *et al.* (1988). It is somewhat unusual because it has a substantial interpulse that emits about 20% of the total pulsed energy. This interpulse is separated from the main pulse by 181° over the frequency range of 408 to 1420 MHz. Also, the upper limit to any 'bridge' of emission connecting the main and interpulse is low, and must have an average flux density $\lesssim$4% of that for the peak. Furthermore, polarization studies have revealed strong right-hand circular polarization of up to 60% in both the main and interpulse.

Studies of the angle subtended by the rotation and magnetic axes (α) strongly suggest that PSR 1702–19 is very nearly an orthogonal rotator. Fitting the observed polarization position angle variation as a function of rotational phase to the Radhakrishnan and Cooke (1969) model resulted in $\alpha \approx 85°$ and 80° in the work of Malov (1986) and Lyne and Manchester (1988), respectively. Using a different method based on the geometry of core (those aligned about the magnetic axis) components, Rankin (1990) calculated $\alpha \approx 85°$.

2 Observations and Analysis

The observations described here were obtained using the Parkes 64-m telescope in 1981 May at 645 MHz with the system described by Biggs *et al.* (1985). Two dedispersed time series of 1070 and 2560 pulse periods duration were acquired for PSR 1702–19 with 400 samples per pulse period resolution. The receiver employed was sensitive to right-hand circular polarization only, and, given the polarization characteristics of PSR 1702–19, (Biggs *et al.* 1988) implies that the pulses studied here are more weighted to their circularly polarized components.

A detailed study of these individual pulses from PSR 1702–19 was not really possible due to the low signal-to-noise ratio of the relatively small datasets acquired. However, we attempted some analyses because of the significant interpulse, a feature that is relatively rare in average pulse profiles.

Average pulse profiles were obtained for the two datasets (see Fig. 1); they are very similar to those presented by Biggs *et al.* (1988). Thus the polarization characteristics of the receiver were not too detrimental to the observations. These profiles were used to define phase intervals corresponding to the main pulse and interpulse. Time series were formed from the total energy in the main and interpulse for each pulse period and were used in the analysis described below.

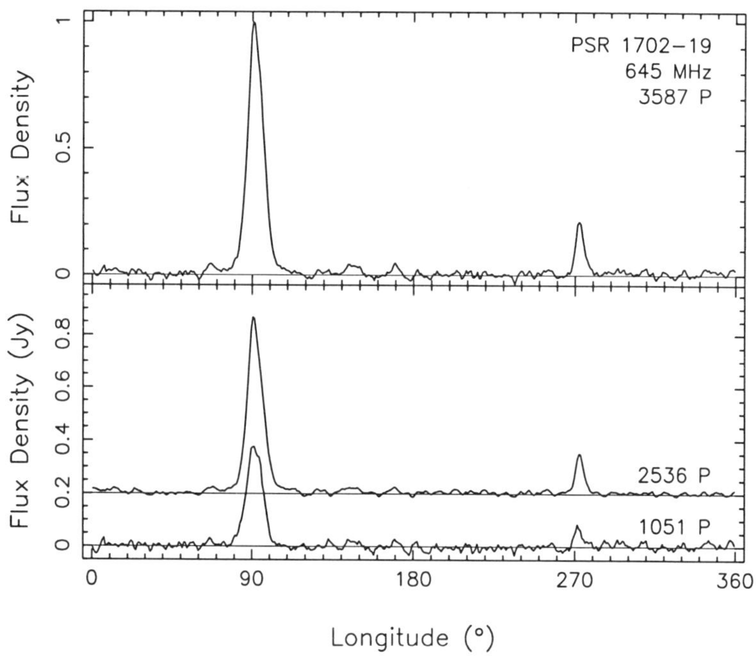

Fig. 1—Lower panel: Average profiles for the two 645 MHz observations (start times 81/05/13 12:05 and 81/05/16 13:49), offset for clarity. Upper panel: The average profile derived from the addition of the two normalized profiles weighted by the number of their pulses. All profiles are aligned so that their peak flux density occurs at 90° longitude. The sample time is 0.7 ms (0.9°), the time constant smoothing is 0.5 ms, and the dispersion broadening is much less than this.

The temporal relationship between the energy emitted from the main and interpulse was examined using the time-lagged Spearman's rank cross-correlation coefficient. A significant correlation was found around 0 P lag (see Fig. 2) for the longer observation. This correlation implies that a strong main pulse is generally followed by a strong interpulse. No significant correlations were found for the shorter observation.

A significant periodicity was noticed in the time-lagged cross-correlation coefficient for the longer observation, so a fluctuation spectral analysis was undertaken by applying a FFT to each time series. A significant quasi-periodic oscillation near 11 P with Q ($=\nu/\Delta\nu$) $\approx$ 1.3 was detected in the main pulse energy time series from both observations and is displayed in Figure 3. No significant signals were found in the analysis of the interpulse energy time series. The 11 P quasi periodicity in the main pulse energy most probably accounts for the structure of the time-lagged cross-correlation coefficient for the longer observation, and its low Q implies that it is probably associated with core emission (Rankin 1986).

3 Discussion

Biggs *et al.* (1988) have argued on the basis of the phase separation of the major components of the pulse profile, the lack of a bridge of emission between main and interpulse, and the relatively large proportion of circular polarization in its average profile that the main and interpulse of PSR 1702–19 are associated with the opposite magnetic poles of this pulsar. This hypothesis is strongly supported by the α estimates of Malov (1986), Lyne and Manchester (1988), and Rankin (1990).

The correlation analysis suggests that activity in the main pulse is correlated with that in the interpulse of PSR 1702–19. This correlated activity was only found in the longer observation and is not discernible in the shorter observation probably because of its lower signal-to-noise ratio and smaller number of pulses. Recently, we have obtained more observations of this pulsar in order to address this uncertainty.

If the main pulse-interpulse correlation is verified, then this finding implies a form of communication between the two magnetic poles of PSR 1702–19. Similar correlations have been found for PSR 1055-52 (Biggs 1990) and PSR 1822-09 (Fowler and Wright 1982). The time scale for the correlation is $\approx$ 0.5 P for both PSR 1055-52 and PSR 1702–19. The agency for this communication is unclear, but this time scale is probably too short for communication along the 'open' magnetic field lines that connect the two magnetic poles (Biggs 1990).

The apparent communication between the magnetic poles of pulsars provides information concerning the 'global' nature of the emission mechanism and must have an important bearing on models of pulsar radio emission.

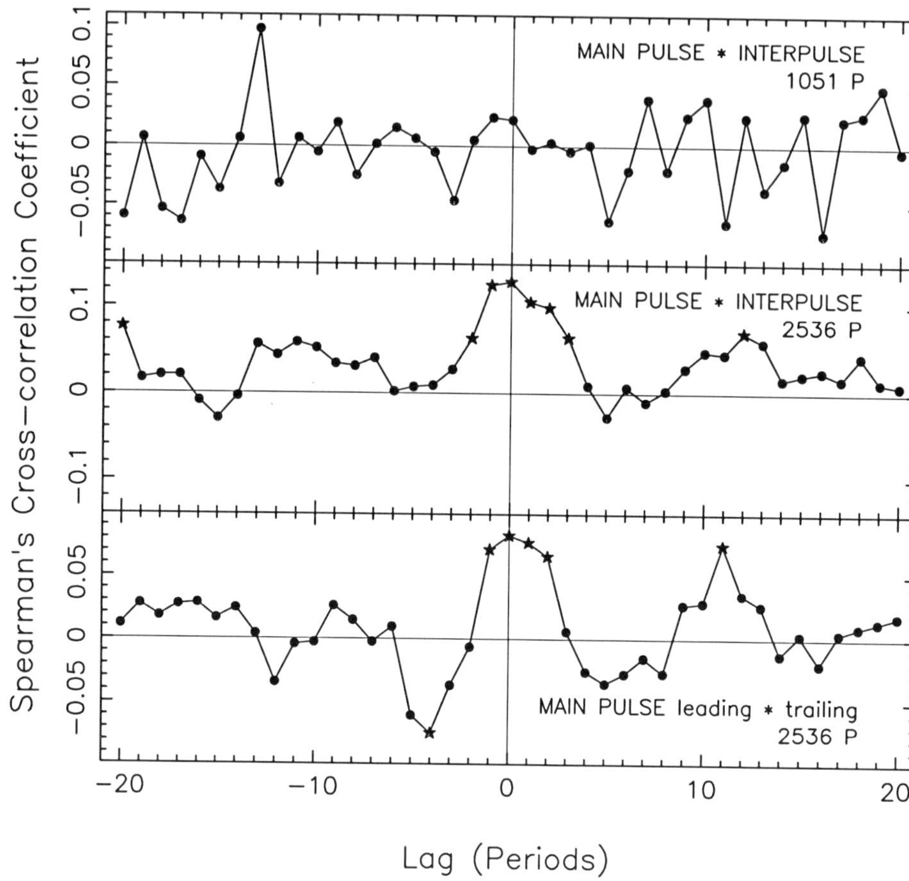

Fig. 2—Upper panel: Time-lagged Spearman's cross-corelation between the main pulse and interpulse for the shorter (and lower quality) observation. Middle panel: Time-lagged Spearman's cross-corelation between the main pulse and interpulse for the longer observation. Lower panel: Time-lagged Spearman's cross-corelation between the leading and trailing edges of the main pulse for the longer observation. The cross-correlation coefficients significant at > 99.9% are indicated by $\star$. Note the lack of significant correlations for the lower quality data (upper panel). Also, note the significant correlation at 0 P lag in the middle panel. The 10-15 P fluctuation in the main pulse is also clear in the middle and lower panels. Zero lag corresponds to aligning the main pulse with the interpulse, and the positive sense of lag brings the main pulse into alignment with later interpulses.

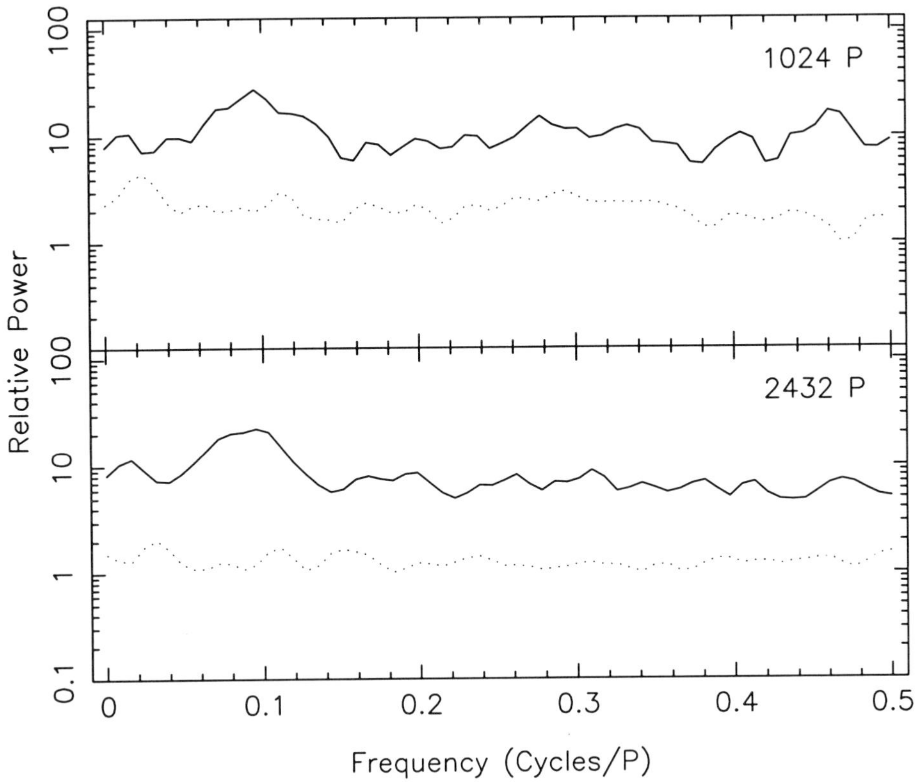

Fig. 3—Fluctuation spectra of the main pulse energy for the shorter and longer observations (upper and lower panels, respectively). Note the significant signal located at ≈ 0.09 cycles/P with Q (=$\nu/\Delta\nu$) ≈ 1.3. The frequency resolution is ≈ 0.008 cycles/P.

References

Biggs, J. D. 1990, *M. N. R. A. S.*, **246**, 341.

Biggs, J. D., Lyne, A. G., Hamilton, P. A., McCulloch, P. M., and Manchester, R. N. 1988, *M. N. R. A. S.*, **235**, 255.

Biggs, J. D., McCulloch, P. M., Hamilton, P. A., Manchester, R. N., and Lyne, A. G. 1985, *M. N. R. A. S.*, **235**, 255.

Fowler, L. A., and Wright, G. A. E. 1982, *Astr. Ap.*, **109**, 279.

Lyne, A. G., and Manchester, R. N. 1988, *M. N. R. A. S.*, **234**, 477.

Malov, I. F. 1986, *Astrofizika*, **24**, 307.

Radhakrishnan, V., and Cooke, D. J. 1969, *Ap. Letters*, **3**, 225.

Rankin, J. M. 1986, *Ap. J.*, **301**, 901.

Rankin, J. M, 1990, *Ap. J.*, **352**, 247.

Self-Consistent Pair-Production Discharges

F. Curtis Michel

Rice University

Abstract

A single electron accelerated in a vacuum gap near a pulsar will produce a cascade terminating in a highly charged bunch capable of intense coherent radio-emission. We report on early results where we have made one-dimensional self-consistent simulations of how these discharges develop. The goal of such simulations is to characterize the discharges and to see how they modulate themselves. Even in one-dimension, the discharges are highly chaotic in nature, suggestive of the micropulse phenomenon.

1 Introduction

We will assume a rudimentary familiarity with the role of nonneutral plasmas in neutron star magnetospheres (Michel 1991a). An important feature of such plasmas is that pulsar magnetospheres can support stable vacuum gaps while the star itself is sheathed with a torus of positively charged nonneutral plasma in the magnetic equatorial zone and a pair of domes of negatively charged nonneutral plasma over the magnetic polar caps. Here the spin axis is assumed to be near the magnetic dipole axis. The opposite sense of rotation would produce oppositely charged torus and domes, and we chose this one for definiteness (Fig. 1).

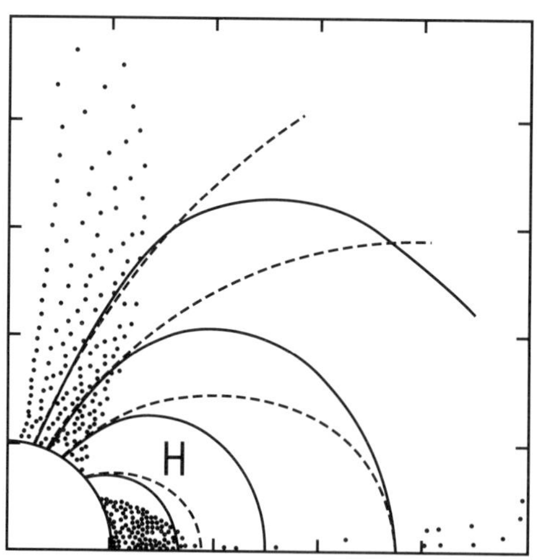

Fig. 1—Distribution of nonneutral plasma in a numerical simulation (Krause-Polstorff and Michel 1985a,b). Dashed lines are dipole magnetic field lines, solid lines are equipotentials, dots above pole (spin and magnetic axes parallel and up) represent negative charge whereas dots in equatorial region represents positive charge. Note discontinuous drop in density to vacuum conditions, leaving a huge gap (H).

Such a configuration is highly charged and electrons moving in such strong electric fields are easily accelerated to such high energies that they lose energy by curvature radiation (emission of gamma rays) as fast as they gain it by crossing equipotential surfaces ("radiation-reaction-limited acceleration"). The gamma rays in turn make increasing angles with the magnetic field as they progress forward and eventually pair-produce. In the meantime, the primary relativistic electron has propagated forward and is essentially along side the secondary pair. The secondary electron is quickly accelerated to the same Lorentz factor as the primary (order 10^8) while the positron halts and is accelerated away from the neutron star. As a result, the downward moving electrons exponentiate in numbers, forming a tight bunch, which is further concentrated by convergence of the magnetic field lines and domes, and we chose this one for definiteness.

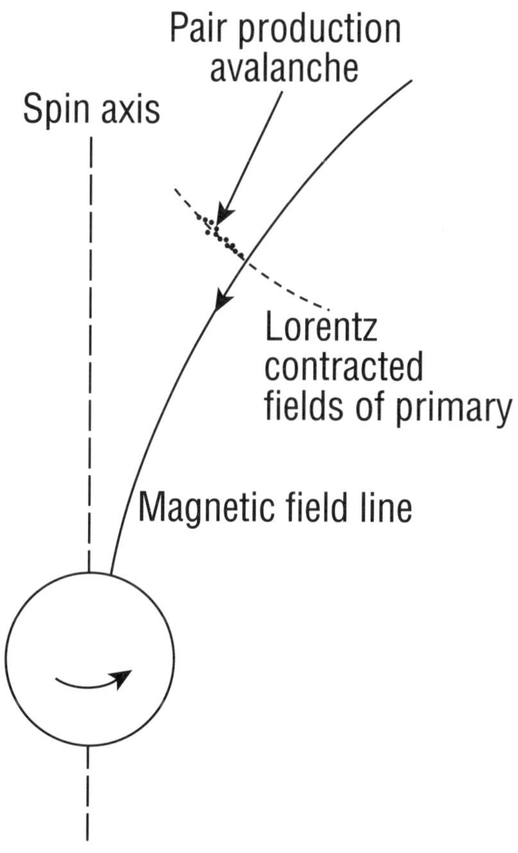

Fig. 2—Pair-production Avalanche. A primary electron following the curved magnetic field line curvature radiates energetic photons. This is represented by the curvature of the Lorentz-contracted fields of the electron, which approximate a sheet. Since the normal to the sheet has a component away from the electron, this portion is radiated. Owing to the angle between the sheet and neighboring magnetic field lines, pairs materialize out of the vacuum, with the secondary electrons continuing forward and the secondary positrons being accelerated oppositely.

A simple picture is that the EM fields of the primary electron are Lorentz contracted into a thin disk (a common model for bremsstrahlung); because the electron follows a curved path along the magnetic field line, this EM disk is also curved (physically representing the emission of photons in curvature radiation). It is essentially in this disk that the secondaries materialize as a result of the nonzero angle it makes with the neighboring magnetic field lines. The resultant avalanche (Michel 1991) is somewhat similar to a terrestrial air show made by a cosmic ray, where a single incident particle triggers the almost simultaneous arrival of a disk of penetrating particles (Čerenkov photons, muons, etc.) at the earth's surface.

2 The One-Dimensional Simulation

Here we report preliminary efforts to simulate the behavior of such a system. Importantly, the upward accelerated positrons will also form bunches, which in turn eject electrons, all of which tends to fill the system with bunches of varying sizes. The discharge in turn shields the accelerating charge and acts to reduce the accelerating field.

Our efforts here are significantly less ambitious but parallel that of two earlier efforts (Al'bert, Krotova, and Ehjdman 1975; Fawley 1978) Here we follow a "one-dimensional" (the distance along a uniformly curved magnetic field line) time development of electrons and positrons as they move in a uniform accelerating electric field. We already see two idealizations to simplify the problem: uniform curvature and uniform electric field. This system can be digitized to run easily on a small computer (in contrast, the above authors attempted digitalization of the parallel continuum equations). First, each radiation-reaction-limited particle can be taken as moving at uniform velocity c. Second, the spectrum of gamma rays emitted can be idealized into the emission of a single gamma-ray with an energy near the high frequency cutoff, with the rate being equal to the energization rate. In the latter case, there will be a mean distance between emission of successive gamma rays that forms our quantized scale size and defines unit distance in the simulation. For pulsar parameters, the gamma ray will go a large number of scale distances before pair-creation, so the natural choice will be to assume an integer number of scale distances for that distance. Third, the pair formed is taken to be an electron with the same energy as the primary and a positron moving the opposite direction with that energy. All of these simplifications could be relaxed, but for a first-look analysis it is simplest to keep things as simple as possible.

With these simplifications, one requires a numerical array consisting of the acceleration region divided into D scale steps, with a certain number of electrons and positrons being found at each location at a given instant. The obvious time interval with which to advance the time steps is just the scale step divided by c. Thus the array at a later time is given by shifting up and down the electrons and positron numbers and adding pairs at the appropriate places from the Pth previous array where P is the number of steps to pair-produce. Thus $2 \times P \times D$ storage is required, and by recycling arrays, the system can be propagated arbitrarily far into the future.

Positron trapping can be imposed by fiat simply by freezing the electron number at some predetermined value Q and setting the positron number thereafter equal to zero. The quantity Q thereafter represents not the number of particles but the (fixed) net charge of the bunch.

Figure 3 shows a simulation with $P = 3$, $N = 17$, and $Q = 88$, with distance steps to the left and time increasing down. Here the system has been started with a single electron at the top of the acceleration zone (the pairs of numbers one above

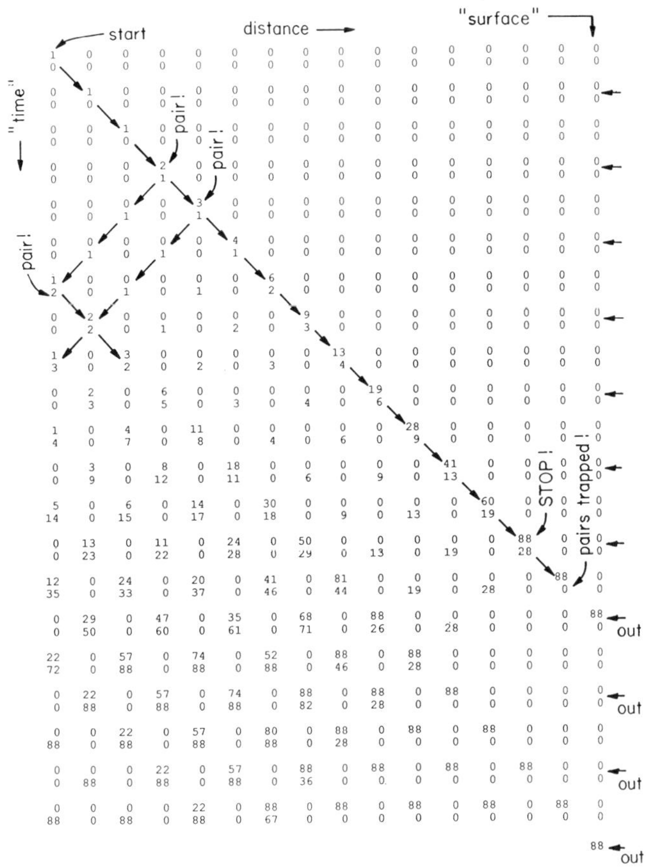

Fig. 3—Detailed program output of paired electron/positron numbers at distinct positions (across) and at distinct times (advancing downward). Electrons move at a 45° angle down to the right while positrons move at a 45° angle down to the left.

the other refer to electron numbers and positron numbers at the indicated time and location). As time progresses, the electron simply moves from site to site until at site 3 the "gamma-ray" emitted at site 1 pair-produces and yields $^{2}_{1}$, indicating 2 electrons and one positron. In the next time step the positron moves left while the electron moves to the right, and the gamma emitted at site 2 pair-produces to give $^{3}_{2}$. Because the number of emitted gammas is equal to the electron number, one gets a "delayed exponentiation": (1, 1, 1, 2, 3, 4, 6, 8, 13, etc.) along the trajectory of the primary electron. When the primary electron approaches the bottom of the acceleration zone, it finally exceeds Q, and the electron value is henceforth fixed at Q. Clearly D, P, and Q are not completely independent because a long acceleration region will always result in bunches of size Q well before exiting the region. Then the remainer of the accelerating region is just a drift region. In contrast, a short one will still produce bunches that grow to size Q after a few transit cycles because the downward electron bunch exiting will yield a positrons bunch moving upwards that continues to exponentiate. If no limiting Q is specified, the system exponentiates without bound even given a small value of D (as long as $D > P$, of course).

3 Results

Although essentially linear, this system is highly chaotic in its dependence on D, P, and Q. The operation can be gauged simply by the consecutive numbers of electrons exiting the acceleration region. One might naively suppose that the electron bunches will all have size Q, but if D is chosen so that exponentiation barely takes place in one pass, more complex behavior is obtained as shown in Figure 4.

0	0	0	0	0	0	0	0	88	0	0	88	88	88	88
88	88	0	0	88	88	88	88	88	88	88	88	88	88	88
88	88	88	88	88	88	88	0	88	88	88	88	88	0	0
88	88	88	88	88	88	88	88	0	0	88	88	88	88	88
0	0	88	88	88	88	0	88	88	88	88	88	88	0	88
88	88	88	88	88	88	88	88	88	88	0	88	88	88	88
88	88	88	88	88	88	88	88	88	88	88	88	88	88	88
88	88	88	88	88	88	88	88	88	88	88	88	88	88	88
88	88	88	88	88	88	88	88	88	88	88	88	88	88	88
88	88	88	88	88	88	88	88	88	88	88	88	0	88	88
88	88	88	88	88	88	88	88	88	88	88	88	88	88	88
88	88	88	88	88	88	88	88	88	88	88	88	88	88	88
0	88	88	88	88	88	88	88	88	88	88	88	0	88	88
88	88	0	88	88	88	88	88	88	88	88	88	88	88	24
88	88	88	88	88	88	88	88	88	88	88	88	88	88	88
88	88	88	88	88	0	0	88	88	88	88	88	0	88	88
88	88	88	88	88	88	88	0	88	88	88	88	88	88	88
88	88	88	88	88	88	88	88	88	88	88	88	88	88	88
88	88	88	88	0	0	88	88	88	88	88	88	0	88	88
88	88	88	88	0	88	88	88	88	88	88	88	88	88	0
88	88	88	88	88	88	88	88	88	88	88	88	88	88	88
0	0	88	88	88	88	0	76	88	88	88	88	88	88	88
88	88	88	88	0	88	88	88	88	88	88	88	88	88	88
88	88	88	88	88	88	88	88	88	88	88	88	88	88	88
88	88	88	88	88	88	88	88	88	88	88	88	88	88	88
88	88	88	88	88	88	88	88	88	88	88	88	88	88	88
88	88	88	88	88	88	88	88	88	88	88	88	88	88	88
88	88	88	88	88	88	88	88	88	88	88	88	88	88	88
88	88	88	88	88	88	88	88	88	88	88	88	88	88	88
88	88	72	88	88	88	88	88	88	0	0	88	88	88	88
88	88	88	88	88	88	88	88	88	88	88	88	0	88	88
88	88	88	88	88	88	88	88	88	88	0	88	88	0	88
88	48	88	88	88	88	88	88	88	88	88	88	88	88	88
88	0	0	0	0	0	0	0	0	0	0	0	0	0	0
0	0	0	0	0	0	0	0	0	0	0	0	0	0	0
0	0	0	0	0	0	0	0	0	0	0	0	0	0	0
0	0	0	0	0	0	0	0	0	0	0	0	0	0	0
0	0	0	0	0	0	0	0	0	0	0	0	0	0	0
0	0	0	0	0	0	0	0	0	0	0	0	0	0	0
0	0	0	0	0	0	0	0	0	0	0	0	0	0	0
0	0	0	0	0	0	0	0	0	0	0	0	0	0	0
0	0	0	0	0	0	0	0	0	0	0	0	0	0	0
0	0	0	0	0	0	0	0	0	0	0	0	0	0	0
0	0	0	0	0	0	0	0	0	0	0	0	0	0	0
0	0	0	0	0	0	0	0	0	0	0	0	0	0	0
0	0	0	0	0	0	0	0	0	0	0	0	0	0	0
0	0	0	0	0	0	0	0	0	0	0	0	0	0	0
0	0	0	0	0	0	0	0	0	0	0	0	0	0	0
0	0	0	0	0	0	0	0	0	0	0	0	0	0	0
0	0	0	0	0	0	0	0	0	0	0	0	0	0	0

Fig. 4—Output of program with just the exiting electrons listed at consecutive time steps (actually every other time step, because the intervening one is always empty) in the usual left-to-right, top-to-bottom reading frame. Thus the sequence of numbers across the top is the same as the arrowed values down the right side of Fig. 3. Note that the bunch size is occasionally less than Q, and in fact terminates for this particular case after a large number of steps.

Here we see that very occasionally either no bunch exits the system or that an incomplete bunch is formed. More unexpected, having been initiated by a *single* electron, the entire discharge clears at the 465th time step!

We have searched some of the D, P, Q parameter space and find that having a discharge "die" is unrepresentative. More commonly, one obtains either a constant electron flux or one that oscillates periodically through a fixed succession of values. The parameter space is actually larger in that the behavior can depend on where the initial electron is introduced! Or whether an electron-positron pair initiates the activity at some level. Thus fixed values of D, P, Q may give substantially different behaviors depending on how the system was excited.

There is some minor uncertainty in determining what happens if positron trapping occurs at the same time that a downward moving electron bunch passes an upward positron bunch, but this uncertainty can be resolved entirely by conserving charge. This compromise can require that upward moving positrons can be incorporated into a downward moving bunch, and vice versa, which is problematical physically, but is largely forced on us unless we also keep track of particle energies. In a three dimensional system, bunches at the same altitudes could be widely separated, as well.

Increasing D would, in principle, always provide enough room for full bunch formation, but test particle bunching is actually somewhat marginal in model pulsar magnetospheres (Michel 1991b) in the sense that D is not huge compared to P. The underlying chaotic behavior is not inconsistent with actual pulsar emission, which is often characterized by intense "micropulses."

References

Al'bert, Ya. I., Krotova, Z. N., and Ehjdman, V. Ya. 1975, Astrfizika, 11, 28 (translated in Astrophysics, 11 (No. 2), 189).
Fawley, W. M. 1978, University of California at Berkeley Ph. D. Thesis.
Krause-Polstorff, J., and Michel, F. C. 1985a, Astr. Ap., 144, 72.
Krause-Polstorff, J., and Michel, F. C. 1985b, M.N.R.A.S., 213, 43p.
Michel, F. C. 1991a, Theory of Neutron Star Magnetospheres (Chicago: University of Chicago Press), p. 241.
Michel, F. C. 1991b, Ap. J., 383, 808

Modeling of Pulsar Magnetospheres

Heinz Herold, Bernhard Hofmann, and Hanns Ruder

Theoretische Astrophysik, Universität Tübingen

Abstract

The magnetosphere of a rapidly rotating, strongly magnetized neutron star with aligned magnetic and rotational axes (parallel rotator) is modeled numerically. Including the radiation of the particles accelerated to relativistic energies as an efficient damping mechanism, we obtain a quasi-stationary self-consistent solution to this classical problem. The numerical simulation, which was started from the well-known vacuum solution, yields a global magnetospheric structure that can be characterized by two regions of oppositely charged particles, which eventually produce a relativistic pulsar wind, separated by a vacuum gap of considerable extent.

1 Introduction

The problem — fundamental to the physics of pulsars — of determining the global structure of the magnetospheres of rapidly rotating, strongly magnetized neutron stars has not yet been solved self-consistently (for a review cf. Michel 1982), but it must be solved before any conclusive interpretation can be given to the large amount of detailed experimental data on pulsar radiation (cf. *e.g.*, Taylor and Stinebring 1986, and references therein). Apart from this reason, it is "*. . . just intolerable that we should not know how a rotating magnetized neutron star comes to terms with its environment* " (Mestel 1981).

We concentrate here on the self-consistent modeling of the magnetosphere of an aligned rotator, where the rotation axis is parallel to the magnetic axis of the neutron star. In this case, the homopolar induction, which should be responsible for populating the magnetosphere with charged particles pulled out from the neutron star surface via field emission (Goldreich and Julian 1969), can be studied in purity, whereas electromagnetic wave effects are neglected. The main questions to be answered are the following: What is the global structure of the electric and the magnetic field? How are the charge and current density distributed that produce these fields (together with the star)? Which particle populations exist in the magnetosphere?

2 Basic Equations

2.1 Maxwell and Vlasov Equations

In the stationary axially symmetric case the electromagnetic fields **E** and **B** in the pulsar magnetosphere can be described (in cylindrical coordinates ρ, φ, z) by the electrostatic potential $\Phi(\rho, z)$, the magnetic flux function $\Psi(\rho, z)$, which determines the poloidal magnetic field, and the toroidal magnetic field $B_\varphi(\rho, z)$: $\mathbf{E} = -\nabla\Phi$ and $\mathbf{B} = (1/\rho)\nabla\Psi \times \mathbf{e}_\varphi + B_\varphi \mathbf{e}_\varphi$. The charge density ρ_e and the current density **j** determine the electric potential via Poisson's equation and the magnetic field via Ampere's law, which here read in suitable units such that all quantities are dimensionless:

$$\Delta\Phi = -\rho_e \; ; \qquad \left(\Delta - \frac{2}{\rho}\frac{\partial}{\partial\rho}\right)\Psi = -\rho j_\varphi \tag{1a}$$

$$\frac{\partial}{\partial z}(\rho B_\varphi) = -\rho j_\rho \; ; \qquad \frac{\partial}{\partial \rho}(\rho B_\varphi) = \rho j_z \tag{1b}$$

The magnetosphere is formed by a collisionless plasma, in which the particles are expected to be extremely relativistic due to the huge electric and magnetic fields (the homopolar induction voltage between pole and equator of the neutron star is typically of the order 10^{17} - 10^{18} V). The charge and current densities are derived from the zeroth and first momentum of the particle distribution function (for the sake of clarity, summing up over the particle species has been omitted here)

$$\rho_e = \int f(\mathbf{r}, \mathbf{p})\, d^3p \; ; \qquad \mathbf{j} = \int \mathbf{v} f(\mathbf{r}, \mathbf{p})\, d^3p \, , \tag{2}$$

where the distribution function $f(\mathbf{r}, \mathbf{p})$ is determined by the Vlasov equation

$$\mathbf{v}\frac{\partial f}{\partial \mathbf{r}} + \frac{\partial}{\partial \mathbf{p}}\left[\left(\mathbf{E} + \mathbf{v} \times \mathbf{B} + \begin{matrix}\text{radiation}\\ \text{damping}\end{matrix}\right) f\right] = 0 \, . \tag{3}$$

The velocity **v** is given by $\mathbf{v} = \mathbf{p}/\sqrt{(\varepsilon^2 + \mathbf{p}^2)}$, where the dimensionless parameter ε is defined by

$$\varepsilon = \frac{2mc^2}{eB_0 a^2 \Omega} \, , \tag{4}$$

i.e., by the ratio between rest mass energy and homopolar induction energy (B_0 is the polar magnetic field strength, a is the radius, and Ω is the angular velocity of the neutron star). Because typical values for ε are extremely small ($\varepsilon \sim -10^{-12}$ for electrons, $\varepsilon \sim 10^{-9}$ for protons), the particles are expected to become extremely relativistic, since their maximum Lorentz factors should be comparable with $|\varepsilon|^{-1}$, at least if the radiation reaction during phases of acceleration can be neglected. Although this is not the case as will be discussed in the following, it is nevertheless convenient to use a scaled Lorentz factor $\Gamma = \varepsilon\gamma$ that never exceeds unity.

2.2 Particle Motion in the Limit of Strong Radiation Damping

Studying the trajectories of particles in realistic pulsar fields, one can get an idea of how strongly the radiation reaction does affect the particle motion.

The Lorentz-Dirac equation of motion in the Landau approximation (cf. Landau and Lifshitz 1975) can be written (for $|\varepsilon| \ll 1$) as

$$\dot{\mathbf{v}} = \frac{1}{\Gamma}\mathbf{F}; \quad \dot{\Gamma} = \mathbf{E}\cdot\mathbf{v} - D_0\Gamma^2\mathbf{F}^2 \tag{5a}$$

with

$$\mathbf{F} = \mathbf{E} + \mathbf{v}\times\mathbf{B} - \mathbf{v}(\mathbf{E}\cdot\mathbf{v})\,. \tag{5b}$$

The dimensionless damping constant D_0, which is defined by

$$D_0 = \frac{2}{3}\frac{e^2}{4\pi\epsilon_0}\frac{\Omega}{mc^3}\frac{1}{\varepsilon^3}\,, \tag{6}$$

characterizes the strength of the radiation reaction. For typical pulsar parameters, its value is of the order of $D_0 \sim -10^{14}$ for electrons, and $D_0 \sim 10$ for protons. Thus, at least for the electrons, the radiation reaction force dominates the particle motion. Large values of $|D_0|$ imply that the factor of D_0 in (5a) always remains very small; this leads us to the assumption, which was confirmed by numerical integration of equations (5) (cf. Herold *et al.* 1985), that during the motion the condition $\mathbf{F} \approx 0$ is fulfilled, which means that the radiation reaction is locally minimized along the trajectory. This is a condition for the velocity and yields, for given **E** and **B** fields

$$\mathbf{v} = \frac{1}{\mathbf{B}^2 + P^2}\Big[\mathbf{E}\times\mathbf{B} + \frac{1}{P}(\mathbf{E}\cdot\mathbf{B})\mathbf{B} + P\mathbf{E}\Big] \tag{7a}$$

with

$$P^2 = \frac{1}{2}(\mathbf{E}^2 - \mathbf{B}^2) + \frac{1}{2}\Big[(\mathbf{E}^2 - \mathbf{B}^2)^2 + 4(\mathbf{E}\cdot\mathbf{B})^2\Big]^{\frac{1}{2}}\,. \tag{7b}$$

Thereby, we have obtained a local velocity field $\mathbf{v} = \mathbf{v}(\mathbf{E},\mathbf{B})$ and thus a fluid-like picture for the particle motions in the magnetosphere. The main characteristics of this "damping-free" motion is that the particles try to come to rest in surfaces where $\mathbf{E}\cdot\mathbf{B} = 0$. How such strong radiation reaction effects influence the particle motions in the case of an *oblique* rotator, has been discussed by Finkbeiner *et al.* (1989).

3 Numerical Approach

Based on these results, the task to determine a self-consistent solution is simpler than before, but still difficult due to the great non-linearity of the problem. Our approach is based on the idea to start from the vacuum solution and to fill up the magnetosphere with the particles that are pulled out from the neutron star surface. This is not a real time-dependent calculation since we assume that the electric field is always described by an electrostatic potential, but for transporting the charge with the velocity (7) we must solve the time-dependent continuity equation

$$\dot{\rho}_e + \mathrm{div}(\rho_e\mathbf{v}) = 0\,, \tag{8}$$

and therefore $\dot{\mathbf{E}}$ cannot be omitted in Ampere's law. Thus equations (1b) have to be replaced by

$$\Big(\Delta - \frac{2}{\rho}\frac{\partial}{\partial\rho}\Big)(\rho B_\varphi) = \rho\Big(\frac{\partial j_z}{\partial\rho} - \frac{\partial j_\rho}{\partial z}\Big)\,. \tag{9}$$

Actually, for each particle species the corresponding continuity equation is solved, as the velocity (7a) depends on the particle charge: the sign of the quantity P ($P = \mathbf{E} \cdot \mathbf{v}$) from (7b) has to be identical to the charge sign. The charge and the current density are calculated by summation over both particle species.

In summary, we solve at each time step the elliptic equations (1a) with Dirichlet boundary conditions for Φ and Ψ on the star — we assume a homogeneously magnetized neutron star — (the change of Φ at infinity has to be derived from Ampere's law) and equation (9) with the Neumann boundary condition $\partial(\rho B_\varphi)/\partial r = -\rho j_\vartheta$ on the star's surface, where for the outcoming current the simple emission law $j = \sigma E_\parallel$ is assumed, with a constant value of σ. The other boundary conditions (at infinity and at the rotational axis) are given either by the rotational symmetry of the problem with respect to the axis of rotation or by Taylor expansion of the field equations. To reduce the computational effort, we use the equatorial plane as another boundary (symmetry for Φ and Ψ, antisymmetry for B_φ).

The method of solution of the three elliptic equations was originally a vectorizable SOR scheme, but because this code was too slow for an extended parameter study, we changed to a multigrid method (cf. Hackbusch 1985). We found that for our problem a linewise Gauß-Seidel scheme for the "smoother" part of the multigrid method works best.

For the continuity equations (8) an explicit discretization in time with 2-dimensional Flux Corrected Transport (cf. Zalesak 1979) is used in order to preserve steep gradients in the charge density. We have developed a Tensor-Viscosity Donor-Cell FCT scheme (TVDC-FCT), which minimizes the numerical viscosity (for the TV part as high-order transport, cf. Dukowicz and Ramshaw 1979). For details the reader is referred to Ertl (1988).

4 Results

The results described in the following were obtained for a rapid rotator with $r_0 = (\Omega a)/c = 0.1$ as ratio of stellar radius and light cylinder radius (cf. also Ertl 1988, Herold *et al.* 1989, 1992). We start with the vacuum fields of an uncharged aligned rotator, *i.e.*, an electric quadrupole field and a magnetic dipole field resulting in $\mathbf{E} \cdot \mathbf{B} \leq 0$ everywhere. In Figure 1 we present the initial configuration of the electric field.

Thus, as long as $\mathbf{E} \cdot \mathbf{B} \leq 0$ at the surface, only negative particles can be emitted and then transported along the magnetic field lines towards the equator where they accumulate. This accumulation causes a change in the electric field such that the $\mathbf{E} \cdot \mathbf{B} = 0$ surface rises from the equator towards the polar field line forcing the negative particles to follow it. As soon as $\mathbf{E} \cdot \mathbf{B} > 0$ at some part of the star's surface, positive particles can enter the magnetosphere.

Then a steadily increasing cloud of positive particles is developing in the equatorial region near the star, whereas the negative particle distribution evolves more and more into a cone-like structure above the magnetic pole. During this phase the electric field changes its structure considerably.

After about one revolution the magnetosphere has developed a structure which already resembles the final situation, although one has to follow the evolution for a

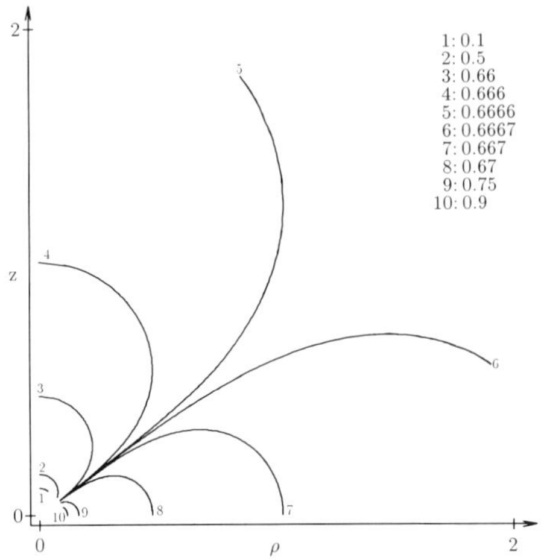

Fig. 1.—Initial phase: contour lines of the electrostatic potential Φ, depicting the quadrupole electric field.

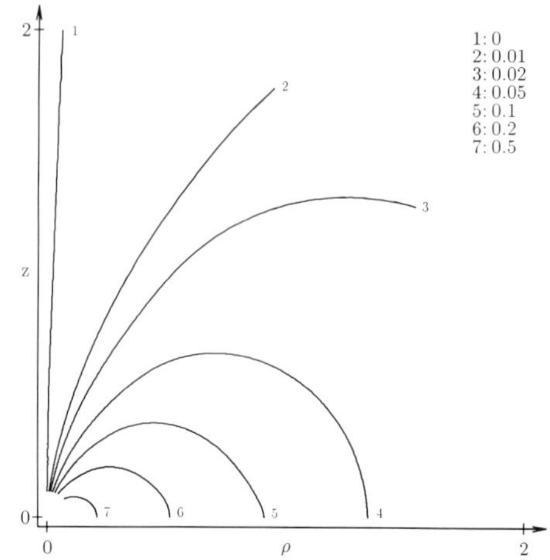

Fig. 2.—Quasi-stationary end phase: contour lines of the electrostatic potential Φ. Note that the field configuration has changed considerably.

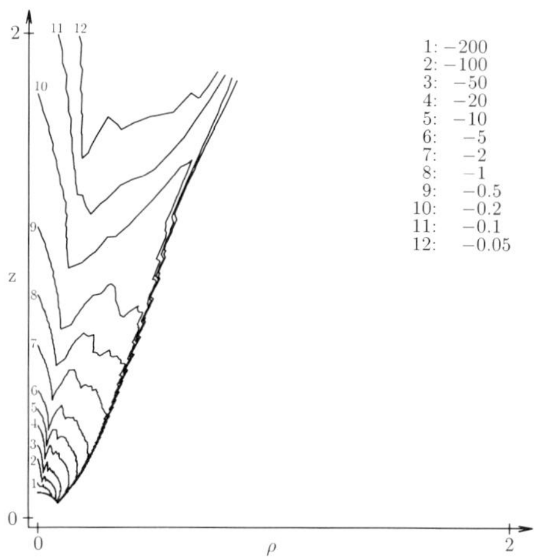

Fig. 3.—Quasi-stationary end phase: contour lines of the density of negative particles.

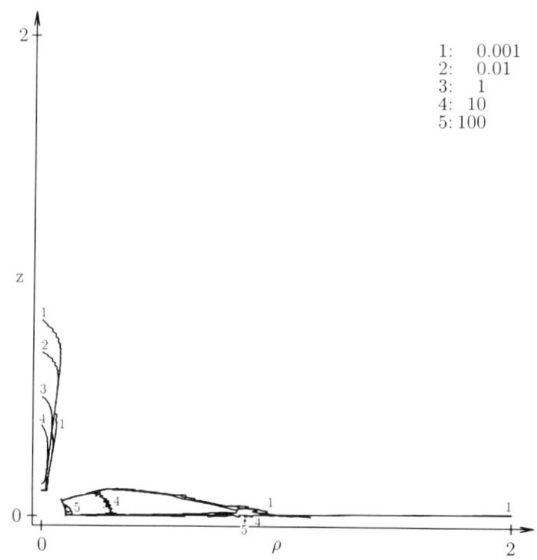

Fig. 4.—Quasi-stationary end phase: contour lines of the density of positive particles.

longer time in order to come close to stationarity. (As indicators for time independence we use various global quantities as the total charge of the star and the total charges of negative or positive particles in the magnetosphere as well as the total particle currents out of the star and at infinity.) This quasi-stationary end phase is depicted (after $\sim$ 30 revolutions; one revolution of the star corresponds to $t = 2\pi$) in Figures 2–6, where the contour lines of the electrostatic potential Φ, of the charge densities, of $\mathbf{E} \cdot \mathbf{B}$, and of $B_\varphi / |\mathbf{B}|$ are shown.

One recognizes two regions of charged particles (the negative ones in a cone around the polar field line, the positive ones around the equatorial plane) separated

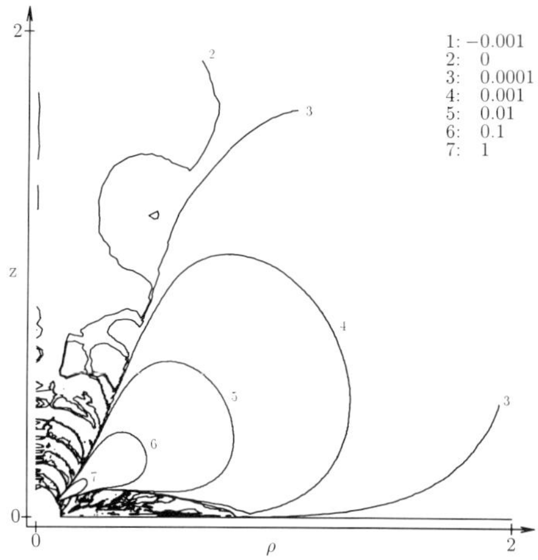

Fig. 5.—Quasi-stationary end phase: contour lines of $\mathbf{E} \cdot \mathbf{B}$. In the plasma regions the parallel electric field tends to become small.

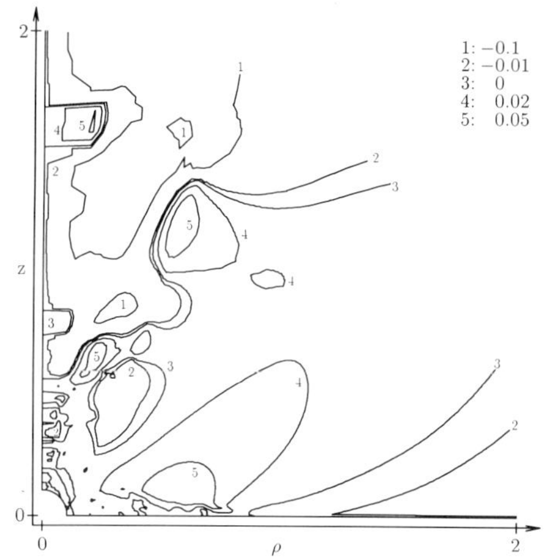

Fig. 6.—Quasi-stationary end phase: contour lines of $B_\varphi/|\mathbf{B}|$. Note that the toroidal component of the $\mathbf{B}$ field is relatively small compared to the poloidal component.

by a vacuum gap in between. (Vacuum gaps in non-self-consistent models are not new; for a more recent model with such a feature, cf. Mestel *et al.* 1985.) In this vacuum gap the parallel electric field $E_\| = (\mathbf{E} \cdot \mathbf{B})/|\mathbf{B}|$ is greatly different from zero (cf. Figure 5), whereas in the regions populated with plasma the electric field (cf. Figure 2), which is now totally dissimilar to the initial quadrupole (cf. Figure 1), has evolved as to achieve $\mathbf{E} \cdot \mathbf{B} \sim 0$ there. The poloidal magnetic field does not change very much, it remains essentially dipolar due to the smallness of the toroidal currents. The neutron star is now positively charged, and the positive and the negative currents out of the star exactly match each other.

From the velocity fields — which are not shown here — one can conclude that the positive particles corotate with the star (to a very good approximation) as long as they are inside the light cylinder, but close to the light cylinder they lose corotation and stream as through a nozzle into the outer parts of the magnetosphere forming a radial wind. A closer inspection of the forces which drive the positive particles near the equator inside the light cylinder across the magnetic field lines reveals that it is the radial component of the electric field which does this job, *i.e.*, that part of the velocity (7a) which is proportional to $\mathbf{E}$ produces the radial drift. Thus, this is a truly relativistic effect. The velocity of the negative particles is dominantly poloidal, and thus, also a negatively charged wind builds up at greater distances from the star.

There are still wave-like fluctuations, especially inside the negative zone, which seem to originate from the tendency of the system to make $\mathbf{E} \cdot \mathbf{B}$ small (or even zero) in the plasma regions. This also explains the strange folded structure of the $\mathbf{E} \cdot \mathbf{B} = 0$ surface and might be an indication that the magnetosphere wants to form an extended region with $\mathbf{E} \cdot \mathbf{B} = 0$; however, this is difficult to achieve with a relativistic particle population. Whether for this end a cool non-relativistic plasma population is necessary or whether such fluctuations are a physically real part of a quasi-stationary configuration is an open question. Globally, our solution represents a stationary structure; for instance, the charges of the star and of the magnetosphere remain, in the end phase, constant with great accuracy.

For the first time in pulsar magnetospheric theory it seems that a stationary self-consistent solution for the aligned rotator has been found by a quasi-time-dependent simulation. Detailed diagnostic investigations of this solution, such as the calculation of the radiation of the accelerated particles, as well as studies of the parameter dependencies (rotation frequency Ω, emissivity σ of the neutron star) are tasks for the future.

References

Dukowicz, J. K., and Ramshaw, J. D. 1979, *J. Comp. Phys.*, **32**, 71.
Ertl, T. 1988, PhD Thesis, Universität Tübingen.
Finkbeiner, B., Herold, H., Ertl, T., and Ruder, H. 1989, *Astr. Ap.*, **225**, 479.
Goldreich, P., and Julian, W. H. 1969, *Ap. J.*, **157**, 869.
Hackbusch, W. 1985, *Multi-Grid Methods and Applications* (Berlin: Springer).
Herold, H., Ertl, T., and Ruder, H. 1985, *Mitt. Astron. Ges.*, **63**, 174.
Herold, H., Ertl, T., Finkbeiner, B., and Ruder, H. 1989, in *Timing Neutron Stars*, ed. H. Ögelman and E.P.J. van den Heuvel (Dordrecht: Kluwer), p. 723.
Herold, H., Ertl, T., Finkbeiner, B., Ruder, H. 1992, in *The Magnetospheric Structure and Emission Mechanisms of Radio Pulsars*, IAU Colloquium No. **128**, ed. T.H. Hankins and J.A. Gil (Zielona Gora: University Press).
Landau, L.D., and Lifshitz, E.M. 1975, *The Classical Theory of Fields* (Oxford: Pergamon Press).
Mestel, L. 1981, in *Pulsars*, IAU Symposium No. **95**, ed. W. Sieber and R. Wielebinski (Dordrecht: Reidel), p. 9.
Mestel, L., Robertson, J.A., Wang. Y.-M., and Westfold, K.C. 1985, *MNRAS*, **217**, 443.
Michel, F.C. 1982, *Rev. Mod. Phys.*, **54**, 1.
Taylor, J. H., and Stinebring, D. R. 1986, *Ann. Rev. Astron. Astrophys.*, **24**, 285.
Zalesak, S. T. 1979, *J. Comp. Phys.*, **31**, 335.

Radiative Ejection in the Magnetosphere of Isolated Pulsars

Anton Chernenko and Igor Mitrofanov

Space Research Institute, Moscow, Russia

1 Introduction

In this paper we consider possible consequences of radiation pressure in the magnetosphere of Neutron Stars. It is shown to be important in modeling (A) acceleration processes in active pulsars since they emit X-rays (Ögelman 1992) and (B) ejection of plasma during cosmic gamma-ray bursts (GRB), which are believed to be generated by dead isolated pulsars. Presence of thermal-like X-ray flux was observed shortly before and after some GRBs (Murakami *et al.* 1992).

2 Radiative Force in the Magnetic Field

Let us consider an electron (or positron) moving along a line of a strong magnetic field $\vec{B}$ at speed $\vec{v}$. In presence of electromagnetic radiation the electron is affected by radiative force:

$$\vec{f} = \frac{1}{c} \cdot \int d\epsilon \int d\vec{n} \left[\vec{n} - \frac{\vec{v}}{c} \gamma^2 (1 - \beta x) \right] (1 - \beta x) \sigma_{SC} (\epsilon' (\epsilon, \vec{n}), \vec{n}' (\vec{n})) W (\epsilon, \vec{n}), \qquad (1)$$

where $W(\epsilon, \vec{n})$ is a spectral density of radiation propagating along a direction $\vec{n}$ inside a unit solid angle, $\sigma_{SC}(\epsilon', \vec{n}')$ is cross-section (C-S) of the electron, $\beta = v/c$, $\gamma = (1 - \beta^2)^{1/2}$, and $x = \cos(\vec{v}, \vec{n})$ (Mitrofanov and Tsygan 1982). Primed values correspond to the electron rest frame (ERF).

In the presence of the magnetic field the C-S near the cyclotron energy $\epsilon_B = \hbar\omega_B / mc^2$ (in this paper all energies will be expressed in units of mc^2) is dominated by cyclotron resonance. If we neglect collision broadening (it is correct if electron number density is $n \ll 2 \cdot 10^{22} B_{12} (kT/1\mathrm{keV})^{3/2}$ cm^{-3}) and thermal broadening, then for non-polarized radiation the resonant C-S takes the form (*e.g.*, see Mitrofanov and Pavlov 1982):

$$\sigma_r(\epsilon', x') = \frac{1}{4} \sigma_T (1 + x'^2) \frac{(\epsilon')^2}{(\epsilon' - \epsilon_B)^2 + (\epsilon'_r/2)^2} \qquad (2)$$

Here σ_T is Thompson C-S, $\epsilon'_r = 4\alpha\epsilon_B{}^2/3$ is the radiative width, $\epsilon' = \epsilon\gamma(1 - \beta x)$, and $x' = (x - \beta)/(1 - \beta x)$.

If in the ERF at ϵ_B there are enough photons, they should result in amplification of the radiative force. The physics of this effect is sketched in Figure 1 for black-body radiation. Maximum amplification occurs if cyclotron energy ϵ_B coincides with the maximum of Plank spectral density ϵ_T. For the electron moving with Lorentz-factor γ

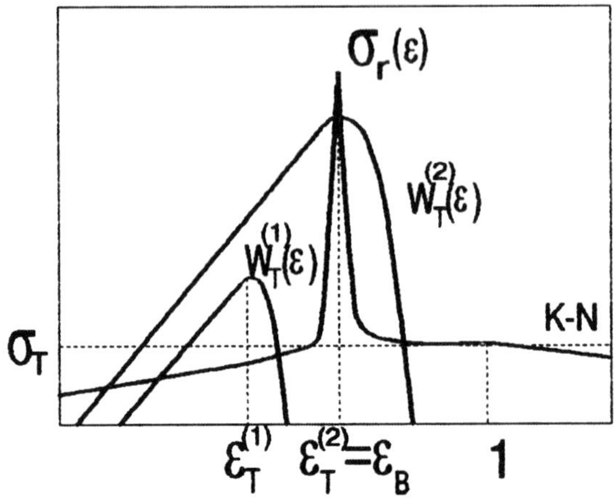

Fig. 1—Resonant magnetic Compton C-S σ_r and spectral density of black-body radiation W are plotted as a function of energy ϵ. The first curve corresponds to Plank spectrum with maximum spectral density at $\epsilon_T^{(1)} = 3kT_1/m_ec^2 \ll \epsilon_B$ whereas the second curve corresponds to Plank spectrum with maximum at $\epsilon_T^{(2)} \sim \epsilon_B$ $(T_1 < T_2)$.

this requires that $\epsilon_B \sim \gamma\, \epsilon_T$. Generally speaking, maximum amplification may occur in two cases provided cyclotron resonance determines the C-S:

- Optically thin plasma rests just above the heated spot on the surface of magnetized NS. Maximum radiative force takes place if the temperature of the spot T corresponds to

$$T_6 \sim 45 B_{12}. \tag{3}$$

  In this case Magnetic radiative force f^B exceeds the one corresponding to field free case $f^T = \sigma_T \sigma T^4/2c \sim 6 \cdot 10^{-16} T_6^4$ erg/cm by factor of $3 \cdot 10^5 T_6^{-1}$ (Mitrofanov and Pavlov 1982). This may lead to supercritical relativistic ejection of the plasma during GRBs in those cases when it would not occur from a NS without a magnetic field.
- Electrons or positrons are accelerated by electric force in presence of soft photons. Maximum radiative force which brakes the particles due to relativistic transformation of radiation angular diagram occurs if

$$\gamma \sim \epsilon_B/\epsilon_T \sim 45 B_{12} T_6^{-1}. \tag{4}$$

  If this condition is satisfied, the amplification factor equals to $2 \cdot 10^4 B_{12}^{-1}$ (Kardashev *et al.* 1984). This effect may influence the equilibrium Lorentz-factor of the particles.

3 Radiative Ejection of Plasma from a Surface of a Dead Pulsar

3.1 Radiative Force vs. Gravity

The first question to be answered regarding the radiative ejection is under which circumstances radiative force dominates gravity. In Table 1 we present the ratios Magnetic radiative force/Gravity force f^B/mg for electron-positron and electron-proton plasma accelerated above the hot surface of an NS. For every B the surface temperature T (K) is set to achieve maximum amplification factor in accordance with condition (3). We presented in brackets also the ratios for these temperatures and the absence

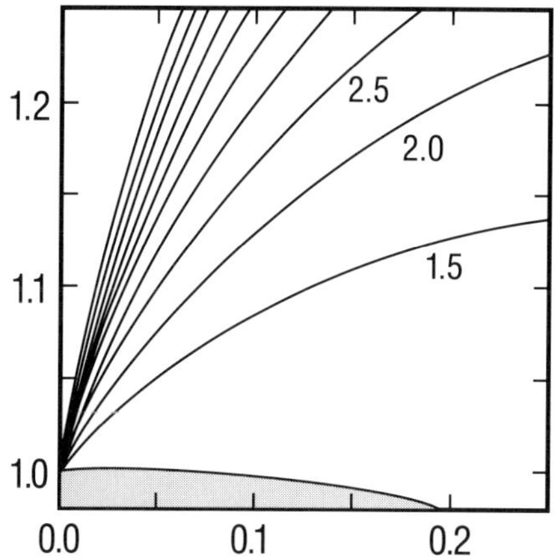

Fig. 2a—Section of axially symmetric surfaces of equal γ_K in one quadrant. Point source of radiation is located at the magnetic pole. Dimensions are in units of R_{NS}.

of magnetic field (f^T/mg) to further illustrate importance of magnetic amplification. Assumed value of surface gravitational acceleration $g \sim 10^{14}\text{cm/s}^2$ corresponds to $R_{NS} = 10^6$ cm and $M_{NS} = 1M_\odot$.

TABLE 1
Force Ratios

B (Gs)	2×10^9	2×10^{10}	2×10^{11}	2×10^{12}
T (K)	10^5	10^6	10^7	10^8
$f^B/m_p g$	10^{-3}	1	10^3	10^6
$(f^T/m_p g)$	(3×10^{-10})	(3×10^{-6})	(3×10^{-2})	(300)
$f^B/m_e g$	2	2×10^3	2×10^6	2×10^9
$(f^T/m_e g)$	(6×10^{-7})	(6×10^{-3})	(60)	(6×10^5)

From this Table one can conclude, that magnetically amplified radiative force can sweep plasma away from NS surface in a large range of surface magnetic field and reasonable surface temperatures. Thus, the dynamics of supercritical ejection are an important issue for modeling the GRBs.

3.2 Dynamics of Supercritical Ejection

Let us consider plasma radiatively ejected by a small hot spot ($R_{spot} \ll R_{NS}$) located at the NS magnetic pole. At every point in the magnetosphere one can calculate an equilibrium Lorentz-factor γ_K for which $\vec{f} \cdot \vec{B} = 0$ (radiative force acts perpendicularly to local magnetic field line). In approximation of point source this condition is equivalent to $x' = 0$. In this case $\gamma_K = (1 - x^2)^{-1/2}$.

If $mg \ll f^B$, then at every point of the magnetosphere the plasma will be driven at γ_K which does not depend on the emission spectrum and intensity. In Figure 2a we present the distribution of γ_K over the NS dipole magnetosphere for acceleration by point source at the magnetic pole. Formally, along the $\vec{Z}$ axis $\gamma_K = \infty$.

Let us take a closer look at the region near the hot spot, where the emission has a broad angular diagram. To find γ_K at a given point in this region, one must solve equation $\vec{f} \cdot \vec{B} = 0$. The solution corresponds to quasi-isotropic emission angular diagram in the ERF. It depends on the emission spectrum and angular diagram, but still does not depend on its intensity (see Fig. 2b).

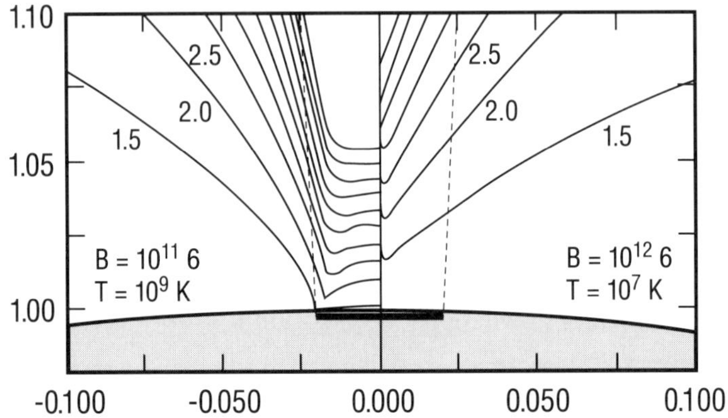

Fig. 2b—Zoom-in of Fig. 2a: the axially symmetric distribution of γ_K near the radiation source for $\epsilon_B \ll \epsilon_T$ (left panel) and $\epsilon_B \gg \epsilon_T$ (right panel). The source located near the magnetic pole is denoted by the thick line. In the region beyond the dashed lines the approximation of point source becomes valid.

Three ideas may be extracted from there:

- There are two distinct regions in the magnetosphere which are separated by a dashed line. The distinctive condition is if there exist photons that propagate in parallel to electron velocity. If so, then γ_K strongly depends on the emission spectrum. This occurs above the spot. If not, then spectrum dependence is weak and distribution of γ_K is physically similar to that for point source.
- γ_K increases with height. This happens, since emission angular diagram becomes more narrow with height, so larger γ_K is necessary to make it quasi-isotropic in the ERF.
- In the region above the spot γ_K decreases with the value ϵ_B/ϵ_T if $\epsilon_B/\epsilon_T > 1$. The reason is that photons which accelerate the particle propagate at smaller x and therefore they must have larger ϵ than those which decelerate the electron to fit resonant condition $\epsilon' = \epsilon_B$. If $\epsilon_B/\epsilon_T > 1$, then ϵ_B lies in the Wien part of the emission spectrum. For large γ this results in increasing deficit of hard accelerating photons.

3.3 Compton Scattering on the Ejected Plasma - Absorption in GRB Spectra

An important issue which can be analyzed on the basis of the calculated field of γ_K is how Compton scattering contributes to the observed source spectrum. We will confine ourselves to the case of resonant scattering on optically thin plasma. Let x_N be a cosine of the angle between the line of sight and magnetic field line, then at a given energy ϵ the contribution consists of two parts:

- photons with energy ϵ which go directly from the source to the observer are removed from the spectrum by scattering at the points of the line of sight where ϵ coincides with cyclotron energy in the ERF ($\epsilon' = \epsilon_B$). This mechanism is responsible for 'absorption-like features' in the spectrum.
- spectral density at the energy ϵ will be increased by photons scattered to the observer from the points of the magnetosphere, where the resonance condition is satisfied:

$$\epsilon = \frac{\epsilon_B}{\gamma_K(1 - \beta_K x_N)} \tag{5}$$

This mechanism could, in principle, compensate 'absorption features'.

It is obvious that condition (5) determines for any energy ϵ and x_N an 'emitting surface' from which photons with energy ϵ may be emitted to the observer. It may be characterized by: ϵ_0 - cyclotron energy corresponding to the magnetic pole and the angle between the line of sight and magnetic axis θ_N^Z, provided the geometry of the magnetic field and, consequently, the distribution of γ_K is known. On the other hand, the point where line of sight intersects the emitting surface is that very point, where direct photons are removed from the spectrum. In Figure 3 we present a family of emitting surfaces, for several values of ϵ/ϵ_0. It shows the same region as Figure 2a.

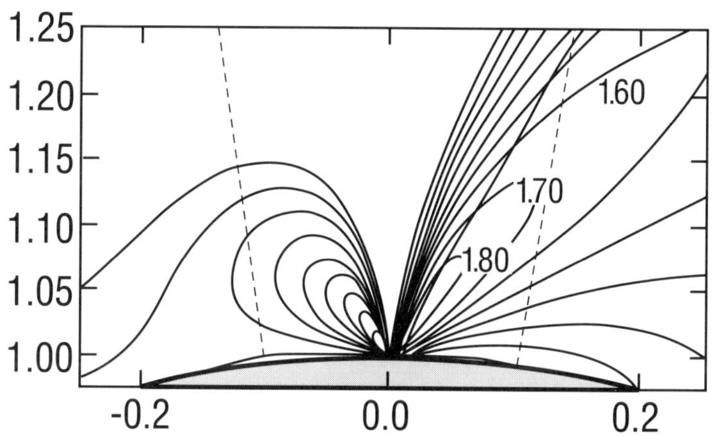

Fig. 3—Section of emitting surfaces for $\theta_N^Z = 30$ for point polar source. The figure is based on γ_K presented in Fig. 2a and the scale is the same. The dashed lines bound the plasma column in which cyclotron absorption will be calculated.

It may be shown that absorption along the line of sight cannot be significantly compensated for by scattering from the emitting surface. Since every type of compensation at given energy is possible only from a corresponding emitting surface, then the absorption corresponding to quantum transitions $0 \to 1$ cannot be distorted also by emission corresponding to transitions like $0 \to 2 \to 1 \to 0$. This matter was a special concern for Miller *et al.* (1991).

In Figure 4 we present absorption lines in the Planck spectrum of point polar source for different θ_N^Z. It was assumed that the scattering plasma is ejected along the magnetic field from a circular region around the magnetic pole with size $0.1 R_{NS}$ (see

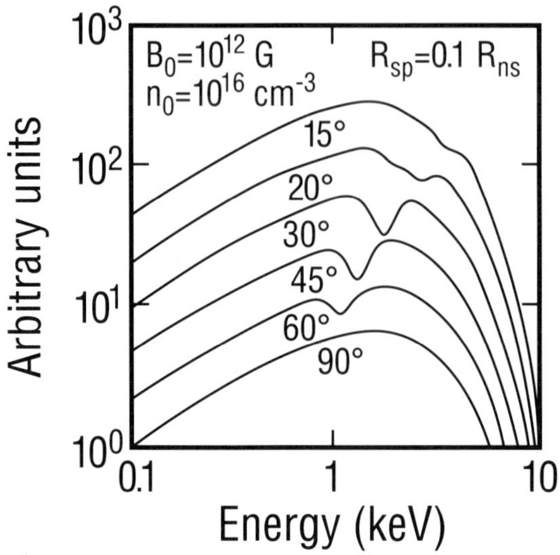

Fig. 4—Profiles of absorption lines for a different observer's angle θ_N^Z. Absorption occurs in plasma column with bottom radius $R_{sp} = 0.1\ R_{NS}$ (see Fig. 3). Assumed values of magnetic field at the magnetic pole B_0 and surface number density of the plasma n_0 correspond to optical depth for resonance scattering $\tau \ll 1$.

Fig. 3). Plasma number density corresponds to optical depth $\tau \le 1$. The spectra were convolved with a typical Gaussian detector response function with FWHM $= 0.3\sqrt{\epsilon}$ and compensation was neglected (Chernenko and Mitrofanov 1992).

One can figure out two interesting points: the line is produced efficiently in rather narrow range of θ_N^Z; both narrow (like those reported in Murakami *et al.* 1988) and wide (like those reported by Mazets *et al.* 1981) lines may be observed depending on θ_N^Z. Another important issue is that since in the ERF photons go perpendicularly to the velocity, there is only second-order Doppler broadening, which is negligible for a few keV plasma. This means that cyclotron line broadening has to be purely geometric.

4 Radiative braking of relativistic electrons in pulsar magnetosphere

Let us now consider how the presence of soft emission may affect acceleration of electrons by electric field in the magnetosphere of an active pulsar. In Figure 5 we present schematically the dependence of the radiative force on γ (Kardashev *et al.* 1984). Cyclotron resonance is clearly visible as well as plateau corresponding to Klein-Nishina slope.

If one fixes some electric force $f_E = eE_\parallel$, then from Figure 5 it is very easy to determine at which γ the acceleration will be restricted by radiative braking. If $f_E < f_B^{max}$ then the acceleration will stop at very moderate $\gamma \sim \epsilon_B/\epsilon_T$ (see eq. [4]). In case $f_B^{max} < f_{K-N}$ which requires $B < 10^{12}$, the acceleration will stop at $\gamma \sim 1/\epsilon_T \sim 100\ (5\mathrm{keV}/kT)$. Condition $f_E < f_B^{max}$ corresponds to $T > 10^{11}\left(E_\parallel/B\right)^{1/2}$ (Kardashev *et al.* 1984).

Let us introduce a very crude estimation of $E_\parallel$ as $E_\parallel = (2\pi R_{NS}/Pc)^\alpha B$, where P is the rotation period and α will be a free parameter different in different models. In Table 2 we present temperatures in K of soft emission necessary to restrict electric acceleration for several values of the rotation period and parameter α.

Taking into account that surface temperature of active pulsars does not strongly exceed 10^6 K, it becomes clear that radiative braking may be expected only for $P \ge$

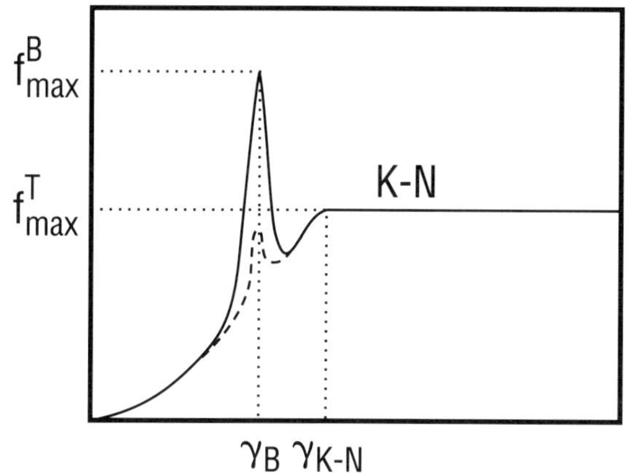

Fig. 5—Dependence of the radiative force on γ-factor of electron. Solid line corresponds to some $B > 10^{12}$, dashed line—to some $B < 10^{12}$. The plateau marked by 'K-N' corresponds to K-N slope in the scattering cross section.

TABLE 2
Soft Emission Temperatures

	$\alpha = 1$	$\alpha = 2$	$\alpha = 2.5$
$P = 0.1$ sec	3.5×10^{11}	2×10^{8}	4×10^{7}
$P = 1.0$ sec	1.3×10^{11}	2×10^{7}	2×10^{6}

1 sec and $\alpha \geq 2.5$. No confirmed detections of X-ray emitting pulsars with $P > 1$ sec have been done so far, but there are several promising candidates including PSR 1524-39 with $P = 2.418$ sec (Ögelman 1992).

Any possible effect must be associated with acceleration process taking place near the radiating surface, because at large distances the radiation is diluted as $\sim 1/R^2$. It is believed that while cone radio-emission comes from remote regions of the magnetosphere, core emission is produced near the surface (Cordes 1992). If so, it may be suppressed by radiative force. Moreover, due to Compton upscattering of the X-ray emission, a hard γ-ray pulse with typical energy $\sim \gamma^2 kT \sim 1.5\ B_{12}^2 T_6^{-1}$ MeV should replace the core pulse. All power of the electric field will be radiated by soft γ-rays.

References

Chernenko, A. M. and Mitrofanov, I. G. 1992, in preparation.
Cordes, J. 1992, these Proceedings.
Kardashev, N. S., Mitrofanov, I. G., and Novikov I. D. 1984, *Russ. A. J.* **61**, 1113
Mazets, E. P. *et al.* 1981, *Nature* **290**, 378.
Miller, G. S. *et al.* 1991, *Phys. Rev. Lett.* **66**, 1395.
Mitrofanov, I. G. and Tsygan, A. I. 1982, *Ap. and Sp. Sci.* **84**, 35.
Mitrofanov, I. G. and Pavlov, G. G. 1982, *MNRAS* **200**, 1033.
Murakami, T. *et al.* 1988, *Nature* **335**, 234.
Murakami, T. *et al.* 1992, in *Gamma-Ray Bursts*, ed. C. Ho, R. Epstein, and E. Fenimore (Cambridge University Press), p. 239.
Ögelman, H. 1992, these Proceedings.
Tsygan, A. I. 1980, *Ap. Space Sci.* **77**, 187.

Small-Scale Texture in the Pulsar Radio Emission Zone

Estelle Asseo

Centre de Physique Theorique, Ecole Polytechnique, Palaiseau, France

Abstract

From a recent study which concerns the development of strong Langmuir turbulence in the emission region of a pulsar, we conclude the possible existence of a lattice of stable microstructures in this region. These microstructures directly radiate sufficient electromagnetic energy to be proposed as responsible for pulsar radio emission. Nevertheless, two other mechanisms, for which I present preliminary results, may also interfere. First, these microstructures, while moving along the dipolar pulsar magnetic field lines, will give rise to curvature radiation. Second, an interaction between the most energetic particles present in the pulsar radio emission zone and the microstructures produces an emission of electromagnetic energy. The radiated power can reach quite a high level if some coherence is maintained between the phase of the particles. I briefly summarize the observable properties associated to these three different emission mechanisms.

1 Physical Conditions in the Pulsar Magnetosphere

All our calculations have been made in the context of the classical Ruderman and Sutherland (1975) polar cap model. This model implies the presence, in the pulsar magnetosphere, of a weakly dense beam of very energetic e^+ (or e^-), with $\gamma_B \approx 10^6$, together with a dense plasma of less energetic e^- and e^+, with $\gamma_p \approx 10^2$–10^3. Both the beam and the plasma are in relativistic motion and can have a mildly relativistic temperature ($T \approx 5 \times 10^{6\circ}$ K). They are immersed in the extremely strong dipolar pulsar magnetic field. Cheng and Ruderman (1977) have shown that the contribution of the very energetic beam to the dispersion features is negligible. In fact, the most efficient two-stream instability arises between the e^- and e^+ of the plasma itself. The fastest plasma particles form the "beam", whereas the slowest ones form the "plasma".

The geometry is defined locally, with respect to the direction of the magnetic field line. The unit vectors are $\vec{l}_\parallel = \vec{B}/|\vec{B}|$, $\vec{l}_\perp = \vec{R}_c/|\vec{R}_c|$ and $\vec{l}_t = \vec{l}_\parallel \wedge \vec{l}_\perp$. $\vec{R}_c$ is the radius of curvature of the magnetic field line.

The unstable plasma mode, excited by the "beam"-"plasma" interaction, is a quasi-longitudinal mode, specific to the physical conditions found in the pulsar magnetosphere. It has non-zero $E_\parallel$, $E_\perp$, and B_t electromagnetic components. In what follows, we study the nonlinear evolution of this mode into Langmuir soliton-like solutions.

2 Stable "Langmuir Microstructures"

The description of the nonlinear evolution of the unstable plasma waves excited by the "beam"-"plasma" interaction has been analyzed in the general case of a beam and plasma immersed in an arbitrary magnetic field. A formal analysis given in Pelletier, Sol and Asseo (1988; hereafter PSA), has been applied to the pulsar case in Asseo, Pelletier and Sol (1990 ; hereafter APS). The final stage is the existence of Langmuir soliton-like structures in the pulsar magnetosphere. $\varepsilon_{\parallel}$, the envelope of the parallel electric field component of one Langmuir wave-packet, slowly varies with space and time. It is the solution of a modified nonlinear Schrodinger equation, written in the plasma rest frame, namely :

$$\Delta_{\parallel}\left[i\partial_t + \frac{1}{2}\Delta_{\parallel} + |\Psi|^2\right]\Psi = \frac{1}{2}s\Delta_{\perp}\Psi. \tag{1}$$

Here $|\Psi|^2 = |\varepsilon_{\parallel}|^2/32\pi n_p k_B T$ and, as usual, ω_c is the gyrofrequency, ω_p is the plasma frequency, n_p the plasma density, and T the temperature. The essential parameter of the problem is the quantity $s = \omega_c^2/(\omega_p^2 - \omega_c^2)$.

The solutions of equation (1) are soliton-like solutions or Langmuir solitons,

$$E_{\parallel} = \varepsilon_{\parallel}\cos\omega_p t = e_{\parallel}\,\mathrm{sech}\,a\left(x_{\parallel} - x_o\right)\cos\Omega_p t$$

Here, $e_{\parallel} = 2\pi a\lambda_D\sqrt{\pi n_p k_B T}$ is the amplitude of the electric field component, $\Omega_p = \omega_p(1 - \lambda_d^2 a_o^2/2)$ is the shifted oscillation frequency of the Langmuir soliton, $\lambda_D = v_{th}/\omega_p$ the Debye length, and v_{th} the thermal velocity.

On the other hand, the perpendicular electric field component of this Langmuir soliton follows a wave-like equation, the parallel part of the electric field acting as a source for its excitation. We obtain $E_{\perp} = E_{\perp}^{+} + E_{\perp}^{-}$, where $E_{\perp}^{\pm} = e_{\perp}\cos\Omega_p(t \pm \frac{x_{\parallel} - x_o}{c})$ and $e_{\perp} = e_{\parallel}\nabla_{\perp}a/a^2$. The transverse magnetic field is easily derived from Maxwell's equation. Together $E_{\perp}$ and B_t, having a propagative character, lead to a radiation process. This process is quite dependent on the presence of the perpendicular irregularities in Langmuir soliton electromagnetic components, through the term $(\nabla_{\perp}a/a^2)^2$.

The corresponding density fluctuations result from the ponderomotive effect $\delta n/n_p \approx -(1/32\pi)\left(\omega_p^2/\omega^2\right)\left(|E_{\parallel}|^2/n_p k_B T\right)(1/\gamma)$. The maxima of Langmuir soliton electric fields thus correspond to density wells.

The stability of Langmuir solitons against perpendicular perturbations depends on the sign of the parameter s. One of our main general results in PSA (1988) is that Langmuir solitons are stable if $s < o$, whereas they are unstable when $s > o$. Obviously, in the region of the pulsar magnetosphere interior to the light cylinder, $\omega_c >> \omega_p$, and consequently Langmuir solitons are stable. We propose to call them "Langmuir microstructures".

Another important result is the fact that we were able to fix the amplitude of the solitons which have zero velocity, $a_o \approx 1/10\lambda_D$. The characteristic dimensions of the Langmuir microstructures depend on a_o and, eventually, on the characteristics of the linear instability. Indeed, their longitudinal extent is $L_{\parallel} \approx 10\lambda_D$ while minimum and maximum values can be determined for their perpendicular extent $L_{\perp}$. We have $L_{\perp} \geq \lambda_{\perp}^{min} = 2\pi k_{\perp\,max}^{-1}$ where $k_{\perp\,max}$ is of the order of $\Delta k_{\perp}$, the perpendicular wavenumber width for the existence of the linear instability. A maximum value for $L_\perp$ is taken as the distance A from a given point on the magnetic field line to the rotation axis. For the last open field line, $A_{\max} \approx r^{3/2} R_{LC}^{-1/2}$ and $L_\perp \leq A_{\max}$. In either case, we assume that the perpendicular and transverse dimensions of the Langmuir microstructures are equivalent, $L_t \approx L_\perp$. Taking APS (1990) pulsar physical parameters $L_\parallel \approx 100$ cm and $L_\perp \approx 10^3$–10^7 cm, at the distance $r = 10^8$ cm.

Obviously, the fact that $L_\perp$ can be determined only within two limits, results in an extremely large latitude for the choice of the soliton volume v_s in the emission region of the pulsar magnetosphere. It has a considerable impact on our numerical estimates. At $r = 10^8$ cm, the presumed distance of the pulsar emission zone, $10^8 \text{cm}^3 < v_s < 8 \times 10^{15} \text{cm}^3$. The whole set of our analytical outcomes is in good agreement with experimental results obtained by Antipov, Nezlin and Trubnikov (1981 and references therein), for the interaction of a monoenergetic beam with a magnetized plasma for which $\omega_c \approx 2 \times 10^{10}$Hz $> \omega_p \approx 3 \times 10^9$ Hz. They observe a stable lattice of solitons in the plasma. These solitons are fixed with respect to the plasma, their electric field maxima are related to their density wells, their longitudinal dimensions are of the order of $6\lambda_D$ and they show concentration in wave energy $W \approx 40 W_{\text{initial}}$.

From all this, we conclude in APS (1990) the probable existence in the pulsar radio emission zone of a lattice of stable Langmuir microstructures, all with the same amplitude, at rest in the plasma frame and certainly regularly distributed in space. These Langmuir microstructures form the small-scale texture of the pulsar radio emission zone.

3 Available Radiated Power from the Langmuir Microstructures Present in the Pulsar Radio Emission Zone

The Langmuir microstructures, being stable, can radiate electromagnetic energy by direct or indirect processes.

3.1 Direct Emission of Electromagnetic Energy by Langmuir Microstructures

Our recent work APS (1990) describes how Langmuir microstructures, perturbed by perpendicular effects, directly emit electromagnetic waves. They contain sufficient electromagnetic energy to provide us with a radiation process if the energy can be transmitted to the observer through the different media. Indeed, the perpendicular irregularities of the Langmuir microstructures induce perturbations of the amplitude a_o but diffuse on a time much longer than any other characteristic time. Thus, these irregularities do persist sufficiently to contribute to the radiation process, the available electromagnetic energy being maximum for $k_\perp \approx k_{\perp\max}$. Considering each Langmuir microstructure as an independent emitter,

$$P_\parallel^{NL} = N_S \frac{c}{4\pi} \varepsilon_\perp B_t \gamma_p^2 S_\perp = 8 \times 10^{27} \text{erg s}^{-1}.$$

This equation represents the Poynting flux of the whole assembly of N_s Langmuir microstructures through a surface $S_\perp$, characteristic of the emission zone. Recent data by Lyne and Manchester (1988) lead us to an estimate of $S_\perp \approx 3 \times 10^{14} \text{cm}^2$ at the distance of the emission zone. We also assume that $N_S \approx 10^3$, a reasonable value since from radio observations one infers that $10^2 < N_S < 10^6$ .Thus, direct emission of electromagnetic energy from the Langmuir microstructures should be sufficient to account for the radio luminosity received from pulsars, $L_{\text{radio}} \approx 10^{26}$–$10^{30}$ erg s^{-1}.

3.2 Curvature Radiation from Langmuir Microstructures

As a first approximation one can consider that the lattice of stable Langmuir microstructures can be represented by a succession of elementary positive and negative charges moving along each magnetic field line. These series describe the alternation of density wells charged with one sign and of interspaces charged with opposite sign, the whole sequence being neutral. A simple estimate for the elementary charge is obtained by means of the charge density fluctuation, $\delta\rho = e\delta n$, derived from the ponderomotive force. An integration over the volume of one Langmuir microstructure fixes the charge estimate between two limits, dependent on the soliton volume. The minimum (maximum) limit corresponds to the minimum (maximum) volume.

$$8 \times 10^{10}(r/R_*)^{3/2}\text{electrons} \leq Q \leq 10^{17}(r/R_*)^{3/2}\text{electrons}.$$

A more refined calculation also provides us with a charge estimate for the Langmuir microstructure. Indeed, the presence of the Langmuir soliton itself in the plasma induces electromagnetic fields, $\delta\vec{\varepsilon}$ and $\delta\vec{B}$, current $\delta\vec{J}$, and charge density $\delta\rho$. These fluctuating quantities are obtained from the relativistic fluid and Maxwell equations, making an ordering in successive powers of the soliton electric field. At lowest order, the presence of the Langmuir soliton induces in the plasma a charge density which is of second order in the soliton electric field. Averaged over wave period,

$$<\delta\rho> \approx \left(3e/4\pi m\omega_p^2\right)\left|\vec{\nabla}\cdot\left(\vec{E}_{\|}^{\text{soliton}}\cdot\nabla\right)\vec{E}_{\|}^{\text{soliton}}\right|.$$

$<\delta\rho>$, depending on $\vec{E}_{\|}^{\text{soliton}}$, varies with the distance to the center of the soliton. It has alternatively negative or positive sign. It corresponds to a positive central charge, Q, and to two negative side charges, $(-Q/2)$, the whole being neutral. Q, in this case, is about 1/30 of the value given in (2). A similar result has been obtained by Melikidze and Pataraya (1983) starting with kinetic theory while we use relativistic fluid and Maxwell's equations. In analogy with Melikidze and Pataraya (1983), we consider that the radiated power P_c^1 associated to one Langmuir microstructure results from the radiation of these three successive charges. Then, at $r = 10^8$ cm

$$10^7\text{erg s}^{-1} \leq P_c^1 \leq 6 \times 10^{21}\text{erg s}^{-1}$$

while at $r \approx R_*$, it is only

$$10^3\text{erg s}^{-1} \leq P_c^1 \leq 6 \times 10^{17}\text{erg s}^{-1}.$$

Let us mention that close to the star, this value can be slightly increased by a factor 10^4 if $R_c \approx R_*$. Consequently, if the Langmuir microstructures are generated sufficiently far from the star surface, at the distance $r = 10^8$ cm, one should require a large enough number of these structures, at least $N_S \geq 10^6$, with all the maximum volume, to reach the high level of pulsar radio radiation. If this mechanism occurs close to the star surface, it seems more difficult to imagine that it can contribute efficiently to the observed pulsar radio radiation.

3.3 Turbulent Radiation from Beam Particles in Interaction with the Langmuir Microstructures

Acceleration of beam particles by the electromagnetic fields of Langmuir microstructures can produce important radiated power if some coherency can be maintained between the phase of the particles. This can be of great consequence, in particular in the case where the perpendicular gradient of Langmuir solitons is too small to induce directly sufficient electromagnetic radiation to account for pulsar radio radiation. Let us notice that when $k_\perp$ is sufficiently smaller than $k_{\perp\,\max}$, $|E_\perp|$ and $|B_t|$ are much smaller than $|E_\parallel^{\max}|$. In this case, we consider that the acceleration suffered by beam particles is only sensible to $E_\parallel$. Then the variation of the momentum of the particle is $\Delta p_\parallel \approx e\varepsilon_\parallel(\pi/2ac)\,\mathrm{sech}(\pi\omega_p/2ac)$. (A similar result is obtained considering the effect of the electrostatic field induced by the presence of the soliton in the plasma). The power radiated by one electron accelerated by one microstructure is $P_{1,1} = 2e^2(ac)^2\left(\Delta p_\parallel\right)^2/3m^2c^3$ or with the APS (1990) set of pulsar physical parameters $P_{1,1} \approx 8 \times 10^{-17}$erg s^{-1} at the distance $r = 10^8$cm.

If one considers that the successive $(n_b V)$ beam charges in interaction with the N_S Langmuir microstructures are quite out of phase, the turbulent incoherent radiated power in the observer's frame is

$$P_{T,I} \approx N_S n_b V \gamma_b^2 P_{1,1} \approx 3 \times 10^8 \text{erg s}^{-1},$$

where $N_S \approx 10^3$, $n_b = 7 \times 10^6$ at $r = 10^8$ cm, V, the volume of the emission region, is $V = LS_\perp \approx 3 \times 10^5 \times 3 \times 10^{14}\text{cm}^3$ and $\gamma_b \approx 3 \times 10^2$. Thus, as explained in APS (1990), turbulent incoherent emission from these accelerated electrons is insufficient to account for L_{radio}. Nevertheless, if some phase coherence can be maintained between the beam particles, the resulting power will be increased. One possibility is that the beam itself, modulated by the unstable wave on a scale length of the order of λ_o, the resonant plasma length, can maintain some phase coherence between beam particles on distances of this order. Such a way to obtain coherence has been recently proposed by Weatherall and Benford (1991) in a different astrophysical context. In this case the radiated power is significantly increased by a coherence factor, equal to the number of beam electrons in one Langmuir microstructure, which can reach very high values. Then the turbulent coherent radiated power in the observer's frame is

$$P_{T,C} \approx \gamma_b^2 \sigma_{Th} c e_\parallel^2 (n_b v_s)^2 (L/\Delta)\left(v_o/\omega_p\Delta\right)\left[1 + 2 \times 10^{-1}\left(\Delta n_b/n_b\right)^2\left(\Delta\omega/\omega_p\right)\right].$$

Here σ_{Th} is the Thomson cross-section, (L/Δ) represents the approximate number of solitons and $(v_o/\omega_p\Delta)$ the number of crossings. The first term in the bracket corresponds to an emission at the plasma frequency ω_p, the second term represents emission at frequencies $\omega \neq \omega_p$. Using the APS (1990) set of parameters and a soliton volume 10^4 times the minimum volume,

$$P_{T,C} \approx 10^{30} \text{erg s}^{-1} \text{ at } \omega = \omega_p \quad \text{but} \quad \approx 6 \times 10^{27} \text{erg s}^{-1} \text{ at } \omega \neq \omega_p.$$

Consequently, coherent turbulent emission, which results from the interaction of beam particles with the Langmuir microstructures, can also be proposed as a possible pulsar radio emission process.

4 Comments

Nonlinear evolution of the beam plasma interaction appears as an efficient mechanism for conversion of the beam kinetic energy into electromagnetic energy. The resultant Langmuir microstructures can give rise to electromagnetic radiated power by three different processes, namely direct electromagnetic radiation, curvature radiation from the Langmuir microstructures themselves, and turbulent coherent emission resulting from the interaction of beam particles with the Langmuir microstructures.

Direct emission produces sufficient radiated power, when $k_\perp \leq k_{\perp\,\max}$, to account for L_{radio}. This process, directly linked to the perpendicular deformations suffered by the Langmuir microstructures, appears as a direct consequence of the obliquity of the propagation. It occurs at the plasma frequency $\omega_p \approx 2 \times 10^9$ Hz and at $r = 10^8$ cm, and is locally narrowband but can account for global broadband emission. The excited modes are 100% linearly polarized perpendicularly to the magnetic field but coupling to the transverse mode could induce a percentage of circular polarization. Also, the associated temporal widths of the directly emitting Langmuir microstructures of the lattice, $\tau_\parallel \approx 10^{-7}$ sec and $\tau_\perp \approx$ few10^{-6} sec, can be related to the observed microstructures. In fact, $\tau_\perp$ arises more naturally as an angular structure for the observed microstructures.

Curvature radiation from the Langmuir microstructures seems more difficult to refer to as a mechanism for pulsar radio radiation, due to the necessity to require the presence of a large number of Langmuir microstructures, with all the maximum volume, to reach L_{radio}. If this mechanism is ever to be invoked, it occurs at radio frequencies, $\omega_c \approx 2 \times 10^7$ Hz. The linearly polarized radiated modes show a percentage of circular polarization, with a possible reversal at the center of the emission. This makes it of interest to interpret emission features in single cone emission profiles, or in triple profiles with both a central core and two conal components, if one can consider that the emission arises at some distance from the star surface. [cf Rankin 1983a, 1983b, 1986, 1990a, 1990b for the observations, Gil (Gil and Snakowski 1990a,b; Gil, Kijak, and Zycki 1991) for the analysis of polarization features]. The corresponding Langmuir microstructure lifetimes, of the order 10^{-5}–10^{-7} sec, could be related to the observed microstructures.

On the other hand, the turbulent coherent emission, which results from the interaction of beam particles with the Langmuir microstructures present in the pulsar emission region, is of interest for interpretation of pulsar radio emission. Radiated frequencies are in the radio domain, with $\omega \approx 2 \times 10^9$ Hz at $r = 10^8$ cm, or close to it.

The polarization is 100% linear, in the direction of the magnetic field line. It is more difficult to connect the observed timescale to microstructure lifetimes.

In conclusion, taken as a whole, the role of Langmuir microstructures, which form the small-scale texture of the pulsar emission region is worth consideration in relation to observed radio emission.

It is a pleasure to thank Helene Sol and Guy Pelletier, and also Georgiu Melikidze for many fruitful suggestions during his three months stay in France in 1991.

References

Antipov, S. V., Nezlin, M. V., and Trubnikov, A. S., 1981, *Physica 3D*, **1&2**, 311.
Asseo, E., Pelletier, G., and Sol, H., 1990, *MNRAS*, **247**, 529.
Cheng, A. F. and Ruderman, M. A., 1977, *Ap. J.*, **212**, 800.
Gil, J. A. and Snakowski, J. K., 1990a, *A&A*, **234**, 237.
Gil, J. A. and Snakowski, J. K., 1990b, *A&A*, **234**, 269.
Gil, J. A., Kijak, J., and Zycki, P., 1991, *A&A* (preprint).
Lyne, A. G. and Manchester, R. N., 1988, *MNRAS*, **234**, 477.
Melikidze, G. I. and Pataraya, A. D., 1984, *Astrofizika*, **20**, 157.
Pelletier, G., Sol, H., and Asseo, E., 1988, *Phys. Rev. A*, **38**, 2552.
Rankin, J. M., 1983a, *Ap. J.*, **274**, 333.
Rankin, J. M., 1983b, *Ap. J.*, **274**, 359.
Rankin, J. M., 1986, *Ap. J.*, **301**, 901.
Rankin, J. M., 1990a, *Ap. J.*, **352**, 247.
Rankin, J. M., 1990b, *Ap. J.*, **352**, 258.
Ruderman, M. A. and Sutherland, P. G., 1975, *Ap. J.*, **196**, 51.
Weatherall, J. C. and Benford, G., 1991, *Ap. J.* **378**, 543.

Preliminary EGRET High-Energy Gamma Ray Observations of the Crab Pulsar

D. J. Thompson,[1] D. L. Bertsch,[1] C. E. Fichtel,[1] J. M. Fierro,[2] R. C. Hartman,[1] S. D. Hunter,[1] G. Kanbach,[3] D. A. Kniffen,[7] Y. C. Lin,[2] J. R. Mattox,[1,5] H. A. Mayer-Hasselwander,[3] P. F. Michelson,[2] C. v.Montigny,[3] H. I. Nel,[6,8] P. L. Nolan,[2] K. Pinkau,[3] H. Rothermel,[3] E. J. Schneid,[4] M. Sommer,[3] P. Sreekumar[6]

[1]NASA Goddard Space Flight Center
[2]W.W. Hansen Experimental Physics Laboratory and Department of Physics, Stanford University
[3]Max-Planck-Institut für Physik u. Astrophysik, Institut für Extraterrestrische Physik
[4]Grumman Aerospace Corporation, Bethpage, NY
[5]Compton Observatory Science Support Center, Computer Sciences Corporation, NASA/GSFC
[6]USRA Research Associate
[7]Department of Physics, Hampden-Sydney College
[8]On leave from Dept. of Physics, Potchefstroom University

Abstract

The Energetic Gamma Ray Experiment Telescope (EGRET) on the Compton Gamma Ray Observatory (CGRO) observed the Crab pulsar three times between April 22 and June 15, 1991. In the energy range from 50 MeV to at least 2 GeV, the pulsar shows the double peak structure seen at other wavelengths. Although the light curve has a larger ratio of first pulse/second pulse than that seen by the SAS-2 observations and the early COS-B observations, the current data show no significant evidence for variability on timescales of days to weeks.

1 Introduction

NASA's Compton Gamma Ray Observatory (CGRO), launched by the Space Shuttle on April 5, 1991, is a large (16,000 kg) satellite designed to study cosmic gamma rays over a broad energy range (30 KeV - 30 GeV). With a design lifetime of at least two years, CGRO will survey the full sky during its first 18 months and then return to targets of particular interest.

Of the four instruments on the Compton Observatory, the Energetic Gamma Ray Experiment Telescope (EGRET) studies the highest energy photons (greater than about 20 MeV). EGRET (shown schematically in Figure 1) is a multi-level spark chamber, triggered by a scintillator coincidence system, and using a large NaI(Tl) Total Absorption

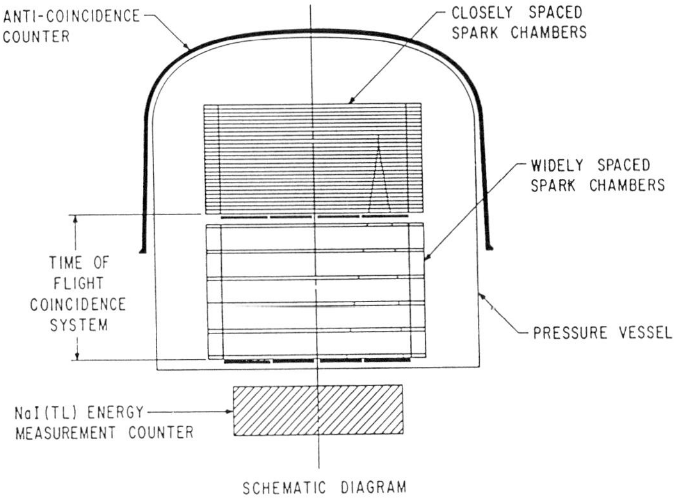

Fig. 1—Schematic diagram of the Energetic Gamma Ray Experiment Telescope.

Spectrometer Crystal (TASC) as a principal energy-measuring device (Kanbach, *et al.* 1988).

The basic principle of operation is to detect gamma rays which interact by pair production within the EGRET spark chamber. An incoming gamma ray gives no signal in the large anticoincidence scintillator surrounding the spark chamber. If pair production of the gamma ray occurs in one of the thin tantalum foils within the upper spark chamber, the electron and positron may trigger the coincidence system, consisting of a 4 by 4 array of plastic scintillator tiles in the middle of the spark chamber and a similar array at the bottom. A coincidence signal in combinations of tiles selected by preset electronics or command, together with a time-of-flight signature indicating downward-moving particles, initiates the spark chamber high-voltage pulse and the readout of the spark chamber and energy data. The recorded spark chamber picture, energy information, gamma ray arrival time, and auxiliary information are transmitted to the ground as one "event". Because there are more events than useful gamma ray detections, the data analysis system must select the subset of all events that are unmistakably recognized as gamma-ray produced pairs, and from these the system extracts the arrival direction and energy of each detected gamma ray. In the middle of its energy range (200–1000 MeV), EGRET has an effective area of about 1500 cm^2 on-axis, with its sensitivity falling to near zero at incident angles greater than 40°. For this same energy range, its energy resolution is about 25% FWHM. Sources can be located to 10 arcmin in general, and to 5 arcmin for strong sources or those away from bright regions of the sky. Each event's arrival time is accurate to 100 microseconds, although an uncertainty in the absolute time of the CGRO clock is presently unresolved.

2 Light Curve of the Crab Pulsar

The Crab Pulsar (PSR0531+21) was observed by EGRET three times early in the Compton Observatory mission: April 22 - May 7, 1991; May 16 - 30, 1991, and June 8

- 15, 1991. The arrival time of each gamma ray from the direction of the pulsar was converted to an equivalent pulsar phase using the timing parameters derived from radio observations (Nice, 1992). Results presented in this paper are preliminary; a more detailed analysis of the EGRET Crab data is given by Nolan *et al.* (1992).

Figure 2 shows the derived light curve of the Crab Pulsar for gamma rays with energy above 50 MeV from the combined April and May observations, for which a single set of radio timing parameters could be used. Several features are noteworthy:

- Two peaks are apparent, separated by the same 0.4 period seen at other wavelengths and previously in this energy range by SAS-2 and COS-B.
- The first peak is narrow, with a full width half maximum of $(4.5 \pm 0.5)\%$ of the period, about 1.5 ms, consistent with the optical pulse width (Biggs, 1992).
- A small, but statistically-significant, excess is visible between the two peaks. Often referred to as "bridge" emission by radio astronomers, this excess is called "interpulse" emission in the gamma-ray literature.

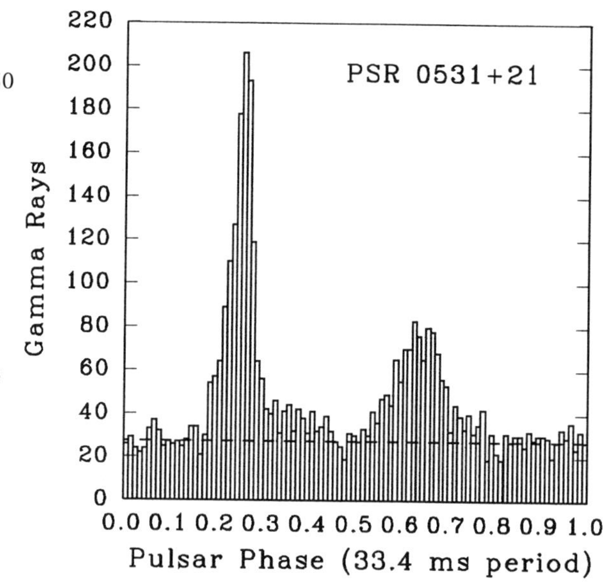

Fig. 2—The high-energy ($E \geq 50$ MeV) gamma-ray light curve of the Crab Pulsar, shown in 100 bins. The dashed line is the average number of gamma rays per bin in the phase ranges 0.0 - 0.15 and 0.75 - 1.0, away from the pulsed emission. These gamma rays come from the diffuse galactic and extra-galactic emissions, as well as any unpulsed emission from the Crab.

3 Energy Spectrum

Figure 3 is the preliminary differential energy spectrum from PSR 0531+21 seen by EGRET during the period May 16-30, 1991. It was derived by dividing the data into 9 energy intervals, constructing a light curve for each interval, and then determining the excess in the pulsed region (0.15 - 0.75) compared to the unpulsed region. The data are consistent with a power law:

$$dN/dE = (6.94 \pm 0.25)\times 10^{-9}(E/227\mathrm{MeV})^{-2.18\pm 0.08} \ \text{photons cm}^{-2}\ \text{s}^{-1}\ \text{MeV}^{-1}, \quad (1)$$

with E in MeV.

The uncertainties shown are statistical only. The energy scale factor, 227 MeV, is chosen so that the statistical errors in the power law index and the overall normalization are uncorrelated. It should be emphasized that this pulsed energy spectrum applies only to the 0.6 period during which the pulse is "on". A time-averaged pulsar spectrum would have the same shape but a normalization 0.6 as large. Most previous gamma-ray results have presented time-averaged spectra.

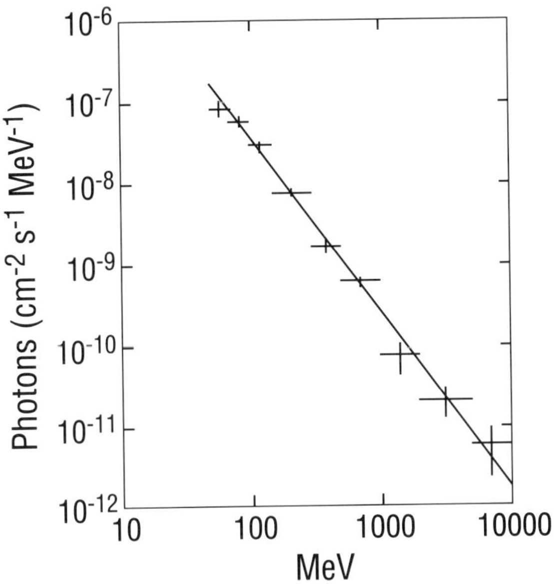

Fig. 3—Differential energy spectrum from the pulsed emission of PSR 0531+21 seen by EGRET during the period May 16-30, 1991. As discussed in the text, a time-averaged spectrum for the pulsar would have a normalization 0.6 as large.

The extrapolation of this power-law spectrum to lower energies is consistent with the COMPTEL (McConnell *et al.* 1992) and OSSE (Ulmer *et al.* 1992) observations of PSR 0531+21. The highest energy point in Figure 3 should be treated as an upper limit. Above 5 GeV, the radiation from the Crab is dominated by an unpulsed component.

4 Time Variations

Over the limited time span of the EGRET observations, no variation in the pulsed emission from the Crab pulsar was detected:

•For each of the 35 days of the observation, a light curve similar to Figure 2 was constructed. The statistics allowed a determination of the pulsed flux for each day with an uncertainty of approximately 40% (1σ). The distribution of values about the mean was consistent with the statistics.

•Light curves for one and two week intervals during the observation all show the same shape—a strong, narrow first pulse and a weaker, broader second pulse.

•Energy spectra for other parts of the observation are consistent with the one shown in Figure 3.

Comparison of the EGRET results with those of SAS-2 (Thompson, *et al.* 1977) and COS-B (summarized by Clear *et al.* 1987) confirms a long-term variation in the shape of the light curve. The 1973 SAS-2 and 1975 COS-B light curves in particular show a much stronger second peak than is visible in Figure 2. Özel and Mayer-Hasselwander (1984), Kanbach (1989) and Özel (1991) suggested that the ratio of counts in the second peak to counts in the first peak (P2/P1) varies nearly sinusoidally in the SAS-2 and COS-B data. For the 1991 EGRET data, the ratio P2/P1 = 0.53 ± 0.05 at Modified Julian Day 8392 (MJD = JD - 2440000.5).

Figure 4 shows the SAS-2, COS-B, and EGRET values for P2/P1 as a function of time. Also shown is the equation:

$$[P2/P1] = 0.9 - 0.5\sin(2\pi(T - 2594)/5110). \tag{2}$$

where T is the time in MJD. Equation (2) is the fit by Kanbach (1989) to the SAS-2 and COS-B data only. The remarkable agreement of the EGRET point with the fit to the older data is intriguing but not conclusive. Observations over the next several years may solidify this result.

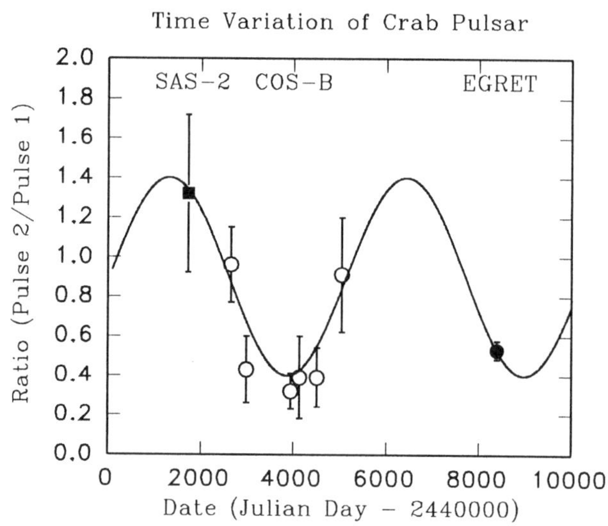

Fig. 4—Time history of the ratio of counts in the second peak to counts in the first peak for the gamma radiation from the Crab Pulsar. The line is a fit to the SAS-2 and COS-B points only, as discussed in the text.

5 Discussion

The detection of pulsed high-energy gamma rays from PSR 0531+21 provided conclusive evidence that the pulsar is accelerating electrons to very high energies. The details and site of the acceleration process, as well as the gamma-ray production mechanism, remain uncertain. The two most prominent models - the polar cap model (Harding, 1981; Daugherty and Harding, 1982; Harding and Daugherty, 1992) and the outer gap model (Cheng, Ho, and Ruderman, 1986a, 1986b; Ho, 1989, 1992) offer substantially different scenarios for the origins of the gamma radiation. The EGRET results, together with results from other telescopes, provide new constraints on these models:

- The shape of the light curve changes as the gamma-ray energy decreases (McConnell *et al.* 1992; Ulmer *et al.* 1992; Wilson, 1992), but the optical light curve (Biggs, 1992) resembles that of the high-energy gamma rays. This result suggests a change in the size, shape, or location of the radiating region as a function of photon energy. The narrow first peak in the light curve indicates a sharp edge or small size for the emitting site.

- The 0.4 period separation and difference in shape of the two peaks is a difficult geometry problem for the configuration of the magnetosphere, as discussed by Chiang (1992).
- The energy spectrum across the five orders of magnitude covered by the Compton Observatory is consistent with a single power law, suggesting a common or closely related production mechanism for all the gamma rays.
- The possibly sinusoidal variation in the ratio of the two pulses seen in high-energy gamma rays might suggest a precession or nutation of the neutron star (Kanbach, 1989; Özel, 1991), but the absence of any indication of such changes at other wavelengths leaves this interpretation uncertain.

6 Summary

Preliminary results from the Energetic Gamma Ray Experiment Telescope on the Compton Gamma Ray Observatory provide new details on the high-energy gamma radiation from the Crab Pulsar, including the shape of the light curve, the energy spectrum, and studies of time variations. PSR0531+21 will be observed again during Phase 2 of the Compton Observatory mission, giving additional information about the high-energy processes in this pulsar.

References

Bertsch, D. L., *et al.* 1989, *Proc. of the Gamma Ray Observatory Science Workshop*, (Greenbelt, MD, GSFC), 2-52.
Biggs, J. 1992, these Proceedings.
Chiang,, J. 1992, these Proceedings.
Cheng, K. S., Ho, C., & Ruderman, M. 1986a, *Ap. J.*, **300**, 500
Cheng, K. S., Ho, C., & Ruderman, M. 1986b, *Ap. J.*, **300**, 522
Clear, J., *et al.* 1987, *A & A*, **174**, 85.
Daugherty, J. K. & Harding, A. 1982, *Ap. J.*, **252**, 337
Harding, A. K. 1981, *Ap. J.*, **245**, 267
Harding, A. K. & Daugherty, J. K. 1992, these Proceedings.
Ho, C. 1989, *Ap. J.*, **342**, 396.
Ho. C. 1992, these Proceedings.
Kanbach, G., *et al.* 1988, *Space Science Reviews*, **49**, 69.
McConnell, M., *et al.* 1992, these Proceedings.
Nice, D. 1992, these Proceedings.
Nolan, P. L., *et al.* 1992, *Ap. J.*, submitted.
Özel, M. E., 1991 *Europhysics Letters*, **14**, 3.
Özel, M. E., & Mayer-Hasselwander, H.A. 1984, in *Int. Workshop on Data Analysis in Astronomy*, Erice, Italy.
Thompson, D. J., *et al.* 1977, *Ap. J.*, **213**, 252.
Ulmer, M., *et al.* 1992, these Proceedings.
Wilson, R. B. 1992, these Proceedings.

COMPTEL Observations of the Crab and Vela Pulsars

M. Busetta,[4] K.Bennett,[4] J. Bloemen,[2] R. Buccheri,[5] W. Collmar,[1] A. Connors,[3] R. Diehl,[1] H. de Boer,[2] J. den Herder,[2] W. Hermsen,[2] L. Kuiper,[2] J. Lockwood,[3] G. Lichti,[1] J. Macri,[3] M. McConnell,[3] D. Morris,[3] R. Much,[1] J. Ryan,[3] V. Schönfelder,[1] G. Simpson,[3] J. Stacy,[3] H. Steinle,[1] A. Strong,[1] B. Swanenburg,[2] B. Taylor,[4] M. Varendorff,[1] C. de Vries,[2] W. Webber,[3] C. Winkler[4].

[1] Max-Planck Institut für Extraterrestrische Physik,Garching
[2] Laboratory for Space Research, LeidenNL
[3] University of New Hampshire, I.S.E.O.S.
[4] Astrophysics Division, Space Science Dept. of ESA, Noordwijk
[5] IFCAI/CNR, Palermo, Italy

1 Introduction

The imaging Compton telescope COMPTEL is one of the four instruments on-board the Compton Gamma Ray Observatory launched on April 5 1991 into a 450 km, 28.5^o orbit. COMPTEL (described in detail by Schönfelder *et al.* 1984) operates in the 0.7 MeV to 30 MeV energy range with a field of view of 1 steradian, an angular resolution of $\sim 1^o$ and an energy resolution of $\leq$ 10% FWHM. COMPTEL measures the time of arrival of gamma-rays to an accuracy of 1/8 ms with respect to UTC.

2 Observation Details

Following 21 days of satellite and instrument check-out, GRO embarked upon a 2-week validation period before beginning the sky survey on May 16, 1991. The validation phase incorporated four 3-day viewings of the Crab Nebula at different aspect angles in the instrument field of view, namely 3° off axis, 9°, 9°-rotated azimuthally and, finally, on-axis. Data from the first period (3° pointing) are not available for scientific analysis as COMPTEL was still undergoing in-flight tuning at that time.

The validation data provided the first in-orbit determination of instrument performance and was destined to be used for confirmation of the instrument parameters by using the Crab as a "standard candle". The Vela pulsar, the second high-energy gamma-ray pulsed source, was also selected as a validation target as it was placed, unlike the Crab, above the orbital plane of GRO.

The first full observation of GRO was a dedicated viewing of the Crab for a standard 14 day interval. By chance a revisit to the same area of sky was made one week later when a Target of Opportunity was declared and GRO was slewed to the sun which was only 11° from the Crab at that time. The observation details are summarized in Table 1.

TABLE 1
Observation Summary

Obs. #	T_{START} dd/mm/yy	MJD	T_{END} dd/mm/yy	MJD	Pointing l	b	View angle (^{o})	Comments
0.2	28/04/91	48374	01/05/91	48377	193.4	-4.3	9.0	Crab
0.3	01/05/91	48377	04/05/91	48380	193.4	-4.3	9.0	Crab
0.4	04/05/91	48380	07/05/91	48383	184.5	-5.9	0.0	Crab
1.0	16/05/91	48392	30/05/91	48406	190.9	-4.7	6.4	Crab
2.1	08/06/91	48415	13/06/91	48420	194.9	-7.3	10.4	Crab
0.6	10/05/91	48386	16/05/91	48392	266.7	-4.7	4.7	Vela
8.0	22/08/91	48490	05/09/91	48504	262.9	-5.7	3.0	Vela

During these early observations, COMPTEL was only operating at about 70% efficiency as several of the detector modules were switched off while being allowed to out-gas. During observation 8, COMPTEL was about 80% efficient owing to one D1- and one D2-cell remaining switched off.

3 Analysis

Since contemporary radio observations of these two pulsars (among many others) were being made by a number of groups, notably the team coordinated by Taylor (Princeton) and Lyne and Pritchard (Jodrell Bank), the radio ephemeredes of Crab and Vela are accurately known for these observations and may be adequately described by a second order polynomial on period: the parameters may be found in the data-base made available by the Princeton group (Taylor 1989).

Transformation of UTC satellite times to the Solar System Barycentre uses software common to all four GRO teams and includes routines used by the radio observatories. Specifically, use is made of part of the established TEMPO package which has been adapted for satellite observations and of the JPL Ephemeris DE200.

As described in Busetta *et al.* (1991), in the case of COMPTEL, period folding of the event arrival times transformed to the solar system barycentre is performed on events selected on energy and scatter angle. It is a property of the Compton Telescope that the arrival directions of the in-falling gamma rays are only known to be somewhere on an "event circle" (Schönfelder *et al.* 1984). For the present analysis events are selected whose event circle passes within 3° of the position of the pulsar. There are many non-source events whose event circles pass near the position of the pulsar under study which, in addition to the unpulsed component of the source emission, provide a background within the light curve. The phase distribution of such background events in the light curve is flat for fast pulsars. This has been verified by phase folding events selected from a symmetric "mirror" position on the opposite side of the detector field of view to the source. The significance of the pulsation has been evaluated by applying the Z_n^2 statistic (Buccheri and Sacco 1985) in an unbiassed way to the un-binned event phases looking for a n-harmonic signal. Three harmonics have been chosen for Crab and five for Vela following a harmonic component analysis of the observed signal.

3.1 The Crab Pulsar

As may be seen in Fig. 1 taken from Strong *et al.* (1992)m the Crab Nebula and contained pulsar shine brightly in the COMPTEL 1-10 MeV view of the Galactic anti-centre.

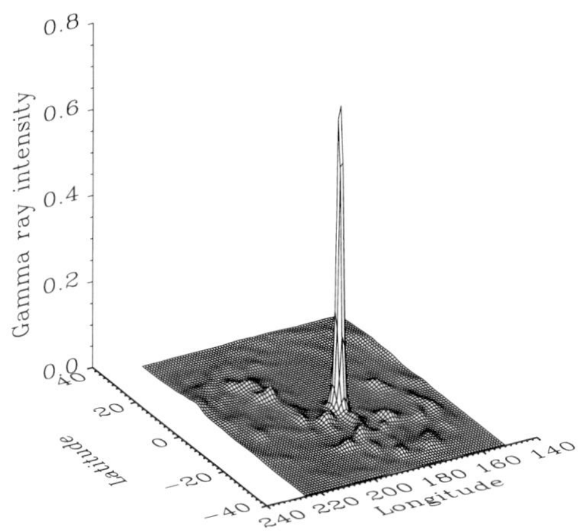

Fig. 1— 1-10 MeV sky-map of the Galactic Anticentre showing the total emission from the Crab.

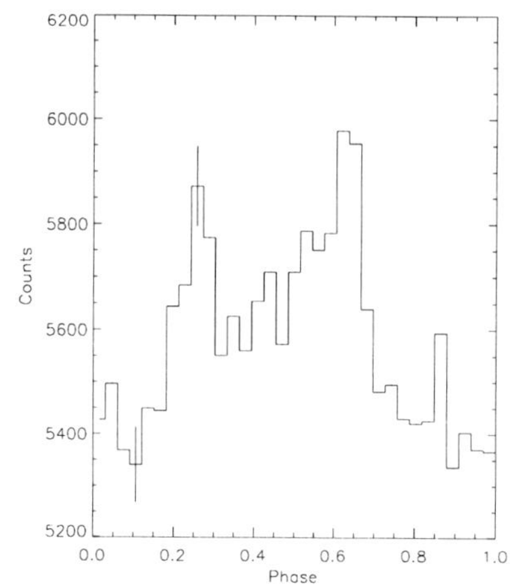

Fig. 2— 1-10 MeV light curve of the Crab Pulsar for three combined observations.

The light curve combined from all the observations is shown for the energy range 1-10 MeV in Figure 2. This graph represents approximately 5400 pulsed counts for a data-selection more appropriate for pulsar studies and which is slightly different from that used by Strong *et al.* The pulsation is judged to be present with statistical significance (from the Z_n^2) of 40 σ. Comparing the shape with that at higher energies measured by COS-B (Clear *et al.* 1987) and EGRET (Thompson *et al.* 1992), it is evident that the interpulse (*i.e.*, between the peaks) emission is proportionally much more intense in the COMPTEL energy range. The light curve resembles more closely that seen at hard X-ray energies by OSSE (Ulmer *et al.* 1992).

To investigate possible evolution within the energy range of COMPTEL, the light-curve is plotted in four energy intervals in Fig.3.

The Crab spectrum combined with the COMPTEL sensitivity results in the bulk of the signal being seen in the 1-3 MeV energy band. Nevertheless pulsed excesses are seen in all energy bands, as is interpulse emission.

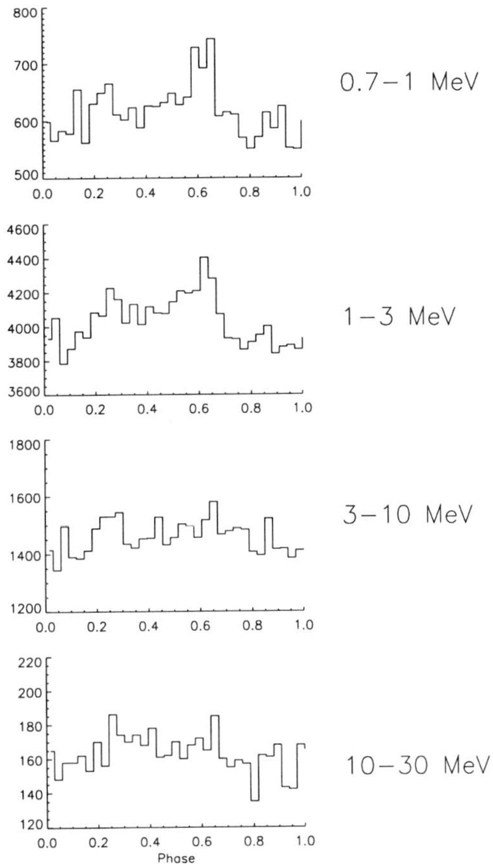

Fig. 3— Light curves of the Crab Pulsar for three combined observations in the four energy bands indicated.

The pulsed flux has been derived by Strong *et al.* (1992). The spectrum is shown in Fig. 4 and is consistent with those of EGRET and OSSE in adjacent energy bands and may be described by an $E^{-2.2}$ power law. The same analysis gives the pulsed ratio to be 25% (at ~ 1 MeV) to 36% (at 30 MeV).

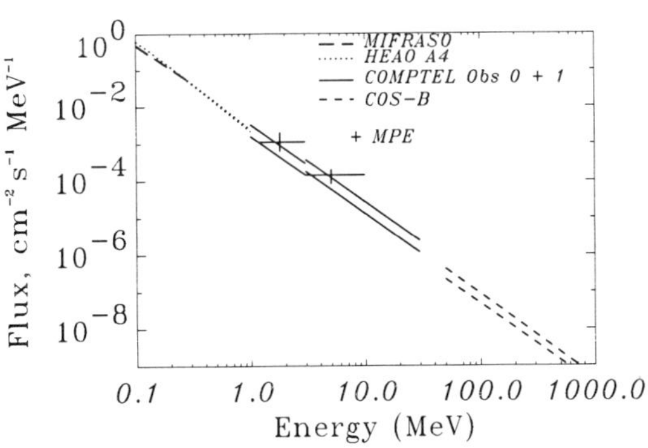

Fig. 4—COMPTEL pulsed spectrum of Crab (from Strong *et al.* 1992)

The complicated pulse structure: two barely resolved peaks, with the first being sharp and the second being a continuation of the interpulse, is rather different compared to the earlier measurements in the same energy range by the U.C.R. (White *et al.* 1985) and MPE (Graser and Schönfelder 1982) groups. Their results show two sharp isolated peaks. Although these balloon flight measurements have poorer statistics than COMPTEL, the difference in the light curves is evidence of a secular change like that reported at $\geq 50\ MeV$ by the COS-B team (Wills *et al.* 1982) and which is is *prima facie* supported by the EGRET findings (Thompson *et al.* 1992). Therefore some variation in the light curve obtained at different epochs may always be expected. The three individual light curves combined to produce Figure 2 and Figure 3 do show marked differences in shape but, given the attendant background of COMPTEL, quantification of these effects is statistically problematic and is subject of on-going analysis.

The phase of the maximum of the primary peak (determined to within 0.5 ms) is in perfect agreement with that derived by the other GRO instruments (private communication). Comparison with the radio phase determination is not yet complete.

3.2 The Vela Pulsar

The light curve from observation 8 is shown for the energy range 10-30 MeV in Figure 5. The optimum energy range for Vela is higher than that for the Crab because of its harder spectrum : the Z_n^2- significance of this signal is 6.6 σ. It should be recalled that Vela has only been seen once at energies below 30 MeV (Tümer *et al.* 1984) while it shines brightly at 100 MeV as seen by SAS-2 (Thompson *et al.* 1975), COS-B (Bennett *et al.* 1977) and Gamma-1 (Akimov *et al.* 1991). The COMPTEL data allow us to derive a phase selected skymap of the Vela region. Figure 6 shows the first image of the Vela pulsar derived in the energy range 10–30 MeV. The COMPTEL light curve (see Fig. 5) is markedly different from that obtained during a 1981 balloon flight by Tümer *et al.* (1984) in a comparable energy range. The U.C.R. results show two narrow pulses at the phase of the two high-energy gamma-ray pulses while the COMPTEL curve shows significant in-fill between the pulses. The harder spectrum of Vela is evident when comparing the energy selected light curves for Vela in Figure 7 with those of the Crab in Figure 3.

Grenier *et al.* (1988) have performed a detailed spectral analysis of several phase intervals (so-called Peak-1, Interpulse-1, -2, Peak-2 and Trailer) using COS-B data accumulated over a six year time span. There is striking evidence for spectral variations and for time variability of these spectra throughout these observations. Furthermore the spectral flickering is most noticeable at the lowest COS-B energies ($\simeq$ 50 MeV): just above the COMPTEL energy range. Akimov *et al.* (1991) used the Soviet-French Gamma-1 to monitor the Vela light curve for long periods (of the order of months) during 1990 and 1991. They confirm spectral variability at most phases.

For COMPTEL the most prominent feature in the observation 8 light curve is Peak-1 whereas for the Validation period, just 16 weeks earlier, Peak-2 dominated. This observation is in agreement with the variation in spectral slope for Peak-2 at high

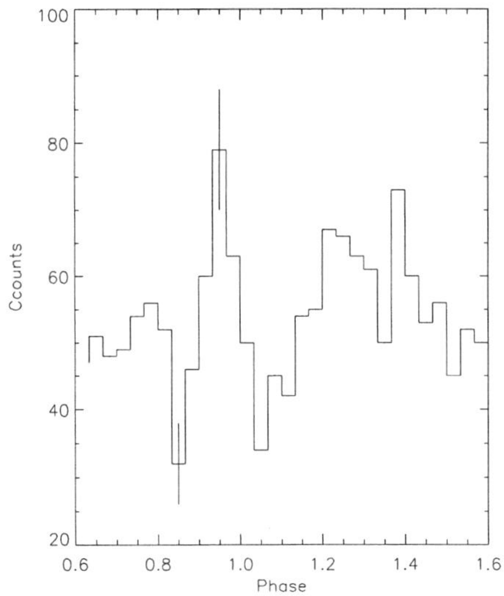

Fig. 5— 10-30 MeV light curve of the Vela Pulsar for GRO obs. 8.

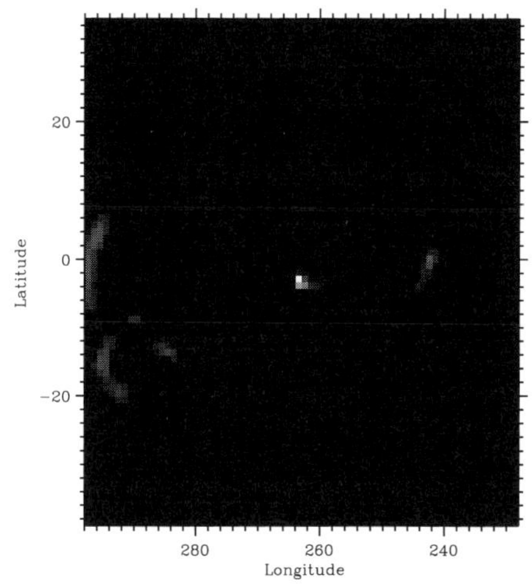

Fig. 6— Phase selected image of the Vela Pulsar in the energy range 10-30 MeV.

energies reported by EGRET (Kanbach *et al.*1992). Note that a glitch occurred in Vela just prior to observation 8 which could have some relevance to this effect. The significance of the variability of the features in the COMPTEL data are still being evaluated, but at least they are consistent with the behaviour at EGRET energies.

The COS-B Interpulse-1 (confirmed by EGRET and Gamma 1) is a region of intense emission just following Peak-1, but this is a region of *low* emission in the COMPTEL light curve. Thus, as was detected by Grenier *et al.*, there is a need for a spectral break in Interpulse-1. On the other hand, the adjacent Interpulse-2 is very intense in the COMPTEL phasogram, but this is consistent with the very soft spectral index (the softest feature of all) in the COS-B data.

The phase of Peak-1 (determined to within 1 ms) is in agreement with that derived by the other GRO instruments (private communication). Comparison with the radio phase determination remains, as for the Crab, the subject of on-going study.

4 Discussion

Rather than Crab and Vela pulsars being standard gamma-ray candles, evidence is mounting from all sides that they are flickering sources whose instability appears most strikingly in the gamma-ray energy range. The synergism of GRO will disentangle the spectral and time variability throughout the 6 decades of energy from 10's of keV to 10's of GeV.

The models, such as Polar-Cap and Outer-Gap must consider the possibility for variability on the time scales of weeks to years as well as spectral differences throughout the pulse.

A revisit to both these pulsars remains a prime goal of future GRO observations following completion of the All Sky Survey phase.

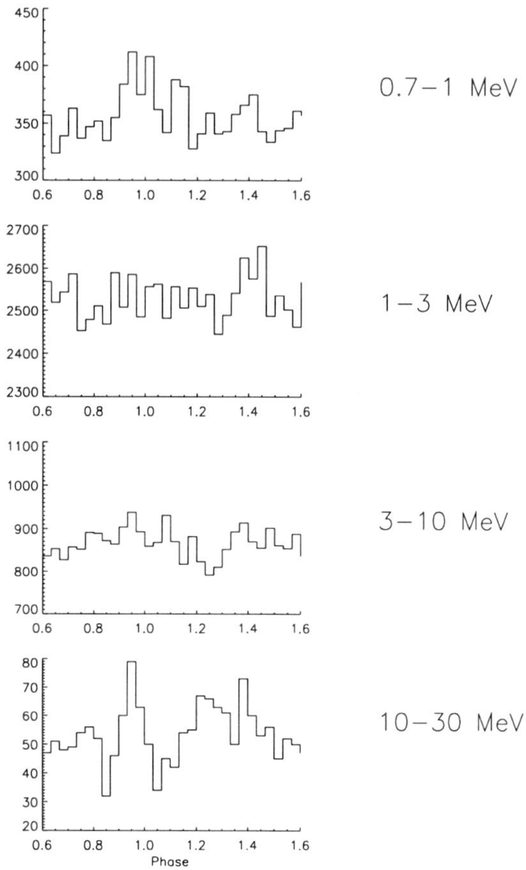

Fig. 7— Light curves of the Vela Pulsar for GRO obs.8 in the four energy bands indicated.

We are grateful to Radio Astronomy colleagues for providing the pulsar ephemeredes and to Miles Kingston for the JPL ephemeris. We benefited from discussions within the GRO Pulsar Group under the chairmanship of Mel Ulmer.

References

Akimov, V. V., *et al.* 1991, *Proceedings 22nd ICRC*, **OG 3.3.1.**
Bennett, K., *et al.* 1977, *Astr. Ap.*, **61**, 279.
Buccheri, R., and Sacco, B. 1985, in *Data Analysis in Astronomy II*, ed. Plenum Press, New York.
Busetta, M., *et al.* 1991, in *Data Analysis in Astronomy IV*, ed. Plenum Press, New York, in press.
Clear, J., *et al.* 1987, *Astr. Ap.*, **174**, 85.
Graser, U. and Schönfelder, V. 1982, *Ap. J.*, **263**, 677.
Grenier, I. A., Hermsen, W. and Clear, J. 1988, *Astr. Ap.*, **204**, 117.
Kanbach, G, *et al.* 1992, This Workshop.
Schönfelder, V., *et al.* 1984, *IEEE Trans. Nucl. Sci.* , **NS-31**, 766.
Strong, A. W., *et al.* 1992, *Astr. Ap. Supp.*, in press
Taylor, J., 1989, in *Proc. GRO Science Workshop.*
Thompson, D. J., *et al.* 1975, *Ap. J. (Letters)*, **214**, L17.
Thompson, D. J., *et al.* 1992, these Proceedings.
Tümer, O. T., *et al.* 1984, *Nature* , **310**, 214.
Ulmer, M, *et al.* 1992, these Proceedings.
White, R. S., *et al.* 1985, *Ap. J. (Letters)*, **299**, L23.
Wills, R., *et al.* 1982, *Nature* , **296**, 723.

Search for Pulsed Gamma-Ray Emission from SN1987A and the Vela Pulsar

A. D. Kerrick,[1,2] F. Ait-Ouamer,[1] T. J. O'Neill,[1] O. T. Tümer,[1] A. D. Zych,[1] R. S. White,[1] and J. Middleditch[3]

[1]Institute of Geophysics and Planetary Physics and Physics Department, University of California, Riverside
[2]Present address: Department of Physics and Astronomy, Iowa State University
[3]Los Alamos National Laboratory

Abstract

Gamma-ray data obtained on 15 and 16 April 1988 with the UCR Compton telescope is analyzed for pulsed emission from SN1987A and the Vela pulsar. No evidence of periodicity was seen in a fast Fourier transform search on SN1987A. Flux limits are $F_p(1.1-30\text{MeV}) < 3.6 \times 10^{-3}\text{cm}^{-2}\text{s}^{-1}$ for frequencies below 453.6 Hz, and $F_p(> 0.85\text{MeV}) < 2.3 \times 10^{-3}\text{cm}^{-2}\text{s}^{-1}$ for frequencies below 97.656 Hz. The Vela pulsar is possibly seen between 2.5 and 5 MeV, where the telescope effective area is maximum. Although the signal is not strong enough to claim a detection, the possible pulsed flux is consistent with that reported previously by UCR. No evidence for Vela is seen in the energy bands 1.1-2.5 MeV, 5-10 MeV, and 10-30 MeV.

1 Introduction

SN1987A has provided spectacular confirmations of theoretical predictions with the detection of neutrinos from core collapse and γ-ray lines from the decay of ^{56}Co. Efforts are still underway to determine if a pulsar was left behind. In searching for a pulsar, γ-rays have the advantage of more easily penetrating the debris created in the explosion, but also the disadvantage of low flux and the uncertainty which comes with only two pulsars (Crab and Vela) having ever been detected at Compton telescope energies. This paper describes an FFT search for pulsed power from SN1987A with data obtained by the UCR Compton telescope on 15-16 April 1988, some 417 days after core collapse.

This paper also describes a search for the Vela pulsar (PSR 0833-45) by folding the γ-ray data using pulsar parameters from radio observations. The Vela pulsar is a strong source above 50 MeV, where it has a double-peaked light curve similar to the Crab pulsar. EGRET, on board the Compton Observatory, is producing beautiful Vela light curves (Thompson 1992). COMPTEL is now seeing a pulsed signal from 10-30 MeV, but not at lower energies (McConnell 1992). A much earlier balloon experiment (Albats 1974) reported detection of a single-peaked Vela light curve at 10-30 MeV, with a different phase than the high energy pulsations.

A previous UCR paper (Tümer *et al.* 1984, hereafter UCR84) reported a double-peaked light curve and plotted a spectrum over the entire 1-30 MeV region. More recently, FIGARO II reported negative results in a search from 0.2–6 MeV (Sacco *et al.* 1990). In those energy regions where UCR84 and FIGARO overlap, the FIGARO upper limits are lower than the fluxes reported by UCR. The FIGARO paper speculates that the discrepancy could be related to glitch activity, as the UCR84 data was taken on 10 November 1981, 31 days after a major Vela glitch, whereas the FIGARO data of 25 November 1988 was removed from any major glitches by more than 3 years.

2 Data Set

The UCR Compton telescope has been described previously (Zych, Tümer, and Dayton 1983; Simone *et al.* 1985; O'Neill 1987; O'Neill *et al.* 1988; Ait-Ouamer *et al.* 1990). It consists of an array of 16 plastic (S1) scintillator bars ($6.35 \times 7.62 \times 103 \text{cm}^3$) situated 1 m above an array of 16 NaI(Tl) (S2) scintillator bars ($4.83 \times 4.83 \times 102 \text{cm}^3$). An incident γ-ray scatters in S1 and is absorbed in S2. The energies and positions of these interactions are recorded and allow reconstruction of the incident event direction to within a circle on the sky. The event time is recorded to within 10 μs. Additionally, "singles" data, the number of counts in the NaI scintillator, is recorded every 5.12 ms, allowing independent pulsar analysis.

The detector was launched on a balloon from Alice Springs, NT, Australia at 21:00 UT on 14 April 1988, reached float altitude ($\approx$ 36 km) at approximately 0:00 UT on 15 April, and was terminated at 6:27 UT on 16 April after more than 30 hours at float altitude. The residual atmosphere at float was less than 4 mbar. The energy thresholds for this flight were 0.25 MeV for S1 and 0.85 MeV for S2. The S2 singles rate was 9800 s^{-1}.

A source is considered in the detector aperture when its zenith angle is less than 60°. SN1987A was observed from 2:30 to 11:25 UT on 15 April, and again from 2:25 to flight termination on 16 April. This data is 417 days after core collapse. Gamma-ray lines from the decay of ^{56}Co were detected during this flight (Ait-Ouamer *et al.* 1992). The time is likely too soon for 1 MeV γ-rays from a pulsar to escape the shell surrounding the supernova (Michel, Kennel, and Fowler 1987; Woosley, Pinto, and Hartmann 1989).

Vela was observed from 5:18 to 14:43 on 15 April, and briefly from 5:17 to termination at 6:27 on 16 April. This brief Vela observation was not included in the analysis.

Telescope efficiencies were determined from Monte Carlo computer simulations with an incident source spectrum of E^{-2}. For the Vela pulsar passage the effective area over the entire 1.1–30 MeV range was 26 cm^2. For energy bands of 1.1–2.5 MeV, 2.5–5 MeV, 5–10 MeV, and 10–30 MeV the effective areas were 24, 36, 26, and 9.4 cm^2, respectively. For the SN1987A passage the telescope effective area was 13 cm^2. The effective area for NaI singles was 2.4×10^3 cm^2.

3 Search for SN1987A

Events (a total of 30,803) with circles within 10° of SN1987A from those times when the supernova zenith angle was less than 60° were selected for input to the fast

Fourier transform program. A single series of data was created by inserting zeros between the data for 15 and 16 April. The time series consisted of $2^{27} = 134,217,728$ points and had a Nyquist frequency of 453.6 Hz. Seventy-five transforms covered values of $\dot{f}_a$ from -9.9×10^{-9} to $+9.9 \times 10^{-9}$ Hz/s for an assumed pulsar frequency $f_a = 200$ Hz. In the binning process, a pulsar with *arbitrary* frequency f would be subjected to an given by $\dot{f}(f) = \dot{f}_a \times (f/f_a)$. The step-size $\delta\dot{f}$ between transforms is chosen so that an mismatch of half the step-size would degrade the pulsed power by 10%.

The highest power was seen at $f = 242.80939$ Hz and $\dot{f} = +4.2 \times 10^{-9}$ Hz/s. The power of 27.05 ± 7.36 is significant at the 99% level, but a true pulsar should show harmonic structure. When the data was rebinned with a higher Nyquist frequency, the power at 242.8 Hz was somewhat reduced and there was no evidence for harmonics. This is not a pulsed signal.

No other bins had power significant at the 90% level, which sets an upper limit of $F_p(1.1\text{MeV} \leq E \leq 30\text{MeV}) < 3.6 \times 10^{-3}\text{cm}^{-2}\text{s}^{-1}$.

S2 singles data were taken between 2:35 and 11:31 UT on 15 April 1988 and 2:35 and 6:27 UT on 16 April 1988, corresponding to the approximate times during which SN1987A was at a zenith angle of less than 60° with respect to the telescope. The ≈ 9 million points were padded and rebinned into a time series of $2^{25} = 33,554,432$ points for transforming. The barycenter corrected sampling frequency gives a Nyquist frequency of 97.656 Hz for this series. Seventeen transforms covered $\dot{f}_a$ from -2.08×10^{-8} to $+2.08 \times 10^{-8}$ Hz/s for an assumed pulsar frequency $f_a = 97$ Hz.

The power spectrum is estimated midway between consecutive pairs of Fourier lattice points with a two-element interpolation in order to avoid missing significant power which occurs between lattice elements. The large gap (≈ 15 hours) in the middle of the time series transformed here complicates the situation by making the peaks even narrower than normal. The spacing of the Fourier lattice was similarly narrowed by rebinning the data to only ≈ 25 million points and then filling the remainder of the 2^{25} points with the mean value of the data.

None of the bins in the power spectrum showed pulsed power significant at the 90% level. This limits the pulsed γ-ray flux above 0.85 MeV to $F_p(> 0.85\ MeV) < 2.3 \times 10^{-3}\text{cm}^{-2}\text{s}^{-1}$. Assuming a (Crab-like) $E^{-2.2}$ energy spectrum, this limiting flux corresponds to an emitted power of $5 \times 10^{39}\text{erg s}^{-1}$ in γ-rays above 0.85 MeV. Arnett and Fu (1989) applied theoretical models to the light curve for days 20 to 1500 and concluded that any pulsar must have a luminosity less than $2 \times 10^{39}\text{erg s}^{-1}$.

4 Search for the Vela Pulsar

Data with event rings coming from within 10° of the Vela pulsar (22,276 events) were folded with parameters provided by Manchester and Siegman (1988) who fit data from 18 December 1987 to 23 July 1988. These parameters were $P = 0.08928188760(10)$ s and $\dot{P} = 124.346(10)\times 10^{-15}$, at Julian day 2447148.1349. 0

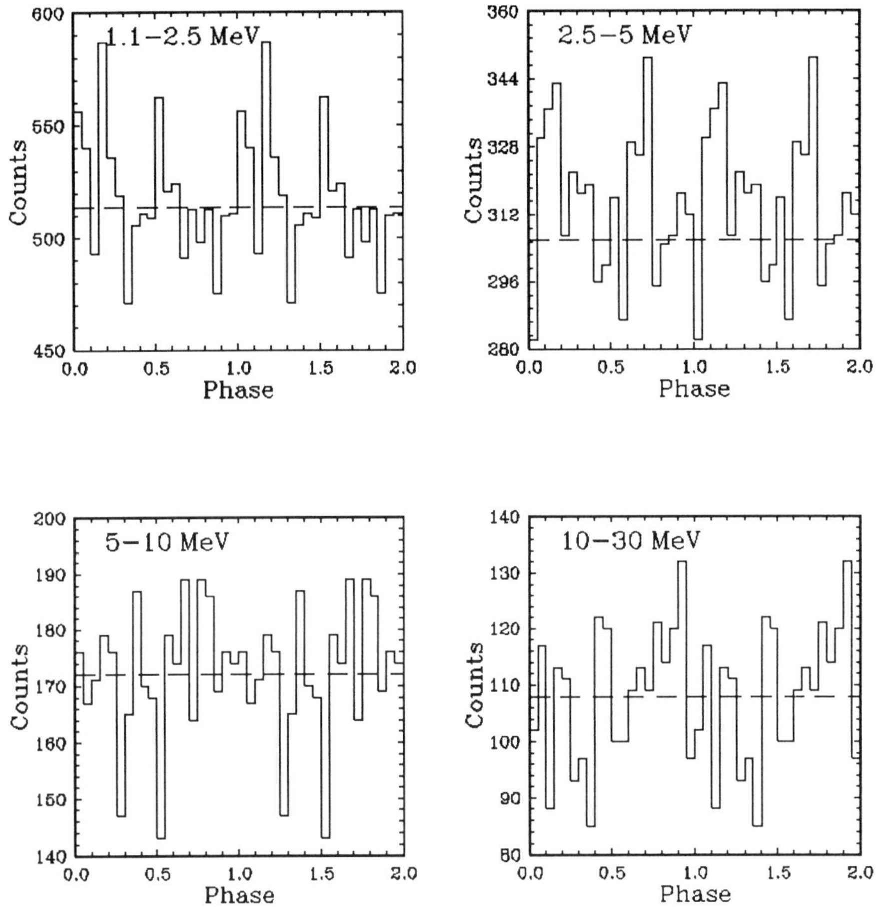

Fig. 1.—Vela pulsar light curves in different energy bands.

hr UT on 15 April 1988 was Julian day 2447266.5. Note that there is insufficient data to locate the absolute positions of any γ-ray peaks relative to the radio peak.

Light curves for energy bands of 1.1-2.5 MeV, 2.5-5 MeV, 5-10 MeV, and 10-30 MeV are shown in Figure 1. A complete light curve goes from 0 to 1 in phase; the histogram bins from 1 to 2 repeat the data from 0 to 1 and are inserted to make peaks near 0 phase easier to recognize. In most of these light curves significances of 3σ can be obtained by combining some of the large peaks, but in no two of the bands are the bins for these "peaks" the same.

The most interesting light curve is for 2.5 to 5 MeV. Interestingly, this is the energy band with by far the highest effective area. It shows two clear peaks with a phase separation of 0.45 ± 0.03, which compares well with the value of 0.43 ± 0.02 reported by UCR84. The peak widths of 0.15 in phase are considerably wider than those of 0.03 of UCR84. Identifying the main peak as bins 13 to 15 and the second peak as bins 2 to 4 gives peak significances of 2.7σ and 2.6σ. Combined, the two peaks have a significance of 3.4σ. Because of the lack of absolute phase information,

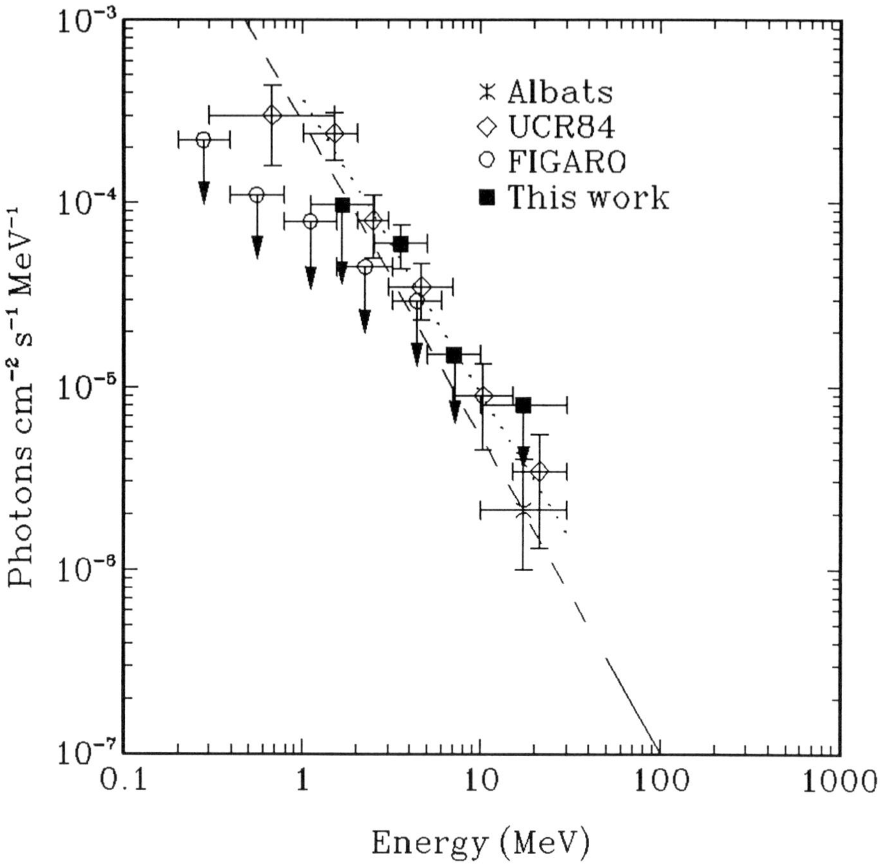

Fig. 2—Vela pulsar energy spectrum

this must be considered only as a possible detection. If it is truly a signal, the flux from 2.5 to 5 MeV is $(1.5 \pm 0.4) \times 10^{-4}\text{cm}^{-2}\text{s}^{-1}$, where the error is only statistical. Integrating the UCR84 fit from 2.5 to 5 MeV gives a flux of $(1.2 \pm 0.4) \times 10^{-4}\text{cm}^{-2}\text{s}^{-1}$.

The 2.5-5 MeV "light curve" was used as a template to analyze the other energy bands. The number of counts in these phase bins was 63 ± 66, 13 ± 37, and 0 ± 25 for 1.1–2.5, 5–10, and 10–30 MeV, respectively. Upper limits (2σ) for each of these energy bands, along with the possible flux from 2.5 to 5 MeV, are plotted in Figure 2, along with the fluxes and curve fit from UCR84, the upper limits of FIGARO II, the flux of Albats (1974), and the extrapolation of the COS-B spectrum measured from 50 to 300 MeV (Grenier, Hermsen, and Clear 1988).

The light curve for all data is given in Figure 3. Applying the template to the light curve for the combined 1.1–30 MeV data gives 256 ± 98 counts as an excess over background. This count corresponds to a flux of $(2.9 \pm 1.1) \times 10^{-4}\text{cm}^{-2}\text{s}^{-1}$, in agreement with UCR84.

Efforts are continuing to locate accurate absolute timing data for Vela, which would allow comparison of the positions of the possible peaks presented here with the location of the peaks of COS-B and UCR84.

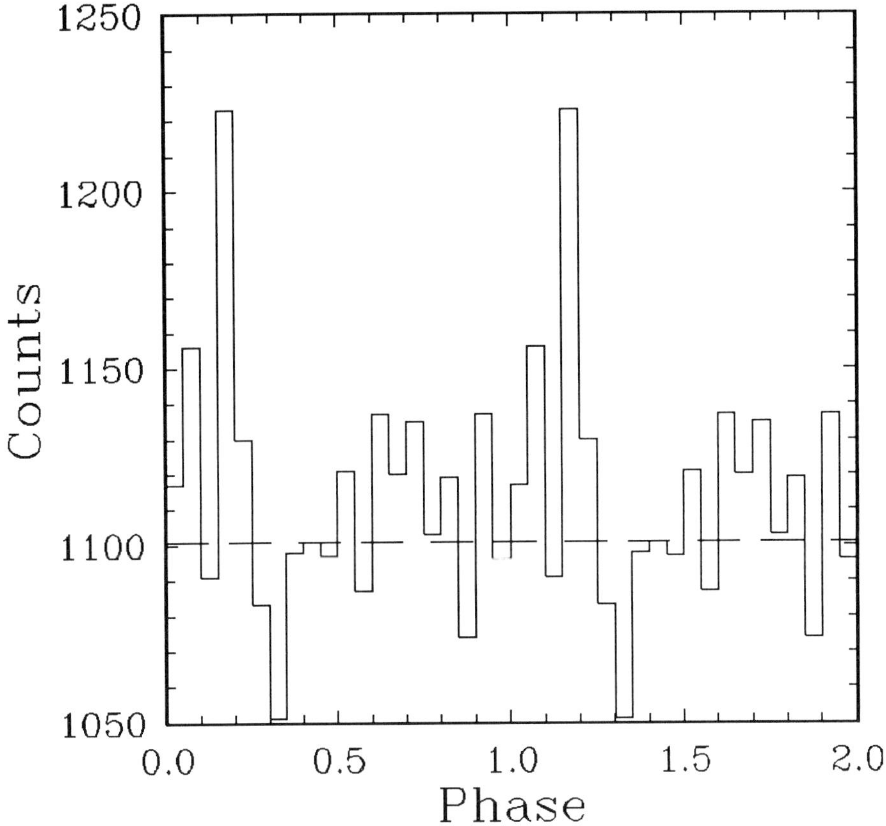

Fig. 3—Vela pulsar light curve ffrom 1.1 to 30 MeV. The dotted line is the fit to the UCR84 data; the dashed line is an extrapolation of the COS-B spectrum.

5 Conclusion

An FFT search has found no indication of pulsed γ-ray emission from SN1987A. The 90% upper limits are $F_p(1.1-30\text{MeV}) < 3.6\times10^{-3}\text{cm}^{-2}\text{s}^{-1}$ for frequencies below 453.6 Hz, and $F_p(> 0.85\text{MeV}) < 2.3\times10^{-3}\text{cm}^{-2}\text{s}^{-1}$ for frequencies below 97.656 Hz.

The Vela pulsar is possibly seen between 2.5 and 5 MeV, where the instrument effective area is maximum, but this detection cannot be claimed until more accurate timing information is obtained. The possible detection is consistent with levels measured by a previous UCR experiment.

We thank NASA for support through grants NGR 05-08-022 and NAGW-1996, and NSBF for launch, data transmission, and recovery operations.

References

Ait-Ouamer, F., *et al.* 1990, *IEEE Trans. Nuc. Sci.*, **37**, 535.
Ait-Ouamer, F., *et al.* 1992, *Ap. J.*, **386**, 715.
Albats, P., *et al.* 1974, *Nature*, **251**, 400.
Arnett, W. D., and Fu, A. 1989, *Ap. J.*, **340**, 396.
Grenier, I. A., Hermsen, W., and Clear, J. 1988, *Astr. Ap.*, **204**, 117.
Manchester, R. N., and Siegman, B. C. 1988, private communication.
McConnell, M. *et al.* 1992, these Proceedings.
Michel, F. C., Kennel, C. F., and Fowler, W. A. 1987, *Science*, **238**, 938.
O'Neill, T. J. 1987, Ph.D. Dissertation, University of California, Riverside.
O'Neill, T. J., Sweeney, W. E., Tümer, O. T., Zych, A. D., and White, R. S. 1988, *IEEE Trans. Nuc. Sci.*, 35, 511.
Sacco, B., *et al.* 1990, *Ap. J.*, **349**, L21 (FIGARO).
Simone, J., O'Neill, T. J., Tümer, O. T., and Zych, A. D. 1985, *IEEE Trans. Nuc. Sci.*, **32**, 124.
Thompson, D. J. *et al.* 1992, these Proceedings.
Tümer, O. T., *et al.* 1984, *Nature*, **310**, 214 (UCR84).
Woosley, S. E., Pinto, P. A., and Hartmann, D. 1989, *Ap. J.*, **346**, 395.
Zych, A. D., Tümer, O. T., and Dayton, B. 1983, *IEEE Trans. Nuc. Sci.*, **30**, 383..

OSSE Observations of the Crab Pulsar

M. P. Ulmer,[1] S. M. Matz,[1] R. A. Cameron,[2] D. A. Grabelsky,[1] J. E. Grove,[3] W. N. Johnson,[4] G. V. Jung,[2] R. L. Kinzer,[4] J. D. Kurfess,[4] M. D. Leising,[5] W. R. Purcell,[1] and M. S. Strickman[4]

[1]Northwestern University
[2]Universities Space Research Association
[3]NRC/NRL Resident Research Associate
[4]Naval Research Laboratory
[5]Clemson University

Abstract

We present preliminary results of the *Compton Gamma Ray Observatory* OSSE observations of the Crab pulsar. The pulsar energy spectra and light curves are in general agreement with previous observations, validating the OSSE pulsar data acquisition modes and data analysis algorithms. The data suggest that the spectrum of the pulsar varies throughout the light curve. The "interpulse" region has a slightly flatter spectrum in the $\sim$ 60–250 keV region and a slightly steeper spectrum at higher energies than the two main pulses. We find no evidence for any lines in the spectra with a typical sensitivity of about 10^{-4} photons cm^{-2} s^{-1}.

1 Introduction

The Oriented Scintillation Spectrometer Experiment (OSSE; Johnson *et al.* 1989) observed the Crab pulsar during instrument activation and also during observation period 1. In this paper we report on the preliminary analysis of the data acquired during viewing period 1 (17 May 1991 to 30 May 1991).

The OSSE pulsar data modes permit acquisition of time-tagged gamma-ray energy losses for the study of fast pulsars. The entire event stream for the detectors cannot be accommodated on an event-by-event basis in telemetry; consequently, the pulsar processing includes event selection and compression for telemetry formatting. The pulsar processing permits the definition of up to eight energy bands to be included in the transmitted pulsar data. These energy bands, as well as the rest of the pulsar data collection configuration, can be defined by OSSE mission operations activities and uploaded into the experiment via command. The pulsar data can therefore be optimized to the specific observing strategy and energy range of particular interest while limiting the event rate to that which can be handled in the OSSE telemetry.

Gamma-ray events qualified as being in one of these eight energy bands are then processed in one of two modes: (1) event-by-event mode, where selected events are time-tagged, and both energy loss and arrival time of the event are transmitted in the telemetry, or (2) rate mode, where high time resolution rate samples are taken in each of the eight energy bands.

The event-by-event mode of pulsar data provides the highest time resolution for the study of fast pulsars. This mode time-tags events accurate to 0.125 milliseconds at its highest resolution, and transmits the arrival times, detector identifications, and encoded energy losses. Depending on the OSSE telemetry format, a maximum of ~ 290 events per second is supported in the event-by-event pulsar mode.

The pulsar rate mode can accommodate a much higher event rate but at the sacrifice of spectral resolution. This mode records the number of events in each of the defined energy bands at a specified sample frequency. The highest sample rate in this mode provides a resolution of 4 milliseconds. Sample times from 4 msec to 512 msec can be selected. In this mode OSSE can achieve its best sensitivity to a continuum flux.

The data presented here were collected in the 1 millisecond event-by-event (or EBE) mode, and various energy ranges were selected at different times for different detectors in order to cover nearly the entire 50 keV–10 MeV range with optimal energy resolution.

TABLE 1
Crab Pulsar Ephemeris

ν	29.9493820573442 s^{-1}
$\dot{\nu}$	-3.77660×10^{-10} s^{-2}
$\ddot{\nu}$	7.45×10^{-21} s^{-3}
T0	8367.0 (TJD)
RA	$5^h34^m31\overset{s}{.}973$
Dec	$22°00'52\overset{''}{.}06$

2 Analysis method

Using a version of TEMPO (Taylor and Weisberg 1989) and a position of the Crab pulsar based on an ephemeris provided by Taylor and Nice (private communication), we corrected the arrival time of the time-tagged photons to the solar system barycenter. Next, we epoch-folded the data using the relative phases of these events calculated from the ephemeris provided by Taylor and Nice (Table 1). The resulting light curve for the ~ 60–246 keV energy range integrated over the observations made from TJD 8393.6 to 8399.5 is shown in Figure 1. In order to compare the phase of our pulse with that seen at radio frequencies, further phase corrections must be applied to the light curve (Taylor, private communication). However, even after applying these corrections, there is still a net offset of approximately 6 milliseconds between the two light curves. This offset is almost certainly not real, but is probably caused by insufficient accuracy in extrapolating the radio phase to infinite frequency. Further study of this problem is required.

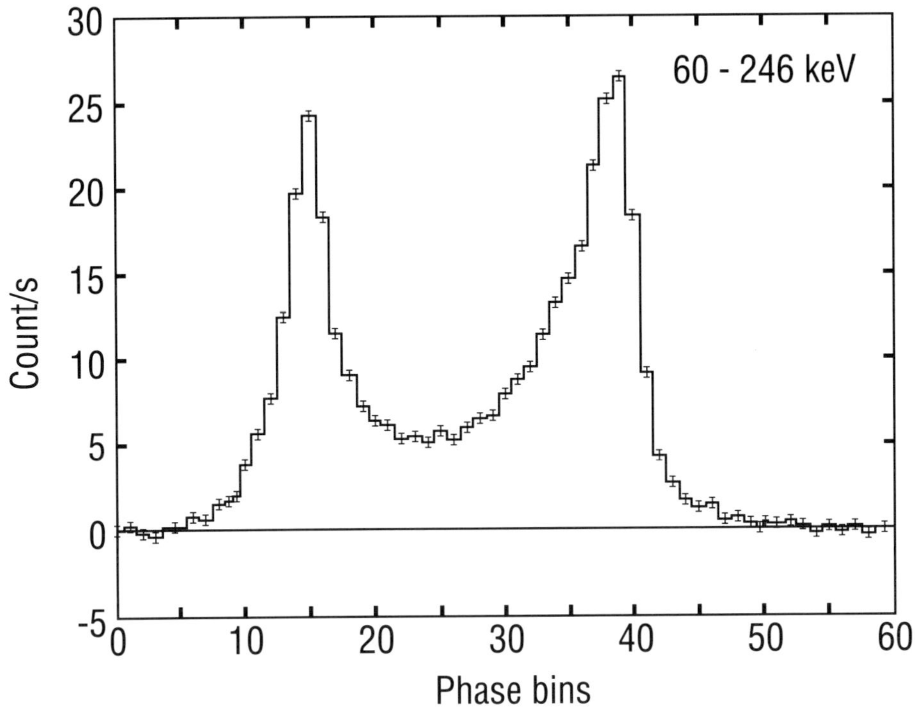

Fig. 1—The average Crab pulsar light curve.

3 Preliminary results

In Figure 2 we show the light curves in several broad energy channels and in Figure 3 we show the hardness ratio as a function of phase. From these two figures we see that in the lower energy range, the spectrum of the interpulse region (phase bins 20 to 29) is flatter than the spectra of the pulse peaks (phase bins 14 to 19 and 35 to 41). This is in basic agreement with Knight (1982), Mahoney *et al.* (1984), and Agrinier *et al.* (1990). We have produced several preliminary spectra, but it is too early to report a detailed spectral comparison between the different regions.

The average spectrum is shown in Figure 4 where we see very good agreement with previous experiments. Note, this is only an *approximate* photon spectrum since we used a simplified instrument response model which ignored Compton scattering.

We cannot conclusively exclude the existence of the 440 keV line in the spectrum of the second peak reported by the FIGARO group (Agrinier *et al.* 1990, Massaro *et al.* 1991) of $0.86 \pm 0.33 \times 10^{-4}$ photons cm^{-2} s^{-1}. The FIGARO result is based on combining data from balloon flights made on 11 July 1986 and 9 July 1990. In our data, we find that no features are obvious at 440 keV or elsewhere. Our 2σ upper limit to a line that is intrinsically much narrower than OSSE's ~ 8.5% detector resolution at 440 keV is $\sim 0.8 \times 10^{-4}$ photons cm^{-2} s^{-1}. On average, between 200 and 500 keV, our 2σ sensitivity to a narrow line for this observation is $\sim 10^{-4}$ photons cm^{-2} s^{-1}.

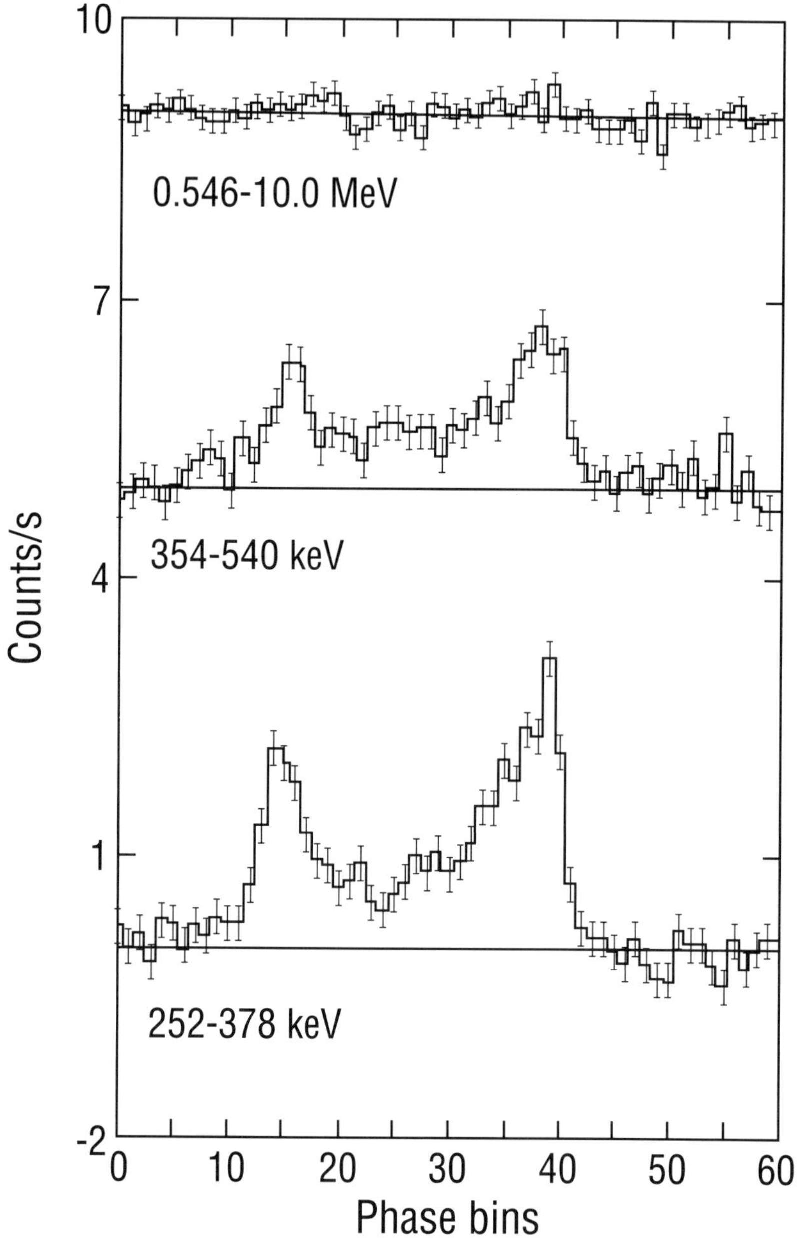

Fig. 2—The Crab pulsar light curve in different energy ranges.

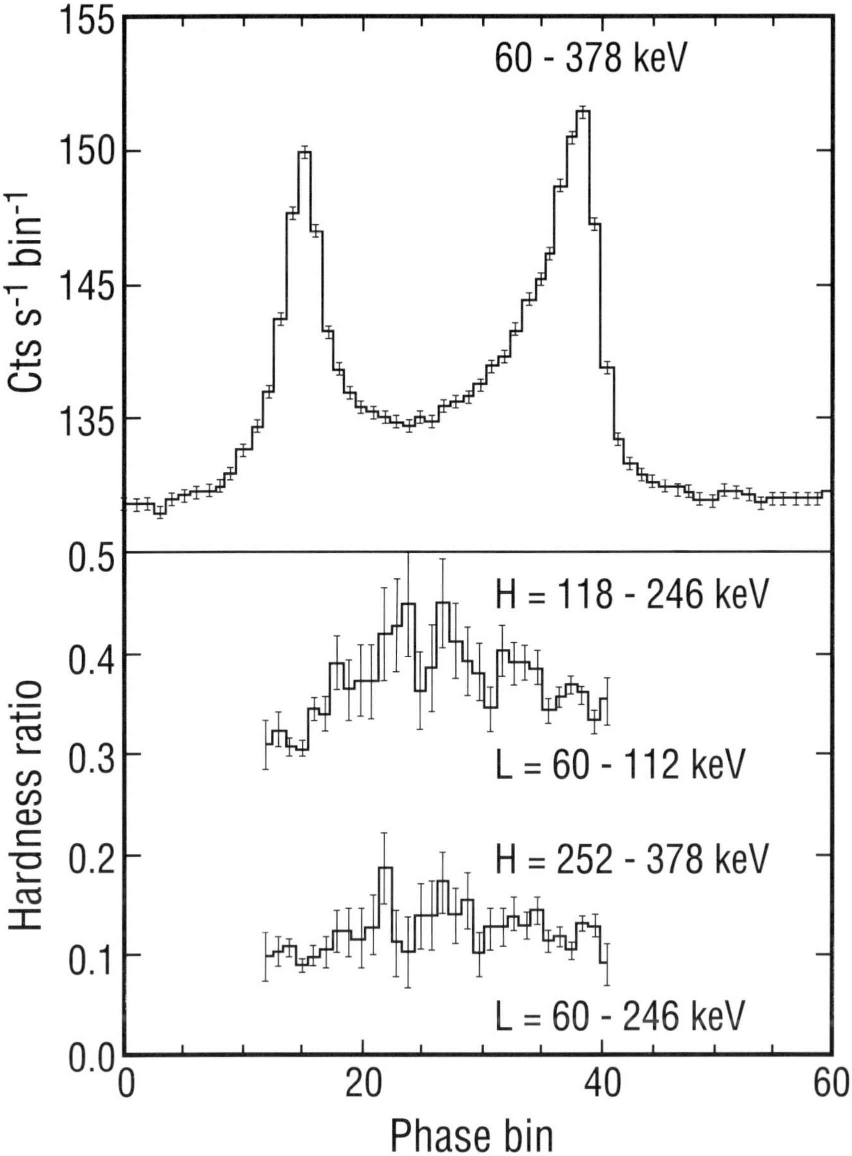

Fig. 3—The Crab pulsar light curve compared with the hardness ratio, plotted as the ratio of the higher energy channel counts over the lower energy channel.

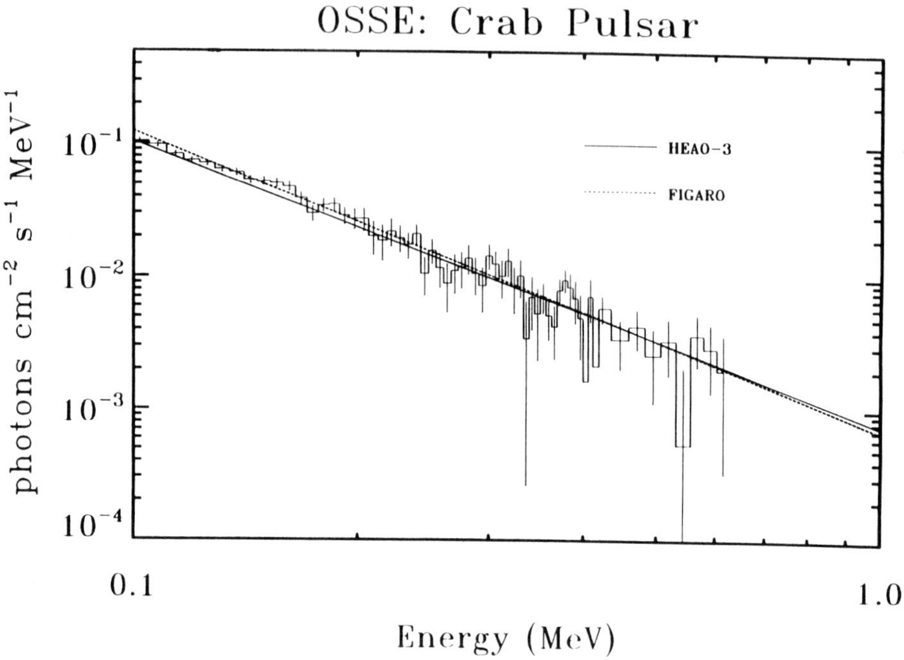

Fig. 4—The average OSSE Crab pulsar spectrum, unfolded using only the instrument photopeak response, compared with the spectra measured by HEAO-3 (Mahoney *et al.* 1984) and FIGARO (Agrinier *et al.* 1990)

4 Conclusions

The OSSE pulsar mode is performing as designed. For the Crab pulsar, we find general agreement with previous experiments in terms of average spectra and pulse shape. We do not yet have any evidence for line features in the spectra, but further analysis must be done. The difference in spectral shape between the interpulse region and the peaks exhibited in our data is consistent with that reported by Knight (1982). This spectral difference suggests that the interpulse emission is coming from a different site than the peak emission. Comparison with concurrent radio observations made by Phase I Guest Investigators from Cornell University may help elucidate the nature of the emission sites (Lundgren and Cordes 1991).

References

Agrinier, B., *et al.* 1990, *Ap. J. (Letters)*, **355**, 645.
Johnson, W. N., *et al.* 1989, in *Proc. of the GRO Science Workshop*, ed. W. N. Johnson, p. 2-22.
Knight, F. K. 1982, *Ap. J.*, **260**, 578.
Lundgren, S. C., *et al.* 1992, in *Proc. of the Compton Observatory Science Workshop*, ed. C. Shrader, N. Gehrels, and B. Dennis, p. 260.
Mahoney, W. A., Ling, J. C., and Jacobson, A. S. 1984, *Ap. J.*, **278**, 784.
Massaro, E., *et al.* 1991, *Ap. J. (Letters)*, **376**, L11.
Taylor, J. H., and Weisberg, J. M. 1989, *Ap. J.*, **345**, 434.

BATSE/CGRO Observations of Isolated Pulsars

Robert B. Wilson,[1] Mark H. Finger,[2] Geoffrey N. Pendleton,[3]
Gerald J. Fishman,[1] Charles A. Meegan,[1] William S. Paciesas[3]

[1]Space Science Laboratory, NASA/Marshall Space Flight Center
[2]Computer Sciences Corporation, NASA/MSFC
[3]Dept. of Physics, University of Alabama in Huntsville

Abstract

The uncollimated BATSE large area detectors (LADs) on the *Compton* Gamma Ray Observatory are capable of nearly continuous monitoring of the stronger hard x-ray pulsars, for objects with periods greater than ~4s. Observation of isolated pulsars, expected to have much shorter periods, is possible using on-board epoch folding. Preliminary pulse profiles and spectra are presented here for the Crab pulsar and for PSR1509-58, detected for the first time at energies as high as 1.8 MeV. An upper limit for the Vela pulsar is also obtained.

1 Introduction

The full-sky, uncollimated field of view of the BATSE detectors enables nearly continuous viewing of objects which can be distinguished from the background by intrinsic source periodicity. For periods greater than a few seconds, data for this purpose is contained in all telemetered data packets. For the objects considered in this paper, however, data was collected for an optimal detector combination and folded on-board.

Although the shielded detectors of OSSE are intrinsically more sensitive for this type of observation, the much longer exposure times available to BATSE, regardless of GRO orientation, permit a useful contribution to be made to searches for γ-ray emission from young radio pulsars.

2 Instrumentation

The Large Area Detectors (LADs) of the BATSE instrument consist of 8 uncollimated NaI(Tl) detectors, each ~2000 cm^2 in area, 1.27 cm thick, placed on the corners of CGRO, parallel to the faces of a regular octahedron. They have an approximately cosine response as a function of angle to the normal of a detector, with an energy resolution of approximately 28% at 88 keV. The low-energy response falls off below ~25 keV due to the presence of a plastic scintillator charged particle veto detector placed in front of each LAD.

For the results presented here, only the scheduled, folded-on-board data types have been used. Data collection for these data occurs in the following way: a schedule entry at a suitable time in the orbit starts the folding hardware and stores the time by latching the CGRO spacecraft clock. Data selected from the desired source-facing detectors only are folded for an integral number of periods, usually selected to complete in less than 8.192 s.

The sensitivity obtained by BATSE for a Crab-like pulse profile is shown in Figure 1. Typically, data with this sensitivity are collected for five sources during each fixed CGRO pointing.

Fig. 1—BATSE pulsed source detection sensitivity for a Crab-like pulse profile. The values, in units of keV-cm^{-2}-sec^{-1}, are the 3σ upper limits for a $10^5 s$ live-time observation. The dashed curve is the Crab pulsar spectrum of Mahoney *et al.* (1984).

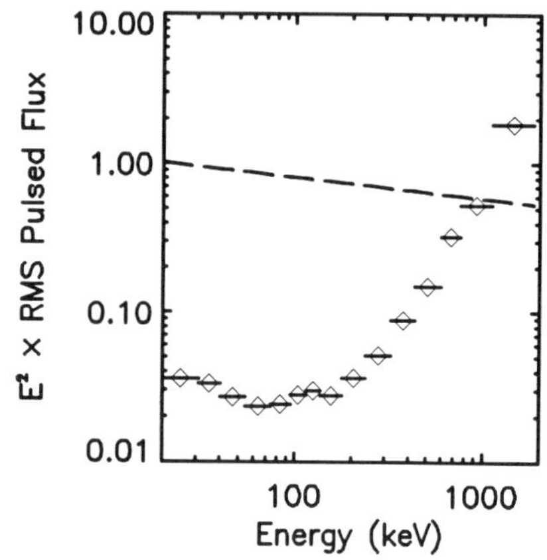

3 Data Analysis

Using the JPL DE200 ephemeris and the CGRO position at the midpoint time of each 8.192 s accumulation (known to better than 30 km), the direction to the Solar system barycenter (SSB) at the midpoint of the accumulation is determined. By using the source ephemeris either provided by Taylor *et al.* (1992) or obtained for a non-radio source by iteration, the phases of each of the 64 bin boundaries, centered on the middle of the accumulation, are determined. The counts are then summed into 64 phase bins fixed in the SSB, with counts split linearly into each output bin if the input bins span adjacent output bins. Additional details on the procedure are contained in Wilson *et al.* (1992).

4 Preliminary Results

4.1 Crab Pulsar

The pulse profile obtained for a representative Crab exposure is shown in Figure 2. This observation is consistent with previous results, as found by Mahoney *et al.* (1984) and Wilson and Fishman (1983). The main peak appears somewhat softer than the interpulse, with the "bridge" region noticeably harder than either peak. Results of a search for variability of the pulse profile over time are contained in Wilson *et al.* (1991) and Wilson *et al.* (1992).

At present it is not possible to provide absolute phase comparisons between any CGRO observations and radio measurements, because of an uncertainty in the knowledge of the CGRO clock absolute time.

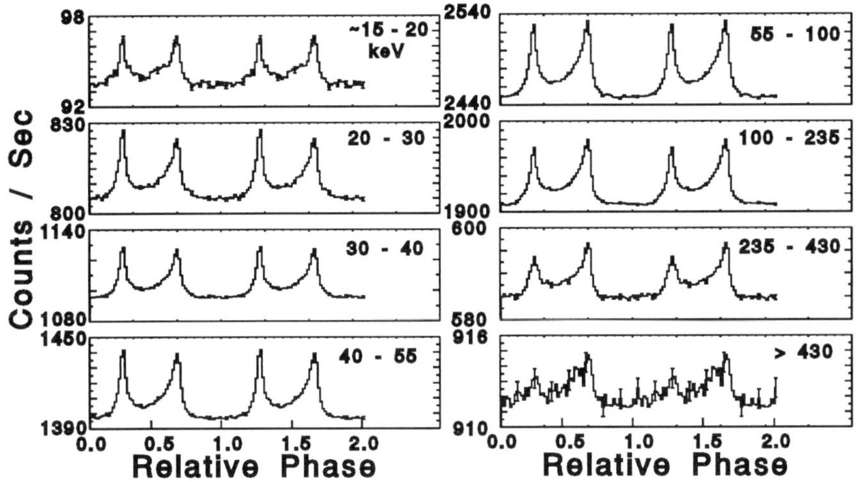

Fig. 2—Pulse profile of the Crab pulsar obtained for a 2-week CGRO pointing from TJD 8490 (August 22, 1991) to 8504, for the indicated energy ranges. The livetime was 144,800 s. The data were combined on board from two detectors, at angles of 34° and 54° from the detector normals.

Deconvolved phase-averaged photon spectra have been obtained using direct inversion of detector response matrices developed by Pendleton *et al.* (1989), based on pre-flight calibration measurements and Monte Carlo simulations. These spectra are shown in Figure 3. All data collected by BATSE on the Crab pulsar have been processed in similar fashion, with comparable results obtained in all viewing periods.

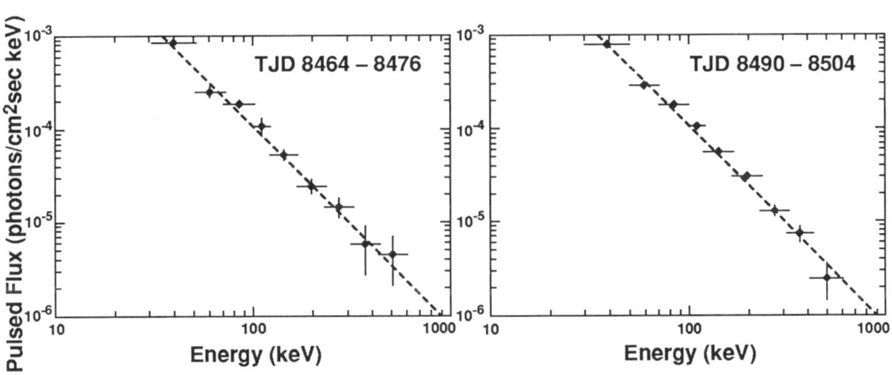

Fig. 3—Phase averaged photon spectra for the Crab pulsar, for two different CGRO pointing intervals, using different BATSE detector combinations. The dashed curve in each frame is the spectral model of Mahoney *et al.* (1984).

The agreement with the Mahoney *et al.* (1984) spectrum is good. At energies below 30 keV, the inversion technique produces results that follow a much softer spectral index, principally because of uncertainties in the energy calibrations where the effective area is rapidly decreasing. It is planned to use the Crab as a "standard candle" to help improve understanding of the detector response in this region.

For the exposure from Truncated Julian Day (TJD) 8490 - 8504, spectral indices have been obtained as a function of phase throughout the pulse, using 8 equal phase intervals during the pulse, with the off-pulse background interval defined as the relative phase interval 0.391-0.609 (see Figure 2). The index was obtained by a forward folding of a trial power-law spectrum through the detector response, using the energy range 50–425 keV. It should be noted that statistically good fits are not obtained, as determined by χ^2 per degree of freedom. The results obtained are shown in Table 1. The qualitative behavior as a function of phase agrees with the results of Knight (1982), and thus disagrees with the measurements of Agrinier *et al.* (1990), which are obtained over a somewhat higher energy interval of 0.2–6 MeV.

TABLE 1
Crab Pulsar Spectral Index vs. Phase[a]

Relative Phase[b]	α
0.609–0.703	2.65±0.40
0.703–0.797	2.19±0.03
0.797–0.891	2.12±0.02
0.891–0.984	1.98±0.04
1.078–1.172	2.01±0.03
1.172–1.265	2.07±0.02
1.265–1.359	2.08±0.02
1.359–1.453	2.24±0.21

[a]Phase origin as shown in Figure 2.
[b]Off-pulse background phase interval is 0.391-0.609.

One contributor to the small systematic departures of these results which prevent reduction of errors to the statistical level is atmospheric scattering of source photons into the uncollimated BATSE detectors. Work is in progress to account for these contributions to detector response, by computing scattered response matrices for a particular source-geocenter-spacecraft geometry, then weighting them according to the time spent during a particular observation in that configuration.

4.2 PSR1509-58

Starting on TJD 8561, data were collected for PSR1509-58. This object is the third youngest radio pulsar, after the Crab and Vela pulsars, and has the highest estimated dipole magnetic fields of any pulsar. It has been observed at x-ray energies by the Ginga spacecraft, as reported in Kawai *et al.* (1991). To date, data up to TJD 8644 have been processed. The pulse profiles obtained are the first detection of the source above ~60 keV and are shown in Figures 4 and 5. The single-peaked profile resembles that obtained by Kawai *et al.*, but is somewhat narrower. There is an evident decrease of the pulse width as a function of increasing energy deposit in the detector. At the highest energies detectable in the LADs, the integral channel registering energy

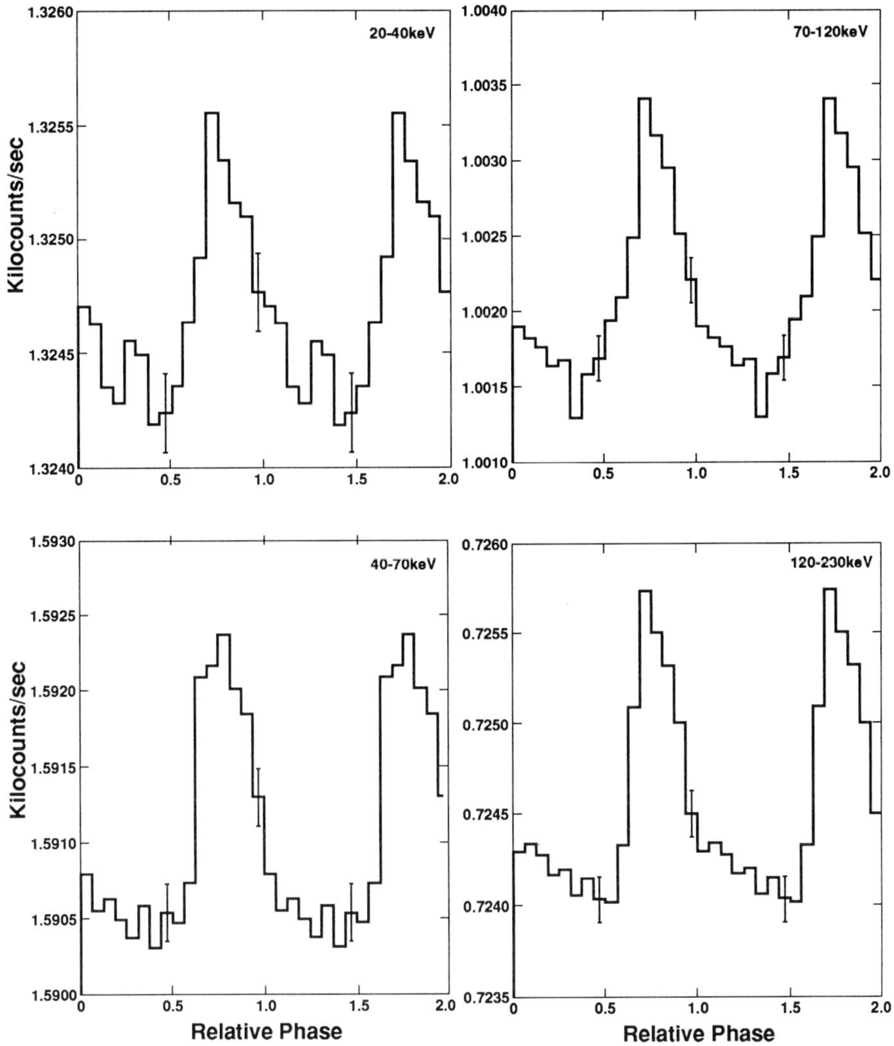

Fig. 4—Pulse profiles for PSR1509-58, obtained using a 723,000 s livetime exposure obtained from TJD 8602-8644.

deposits >1.8 MeV, some evidence of a narrow feature in the two highest bins can be seen in Figure 5. The excess rate in those combined bins is 3.3σ above background taken over the rest of the phase interval and is aligned in phase with the main peak at lower energies.

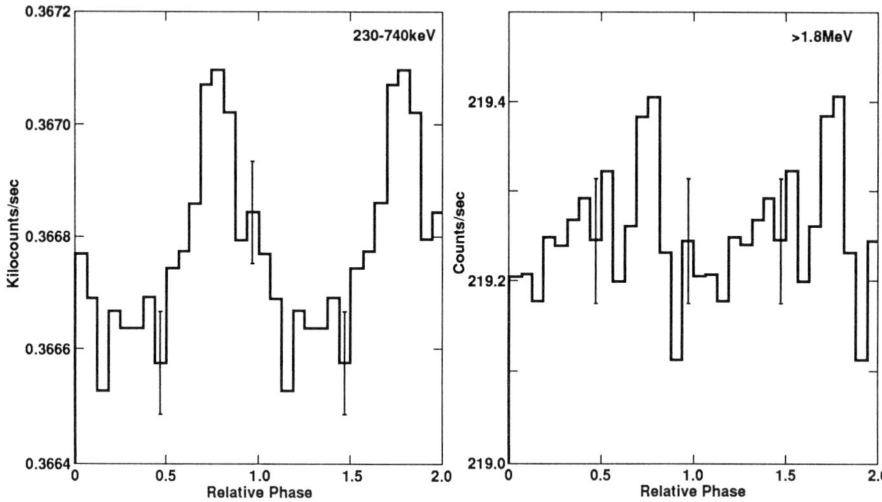

Fig. 5—Pulse profiles for PSR1509-58 at energies above 230 keV, from the same exposures as in Fig. 4.

Spectral deconvolution was performed by direct inversion of the detector response matrices for each of the observations that were used to produce the above pulse profiles. A weighted average was then calculated for each energy range separately, to obtain the spectrum shown in Figure 6. The spectrum is clearly harder than that of the Crab pulsar, shown for reference. The spectral index obtained using forward folding of a simple power law model is 1.8 ± 0.1, although a statistically good fit is not obtained.

Spectral deconvolution using the integral channel is difficult, since the energy distribution of the photons registering there is not known, and the predominant photon interaction in the detector changes from Compton to pair-production.

4.3 Vela Pulsar

Data was collected for the Vela pulsar with a livetime of 798,000 s, with no signal observed. The 3σ upper limits for this source thus lie a factor of 2.8 below the points in Figure 1.

5 Conclusions

BATSE results on the Crab pulsar are consistent with previous observations and show good agreement over different CGRO pointings. Results below 30 keV indicate

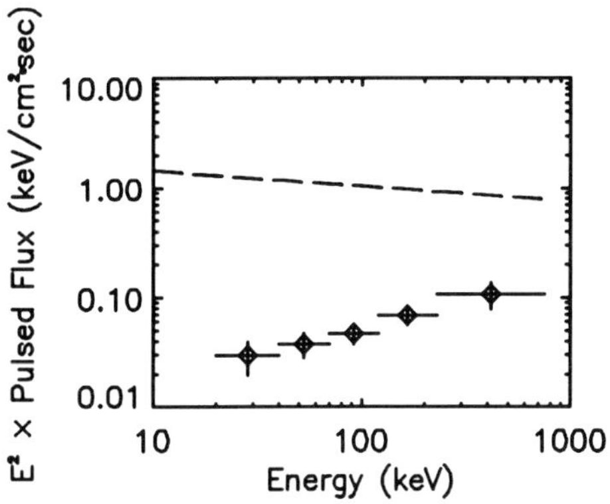

Fig. 6—Deconvolved spectrum for PSR1509-58, as obtained using a 723,000 s livetime exposure obtained from TJD 8602-8644. The fluxes have been multiplied by E_γ^2 to flatten the spectrum. The dashed line is the spectrum of the Crab pulsar of Mahoney *et al.* (1984)

a need for improvements to the energy calibration and BATSE LAD detector response matrices, which may well be provided using these observations.

This observation of PSR1509-58 represents the highest energy range detection of this source and only the third isolated radio pulsar observed in the hard x-ray - γ-ray range. The harder spectrum and varied pulse profile compared to the Crab and Vela pulsars provides a third system in which to test models for isolated pulsars.

We thank the entire BATSE operations team for timely completion of the initial data processing. The contributions of the programming staff, especially Lisa Gibby and Rogers Hunt, have been most important in these analyses.

References

Agrinier, B., Masnou, J. L., Parlier, B., Niel M., Mandrou, P., Vedrenne, G., Gerardi, G., Mineo, T., Sacco, B., Scarsi, L., Costa, E., Massaro, E., Matt, G., Salvati, M., and Lyne, A. G. 1990, *Ap. J.*, **355**, 645.

Kawai, N., Okayasu, R., Brinkmann, W., Manchester, R., Lyne, A. G., and D'Amico, N. 1991, *Ap. J. Lett.*, **383**, L65.

Mahoney, W. A., Ling, J. C., and Jacobson, A. S. 1984, *Ap. J.*, **278**, 784.

Pendleton, G. N., Paciesas, W. S., Lestrade, J. P., Fishman, G. J., Wilson, R. B., and Meegan, C. A. 1989, in *Proceedings of the Gamma Ray Observatory Science Workshop* 1989, ed. Johnson, W. N., p. 4-547.

Taylor, J. H., 1992, public database: anonymous login via **ftp** to **puppsr.princeton.edu**.

Wilson, R. B., Finger, M. H., Fishman, G. J., Meegan, C. A., and Paciesas, W. S. 1992, these Proceedings

Wilson, R. B., and Fishman, G. J. 1983, *Ap. J.*, **269**, 273.

Wilson, R. B., Harmon, B. A., Finger, M. H., Fishman, G. J., Meegan, C. A., and Paciesas, W. S. 1991, in *Proceedings of the Compton Observatory Science Workshop*, ed. Shrader, C. R., Gehrels, N., and Dennis, B., p. 35.

HST High Speed Photometer Observations of the Crab Pulsar

J. D. Biggs,[1,2] J. W. Percival,[3] R. C. Bless,[3] J. F. Dolan,[2] J. L. Elliot,[4] M. J. Nelson,[3] E. L. Robinson,[5] M. J. Taylor,[3] and G. W. van Citters[6]

[1]Universities Space Research Association
[2]NASA/Goddard Space Flight Center
[3]University of Wisconsin,
[4]MIT
[5]University of Texas
[6]NSF

Abstract

We present observations of the Crab Pulsar using the High Speed Photometer (HSP) aboard the *Hubble Space Telescope* (*HST*). So far, we have obtained 4 half-hour observations at visual wavelengths, and one half-hour observation at UV wavelengths, each with 10.7μs time resolution. The UV observations are particularly interesting because the Crab Pulsar has not been photometrically observed in this wavelength region before. In the preliminary analysis completed to date, we have found that the UV main pulse is narrower, and the pulse component separation is smaller, than in the optical. Also, we calculate that the pulsed emission in the lowest part of the optical average profile was $\lesssim$0.5% than that of the main pulse. A study of the individual optical pulses failed to find any correlation between emission between any two phase intervals of the average profile.

1 Introduction

The pulsar in the Crab Nebula (PSR0531+21) was the first radio pulsar to be detected at optical wavelengths, and to date only two other pulsars have optical detections. The average pulse profile has a very similar shape across a large portion of the electromagnetic spectrum, from the infra-red to very high energy γ-rays (Lyne and Graham-Smith 1990, and references therein). Because of the relative ease with which observations can be acquired in the optical, as compared to other wavelengths, the optical average pulse profile for the Crab pulsar has received the most scientific study. In our study, we hope to obtain further clues to the high energy emission mechanism of radio pulsars by a comparison between the optical and UV average profiles.

The instrument we have used to acquire our observations is the HSP. It is located aboard the *HST* and can undertake measurements covering the UV and optical spectrum (1200 to 7500Å) with sample times as short as 10.7μs. The HSP has many advantages over similar ground based instruments because of its access to the UV, its smaller apertures that afford higher spatial resolution and reduction in background, and the lack of an atmosphere that absorbs and scintillates starlight. For a full description of its operation consult the HSP Handbook (White 1990), and for details of the results from early observations with this instrument see Percival, Bless and Nelson (1991).

2 Observations

Observations of the Crab pulsar were undertaken for about 30min on five occasions with a sampling time of 10.7μs (see Table 1). The first four observations were acquired using a broad band optical filter (3600 to 6000Å) and were spaced over about four days. A broad band UV filter (1600 to 3000Å) was used for the last observation which was acquired about 3 months later. Because of the spherical aberration in the *HST* a 1″ aperture was employed for all observations to maximize the detected signal.

TABLE 1
Crab Pulsar Observations

UT	no. P[a]	count rate[b] (s^{-1})
91/10/17 16:09	64636	354.0±0.4
91/10/18 06:39	53864	490.1±0.5
91/10/19 06:49	53863	378.1±0.4
91/10/20 07:00	53864	346.6±0.4
92/01/21 07:58	50259	43.6±0.2

[a]Number of pulse periods (P) observed.
[b]Total count rate, includes emission from pulsar and nebular background.

3 Data Reduction and Analysis

The first stage of the data reduction was to assign a time of arrival at the solar system barycentre to each sample in order to be able to recover the individual pulses from the pulsar. This is a more complicated process than is required for ground based observations because of the *HST* orbit. Barycentric arrival times were calculated for each sample from its arrival time at the telescope by using the definitive ephemeris from the Flight Dynamics Facility at NASA/GSFC to calculate the position of the *HST* with respect to the earth, and then the JPL DE-200 planetary ephemeris was used to determine the position of the earth with respect to the solar system barycentre. The pulse period was calculated using the Crab pulsar ephemeris (Lyne and Pritchard 1992) and divided into 3000 phase intervals, then the number of photons in each 10.7μs sample were assigned to each interval depending on its barycentric time of arrival. The resulting phase resolution was equivalent to 11.1μs for all the observations.

3.1 Average Profiles

Average profiles were computed for each observation by adding the number of photons that arrived in each phase interval. A composite average profile was created from the four separate optical average profiles by visually aligning the first, third

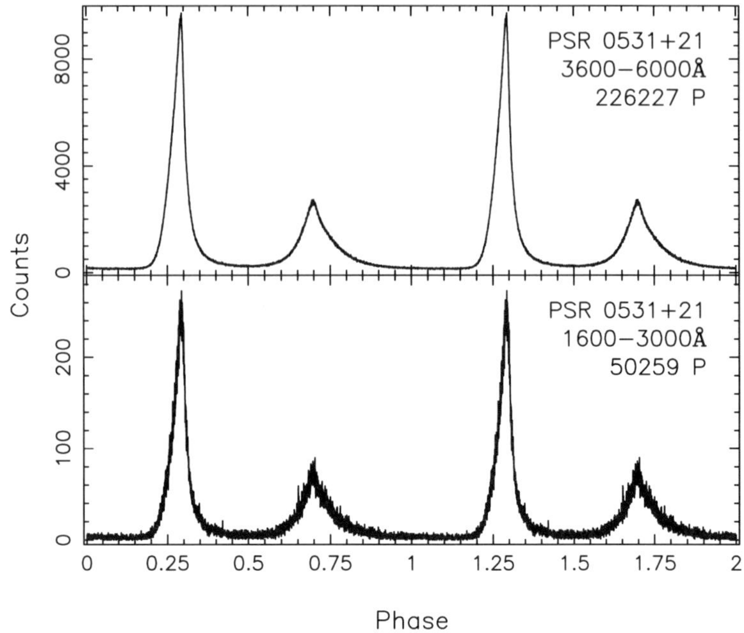

Fig. 1—Upper panel: Composite Crab pulsar average profile calculated from the broad band optical HSP observations (effective time resolution ≈ 12.4μs). Lower panel: Crab pulsar average profile calculated from the broad band UV HSP observations (time resolution = 11.1μs). Note that the profiles are repeated for two periods in order to show the continuity of the pulse profile around phase 1.0 and phase 0.0 is arbitrary.

and fourth observations with the second (this observation had the largest number of counts). The final phase alignment of ±1 phase interval was performed by choosing the composite average profile that was the narrowest. We estimate that this procedure degraded the effective time resolution to 12.4μs. The resultant two-component profile shown in Figure 1 has a shape similar to that obtained by ground-based high speed photometry. However, we find that the count rate at the lowest part of the pulse is substantially reduced for the HSP observations than for ground-based observations. This clearly illustrates the benefits of a space-based optical observatory because the background can be reduced without a significant loss of signal by the use of smaller apertures. In the following, we shall use the terminology main pulse, interpulse and 'bridge' for the high narrow peak, the low broad peak, and the region just after the trailing edge of the main pulse and before the leading edge of the interpulse, respectively.

Using the AAT, Smith *et al.* (1978) were able to resolve the peak of the Crab pulsar main pulse because they found that the five highest phase intervals (of 20μs width) were relatively flat. Our observations are in agreement with theirs, because a χ^2 test we performed on the 13 highest and contiguous samples (phase 0.291 to 0.295) of

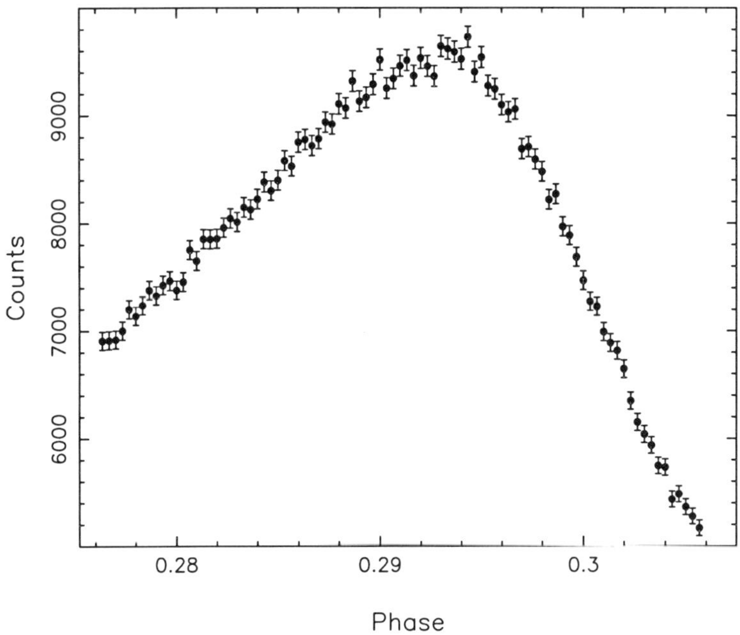

Fig. 2—Expanded plot of the peak of the Crab pulsar optical average profile shown in Figure 1. The error bars represent ±1 standard deviation errors from counting statistic alone.

the peak shown in Figure 2 suggests that these samples are statistically consistent (probability > 99%) with having the same value, *i.e.*, the peak is flat.

Qualitatively, the UV average profile for the Crab pulsar, shown in Figure 1, is very similar to the optical when account is taken of its lower signal-to-noise ratio. This similarity is to be expected because the centre frequency of the observations are only separated by a factor of two. The low count rate of the UV observations does not facilitate the resolution of the main pulse in the average profile as is possible in the optical.

Differences between the optical and UV profiles were studied by calculating the widths of their components, their separation, and their relative heights. The autocorrelation function (ACF) was used in this analysis because it has the virtue of being independent of any modelling of the pulse structure. However, we also note that the widths determined in this manner will be systematically larger than other measures such as the full width at half maximum. To remove any bias in the ACF calculation of widths, the 'background' level was removed from the average profiles. The widths were determined from the phase lag at which the ACF had fallen to half of its 0 lag value. Only the first half of the pulse (see Figure 1) was used in the calculation for the main pulse, and the second half of the pulse was used likewise for the interpulse. The phase lag of the second peak in the ACF of the whole pulse was used to calculate the separation of the main and interpulse. Table 2 contains the widths and separations calculated as outlined above. The peak heights of the main pulse and interpulse were

TABLE 2
Parameters of the Average Pulse Profiles

	width (ACF=0.5)[a] main pulse	width (ACF=0.5)[a] interpulse	separation[a]	$\frac{N_{\text{peak main pulse}}}{N_{\text{peak interpulse}}}$	$\frac{\sigma_{\text{off pulse}}}{N_{\max}}$
Optical	0.0457±0.0001	0.1136±0.0009	0.4167±0.0004	3.78±0.11	< 0.5%
	16.44±0.04°	40.89±0.32°	150.01±0.16°		
UV	0.0442±0.0001	0.1111±0.0010	0.4139±0.0005	3.63±0.14	< 2%
	15.91±0.05°	39.98±0.36°	149.00±0.18°		

[a]The upper values are measured in phase and the lower values are measured in degrees.
Note—one pulse period = 360°, or 1.0 in phase.

calculated by a quadratic fit to the peaks of these components, and the ratio of these were calculated and are also given in Table 2.

The ACF analysis clearly shows that the main pulse is significantly narrower in the UV than in the optical, and the main pulse-interpulse separation is significantly smaller in the UV. There is also some indication that the interpulse may be narrower in the UV, and relatively higher with respect to the main pulse. In Figure 3 we have overlaid the UV and optical average profiles that are plotted in polar coordinates, and the difference in width of the main pulses is clearly visible. However, the difference in main pulse-interpulse separation is masked by the relatively low signal-to-noise ratio of the UV profile.

Because of the manner in which the HSP targets are acquired, we can also obtain some information on the background (nebular emission) close to the pulsar. This background can then be used to determine its contribution in the observation aperture and place limits on the pulsar emission at the phase interval of the lowest count rate region of the average pulse profile. The background near the Crab pulsar was estimated from the finding aperture maps that are acquired in order to locate the pulsar, and facilitate its centering in the observation aperture. We fitted a two-dimensional quadratic function to the map, weighting the pixels within a 2″ radius around the pulsar to 0, and then interpolated the background count rate at the position of the pulsar resulting in values of 63±11 and 4.5±0.7 counts/s, in one phase interval, for the optical and UV observations, respectively. These values compare well to the minima of the average profiles (located at about 0.15 in phase before the peak of the main pulse) that were determined by fitting a quadratic function to the lowest part of the average pulse and have values of 58.9±0.3 and 4.6±0.3 counts/s for the optical and UV, respectively. The agreement of the background and lowest average pulse count rate are shown in Figure 4, and this suggests that the pulsar may 'turn off' to the level of our uncertainties. The upper limits (ratio of the uncertainty in the 'off-pulse' region with the count rate at the pulse peak) to emission in the off-pulse region are also given in Table 2. Figure 4 also shows that the ratio of the count rate at the centre of the bridge region is ≈ 1% of the peak for both the optical and UV observations.

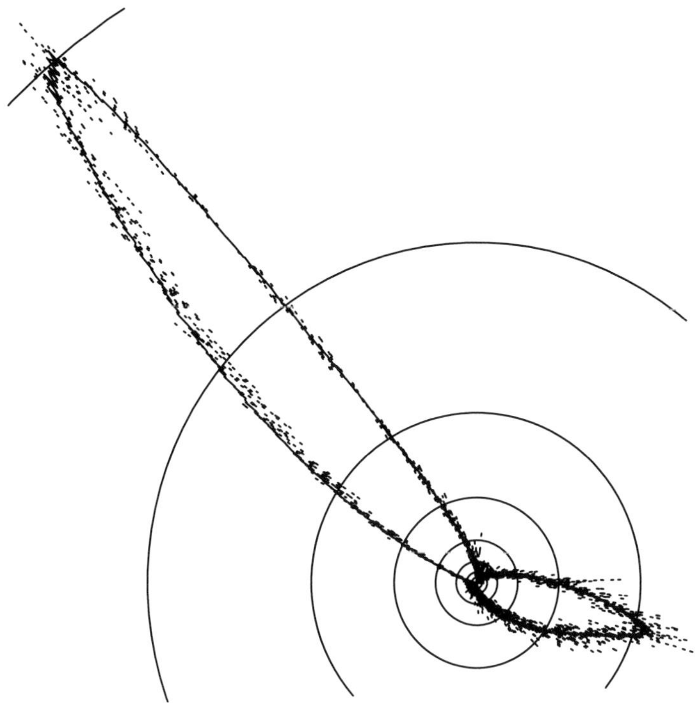

Fig. 3—Comparison of the normalized and background-subtracted optical (thick line) and UV (dotted line) Crab pulsar average profiles plotted in polar coordinates. The count rate and phase are plotted as the radial and azimuthal coordinates, respectively. An arc of radius unity is shown at the highest part of the profile and the successively smaller arcs and circles have radii that decrease by factors of 2. Note the preponderance of dots inside the optical main pulse (the largest lobe) which shows that the UV main pulse is narrower than in the optical.

3.2 Individual Pulse Studies

In the optical and UV observations only about 12 and 1.5 photons, respectively were detected in each pulse period (see Table 1). For all observations, the distribution of the number of photons per pulse was consistent with Poisson statistics. Time series were constructed from the number of photons detected in each pulse and these were searched for periodicities. FFTs and ACFs were used in this analysis but no significant periodic signals from the pulsar were found.

A correlation analysis was conducted in order to search for related activity in various parts of the optical pulse. Time series were constructed from the number of photons that arrived in the phase intervals corresponding to the main pulse, interpulse and bridge, for each pulse period. All combinations of pairs of these time series were cross-correlated using the standard method (Pearson's) for time lags of -500 to 500 pulse periods lag. No significant correlation was found ($\gtrsim$0.015, 99% confidence level) for any of them. Correlations were also sought between time series composed of the leading and trailing edges of the main and interpulse, for -500 to 500 periods lag,

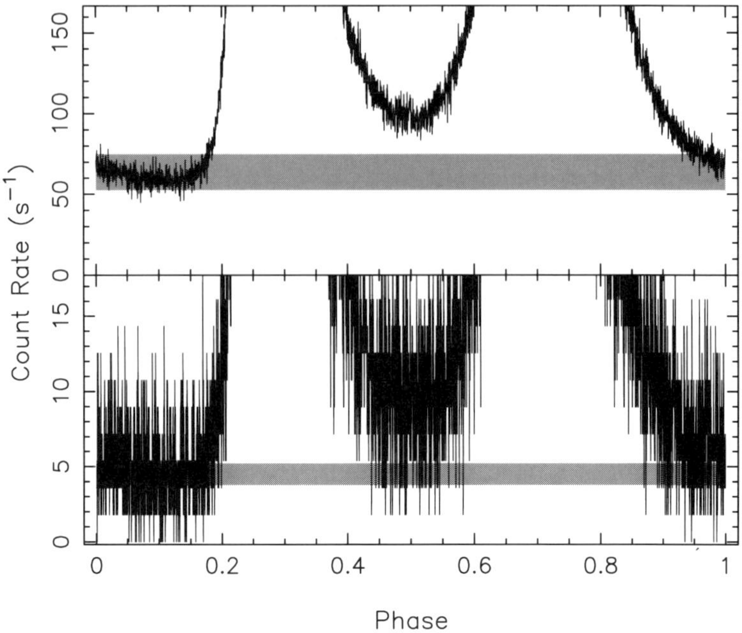

Fig. 4—Crab pulsar average profiles (optical and UV in the upper and lower panels, respectively) suitably scaled in order to highlight their low level emission. The shaded regions represent the background level and its uncertainty calculated from finding-aperture maps (see text).

without success. The average pulse profile was divided into 100 equal phase intervals and correlations within a given pulse were sought between time series of the number of photons in these intervals, again without a positive result. Thus it appears that on the scale of 0.01 pulse periods, the emission is independent, or at most, correlated at the 0.015 level.

The work is based on observations with the *HST*, obtained at the Space Telescope Science Institute, which is operated by AURA, Inc., under NASA contract NAS 5-26555. The research conducted by JDB is supported through NASA contract NAS 530442, and that of the High Speed Photometer instrument definition team is supported through NASA contract NAG 5-1613.

References

Lyne, A. G., and Graham-Smith, F. 1990, *Pulsar Astronomy* (Cambridge University Press).

Lyne, A. G., and Pritchard, R. S. 1992, *Crab Pulsar Ephemeris.*

Percival, J. W., Bless, R. C., and Nelson, M. J. 1991, in *The First Year of HST Observations*, ed. Kinney, A. L. and Blades, J. C. (Space Telescope Science Institute), p. 123.

Smith, F. G. *et al.* 1978, *M.N.R.A.S.*, **184**, 39P.

White, R. E. 1990, *Hubble Space Telescope High Speed Photometer Handbook* (Space Telescope Science Institute).

High-Energy Emissions from Young Pulsars: Perspective from the Outer Gap Model

Cheng Ho

Los Alamos National Laboratory

1 Introduction

Young pulsars such as the Crab and Vela pulsars are the most prominent gamma-ray sources in the Galaxy, powered by the neutron star's rotational energy and the spinning magnetic field. Since the late 1960's, a large amount of data has been amassed for these pulsars. With the recent launches of ROSAT, Compton Gamma-Ray Observatory and other new instruments, many exciting new results have been reported. In this Workshop, we learned about the weak and soft pulsed emission from the Vela pulsar, the possible detection of spectral turn-over at $\sim$ 30 MeV, the detection of gamma-ray emission from PSR1509−58, and observation of pulse profile variability from the Crab pulsar. Soon after the Workshop, the discovery of gamma-ray pulsations from PSR1706−44 (Kniffen *et al.* 1992) and Geminga (Halpern and Holt 1992, Bertsch *et al.* 1992) were reported. All of these contribute to the large pool of hints that theorists need in order to explain the origin of these powerful emissions. In this paper, we discuss results from the outer gap model (Cheng, Ho and Ruderman, 1986*a* and *b*, hereafter CHR*a* and *b*) and comparisons to observations.

Section 2 gives a brief overview of the outer gap model. Section 3 discusses the high-energy spectrum from the Crab, PSR0540−693, and Vela. In Section 4, we discuss the pulse profile for pulsars with Crab-like parameters. Section 5 discusses outer gap evolution.

2 The Outer Gap Model

The outer gap model for high-energy emission from young pulsars was developed with an overall picture of the accelerator formation, the energy balance, the charge production, and the radiation mechanism (CHR*a* and *b*). In this model, a finite region of vacuum (the gap) is formed in the outer magnetosphere within which a strong electric field parallel to the magnetic field accelerates electrons to ultra-high energies and leads to copious gamma-rays production. Detailed considerations of the electrodynamics, radiation mechanisms, and charge production processes of the gap indicate that the Crab and Vela pulsars work differently.

Crab-Type Outer Gap: Relativistic electrons inside a Crab-type gap emit GeV primary photons by means of curvature radiation; these photons are subsequently annihilated to create secondary $e^{\pm}$. The secondary electrons radiate synchrotron radiation below $\sim$ 10 MeV and inverse-Compton-scattered photons above $\sim$ 10 MeV. As described in CHR*a*, the fan-beam pulsed emission at all energies for the Crab pulsar arises from a common population of electrons, and the observed pulse separation is a logical result of the general emission geometry.

Vela-Type Outer Gap: In the outer gap of the Vela pulsar, energy is extracted through inverse Compton scattering of low-energy photons by relativistic electrons with energy $\gtrsim$ TeV. The primary photons are converted through $\gamma - \gamma$ annihilation into secondary $e^{\pm}$ which emit pulsed synchrotron radiation up to the GeV range. The observed absence of X-ray pulsation from the Vela pulsar is caused by the lack of secondary electrons at sufficiently low energy to emit a large number of X-ray photons within the light cylinder.

3 Spectra of Pulsed Radiation From Outer Gaps

3.1 The Crab-like pulsars: Crab and PSR0540−693

Ho (1989*a*) improved the calculation for the Crab pulsar presented in CHR*b*. Besides the fundamental parameters for the pulsar, the spin rate (Ω) and the magnetic dipole moment (μ), this model adopted only one free parameter, f_G, which represents the dimensionless size of the gap. This model generates spectra which have all of the characteristics described in CHR*b*, and are consistent (including overall normalization) with observations. It was pointed out in Ho (1989*b*) that certain spectral features from the Crab pulsar can be tested by CGRO instruments. Work along this line is underway as a part of the CGRO Guest Investigator Program.

The 50 ms pulsar located in the Large Magellanic Cloud (PSR0540−693) has parameters very similar to those of the Crab pulsar and probably produces high-energy emission through a Crab-type outer gap (see Section 5). Figure 1 gives the comparison between the observed pulsed flux and the theoretical prediction. Also plotted is the sensitivity of OSSE on GRO. The plotted sensitivity for OSSE indicates marginal detectability over a nominal two weeks' observation period. It has been estimated that a nine months integration is needed for EGRET to detect the predicted flux (John Mattox, private discussion).

3.2 The Vela-like pulsars: Generic Spectra

The spectral calculation presented in CHR*b* for the Vela-type gap is not as complete as that for the Crab-type gap (Ho 1989*a*). Following a number of simplifying assumptions, CHR*b* suggested that the secondary electron production distribution in Vela-type outer gap would roughly scales like γ^{-1}, where γ is the electron's Lorentz factor. It was also argued that most of the secondary electrons would have similar pitch angle θ with respect to the local magnetic field. As a result, the steady state electron distribution is roughly given by

$$N_e(\gamma) \sim \gamma^{-2} \ln(\gamma_{\max}/\gamma), \tag{1}$$

where $\gamma_{\max}$ corresponds to the maximum energy that an electron can achieve in the Vela-type outer gap. Because of the weakness of the magnetic field in the outer magnetosphere of Vela-type pulsars, an relativistic electron does not radiate away all of

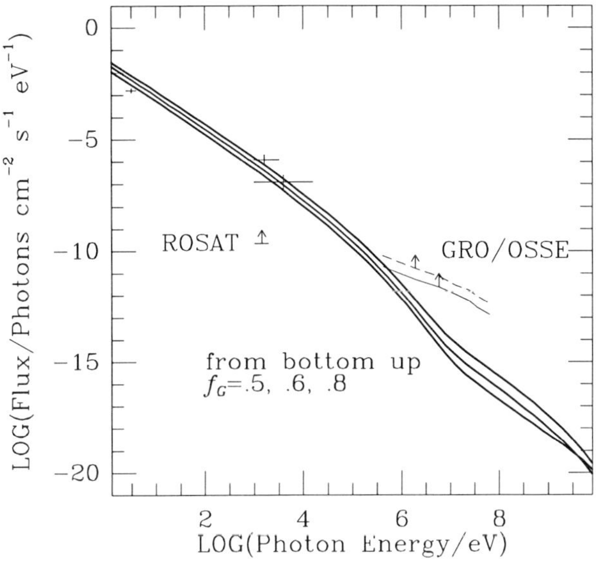

Fig. 1—Observed and theoretical spectrum of the 50 ms pulsar PSR0540−693 in LMC. See also Ho 1989.

its energy through synchrotron loss before it leaves the light cylinder. It is argued that once the electrons leave the light cylinder, they cannot retain the memory of the neutron star's rotation and the ensuing radiation beyond the light cylinder will not be pulsed. Thus, the electron distribution function given by equation (1) is valid only for $\gamma > \gamma_{\rm min}$ where $\gamma_{\rm min}$ is roughly given by

$$\frac{\gamma_{\rm min}^2 \sin^2\theta\,(e^2/c)\,(eB/mc)^2}{mc^2} \sim \Omega \tag{2}$$

where B is the local magnetic field. We expect the electron distribution below $\gamma_{\rm max}$ to be between two extreme ranges:

$$N_e(\gamma) \sim (\gamma\gamma_{\rm min})^{-1}\ln(\gamma_{\rm max}/\gamma_{\rm min}) \qquad \text{for } \gamma < \gamma_{\rm min}, \tag{3a}$$

or

$$N_e(\gamma) = 0, \qquad \text{for } \gamma < \gamma_{\rm min}. \tag{3b}$$

The first case results from the power-law electron production distribution. The second case arises from the case where no electron is created below $\gamma_{\rm min}$. The exact value of $\gamma_{\rm min}$ depends critically on the geometry.

The pulsed radiation spectrum from a Vela-type outer gap is calculated by convolving the electron distribution of equations (1) and (3) with the single particle synchrotron radiation spectra:

$$\frac{dN_{ph}}{dtdE} = A\int_{\gamma_{\rm min}}^{\gamma_{\rm max}} d\gamma\, \mathcal{F}\left(\frac{E}{E_c\gamma^2}\right) \times N_e(\gamma), \tag{4}$$

where A is the overall normalization, $E_c = (3/2\sin\theta)(\hbar eB/mc)$. The function $\mathcal{F}$ is the standard synchrotron radiation spectral function (Jackson 1976). The radiation spectrum given by equation (4) follows a smoothly varying power law between $E_{\rm max} =$

$\gamma_{max}^2 E_c$ and $E_{min} = \gamma_{min}^2 E_c$. The spectrum hardens below E_{min} with $E^{-2/3}$ being the hardest spectrum derived from equation (3b).

The range of validity of equation (4) could be significantly affected by a number of factors. First, the magnetic mirroring is important in the outer magnetosphere for Vela-like pulsars (Ho 1986). Relativistic electrons moving toward the star could be mirrored and produce broadly beamed radiation near the mirroring point. The outgoing electrons will conversely be beamed sharper as they move into the weaker magnetic field. Spectrum and pulse profile could be significantly modified. Secondly, the location along the field line is critical in determining the turn-over of electron distribution at γ_{min}. Work is underway to study these effects. In summary, the Vela-type outer gap emits pulsed radiation with a spectrum specified by three parameters, and a number of complications could affect the exact observed spectral shape. In the following, we discuss the comparison of the theoretical spectrum with observations.

3.3 Vela, PSR1509−58, PSR1706−44 and Geminga

Figure 2 shows the comparison between the theoretical and observed spectrum. The observed fluxes above ~ 100 keV are taken from Grenier *et al.* (1988) and Strickman *et al.* (1992). See CHR*b* for references for previous observations. Due to the possible inconsistency of data between different experiments, data between 1 MeV and 60 MeV are not shown (see Kerrick *et al.* 1992, Busetta *et al.* 1992, and reference therein). The "fit" is obtained through visual examination: no statistical analysis is applied. The theoretical spectrum is calculated using equation (3a) with E_{max}=10 GeV and $\gamma_{min}/\gamma_{max} = 0.056$.

From the theoretical point of view, how the Vela pulsed emission turns off toward low energy is critical to our understanding of the gamma-ray pulsar phenomena. Specifically, if the low-energy (~ MeV) spectra becomes harder than $E^{-2/3}$ (the spectrum predicted by equation [3b]), synchrotron and curvature radiation will be eliminated as the dominant emission mechanism. Through the CGRO Guest Investigator program, we plan to make more detailed comparison of theoretical spectra with observations and study these signatures.

Recently, hard X-ray and γ-ray pulsation has been detected from the 150 ms pulsar PSR1509−58 (Kawai 1991, Wilson 1992). Hints of gamma-ray emission in the COS-B data have been reported (Hartmann *et al.* 1992). Confirmation of this result by EGRET is eagerly awaited. Based on its period and period derivative, it should have a Vela-type gap in the context of the outer gap model. It, however, emits proportionately much stronger X-ray emission compared to the Vela pulsar. In addition, the single pulse profile is rather different from the Vela pulsar. More detailed study of the possible geometry is required to examine whether the observed emission is consistent with the outer gap model.

The pulsars PSR1706−44 and Geminga (1E0630+178) both have strong gamma-ray emissions. The pulsed profile of Geminga is doubly peaked, *i.e.*, Vela-like. Gamma-ray photon statistics for PSR1706−44 are such that we cannot yet preclude the possibility of a double pulse profile with an intensity ratio of, say, 5. The similarity with the Vela pulsars strongly suggested that they should be modeled in the same context as Vela.

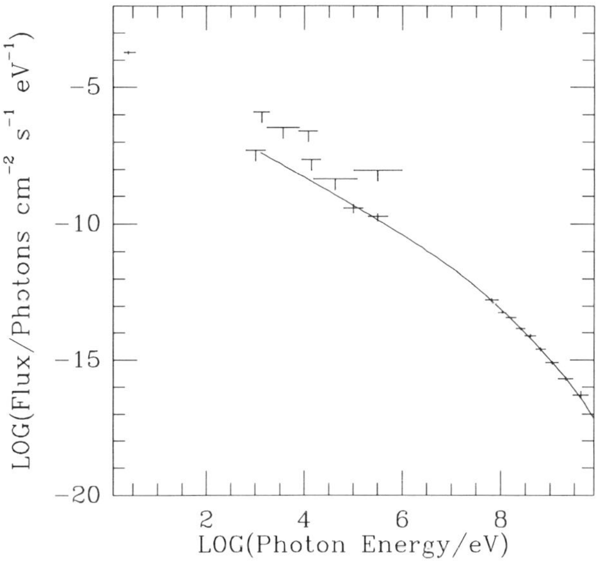

Fig. 2—Spectrum of the Vela pulsar. The theoretical spectrum is calculated using equation (4) along with equation (3a). See text for more details.

4 Pulse Profile and Emission Geometry

In general, the pulse emission geometry can be divided into two components: 1) a population of relativistic electrons distributed on magnetic field lines pointing at the same tangent, and 2) a spread of particles among field lines pointing at different tangents. Current outer gap model gives the particle distribution in momentum space along a set of field lines pointing at the same tangent, under a simplified geometry. Because of the finite size and spread of electron pitch angles, the radiation emitted from these field lines will be spread out over a range of angles. Figure 3 shows this "pencil beaming" pattern as a function of energy. The correlation between pulse sharpness and photon energy is evident.

A more realistic pulse profile calculation has to take into account the distribution of electrons among field lines pointing at different tangents. Such distribution can be divided into the latitudinal and longitudinal components; the former is the spread of field lines at the same rotational phase whereas the latter is the spread across different rotational phases. In the context of the outer gap model, the latitudinal distribution largely arises from various sectors of the same curved field line pointing at different tangents. A straightforward integration of the pencil beam through the a constant latitudinal distribution yields the fan beam pattern shown in Figure 4. The fan beam is broadened from the pencil beam, as expected. The sharpness of the emission pattern at ~ 1 GeV is consistent with the observed sharp pulse profile of the Crab pulsar seen by EGRET (Thompson *et al.* 1992). The predicted optical profile is, however, broader than observation.

Assuming that the fan beam as a function of the rotational phase has the same shape and differs only in the overall normalization, the observations can conceivably be used to derive the longitudinal electron distribution. In particular, the sharp fan beam see at GeV range may be used in conjunction with the pulse profile from the

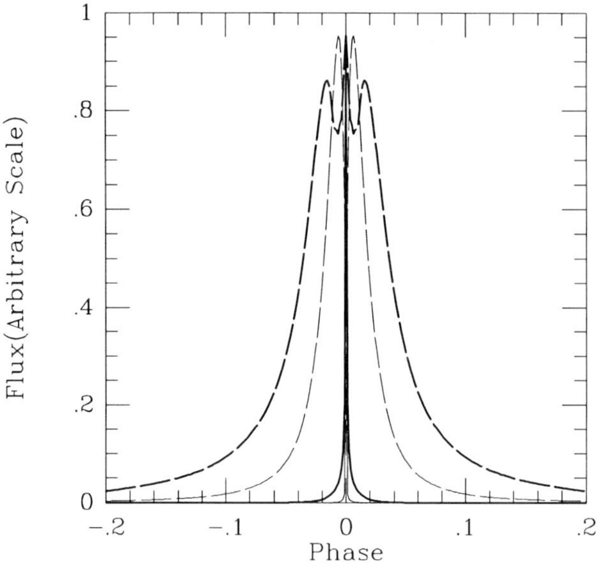

Fig. 3—Pencil beam pulse profile of the Crab pulsar predicted by the outer gap model. The gap fraction f_G is 0.45. The thick dashed, thin dashed, thick solid, and thin solid lines are for eV, keV, MeV and GeV emission respectively. The vertical axis gives the flux per "solid angle", *i.e.*, $dI/d(\cos\theta)$, where θ is the angle of the line of sight with respect to the direction of the magnetic field.

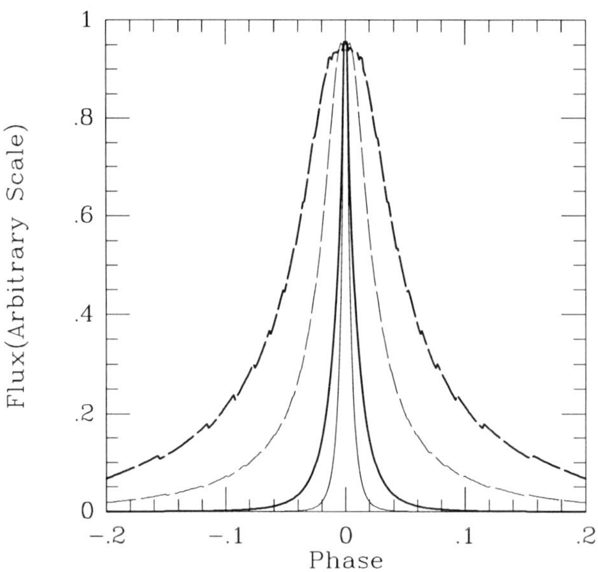

Fig. 4—Fan beam pulse profile of the Crab pulsar predicted by the outer gap model. The gap fraction f_G is 0.45. The thick dashed, thin dashed, thick solid, and thin solid lines are for eV, keV, MeV and GeV emission respectively. The vertical axis gives the flux $dI/d\phi$, where ϕ is the observed phase angle.

Crab pulsar to deduce the distribution of particles in the longitudinal direction. Such distribution can in turn be used to calculate the expected pulse profile at low energies.

5 Evolution of Outer Gaps

Ruderman and Cheng (1988) discussed the overall outer gap evolution scenario. They adopted the arguments in CHR*b* and divided pulsars into Crab-type, Vela-type with Crab-like parameters, Vela-type with Vela-like parameters, and post-Vela (dead outer gap). A revisit to the arguments indicates that the Vela-type outer gap with Crab-like parameters occupies negligible parameter space. Here, we argue that young

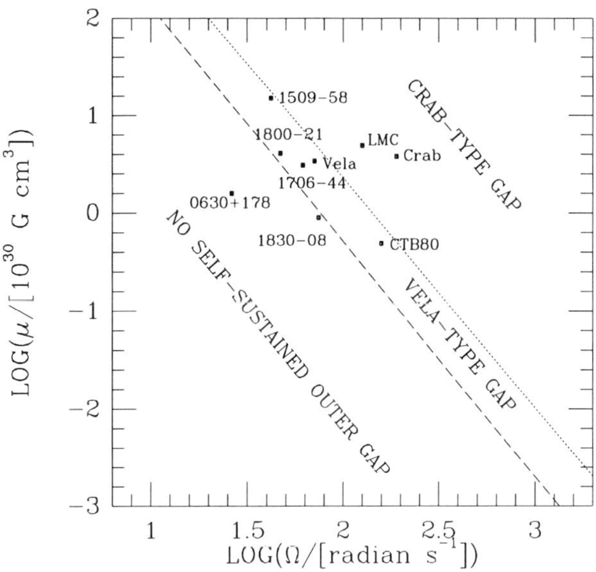

Fig. 5—Outer gap emission mechanism as a function of the pulsar parameters, spin rate Ω, and dipole moment μ. The dotted line gives $\Sigma = 1$ (Eq. [5]) and the dashed line gives $\Xi = 1$ (Eq. [6]).

pulsars can be roughly divided into three classes: Crab-type, Vela-type, and no self-sustained outer gaps. The Crab-type and Vela-type outer gaps are distinguished by the dimensionless parameter Σ:

$$\Sigma = \frac{\mu^2 \Omega^5}{e^4 m^3 c^{11}} \tag{5}$$

where μ is the pulsar's magnetic dipole moment. This criterion determines whether the magnetic mirroring in the outer magnetosphere is strong enough (CHR*b* and Ho 1986). The second criterion separates Vela-type outer gap and dead outer gap. This is parameterized by

$$\Xi = \frac{e^{15} \mu^7 \Omega^{17}}{\hbar m^{10} c^{38}}. \tag{6}$$

Pulsars with $\Sigma > 1$ will have a Crab-type outer gap. Pulsars with $\Sigma < 1$ and $\Xi > 1$ will have a Vela-type gap. For pulsars with $\Xi < 1$, the outer magnetosphere will be largely open without a self-sustained gap. Other areas of charge production are needed if the outer magnetosphere is to maintain simultaneously the Goldreich-Julian density and current flow. Figure 5 shows the locations of various young pulsars and $\Sigma = 1$ and $\Xi = 1$ lines in the Ω-μ parameter space. Note that Geminga lies outside the outer gap parameter space predicted by equation (6).

In arriving at equations (5) and (6), physical quantities such as magnetic field strength and radius of curvature are taken to be those at the light cylinder. As the gap extends a large part of the magnetosphere, typical numbers could be different; these estimates could contain large uncertainties and are strongly geometry dependent. These uncertainties are especially large for Ξ with the strong dependence of B. Recently, Chen and Ruderman (1993) presented a different scaling law for Ξ. Through a modified estimate of the "typical" field strength in the outer gap region, they showed that the outer gap "death line" can be pushed down to include Geminga. Gamma-ray pulsations from Geminga can still be produced through an outer gap.

Conversations with Kaiyou Chen are greatly appreciated. This study was supported in part by the NASA CGRO Guest Investigator Program and in part by the NASA High-Energy Astrophysics Theory and Data Analysis Program. Part of the work was done at the Aspen Center for Physics. This work is performed under the auspices of US Department of Energy.

References

Bertsch, D., *et al.* 1992, *Nature*, **357**, 306.
Busetta, M., *et al.*, 1992, these proceedings.
Chen, K. and Ruderman, M. 1993, *Ap. J.*, in press.
Cheng, K.S., Ho, C., and Ruderman, M. 1986*a*, *Ap. J.*, **300**, 500.
Cheng, K.S., Ho, C., and Ruderman, M. 1986*b*, *Ap. J.*, **300**, 522.
Grenier, I.A., Hermsen, W., and Clear, J. 1988, *Astr. Ap.*, **204**, 117.
Halpern, J. and Holt, S., 1992, *Nature*, **357**, 222.
Hartmann, D., *et al.* 1992, these proceedings.
Ho, C. 1986, *M. N. R. A. S.*, **221**, 523.
Ho, C. 1989*a* *Ap. J.*, **342**, 396.
Ho, C. 1989*b*,, in *Proceedings of the Gamma-Ray Observatory Science Workshop*, ed. N. Johnson, (Greenbelt:GSFC) p. 4-174.
Jackson, J. D. 1976, *Classical Electrodynamics*, (New York: Wiley).
Kawai, N., *et al.* 1991, *Ap. J. (Letters)*, **383**, L65.
Kerrick, A., *et al.* 1992, these proceedings.
Kniffen, D., *et al.* 1992, *IAU Circ.* 5485.
Ruderman, M.A. and Cheng, K.S. 1988, *Ap. J.*, **335**, 396.
Strickman, M., *et al.* 1992, *IAU Circ.* 5557.
Thompson, D., *et al.* 1992, these proceedings.
Wilson, R., *et al.* 1992, *IAU Circ.* 5429; these proceedings.

Pulsar Gamma-Ray Emission in the Polar Cap Cascade Model

Alice K. Harding[1] and Joseph K. Daugherty[2]

[1]Laboratory for High Energy Astrophysics, NASA/Goddard Space Flight Center
[2]Computer Science Department, University of North Carolina-Asheville

Abstract

We present theoretical phase-dependent energy spectra and pulse profiles resulting from pair production cascades above pulsar polar caps. Our treatment applies to the class of models for pulsar gamma-ray emission in which curvature radiation from VHE primary electrons initiates a synchrotron-pair cascade near the neutron star surface. A revised Monte-Carlo calculation includes improved treatments of magnetic pair production and quantized synchrotron emission from the pairs. In addition, the angular distribution of the emitted radiation is used to analyze the behavior of pulse time profiles and to predict pulse-phase variations in the energy spectrum. We will show the dependence of these profiles on the geometric model parameters (inclination of magnetic axis, observer orientation) over various energy ranges.

1 Introduction

The observation of pulsed gamma-ray emission from radio pulsars provides a very powerful source of information on the acceleration of primary particles in the magnetosphere, and thus on the origin of pulsar emission at all other energies. Models for pulsar gamma-ray emission have fallen into two classes. The polar cap model (Sturrock 1971, Daugherty and Harding 1982) assumes that acceleration takes place near the neutron star surface at the magnetic poles, a picture generally assumed from studies of radio emission properties. The outer gap model (Cheng, Ho and Ruderman 1986) assumes that primaries are accelerated in the outer magnetosphere, in vacuum gaps that develop along null charge surfaces. In both cases, electromagnetic cascades develop, creating large numbers of electron-positron pairs which radiate most of the observed gamma-rays. Before the Compton Gamma-Ray Observatory, only the two youngest pulsars, the Crab and Vela, were detected at gamma-ray energies. Although extensive study of the spectra and pulse profiles of these two gamma-ray pulsars was carried out over a decade and a half, the information was not enough to distinguish between the two kinds of models. Compton GRO now has the increased sensitivity which could change this picture, by providing more extensive and detailed measurements of spectra and pulse profiles and, even more important, detecting new gamma-ray pulsars to enlarge the source sample.

In light of the improved observational picture, it is now essential to have more detailed calculations of the expected gamma-ray emission characteristics predicted by each of the models. This paper will describe the basic properties of gamma-ray emission in the polar cap model and present some new calculations of pulse profiles and phase-resolved spectra for the Crab and Vela pulsars. Radio observers have been successful in identifying signatures of polar cap emission in many pulsars through detailed studies of radio pulse morphology and polarimetry (Lyne and Manchester 1988, Rankin 1990). For example, the sweep of polarization position angle through the pulse profile displays the characteristic "S" shape expected for an observer seeing emission from near the magnetic pole. Similar signatures would be very useful for studying pulsar gamma-ray emission, and we will point out several unique gamma-ray signatures of the polar cap model.

2 Polar Cap Cascade Physics

The calculations presented here use an improved version of the Monte-Carlo code developed by Daugherty and Harding (1982) and described in detail in that paper. The code simulates the cascade initiated by energetic primary electrons injected above the polar cap in a dipole magnetic field. Although several polar cap acceleration models have been studied (Ruderman and Sutherland 1975, Arons 1983, Beskin *et al.* 1983), we argue that the details of the acceleration process are not important in determining the final gamma-ray emission as long as the acceleration gap is pair limited instead of radiation-reaction limited. All of the above models result in pair-limited gaps, where the acceleration region is much smaller than the curvature radiation loss scale. Thus, we can ignore any acceleration of primaries (and secondary pairs) after injection. The energy to which primaries are accelerated is somewhat model dependent, but generally in the range $10^{12} - 10^{13}$ eV. We have adopted an initial injection energy of 10 TeV for all calculations, and inject primary electrons uniformly over the polar cap, although this can be changed to model different injection patterns.

The cascade simulation follows each primary as it moves out along a dipole field line in discrete increments of arc length. The curvature radiation from each step is divided into a number of frequency bands, and the cascade from a representative photon at each frequency is calculated, the resulting pairs and photons eventually accumulated with the appropriate spectral weighting. The probability of magnetic pair production is computed for each curvature photon and continually updated along its path until its survival probability has dropped to $1/e$. At the pair conversion point, both members of the pair are assigned the same direction and half the energy of the parent photon. The quantized synchrotron radiation from the electron and positron are calculated, after transforming to the frame where each particle has zero parallel momentum, until each particle has decayed to the ground Landau state. Each emitted synchrotron photon is transformed back to the lab frame and followed either to its pair production point or to escape from the magnetosphere in the same way as the curvature photons. The cascade develops through as many generations as are needed to allow all emitted photons to escape, with each generation of photons calculated by a recursive subroutine call. Each escaping photon is accumulated in a three dimensional

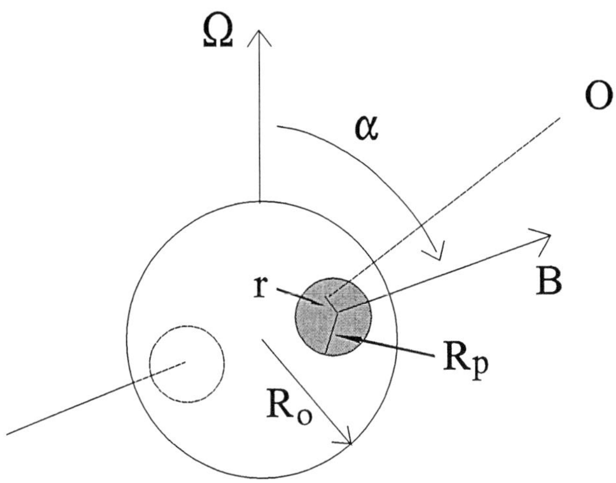

Fig. 1—Polar cap emission geometry, showing pulsar obliquity angle α, polar cap radius R_p and observer "impact parameter" r.

table of energy and polar and azimuthal angles with respect to the rotation axis in the inertial frame, taking into account aberration and light travel time delays.

The spectrum and pulse profile seen by various observers can be determined from this 3D table by specifying the observer's "impact parameter," or the distance of closest approach of the observer's line-of-sight to the magnetic pole at the stellar surface. We define a dimensionless impact parameter, r, as a fraction of the polar cap radius in the standard (Goldreich and Julian 1969) model, $R_p = R_o(\Omega R_o/c)^{1/2}$, where Ω and R_o are the pulsar rotation frequency and radius.

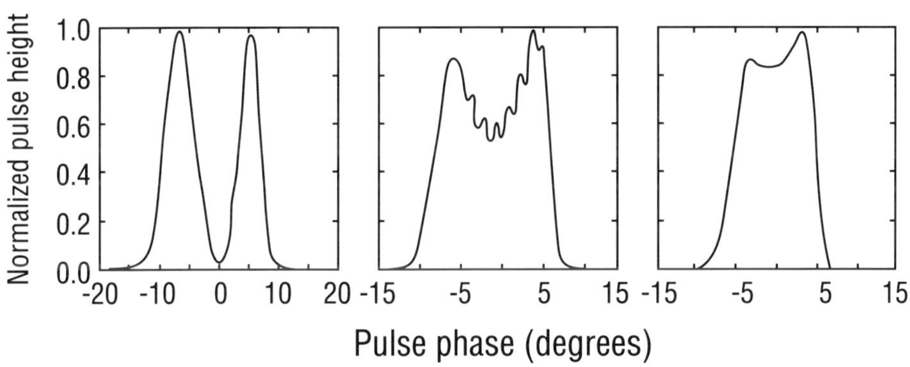

Fig. 2—Pulse profiles for $P = .033s$, $B_o = 10^{12}$ G and $\alpha = 85°$ seen at different observer impact parameters, $r = 0$ (left), $r = 0.5$ (middle), $r = 0.75$ (right). Closest approach to the magnetic pole is 0°.

3 Pulse Profiles

The gamma-ray emission characteristics vary significantly over the polar cap surface. Primary electrons injected near the polar cap edges encounter field lines with

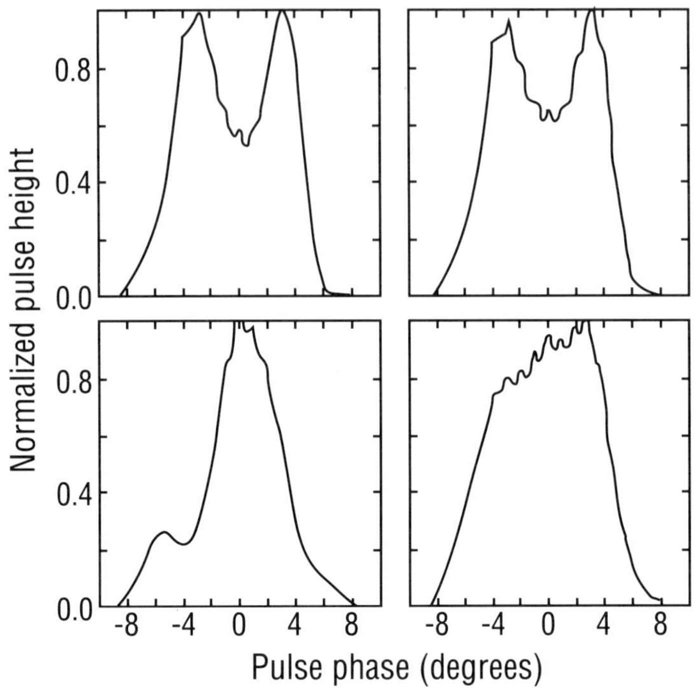

Fig. 3—Pulse profiles for P = .089 s, $B_o = 10^{12}$ G and $\alpha = 85°$ in four different energy bands (clockwise from top left): 1 - 100 MeV, 100 - 500 MeV, 500 MeV - 3 GeV, > 3 GeV.

the smallest radii of curvature, and therefore suffer higher curvature radiation losses, radiating more photons up to higher energies. The cascades generated by these electrons are also more extensive, because the small radii of curvature decrease the pair production mean free paths of both the curvature and synchrotron photons. Electrons injected near the magnetic pole encounter very large curvature radii, and both the number of curvature radiation photons, and the ensuing cascades from these primaries will be comparatively much smaller. Consequently, the change in pulse profile with observer impact parameter, as shown in Figure 2, is quite dramatic. At $r = 0$, the observer's line-of-sight sweeps through the magnetic pole, where the absence of curvature and cascade photons produces a double-peaked profile. At intermediate impact parameters, the central dip in the profile partially fills in, and at large impact parameters, where the observer samples emission only from the outer parts of the polar cap, the profile is single-peaked. It should be noted that $r = 1.0$ is not a maximum impact parameter, as emission from the polar cap may be seen at $r > 1.0$ due to curvature of field lines and the vertical extent of the emission (at least 10 stellar radii). However, at very large impact parameters, the observer will no longer see emission from the synchrotron core of the cascade (which is emitted within a few stellar radii), but only the weaker curvature emission extending to high altitudes.

Because of the variation of emission across the polar cap, the pulse profiles are also energy dependent. Figure 3 shows an example of the change in pulse shape as a function of energy for an intermediate impact parameter. At low energies, the profile shows a distinct central dip that gradually fills in at higher energies, until it becomes single-peaked. The pulse width also decreases with increasing energy.

Since neither the Crab nor Vela shows a double-peaked profile, we take $r = 0.4$ to compute model pulse profiles for these pulsars, shown in Figure 4. At the large angle of obliquity assumed, the observer would also see a second pulse from the other pole at a phase separation of about 180° (Figure 4 shows only half a period). The FWHM of the pulses is $\Delta\phi \simeq 15°$ for Crab and $\Delta\phi \simeq 10°$ for Vela, following the expected $\Delta\phi \propto P^{-1/2}$ polar cap dependence on period. These values are in rough agreement with the observed pulse widths (Thompson *et al.* 1992, Kanbach *et al.* 1992), although the Vela pulse widths are somewhat time variable.

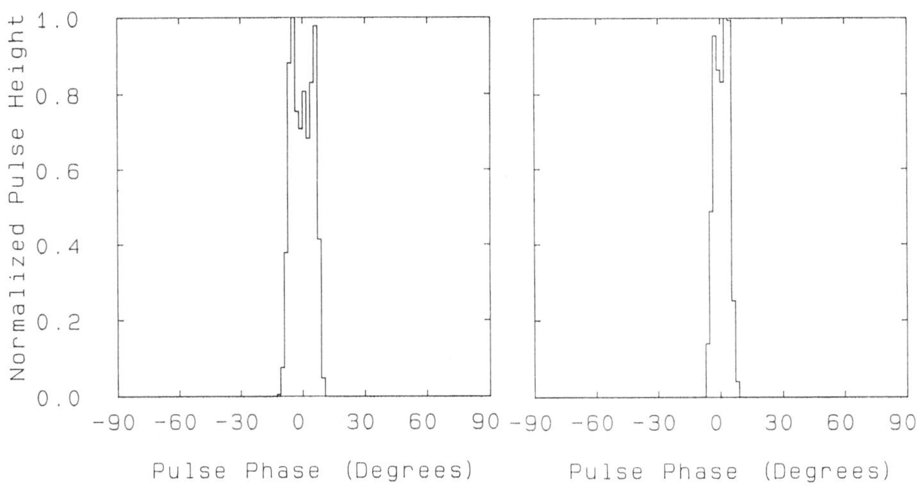

Fig. 4—Model pulse profiles at $E > 100$ MeV for the Crab ($P = .033$ s) and Vela ($P = .089$ s) pulsars with $B_o = 10^{12}$ G, $r = 0.4$ and $\alpha = 85°$, shown over a half period.

4 Spectra

The total spectrum of escaping photons from cascades produced by primary electrons injected over the polar cap surface consists primarily of synchrotron photons and is steeper (index ~ -2) than the primary curvature radiation spectrum (index $= -5/3$). The observed spectrum will depend on the observer impact parameter and could contain different mixtures of curvature and synchrotron photons, producing different spectral indices. Smaller impact parameters ($r < 1.0$) will have a higher proportion of synchrotron photons in the total pulse spectrum, whereas large impact parameters ($r > 1.0$) may have a pure curvature spectrum. The observed spectral indices of the Crab and Vela pulsars are both near -2, indicating within the context of the polar cap model that $r < 1$. Chiang and Romani (1992) have shown that computed cascade spectra for the Crab and Vela assuming $r > 1.0$, based on radio and optical polarization data, are much flatter than those observed.

Figure 5 shows model spectra for the Crab and Vela pulsars assuming $r = 0.4$. Both spectra are power laws and exhibit pronounced steepening above a few GeV due to magnetic pair production attenuation. This feature is a characteristic of the

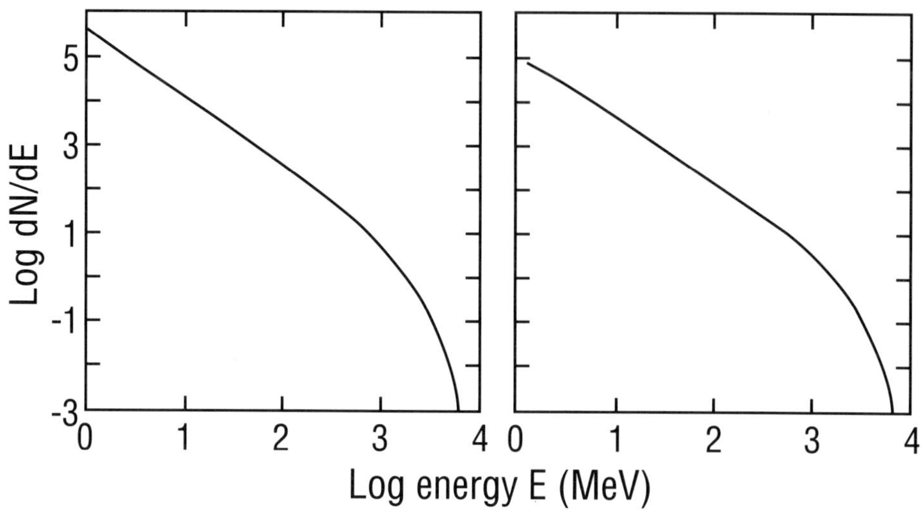

Fig. 5—Total pulse spectra for $B_o = 10^{12}$ G, $\alpha = 85°$, $r = 0.4$ and $P = .033$ s (left), $P = .089$ s (right).

polar cap model, when the pulsar surface field is near or above 10^{12} G. Higher field strengths will produce steepening at lower energies. Pulsar period also has an effect on the cutoff energy because aberration and field-line slippage tend to decrease pair production mean free paths (Harding *et al.* 1978), and this can be seen in Figure 5.

The change in pulse profile with energy, discussed in Section 3, leads to a change in spectrum across the pulse. Figure 6 shows phase-resolved spectra at the leading and trailing edges of the pulse and at pulse center for an extreme case where the observer line-of-sight sweeps through the magnetic pole. The softer spectra near the pulse edges reflect the enhanced cascades produced by smaller field line curvature, whereas the spectrum near pulse center is a pure, nearly unattenuated curvature spectrum. Near the magnetic pole, the radius of curvature is so large that the electron radiative losses are small, producing a spectrum with index $-2/3$ of the curvature emissivity, whereas over other parts of the polar cap the curvature spectrum has index $-5/3$ due to electron energy loss. Therefore, phase-resolved spectra at other impact parameters will change less dramatically. There is a slight difference in cutoff energies of spectra in leading and trailing edges. At the leading edge, the field lines rotate toward the direction of curvature such that photons see larger angles to the field, increased pair production probabilities, and a lower cutoff energy. On the trailing edge, where field lines rotate away from the direction of curvature, the situation is reversed and the cutoff energy is higher.

5 Summary

Polar cap models based on the standard dipole field, oblique rotator make a number of predictions:

- The cascade spectra are power laws with high-energy turnovers around a few GeV.

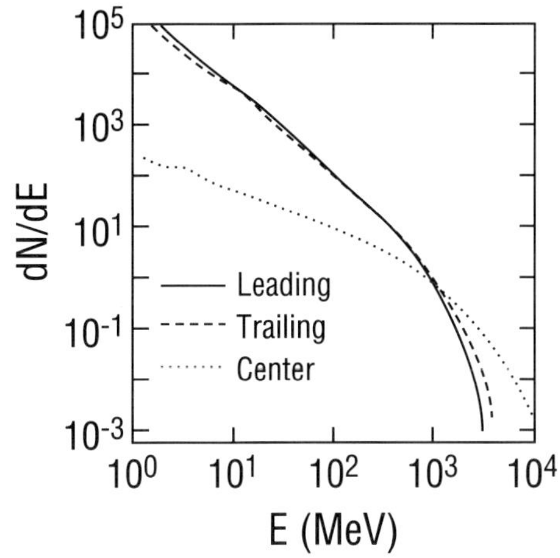

Fig. 6—Phase resolved spectra for the $r = 0$, $\alpha = 85°$, and $P = .033$ s pulse profile shown in Fig. 2.

- Pulse profiles may be double peaked or show central dips, and have widths which decrease with increasing energy and also with increasing period as $\Delta\phi_{FWHM} \propto P^{-1/2}/\sin\alpha$.
- Phase-resolved spectra are harder near pulse center and have higher turnover energies at the leading edge.

Several of these predictions seem to agree with the observed characteristics of pulsar gamma-ray emission. The predicted power law spectra are in rough agreement with the observed spectra of the Crab and Vela pulsars, at least above 10 MeV. High-energy turnovers have not been confirmed, although EGRET detects only unpulsed flux from the Crab above 5 GeV (Thompson *et al.* 1992) which may suggest the presence of a turnover in the pulsed spectrum. EGRET also reported that the first Vela pulse disappears above 2 GeV, but emission from the second pulse extends to 10 GeV. The observed pulse widths also decrease with period as roughly $P^{-1/2}$, assuming both pulsars have large obliquity, and the Vela pulse widths are observed to decrease with increasing energy (Kanbach *et al.* 1992; Grenier *et al.* 1992).

However, the polar cap model does have problems accounting for some observed features. Most notably, the observed 140° phase separation of the pulses in both the Crab and Vela light curves is hard to explain in the standard dipole model which predicts a 180° phase separation. Since neither aberration nor light travel time effects are large enough to account for the difference, one must invoke off-centered dipoles or quadrupole components. A subtler but potentially more serious problem is that the standard Golreich-Julian current flow over a polar cap area falls short of accounting for the total energy radiated between optical and GeV gamma-rays in the Crab pulsar. While the cascade spectrum fits the observed spectrum very well down to $\sim$ 10 MeV, the predicted spectrum flattens at lower energies as the cascade runs out of energy. Inverse Compton scattering of thermal photons from the neutron star surface by cascade secondaries may add to the spectrum at lower energies (Daugherty and Harding 1989). The Vela spectrum, on the other hand, does flatten below $\sim$ 10 MeV as the polar cap cascade model would predict.

References

Arons, J. 1983, *Ap. J.*, **266**, 215.

Beskin, V. S., Gurevich, A. V. and Istomin, Ya. N. 1983, *Sov. Phys. JETP*, **58**, 235.

Cheng, K. S., Ho, C. and Ruderman, M. A. 1986, *Ap. J.*, **300**, 500.

Chiang, J. and Romani, R. W. 1992, in *Proc. of 2nd GRO Workshop*, ed. E. Chipman, in press.

Daugherty, J. K. and Harding, A. K. 1982, *Ap. J.*, **252**, 337.

Daugherty, J. K. and Harding, A. K. 1989, *Ap. J.*, **336**, 861.

Goldreich, P. and Julian, W. H. 1969, *Ap. J.*, **157**, 869.

Lyne, A. G. and Manchester, R. N. 1988, *M. N. R. A. S.*, **234**, 477.

Grenier, I. *et al.* 1992, this Workshop.

Harding, A. K., Tademaru, E. and Esposito, L. W. 1978, *Ap. J.*, **225**, 226.

Kanbach, G. *et al.* 1992, this Workshop.

Thompson, D. J. *et al.* 1992, these Proceedings.

Rankin, J. M. 1990, *Ap. J.*, **352**, 247.

Ruderman, M. A. and Sutherland, P. G. 1975, *Ap. J.*, **196**, 51.

Sturrock, P. A. 1971, *Ap. J.*, **164**, 529.

Gamma Radiation from Pulsar Magnetospheric Gaps

James Chiang and Roger W. Romani

Department of Physics, Stanford University

1 Introduction

With the launch of the Compton Gamma Ray Observatory, new opportunities for distinguishing between models of gamma ray emission from pulsars have arisen. In order to test the predictions of these models more thoroughly, we have performed computations based on the polar cap model (Daugherty and Harding 1982) and the outer gap model (Cheng, Ho, and Ruderman 1986) using a fully three-dimensional model of the pulsar magnetosphere. We have also constrained the magnetosphere geometry by inclination and viewing angle information available from radio and optical observations of the Crab and Vela pulsars. Use of the specific viewing and inclination angles from lower frequency observations has allowed us to determine "predicted" light curve shapes and phase-resolved spectra for the Crab and Vela pulsars which can be directly compared with the data.

For the Crab pulsar, we use the well established values of the period and period derivative to determine dipole magnetic field strength. We also use the values for the dipole-to-rotation axis inclination angle, $\alpha = 86°$, and viewer impact parameter, $\beta = -9.6°$, obtained from optical polarization studies (Narayan and Vivekanand 1982). Similarly, for the Vela pulsar, we use viewing angles obtained from radio polarization studies (Lyne and Manchester 1988): $\alpha = 90°$ and $\beta = 6.4°$. Figure 1 illustrates how the angles are defined.

2 Polar Cap Calculations

Our polar cap calculation follows closely the work of Daugherty and Harding (1982). We inject electrons with energies of $\simeq 10^{13}$ eV at the pulsar polar cap; this energy is appropriate for a Ruderman-Sutherland (1975) type gap. In addition, we have scaled our computations so that the total charge flow is consistent with a Goldreich-Julian (1969) current density at the neutron star surface. The main emission region is defined by the open field line region, that is, those field lines which do not close within the speed-of-light cylinder (see Fig. 1). The highly relativistic electrons flow out along the open field lines emitting curvature radiation. For each step in our simulation along the electron trajectory, we compute the emissivity to curvature radiation for energies from 10 MeV and above; we also compute the optical depth, τ, to $\gamma - B$

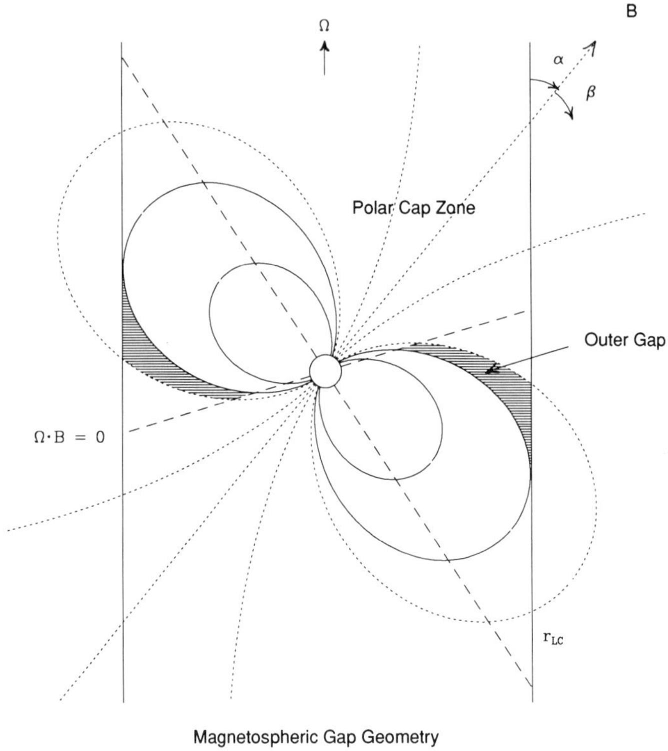

Fig. 1—Pulsar viewing geometry, in the $\vec{\Omega}$-$\vec{B}$ plane. Polar cap emission propagates in beams from the north and south open field line regions. Outer gap photons are directed in outward (and/or inward) beams along the shaded gap field lines above the null charge surface.

pair-production and attenuate the flux by $e^{-\tau}$. For each photon, a pair conversion point is Monte Carlo drawn along its initial trajectory. The resulting electron-positron pair synchrotron radiate, and an electromagnetic cascade ensues. At each generation, we add the properly weighted contribution of the photon flux to our final light curves and spectra.

Figure 2 is an image of the Crab pulsar in gamma rays projected onto the sky. Darker regions indicate greater flux. The darkest areas correspond to the synchrotron emission near the neutron star surface where the magnetic fields are the strongest. Because of the rapid rotation of the pulsar, photon aberration plays an important role in the distribution of the gamma ray photons on the sky. The effect of aberration can clearly be seen as it tends to project all photons forward in phase. The horizontal axis of the plot gives pulsar phase, and the vertical axis indicates viewing angle–a single cut in latitude on figure 2 corresponds to a particular line-of-sight.

For the viewing angles prescribed by the optical polarization sweep analysis, the light curves in Figure 3 obtain. The solid curve is for photon energies $E > 100$ MeV, and the dot-dashed curve is for $E > 500$ MeV. The dashed line is the corresponding radio pulse phase (at infinite radio frequency) as a function of emission height above the polar cap surface.

At first glance, the polar cap light curve has some promising-looking features. The relative height of the main pulse and interpulse compares favorably with the pulse

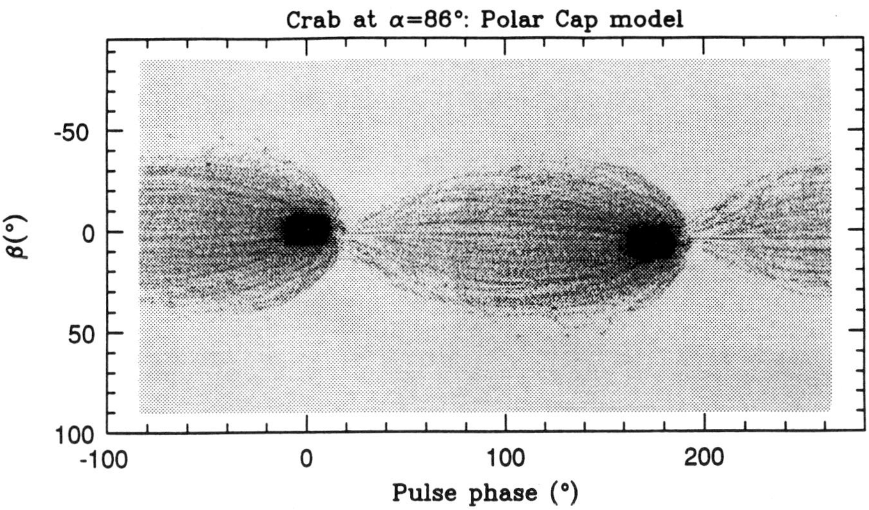

Fig. 2—Phase map of the Crab pulsar in the polar cap model.

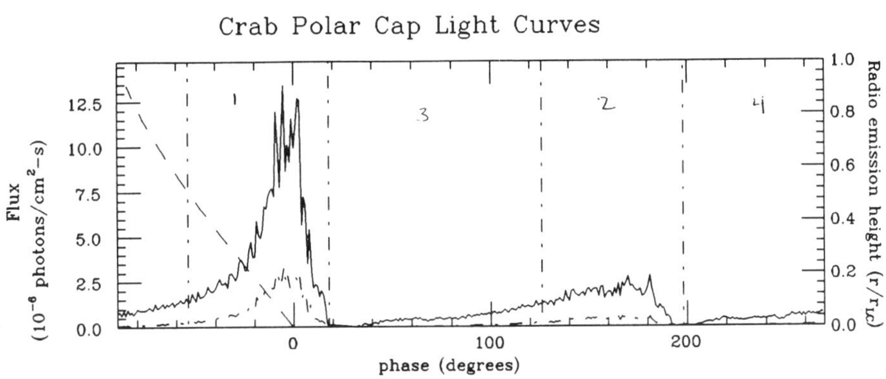

Fig. 3—Light curves for the Crab pulsar in the polar cap model.

heights found by COS-B (Clear *et al.* 1987). Because the line-of-sight passes close to the synchrotron core of the corresponding pole, the main gamma-ray pulse has a sharp cusped peak. Furthermore, for reasonable emission heights above the star's surface, the relative position of the radio main pulse seems to agree roughly with the position of the observed radio precursor pulse when compared to COS-B phases. Additionally, when the total luminosity is scaled by an appropriate beaming fraction incident on the earth, the resulting flux values agree fairly well with what is observed: we obtain flux values of $\sim 3 \times 10^{-6}$ for $E > 100$ MeV.

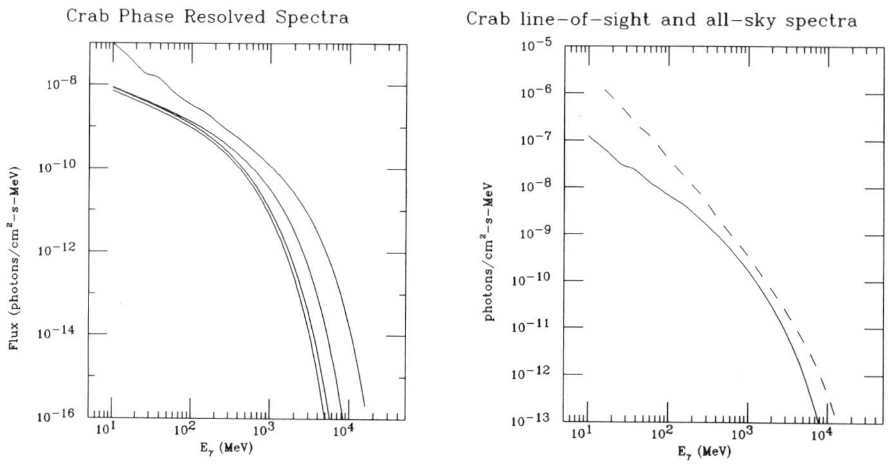

Fig. 4—Polar cap phase-resolved spectra for the Crab pulsar.

Unfortunately, there are still several problems with the polar cap model, at least in this implementation of it. Since we are employing a strict vacuum dipole field geometry, the phase difference between the pulse and interpulse will always be identically 180°; whereas the observed phase difference is approximately 140°. However, with more realistic field geometries, including the distortion of field lines due to effects such as the inertial loading of the co-rotating charges, this phase difference may be subject to change. The effect of aberration is more serious. Since aberration will always project photons to earlier phase, the emission from each pole will have a "ramp" of flux on the leading side and a sharp drop-off of flux at the trailing side. By contrast, the COS-B light curves for the Crab show substantial "bridge" emission between the two pulses and sharper drop-offs in flux at the leading edge of the main pulse and trailing edge of the interpulse.

There are also disagreements when the phase-resolved spectra are considered. Figure 3 shows how we have divided the Crab light curve into four regions in phase. They correspond roughly to the phase regions defined by Clear *et al.* in their analysis of COS-B data for the Crab: pulse (1), interpulse (2), bridge (3), and trailer (4). The spectra for these four regions is shown in the left panel of Figure 4. The spectrum for the pulse has the correct spectral index of $\simeq -2$ for energies below a few GeV, but shows a cut-off at higher energies. The situation is worse for the interpulse spectrum. Since the observer line-of-sight misses the synchrotron core for these regions, there is a deficit of low-energy photons resulting in a ~ -1.7 spectral index. The observed spectral index for the interpulse is closer to -2. Moreover, when the entire line-of-sight spectrum is computed and compared to the all-sky averaged spectrum, the problems which arise when the radio or optical viewing angles are used become more apparent. As with previous polar cap calculations, the all-sky averaged spectrum and flux agree fairly well with what is actually seen from the Crab pulsar. However, when the measured viewing angles from the optical are used, a significant deficit of synchrotron photons at low energies results.

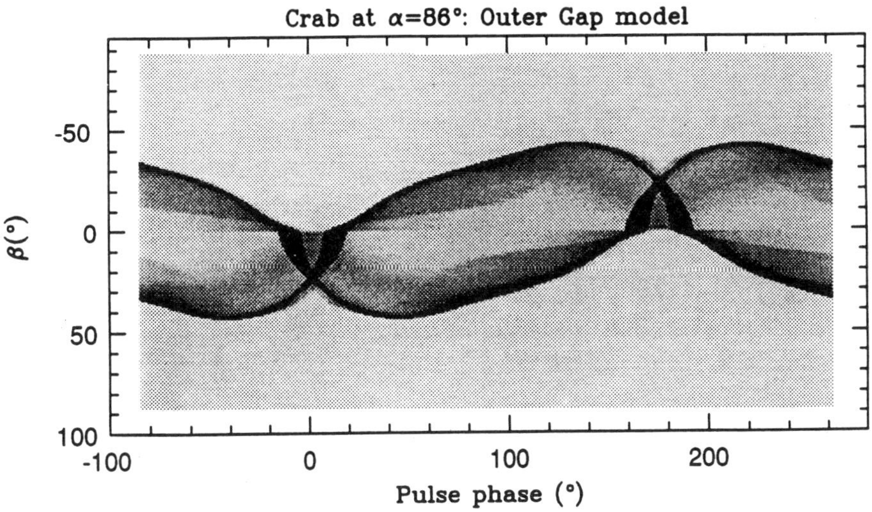

Fig. 5—Phase map of the Crab pulsar in the outer gap model.

3 Outer Gap Calculations

In the outer gap case, we performed light curve calculations based only on the geometry of the magnetosphere. (No spectral calculations were performed, though we plan to include these calculations in the future.) One of the major differences between the outer gap and polar cap models is the location of the acceleration and emission regions. In the polar cap model, the emission region consists of the entire open field line zone; however, all of the acceleration and most of the photon production occurs relatively close to the neutron star surface. In the outer gap model, the acceleration zone is farther out in the pulsar magnetosphere-extending, in a thin gap, from the null charge line out to near the light cylinder. For our calculation, we have defined the outer gap emission region to consist of the set of last closed field lines bounded on the inside by the null charge line and on the outside by the light cylinder. The outer gap as defined by previous authors (for example, Ho 1989) consists only of a smaller subset of these field lines, in particular, the "most favorably" curved field lines. However, since the Crab and Vela pulsars are both nearly counter-aligned rotators (*i.e.*, $\alpha \simeq 90°$), the curvature of the last closed field lines does not vary much as a function of azimuthal angle about the dipole axis. Lacking a compelling reason to prefer one set of last closed field lines to another, we have chosen to use all of them.

In illuminating the outer gap region, we have propagated photons in both the inward and outward directions tangential to the field lines. Since charges are accelerated both away (outward) and toward (inward) the neutron star, radiation will be produced in both directions. After performing the same aberration calculation as in the polar cap case, the sky image in Figure 5 obtains. Again, a horizontal cut in latitude corresponds to a particular observer line-of-sight.

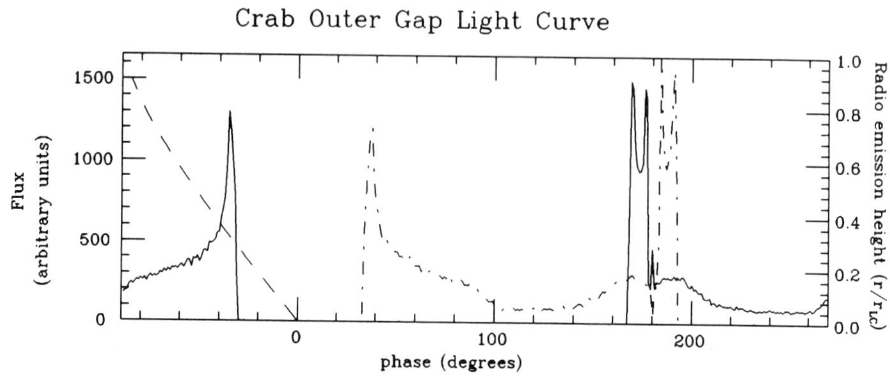

Fig. 6—Crab outer gap light curve.

The outer gap light curve for the Crab is shown in Figure 6. We display emission from both the outward-going photons (solid line) and the inwards-going photons. The outer region of this gap contributes strongly only to the interpulse cusp of the emission; for angles β not close to 0, even this second cusp arises from further inside the magnetosphere. For this particular line-of-sight, the radio phase does not have the observed ($\sim$ a few tens of degrees) lead of the main pulse. This is because the outward photons are coming from the opposite hemisphere from which the main radio pulse originates for these viewing angles. To get the radio phase to lead the gamma-ray phase of the outward photons by the expected amount, emission at both frequencies must occur from the same hemisphere. Furthermore, a simple double pulse structure will in general obtain only if either the outward or inwards beam is illuminated - the outer gap emission prefers a single pole model. It should be noted that the parameters determined by the optical observations are based on the model for two-pole emission originating near the stellar surface; for the outer gap model, since emission occurs farther out in the magnetosphere, the corresponding viewing angle may quite different. We are re-examining the polarization sweep data in this context but at present retain the parameters based on cap emission model for consistency.

Generally, the light curves of both models present some successes. The relatively sharp pulse emission, the weaker second pulse in the Crab case, and the presence of interpulse emission all show rough agreement with the observations. However, in both models, the observed pulse separation of $\sim 140°$ and the sharpness of the observed peaks (particularly for Vela) require some departures from the generic dipole geometry and viewing angles prescribed by the radio and optical observations. As an example, Figure 7 shows an outer gap light curve for Vela in which only the parameter β has been modified to give the observed pulse-interpulse separation ($\beta \sim 25°$). This one adjustment produces a light curve with remarkable similarity to the profile observed by EGRET (including the third peak in the interpulse region) if only one beaming direction (outward) is illuminated. The emission plotted comes from within 80% of distance to the light cylinder and so should not be strongly affected by the field line

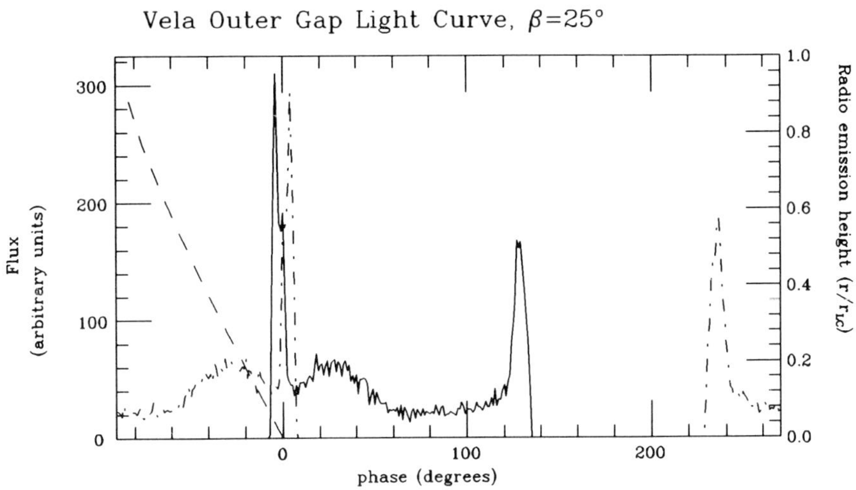

Fig. 7—Vela outer gap light curve for $\beta = 25°$.

distortion due to inertial loading. Clearly, good fits to the observed light curves are possible with minor departures from the standard geometry, especially for the outer gap model, but further tests of the models are needed, particularly with more realistic field geometries and magnetospheric currents. More complete descriptions of our polar cap and outer gap calculations can be found elsewhere (Chiang and Romani 1992).

4 Conclusions

When normalized to a vacuum co-rotation charge density flowing through the cap surface, polar cap models roughly reproduce the overall flux and spectrum of observed gamma ray pulsars at energies ≥10 MeV. These cap models produce beams of secondary synchrotron photons tightly collimated to the dipole axis. It is clear that models with lines-of-sight at large impact parameters to the field axis will not reproduce the cusp-like profiles of Crab and Vela. For the angles determined from low-frequency observations, the match of the models to observed profiles is reasonable at high (EGRET) energies but does not fit in detail. Furthermore, models observed at these angles have worse agreement with present spectral data than simple sky averaged models.

Outer gap light curves offer some hope for detailed reproduction of observed pulse morphologies but probably require a re-appraisal of the radio polarization data. In the outer gap case, as opposed to the polar cap emission, the double pulse light curve is the result of radiation from a single magnetic pole. The best hope for discrimination of these models arises from the phase-resolved spectra–in the polar cap case, lowered fields off the pulse center cause harder curvature spectra, while simultaneously lowering the high-energy cut-off. Comparison of spectral variations we derive here for the polar cap model with similar calculations for gaps in the outer magnetosphere will thus illuminate both the geometry and the emission zone physics. As more examples of gamma ray pulsars are observed by the Compton Observatory with different viewing geometries and spin periods (perhaps even including millisecond pulsars), we expect that the wider range observed parameter space will give strong tests of this modeling.

References

Cheng, K. S., Ho, C. and Ruderman, M. 1986, *Ap. J.*, **300**, 522.
Chiang, J. and Romani, R. W. 1992, *Ap. J.*, submitted.
Clear, J., *et al.* 1987, *Astr. Ap.*, **174**, 85.
Daugherty, J. K. and Harding, A. K. 1982, *Ap. J.*, **252**, 337.
Goldreich, P. and Julian, W. H. 1969, *Ap. J.*, **157**, 869.
Ho, C. 1989, *Ap. J.*, **342**, 396.
Lynn, A. and Manchester, R. M. 1988, *MNRAS*, **234**, 477.
Narayan, R. and Vivekanand, M. 1982, *Ast. Ap.*, **113**, L3.
Ruderman, M. A. and Sutherland, P. G. 1975, *Ap. J.*, **196**, 51.

Unpulsed High-Energy Gamma-Ray Emission from the Crab Pulsar

J. R. Mattox,[1,2] O. C. De Jager,[3] and Alice K. Harding[1]

[1]Laboratory for High Energy Astrophysics, NASA/Goddard Space Flight Center
[2]Compton Observatory SSC, NASA/Goddard Space Flight Center
[3]Physics Department, Potchefstroom University, Potchefstroom, South Africa

Abstract

An analysis by Clear *et al.* (1987) of COS-B emission from the Crab pulsar reported a 53 ± 15% unpulsed component (50 - 3000 MeV), whereas an earlier analysis by Bennett *et al.* (1977) reported a lower limit of 78% for the Crab pulsed fraction. Because the earlier analysis used only the first month of COS-B observation and the 1987 report used all the data together, we have reanalyzed the COS-B emission from the Crab pulsar in the six separate observing periods to investigate the possibility that the unpulsed emission is variable. A likelihood ratio test is done using a model of the diffuse gamma-ray emission obtained from radio observations of neutral H, CO and electron synchrotron emission. We find no significant variability in the unpulsed emission in the energy range 150 MeV - 3 GeV, but we confirm the overall unpulsed component reported by Clear *et al.* (1987).

1 Introduction

The pulsed emission from the Crab pulsar and the unpulsed emission from the Crab nebula are easily differentiated at radio through X-ray frequencies, where the extended nebular emission is clearly resolved. At gamma-ray energies, the situation is more complicated as both pulsed and unpulsed components appear as point sources to detectors with poorer resolution. In addition, the extent of the nebular emission decreases with increasing frequency, probably due to synchrotron losses of centrally injected particles, making resolution of the nebula at high energies even more difficult. The nature and spectrum of the two components at gamma-ray energies is thus not well determined.

The synchrotron nature of the nebula from radio through X-ray energies has been established through 1) the power law spectrum between 10^7 Hz and 10 MeV, 2) strong linear polarization, and 3) the observed decrease in scale size of the nebular images with increasing frequency. The scale size is roughly proportional to the lifetime of electrons radiating at a frequency ν times the radial velocity of the bulk motion. This picture requires a continuous injection of relativistic particles near the center of the nebula, presumably from the pulsar, whose spin-down power is about five times the radiated power of the nebula. Although the overall energetics makes sense, the mech anism by which the pulsar transfers its spin-down energy to relativistic particles is still not understood.

There is an indication that the synchrotron spectrum of the Crab nebula extends to gamma-ray energies. An unpulsed component at 1 - 20 MeV was suggested by Graser and Schonfelder (1982) and recently confirmed by COMPTEL (McConnell *et al.* 1992). At higher energies, SAS-2 reported an unpulsed component above 30 MeV (Thompson *et al.* 1977) which appeared to be an extension of the spectrum at lower energy. COS-B has also reported detection of an unpulsed flux from the Crab above 50 MeV, but the several analyses reported give different results. Bennett *et al.* (1977) analyzed only the first COS-B observation of the Crab (Sept. 1975) and reported a lower limit on the pulsed fraction above 50 MeV of 78%, which is consistent with no unpulsed emission. Lichti *et al.* (1980) reported an unpulsed flux of 45 ± 10% in the first two COS-B observations. A more sophisticated analysis by Clear *et al.* (1987) of the entire COS-B Crab data base revealed that 53 ± 15% of the emission (from 50 - 3000 MeV) is unpulsed. These analyses of COS-B emission taken together suggest that the flux between 50 and 500 MeV could be variable.

There also exists a theoretical reason for suspecting variability at high energy gamma-ray energies. If the emission up to 500 MeV is an extension of the synchrotron spectrum, then the radiating electron spectrum must extend to $\sim 10^{16}$ eV, where the synchrotron lifetime is around a month. The nebula is also likely to be quite compact in gamma-rays, so that fluctuations in either the injection or acceleration on timescales longer than a few months could produce variability in the emission at the upper end of the spectrum. De Jager and Harding (1992) have shown that the inverse-Compton component due to scattering of the synchrotron radiation by the relativistic electrons predicted by Gould (1965) cannot account for any significant part of the 50 - 500 MeV unpulsed flux detected by COS-B. Thus, it seems that the unpulsed component detected by COS-B is predominantly synchrotron emission from the highest energy electrons accelerated in the nebula.

In Section 2, we discuss the importance of unpulsed emission between 10 MeV - 1 GeV in the context of the synchrotron-self Compton model of the nebula. In Section 3, we present our reanalysis of the COS-B Crab data to confirm the presence of an unpulsed component and to search for variability in the unpulsed emission.

2 Synchrotron-Self Compton Model of the Crab Nebula

Gould (1965) predicted a flux of TeV γ-rays from inverse Compton (IC) scattering of the synchrotron radiation by the relativistic electrons. Although the observations to confirm this prediction have been inconclusive for several decades, the development by the Whipple group (Weekes *et al.* 1989, Vacanti *et al.* 1991) of an imaging technique to eliminate a large fraction of the background hadron-initiated showers has finally produced a solid ($\approx 20\sigma$) detection of unpulsed γ-rays from the Crab nebula with a spectral index of 2.4 ± 0.3. Since the Crab nebula is optically thin to absorption and scattering, the calculation of the IC spectrum is straightforward with the aid of detailed maps and photon spectra, provided that the field distribution is known.

The treatment of the field distribution is most crucial in synchrotron self Compton (SSC) estimates, since the scattered flux depends on the field strength B as $B^{-(\alpha+1)/2}$ (given an electron spectral index of α). Since the earlier calculations of the IC flux of the Crab, more sophisticated models have been developed for the field distribution in the nebula, based on the pulsar wind model first suggested by Rees and Gunn (1974). They argued that the spindown power of the pulsar would be transported in the form of a relativistic MHD wind and showed that a shock would form in the pulsar wind due to the confinement by the nebula expanding at $v \ll c$. They estimated r_s as the distance from the pulsar where the ram pressure in the pulsar wind equals the total nebular pressure, giving $r_s \approx 10''$. In the MHD models investigated by Kennel and Coroniti (1984), the spindown luminosity L of the pulsar upstream of the shock is divided between particle and electromagnetic energy as $L = 4\pi n_1 \gamma_1 u_1 r_s^2 mc^3 \times (1 + \sigma)$, where n_1 is the proper density and u_1 is the radial three speed (such that $\gamma_1^2 = 1 + u_1^2$). The only free parameter in their model is $\sigma \equiv B_1^2/4\pi n_1 u_1 \gamma_1 mc^2$, or the ratio of magnetic to particle energy density just ahead of the shock. The best estimate of σ, obtained by matching the wind flow solutions to the observed velocity of $\approx$ 2000 kms^{-1} at the outer boundary of the nebula, gives a value of $\sigma \approx 0.003$. For such a small σ the magnetic field strength at r_s, $B_s = 8 \times 10^{-5}$ G, is a minimum and increases as $B \propto r$ for $r > r_s$ until equipartition is reached, thereafter decreasing until the nebular radius r_N is encountered.

De Jager and Harding (1992) have recently re-examined the SSC model of the Crab nebula, incorporating the nebular magnetic field distribution derived from MHD flow models (Kennel and Coroniti 1984). They showed that the optimal field distribution as derived by Kennel and Coroniti provides a TeV flux and spectral index which is consistent with the observations and also gives the required $\bar{B} \approx 3 \times 10^{-4}$ G calculated from the spectral break between radio and optical. However, as shown in Figure 1 the extension of this IC spectrum below GeV energies cannot account for the 50 - 500 MeV unpulsed flux reported by Clear *et al.* (1987). These results indicate that the IC spectrum is much flatter and much lower in flux than the observed COS-B unpulsed spectrum between 50 MeV and 500 MeV, so that it would be difficult to associate the COS-B flux with an inverse Compton component in the nebula.

Interpreting this spectrum as synchrotron radiation at the pulsar wind shock allows an estimate of the maximum electron energy at the shock. If this emission is fit with an extension of the synchrotron spectrum, assuming the same nebular field distribution used to compute the IC spectrum, then an electron spectrum extending to $\sim 10^{16}$ eV is required to produce this radiation. In Figure 1 we have calculated the synchrotron component at high energies for $B_S = 8 \times 10^{-5}$ G ($\sigma = 0.003$) for various values of E_{max}. The spectrum for $E_{max} = 10^{16}$ eV agrees best with the COS-B spectrum and is consistent with the spectrum extrapolated from the low energy γ-ray range. Reducing E_{max} by a factor of about 3 (*e.g.*, due to variability) can turn off the detection of unpulsed γ-rays (from the shock) between 50 MeV and 1 GeV, but the emission above 1 GeV is expected to be unaffected because it is due primarily to IC scattering of radio and optical photons by much lower energy electrons (GeV to TeV) with long synchrotron lifetimes.

The electron energy $\sim 10^{16}$ eV required to account for the COS-B unpulsed flux approaches the full vacuum potential across the pulsar polar cap and is at the maximum theoretical limits of acceleration at the pulsar wind shock (De Jager and Harding 1992). Detailed study of this emission, to determine where the upper end of the synchrotron

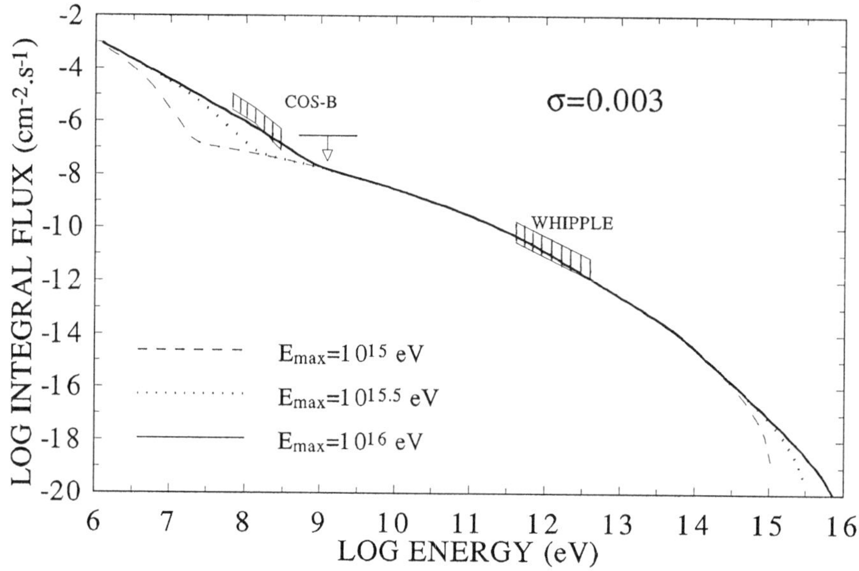

Fig. 1—Combined synchrotron and inverse Compton γ-ray spectrum of the Crab Nebula for σ = .003 and various maximum electron energies E_{max} (from De Jager and Harding 1992).

spectrum lies and whether the emission is variable, is important in uncovering clues to the nature of the acceleration mechanism in the nebula.

3 Reanalysis of the Crab COS-B Data

To investigate the possible variability of the unpulsed emission from the Crab, we performed a new analysis of all the available data from 1975 - 1982. We have derived the unpulsed flux between 150 MeV and 5 GeV in each of the six observation periods separately. At this stage, we have not included the 70 - 150 MeV data in the analysis because the detector background is not proportional to exposure at these energies as it is above 150 MeV. Two sets of counts and exposure maps were constructed for each of the two energy intervals, 150 - 300 MeV and 0.3 - 5 GeV, one containing all events and one containing only unpulsed events, *i.e.*, those outside the pulse phase interval 0.1 - 0.62 (phase calculated with the ephemerides of Clear *et al.*, 1987). A likelihood analysis (Mattox, 1992, in preparation) was then done (using the galactic diffuse emission model of Strong *et al.* [1988] at a fixed intensity) to simultaneously fit the intensity of the Crab and Geminga emission, and an isotropic term (which comprises instrumental background and a possible extragalactic flux) for maps of all events. The unpulsed maps were then evaluated with likelihood (with the isotropic term fixed at the value obtained from the map of all events; which afforded better determination of this parameter due to improved statistics). The unpulsed flux thus obtained for each observation period is shown in Table 1. The uncertainties given are approximate 1 σ errors. There is no significant evidence for variability in the unpulsed emission component in either energy range. Figure 2 shows a profile of the result of this likelihood analysis for all observation periods combined (150 - 300 MeV). The

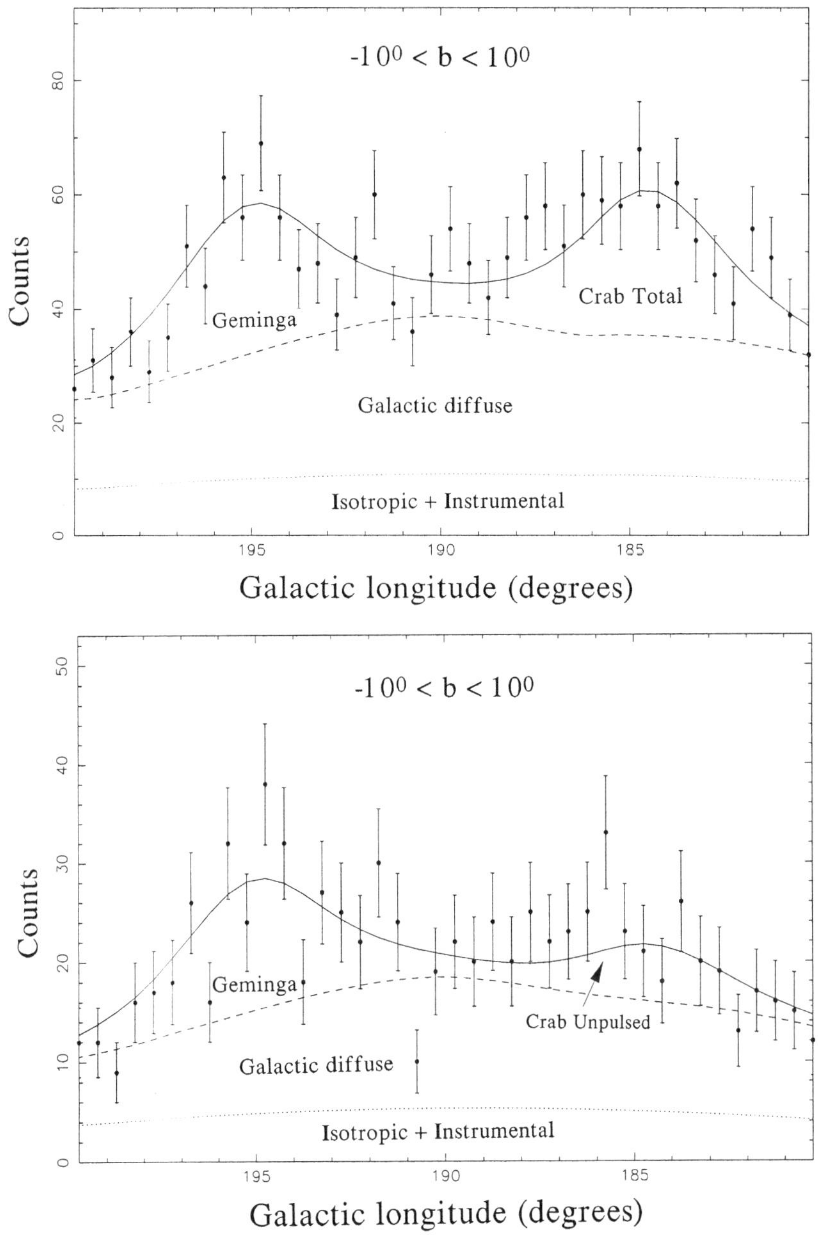

Fig. 2—COS-B longitude profiles of the Crab and Geminga region, showing the background and point spread function fits. Top profile includes all Crab events, pulsed and unpulsed. Bottom profile includes only unpulsed events (phases 0.62 - 0.10). Vertical axes are counts in $5^0 \times 20^0$ bins.

TABLE 1
Crab Unpulsed COS-B Flux

OBSN	Date	FLUX (10^{-7} cm^{-2} s^{-1}) $150 < E < 300$ MeV	$300 < E < 5000$ MeV
0	75/9	5.9 ± 2.5	1.2 ± 1.2
14	76/10	2.2 ± 3.5	0.7 ± 2.7
39	79/3	7.2 ± 3.6	3.3 ± 2.4
44	79/9	7.7 ± 5.3	4.3 ± 3.2
54	80/10	0.0 ± 4.7	3.4 ± 2.2
64	82/3	5.2 ± 2.9	1.6 ± 1.6
ALL	75-82	4.7 ± 1.4	2.2 ± 0.8

fraction of unpulsed flux is 0.39 ± 0.11. A similar analysis for 300 - 5000 MeV gives a fraction of unpulsed flux of 0.26 ± 0.10. This is consistent with the result of Clear *et al.* (1987).

Figure 3 shows the results plotted for the lower energy band, 150 - 300 MeV. It is clear that, given the size of the errors in the fluxes, there is no significant evidence for variability in the unpulsed emission component. The data is consistent with a constant flux level of $(4.7 \pm 1.4)\times10^{-7}$ cm^{-2} s^{-1} , shown by the dashed line in the figure. We therefore confirm a significant overall unpulsed flux from the Crab as reported by Clear *et al.* (1987). The pulsed fraction for the initial observation period is 0.49 ± 0.21 (150 - 300 MeV). Although we have not analyzed the low energy gamma rays, this appears to be inconsistent with the upper limit on unpulsed flux of 22% of Bennett *et al.* (1977) for E > 50 MeV.

4 Summary

Though analysis of the Crab unpulsed data from COS-B does not yield a clear answer to the question of variability of the emission in this energy range, data from EGRET on the Compton Observatory will be able to shed new light on this issue. In particular, the likelihood analysis we have presented here can easily be applied to EGRET data to derive the Crab unpulsed component in different energy ranges and observing epochs with more significant results. COMPTEL has already reported detection of an unpulsed component from the Crab up to an energy of 30 MeV (McConnell *et al.*) 1992). It will be critical to examine the low EGRET energy range (20 - 100 MeV) to determine both whether the synchrotron spectrum extends to these energies and whether it is present at the same flux level as was seen by COS-B a decade earlier. The sensitivity of EGRET extends to at least 10 GeV, where unpulsed emission from the inverse Compton component is expected to make a significant contribution.

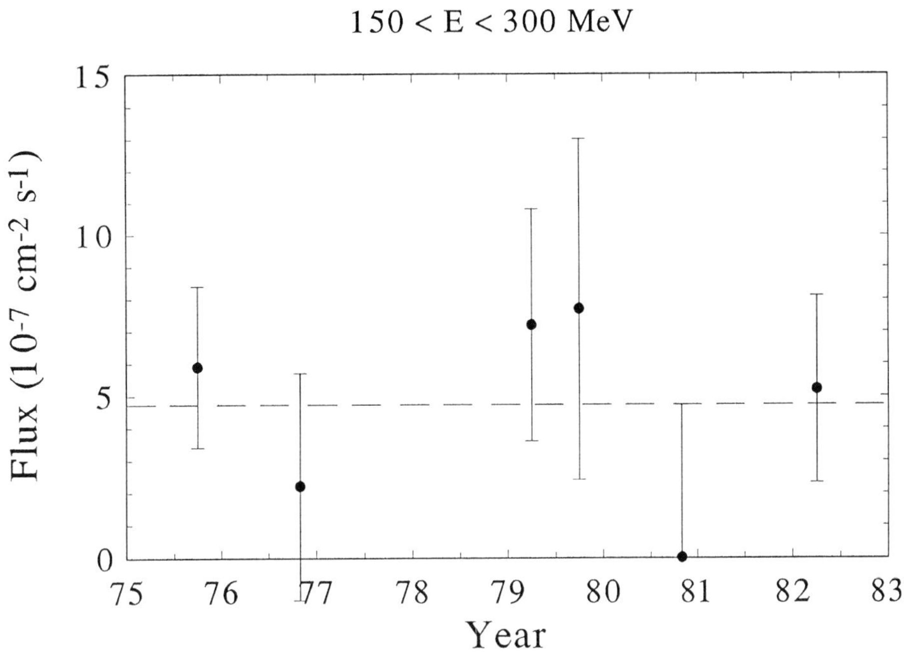

Fig. 3—Unpulsed flux from the Crab pulsar in six observation periods (cf. Table 1). Dashed line is best fit constant flux level.

References

Aschenbach, N. and Brinkmann, W. 1975, *Astron. Astrophys.*, **41**, 147.

Bennett, K. *et al.* 1977, *Astron. Astrophys.*, **61**, 279.

Clear, J. *et al.* 1987, *Astron. Astrophys.*, **174**, 85.

De Jager, O. C. and Harding, A. K. 1992, *Ap. J.*, in press.

Gould, R. J. 1965, *Phys. Rev. Lett.*, **15**, 577.

Graser, U. and Schonfelder, V. 1982, *Ap. J.*, **263**, 677.

Kennel, C. F. and Coroniti, F. V. 1984, *Ap. J.*, **283**, 694.

Lichti, G. G. *et al.* 1980, *Non-Solar Gamma Rays*, ed. R. Cowsik and R. D. Wills (Pergamon: Oxford), **7**, 49.

McConnell, M. *et al.* 1992, these Proceedings.

Rees, M. J. and Gunn, J. E. 1974, *Ap. J.*, **167**, 1.

Strong, A. W., Bloemen, J. B. G. M., Dame, T. M., Grenier, I. A., Hermsen, W., Lebrun, F., Nyman, L. A., Pollock, A. M. T., Thaddeus, P. 1988, *Astron. Astrophys.*, **207**, 1.

Thompson, D. J., Fichtel, C. E., Hartmann, R. C., Kniffen, D. A. and Lamb, R. C. 1977, *Ap. J.*, **213**, 252.

Weekes, T. C. *et al.* 1989, *Ap. J.*, **342** 379.

Vacanti, G. *et al.* 1991, *Ap. J.*, **377**, 467.

Observations of Plerions with the Whipple Imaging Telescope

T. C. Weekes,[1] C. W. Akerlof,[2] M. F. Cawley,[3] M. Chantell,[1] D. J. Fegan,[4] S. Fennell,[1,4] A. M. Hillas,[5] R. C. Lamb,[6] M. A. Lawrence,[1] D. A. Lewis,[6] D. I. Meyer,[2] G. Mohanty,[6] K. S. O'Flaherty,[4] M. Punch,[1,4] P. T. Reynolds,[6] A. Rovero,[1] M. S. Schubnell,[2] T. Whitaker[1].

[1]Harvard-Smithsonian Center for Astrophysics
[2]University of Michigan
[3]St. Patrick's College, Maynooth
[4]University College, Dublin
[5]University of Leeds
[6]Iowa State University

Abstract

The significance of the observation of TeV gamma rays from pulsars and plerions is discussed. The advantages of the atmospheric Čerenkov technique is contrasted with the problems inherent in the technique. Only the detection of the Crab Nebula can be regarded as well-established, but new telescopes offer the possibility of significant advances in the near future.

1 Motivation

The TeV gamma-ray sky is still largely a matter of conjecture. Nonetheless, since their discovery, radio pulsars have been considered to be major constituents and hence prime targets of investigation by ground-based TeV gamma-ray astronomers. Before briefly discussing the results of such investigations, it is worth considering what it is about the pulsars that makes them such attractive targets.

The prime motivation is the source models, almost all of which consider the presence of electrons with energies up to TeV (and beyond), as intrinsic to the mechanics of the sources. Despite the fact that there is no universally accepted model of pulsar acceleration, there does seem to be general acceptance that there is copious production of TeV electrons (and positrons).

In several cases the so-called "isolated pulsars" are surrounded by bubbles of relativistic electrons which are observed to emit synchrotron radiation from radio to x-ray wavelengths. These are called plerions, and they are widely believed to be continually fed from the pulsar. Hence, not only is the pulsar itself a candidate for *pulsed* gamma-ray emission, but the surrounding synchrotron nebula is also a possible source of *steady* gamma-ray emission.

Prior to the launch of the Compton GRO, only a few sources at GeV energies were positively identified; two of these were the Crab and Vela pulsars. Their measured spectra indicated that a reasonable extrapolation might lead to fluxes at TeV energies that would be easily detectable with existing ground-based telescopes. For this reason alone, pulsars would have been prime targets for TeV astronomers.

While it is usual to assume that where there are relativistic electrons, there will be gamma rays (as a result of synchrotron or curvature radiation, bremsstrahlung or Compton scattering), there is the strong possibility that gamma rays cannot escape from the acceleration region without undergoing further interactions which downgrade their energy. In particular, in the intense magnetic fields close to the surface of neutron stars, pair production is very probable. Hence the absence of pulsed TeV gamma radiation can be easily explained and is not a serious constraint on existing models.

The definite detection of pulsed radiation would confirm that particle acceleration does take place at the pulsar; the details of the emission (time variability, energy spectrum) might assist with discrimination between different pulsar models. It seems likely that pulsars play a role in the production of cosmic ray electrons in the Galaxy; the measurement of the associated gamma- ray spectrum might allow us to determine just what this role is.

There is another pragmatic reason why pulsars are attractive to ground-based observers: the observation of a steady source is difficult and prone to systematic effects. A pulsed signal has a far more definite character and can increase confidence in the detection; in fact some observers feel that they should only believe signals that have a periodic character! (With this definition the moon would not be accepted as an optical source.) This observational bias should not detract from the fact that gamma radiation from pulsars may be unpulsed if it originates outside the light-cylinder or in the associated plerion. In fact, as we will see below, such steady emission may the strongest evidence for gamma-ray emission from pulsar systems.

2 Techniques: Limitations and Possibilities

The only feasible method for the study of TeV radiation from cosmic sources involves the use of the atmospheric Čerenkov technique (Weekes, 1988). Even the large spark chamber telescope on the Compton Gamma Ray Observatory (EGRET) has little effective sensitivity above 10 GeV; hence, in the near future only ground- based measurements are anticipated.

This is not the place to review the basic atmospheric Čerenkov technique or to describe the recent advances that have been made; however, we should mention the most significant development, namely that it is now possible to effectively discriminate against the charged cosmic ray background (99% rejection) in the energy range from 0.3 to 10 TeV (Vacanti *et al.* 1991; Baillon *et al.* 1991) and this development should dramatically change the sensitivity of all TeV measurements of pulsars and other sources.

There are some special considerations favoring the observation of pulsars with ground-based atmospheric Čerenkov telescopes.

(1) Since ground-based measurements are independent of space vehicles, the atmospheric Čerenkov technique is well-suited for long-term monitoring of specific objects.

(2) The large collection area of even the simplest telescopes allows the detection of short outbursts.
(3) A periodic signal can easily be picked out against a noisy or variable background e.g under poor sky conditions.

On the other hand there are several reasons why such observations are difficult and why reports of TeV emission should be treated with caution.

(1) To get maximum sensitivity for the detection of a steady (but weak) periodic signal, long exposures are required; because of their nature these measurements are interrupted by the sun, moon and weather conditions.
(2) Pulsar observations require precise time standards; many gamma-ray observatories lack the facilities necessary to phase link observations over long intervals. Pulsars are not the only sources under investigation, and hence, TeV astronomers are not full-time professional pulsar observers. Mistakes can (and have) been made in barycentering, etc.
(3) New statistical procedures are often used to detect the weak emission; the exact interpretation of such procedures only comes with experience. It is almost impossible to rigorously treat the statistics of transient phenomena.
(4) The signal-to-noise ratio of these experiments is always low; there are no standard candles and no assurance that the correct observational techniques or analysis procedures are being applied.
(5) Electronic pickup of spurious periodic signals is a hazard of any pulsar experiment. In any atmospheric optical technique there is the added possibility of the detection of spurious background *optical* signals giving spurious periodicities. Possible sources include satellites, aircraft, beacons, and other manmade light sources.

Whereas most of these pitfalls are now recognized, this was not the case in the early days of TeV pulsar observation. The availability of time standards such as GPS and the use of the optical signal from the Crab pulsar as a test of analysis procedures have increased confidence in recent TeV observations.

3 Observations

Following the discovery of pulsars there were the usual crop of tantalizing results in TeV astronomy; these bordered on statistical significance but were not verified by subsequent, more sensitive, observations. These pulsar observations have been reviewed elsewhere (Weekes, 1988; Chadwick *et al.* 1990): recent observations are discussed in a companion paper at this workshop (Punch *et al.* 1992). Here we discuss only the recent observation of plerions.

3.1 Crab Nebula

Supernova remnants were a prime target for early experimenters in TeV astronomy. The Crab Nebula has played a key role in the development of most astronomies. This interest was inspired by a very basic Compton-synchrotron model (Gould 1965) for plerions; the Crab is the prime example of a supernova remnant with centrally-filled morphology. There was some evidence for a detection of the Crab in a first generation atmospheric Čerenkov experiment (Fazio *et al.* 1972). The first clear detection of the Crab Nebula as a TeV gamma-ray source was reported in 1989, based on observations taken with the 37 pixel camera on the Whipple Observatory 10 m optical reflector in 1986-88 (Weekes *et al.* 1989). That camera was replaced in 1988 with a higher resolution 109 pixel camera (Cawley *et al.* 1990). Observations of the Crab Nebula with this camera over the epoch 1988-91 have been reported (Vacanti *et al.* 1991) and are updated here; a more complete report on observations of the Crab and the other sources is in preparation (Reynolds *et al.* 1992).

The 1988-89 database consisted of 65 ON/OFF pairs, comprising some 30 hours of ON source observations. The observations were taken under optimum conditions (clear sky, newly-coated mirrors, low zenith angles, etc.). The first analysis was on lines employed in the initial detection (Weekes *et al.* 1989) which had given a 9 sigma detection based on 60 hours of ON source observation. Candidate gamma-ray images were selected based on their orientation and shape; at the zenith this selection rejected 97% of the background. These selection criteria were determined a priori from Monte Carlo simulations of images from hadron and photon-initiated showers. An excess of 5 sigma is seen in the raw data and 20 sigma in Azwidth-selected data (Vacanti et al. 1991).

Clearly the new camera is more sensitive than the earlier version, and the early detection is confirmed with high significance with the improved detector (Fig. 1). The measured flux is $70 \times 10^{-12} \mathrm{cm}^{-2} \mathrm{s}^{-1}$ above 0.4 TeV. Using this single parameter discriminant (Azwidth), the factor of improved flux sensitivity is 2–3.

In the above analysis, the data (events) were calibrated and prepared for analysis essentially using the methods outlined in previous papers (Weekes *et al.* 1989; Vacanti *et al.* 1991). A more efficient method of event treatment has now been developed involving sophisticated treatments of pedestals, noise, and gains (Punch *et al.* 1991). Previously the gamma-ray selection was based principally on the Azwidth parameter using gamma-domain boundaries which were predetermined by shower simulations (Hillas 1985). Here a new multi-parameter selection was used in which the domain boundaries, defined by the shape and orientation parameters, was optimized on the data: the 1988–89 season of Crab observations. With this new procedure (Supercuts), the significance of a source detection is increased by a factor of 1.75 (Table 1) (Lang *et al.* 1991). The efficiency of the procedure is demonstrated by the application of Supercuts to new observations of the Crab Nebula in 1989-91 (see below).

TABLE 1
Supercuts 1988–89

	Raw	Shape	Orientation	Supercuts
ON	498,426	14,218	43,099	4,452
OFF	493,434	11,216	40,413	1,766
Diff.	+4,992	+3,002	+3,686	+2,686
Excess (sigma)	+5.0	+18.8	+12.7	+34.1

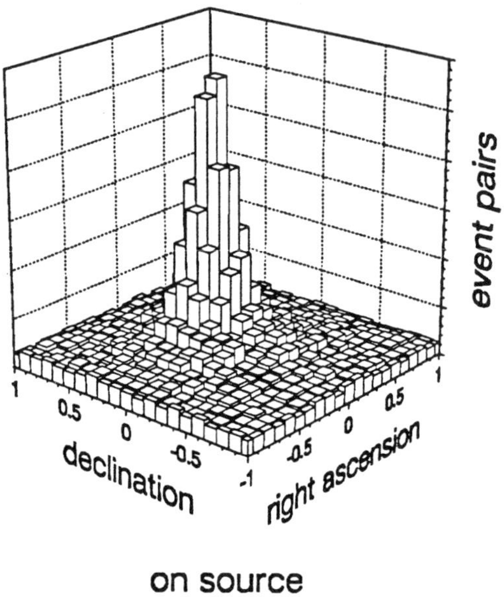

Fig. 1—The "picture" of the Crab Nebula in TeV gamma-rays made by plotting the locus of intersections of the major axes of images in the field of view of the camera. These images have been preselected by their image widths and lengths. This picture is made by subtracting the "OFF" picture from the "ON" source images (Akerlof *et al.* 1991a). It is not possible yet to distinguish a point source from a source of several arcminute diameter.

The same procedure has been used in analyzing data from several other sources (plerions: below; pulsars: companion paper (Punch *et al.* 1992; binary sources: Reynolds *et al.* 1991, extragalactic sources: Akerlof *et al.* 1991b). In these cases we can derive upper limits (or fluxes) *for gamma rays* relative to the Crab Nebula *provided the source spectra are similar to that of the Crab*; if significantly different, the sensitivity may be less.

TABLE 2
Supercut Analysis 1988–91
Epoch: 1988-91 Time: 4105 min Elevation: 71.3 deg

	Raw	Shape	Orientation	Supercuts
ON	968,038	26,747	83,823	8,230
OFF	958,970	21,542	76,956	3,339
Diff.	+9,068	+5,205	+6,867	+4,891
Sigma	+6.5	+23.7	+17.1	+45.5

Using Supercuts the detection level has been increased to a nominal 34 sigma. The same Supercuts selection is now applied ("a priori") to the Crab observations taken in 1989-90 and 1990–91. The results are shown in Table 2. Note the following:

i) the signal is seen every year with no evidence for variability; the results of the search of this data for evidence of periodicity is reported elsewhere (Punch *et al.* 1992);
ii) the detection does not have the full significance associated with the 45.5 sigma detection since it was optimized on the first year of data;

iii) the signal is seen strongly in raw data and in selection by shape or orientation and shape and orientation; the signal has all the characteristics expected of a gamma-ray shower and hence confirms that there are no unexpected features in shower development at these energies;

iv) with Supercuts, a signal has been isolated that is 59% gamma rays, 41% background; the gamma-ray signal is detected at a rate of 1.2/minute, and the number of gamma rays recorded from this source is 4,891;

v) in a recent paper (Akerlof *et al.* 1991a) it is shown that the source location capability (angular resolution) of the technique for a source of this strength is a few arc-minutes (similar to EGRET);

vi) the signal is about 0.5% of the cosmic ray background; for a conventional camera which triggered on cosmic ray showers this might be about 0.2% of the background.

3.2 Plerions

Apart from the detection of the Crab Nebula (Fazio *et al.* 1973; Weekes *et al.* 1989; Akerlof *et al.* 1989; Vacanti *et al.* 1991; Baillon al. 1991; Grenier *et al.* 1992), there have been no clear reports of a TeV signal from any supernova remnant. Three other plerions were observed by the Whipple experiment. The results of these observations are given in Table 3; there is no evidence for emission at the level of 50% of the Crab Nebula. Based on their x-ray luminosities, ages and distances, none of these could be expected to have the intensity of the Crab Nebula.

TABLE 3
Plerions

Source	Time (min)	Supercuts ON	Supercuts OFF	Supercuts Excess	Energy (TeV)	Flux < $\times 10^{-12}$cm^{-2}s^{-1}
3C58	403	484	446	+38 (+1.2)	0.55	32
SS433	191	280	279	+1 (+0.0)	0.55	18
PSR0656	84	191	200	-9 (-0.5)	1.0	34

4 New Telescopes

It is clear that definitive results in these fields await the development of more sensitive imaging telescopes. One such telescope that has recently come on-line is the stereo imaging system at the Whipple Observatory. There the 10 m optical reflector has been supplemented by a second large optical reflector of nominal aperture 11 m. The 11m reflector is actually a converted solar energy device which was acquired in 1991. The 364 light-weight mirrors are of 45 cm diameter. Situated 140 m away from the 10 m reflector, the second telescope is also equipped with a photomultiplier camera. This camera initially has just 37 pixels but is otherwise identical. The two

Fig. 2—The Whipple Observatory Stereo Imaging Telescopes.

cameras (Fig. 2) are designed to operate independently and together to give stereoscopic imaging of air showers; the combination of images improves the sensitivity of the system (by a factor of three in energy threshold and a factor of ten in flux sensitivity).

A similar system (with somewhat smaller reflectors but with cameras with more pixels) is under construction at Woomera, Australia; this is a joint project of the Institute of Cosmic Ray Research, Tokyo, Japan and the University of Adelaide, Australia (Ebisuzaki *et al.* 1991). When this telescope comes on-line, it will permit the study of southern sources with greater sensitivity than has hitherto been possible.

These two reflector systems are now considered prototypes of a large array of Čerenkov imaging telescopes which may be part of an international project to be built in the next decade. In principle it is possible to reduce the energy threshold of the atmospheric Čerenkov technique down to 10 GeV giving complete overlap with space gamma-ray telescopes and to reduce the flux sensitivity down to the cosmic electron background limit.

References

Akerlof, C. W. *et al.* 1989, *Proc. GRO Science Workshop (Goddard)*, 4-49.
Akerlof, C. W. *et al.* 1991a, *Ap. J. Lett.*, **377**, L97.
Akerlof, C. W. *et al.* 1991b, *Proc. 22nd ICRC.*, (Dublin), **1**, 324.
Baillon, P., *et al.* 1991, *Proc. 22nd ICRC, (Dublin)*, **1**, 220.
Chadwick, P. M.. Macomb, T. J. L. and Turver, K. E., 1990, *J. Phys. G.: Nucl. Part. Phys.*. **16**, 1773.
Cawley, M. F. *et al.* 1990, *Exper. Ast.*, **1**, 173.
Ebisuzaki, S. *et al.* 1991, *Proc. 22nd ICRC, (Dublin)*, **1**, 611.
Fazio, G. G., Helmken, H. F., O'Mongain, E. and Weekes, T. C., 1972, *Ap. J.*, **175**, L117.
Gould, R. J., 1965, *Phys. Rev. Lett.*, **15**, 511.
Grenier, I., 1992, this workshop.
Hillas, A. M., 1985, *Proc. 18th ICRC, (La Jolla)*, **3**, 445.
Lang, M. J., *et al.* 1991, *Proc. 22nd ICRC, (Dublin)*, **1**, 204.
Punch, M., *et al.* 1991, *Proc. 22nd ICRC, (Dublin)*, **1**, 464.
Punch, M. 1992, these Proceedings
Reynolds, P. T., *et al.* 1991, *Proc. 22nd ICRC, (Dublin)*,**1**, 392.
Reynolds, P. T., *et al.* 1992, in preparation.
Vacanti, G. *et al.* 1990, *Proc. 21st ICRC, (Adelaide)*, **2**, 329.
Vacanti, G. *et al.* 1991, *Ap. J.*, **377**, 467.
Weekes, T. C. 1988, *Physics Reports*, **160**, 1 and 2.
Weekes, T. C. *et al.* 1989, *Ap. J.*, **342**, 379.
Note: ICRC = International Cosmic Ray Conference

Observations of Pulsars with the Whipple Imaging Telescope

M. Punch,[1,4], C. W. Akerlof,[2] M. F. Cawley,[3] M. Chantell,[1] D. J. Fegan,[4] S. Fennell,[1,4], A. M. Hillas,[5] R. C. Lamb,[6] M. A. Lawrence,[1] D. A. Lewis,[6] D. I. Meyer,[2] G. Mohanty,[6] K. S. O'Flaherty,[4] P. T. Reynolds,[6] A. Rovero,[1] M. S. Schubnell,[2] T. C. Weekes,[1] T. Whitaker[1]

[1] Harvard-Smithsonian Center for Astrophysics
[2] University of Michigan
[3] St. Patrick's College, Maynooth
[4] University College, Dublin
[5] University of Leeds
[6] Iowa State University

1 Introduction

The Whipple Observatory's High Resolution Imaging Čerenkov Telescope has been used in the search for TeV emission from pulsars and other objects. Results of these observations are presented on PSR's 0355+59, 0531+22, 0950+08, and 1855+09 for steady emission and emission at the pulsar period.

2 Method of Observation

The sources were observed using the Whipple's High Resolution Imaging Telescope (Cawley *et al.* 1990) during the period May 1988 to June 1991. All observations were taken on moonless nights, using ON/OFF scans under clear skies and tracking scans under less than optimal conditions. The telescope's 10m optical reflector was used to observe the Čerenkov flashes from showers of particles in the atmosphere arising from TeV γ-rays and hadrons. Images of the showers were obtained using an array of 109 photomultipliers at the focal plane of the reflector coupled to fast ADCs and data acquisition electronics. The events are time-tagged with a relative accuracy of $1\mu s$, and absolute minute markers are inserted in the data stream from WWVB and Rubidium clocks to give an absolute accuracy of 0.5 ms.

3 Previous Observations

3.1 PSR0355+59

This pulsar is unusual in that large glitches in the radio period have been reported. TeV emission was reported by the Tata group after a large glitch (Acharya *et al.* 1991); no subsequent emission has been seen.

3.2 PSR0531+22

The most convincing observations of pulsed radiation from a pulsar were reported by the Durham group based on observations taken between September 1982 and November 1983 (Dowthwaite *et al.* 1984). A sharp peak was observed in coincidence with the main peak of the light-curve, and the emission was apparently steady. The probability of the detection being a chance fluctuation was estimated as 10^{-5} and the data were internally consistent. However, since that observation there has been no convincing confirmation of this emission.

3.3 PSR0950+08

This pulsar was one of the first discovered and is one of the closest; no evidence has been found for periodic emission at γ-ray energies.

3.4 PSR1855+09

More recently the Durham group have reported emission from this nearby millisecond pulsar (Bowden *et al.* 1991). Emission was seen at the second harmonic throughout the three year period of observations. There was also evidence for transient outbursts. A detection was reported by the Potchefstroom group but with somewhat lower statistical significance (de Jager *et al.* 1990).

4 Method of Analysis

For each event the time is corrected to the Solar-System barycenter as follows:

1. Conversion to UTC (Coordinated Universal Time) by addition of WWVB propagation delay to Mt. Hopkins.
2. UTC to TAI (International Atomic Time) by addition of appropriate number of leap seconds.
3. TAI to TDT (Terrestrial Dynamic Time) by addition of 32.184 s.
4. TDT to TDB (Barycentered Dynamic Time) using program supplied by Jodrell Bank.
5. Earth barycenter to Solar-System barycenter correction using JPL-DE200 ephemeris.
6. Time-of-flight correction due to relativistic effects.

The barycentring procedure was checked by optical observations of the Crab pulsar using a photomultiplier at the focus of a 48" reflector connected to the data acquisition system of the HRC. The Crab lightcurve is easily observed in these data, with the peak and interpulse at the expected phase.

The images of each event are parameterized, and the γ-like candidates are selected using the "Supercuts" procedure (Punch *et al.* 1991). For the ON/OFF data, the data is examined for any steady excess from the source region.

The phase of each event is calculated from the corrected time. (For PSR1855+09, corrections for binary motion were also applied, and the data was also tested for emission at the second harmonic.) The ON and tracking data is examined for pulsed emission both in the overall phase-linked datasets and on a run-by-run basis. The tests used on the overall datasets were the Rayleigh and Pearson tests. For the individual runs, the Protheroe test was also applied. These tests are described below.

Pearson test: Phases are binned into M bins, sample size n. The resulting lightcurve is tested for deviation from a uniform lightcurve.

$$S_{m-1} = \sum_{i=1}^{n} \frac{\left(x_i - \frac{n}{M}\right)^2}{\left(\frac{n}{M}\right)^2},$$

where x_i is the number of events in bin i and S_{m-1} is distributed as χ^2_{n-1} for $n/M > 5$.

Rayleigh test: Phase vectors for each event are summed. The magnitude of the resultant gives a measure of the probability.

$$2n\bar{R}^2 = 2n\left(\hat{\alpha}^2 + \hat{\beta}^2\right),$$

where

$$\hat{\alpha} = \frac{1}{n}\sum_{i=1}^{n} \cos\theta_i, \quad \hat{\beta} = \frac{1}{n}\sum_{i=1}^{n} \sin\theta_i,$$

θ_i is the phase in radians, $2n\bar{R}^2$ is distributed as χ^2_2, and $p = \exp(-n\bar{R}^2)$ for $n > 100$.

Protheroe test: Probabilities must be found from Monte Carlo simulations or estimated from tables derived from simulations.

$$Y_n = 2[n(n-1)]^{-1} \sum_{m=1}^{n-1} \sum_{k=m+1}^{n} \left(\Delta_{mk} + \frac{1}{n}\right)^{-1},$$

where $\Delta_{mk} = 0.5 - ||\phi_m - \phi_k| - 0.5|$ and $\phi_i = \theta_i/2\pi$.

The Pearson and Protheroe tests are good for narrow-peaked lightcurves, the Rayleigh test for broad lightcurves.

5 Results

Flux upper limits were derived based on a 10% duty-cycle and assuming energy threshold and spectrum similar to the Crab Nebula. The results of the analysis on the combined datasets for each pulsar are shown in Table I.

TABLE 1
Overall Periodic Emission from Pulsars

Pulsar	Time observed (Sidereal mins)	Flux $10^{-12}\text{cm}^{-2}\text{s}^{-1}$	Ephemeris
PSR0355+54	2328	1.3	Ramana-Murthy (1990)
PSR0531+22	7390	1.2	Lyne and Pritchard (1992)
PSR0950+08	1435	1.4	Gullahorn and Rankin (1978)
PSR1855+09	419	3.4	Rawley *et al.* (1988)

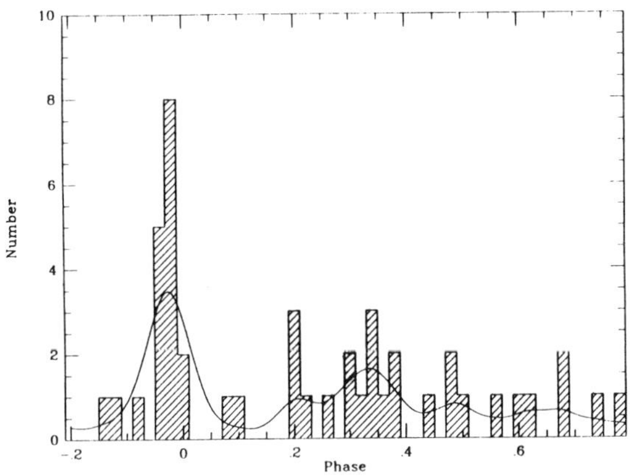

Fig. 1—Lightcurve from anomalous scan on Crab Pulsar, showing number of events per phase bin. The curve is a kernel density estimator of this lightcurve.

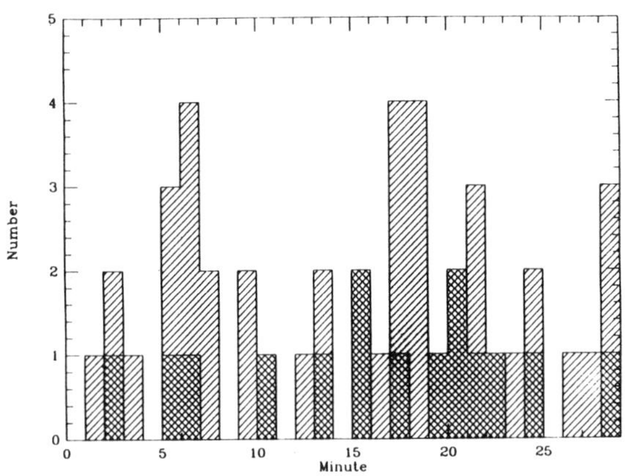

Fig. 2—Distribution of events within the anomalous scan. Double hatching indicates that the event was in the phase range -0.05 to +0.05.

In the examination of the individual runs, one anomalous scan was seen from the Crab Pulsar (0531+22). This scan was a tracking run taken on MJD 48267 (11 January, 1991), starting at 0402 UT and lasting for 29 sidereal minutes. The scan showed a Crab-like lightcurve (Fig. 1) with a probability (taking into account that there were 250 scans on the Crab) of 0.22% from the Pearson test and 3.8% from the Protheroe test that it was a random occurrence. These probabilities do not take into account the fact that the peak of the lightcurve is close to the expected position (within $\simeq$ 0.87 ms).

Since the data were taken in tracking mode, it is not possible to find a steady excess for this scan. However, the scan is bracketed by several contiguous tracking scans on the Crab. There is no significant deviation in the number of events which pass the cuts for the anomalous scan and its bracketing scans. Also, the parts of the bracketing scans closest to the anomalous scan were examined for evidence of a periodic signal, but none was found. The distribution of events within the scan (Fig. 2), does not indicate any localization of the emission within the scan.

As a check on the database, the analysis on the Crab was repeated at two arbitrary frequencies (10π and 10e Hz). No significant runs were found in this check.

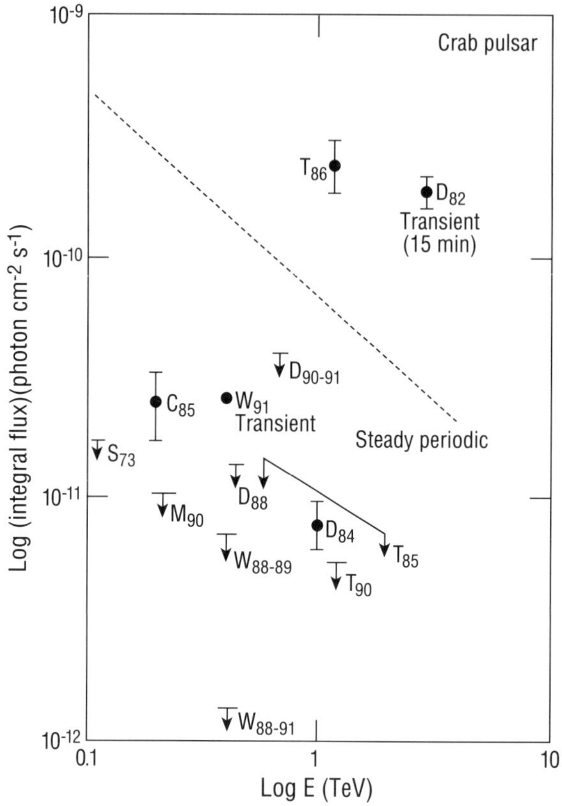

Fig. 3—-Some of the principal VHE measurements reported on the Crab Pulsar. Points D_{88}, D_{90-91} are from Bowden *et al.*(1991). Points W_{88-91}, W_{91} are from this work. Other references are found in Weeks *et al.* (1992).

Taking into account the number of tests on the data, the probability which can be assigned to this scan is less than 0.66% (since the tests are not independent). The flux from the main peak is calculated to be $2.6 \times 10^{-11} \mathrm{cm}^{-2} \mathrm{s}^{-1}$. In Figure 3, the pulsed upper limit and the transient flux from the Crab are plotted for comparison with results reported by other groups.

6 Conclusions

No overall pulsed emission was seen from any of the four pulsars considered here; upper limits are presented. In the run-by-run examination of the data, one anomalous scan from PSR0531+22 with a Crab-like lightcurve was seen. However, such an isolated instance of tranisient emission is difficult disprove without a large database of subsequent observations.

References

Acharya, B. S. *et al.* 1991, *AIP Conf Proc.*, **220**, 87.
Bowden, C. C. G. *et al.*1991, *Proc. 22nd ICRC, (Dublin)*, **1**, 396.
Cawley, M. F. *et al.* 1990, *Exper. Ast.*, **1**, 173.
deJager, O. *et al.* 1990, *Nucl. Phys. B*, **14A**, 169.
Dowthwaite, J. C. *et al.* 1984, *Ap. J. Lett.*, **286**, L35.
Gullahorn, G. E. and Rankin, J. M. 1978, *Astron. J.*, **82**, 1219.
Lyne, A. S. and Pritchard, R. S. 1992, Jordrell Bank Crab Pulsar Timing Results (private communication).
Punch, M., *et al.* 1991, *Proc. 22nd ICRC, (Dublin)*, **1**, 464.
Ramana-Murthy, P. V. 1990, (private communication).
Rawley, L. A., Taylor, J. H., and Davis, M. M. 1988, *Ap. J.*, **326**, 947.
Weekes, T. C. 1992, *Space Science Reveiw*, **59**, 315.

Observations of the Crab Pulsar with the Asgat Atmospheric Čerenkov Telescope

Giuseppe Vacanti,[1] Philippe Goret,[1] Isabelle Grenier,[1,2]
Thomas Palfrey,[3] and Andre Tabary[1]

[1] DAPNIA, SAp, CE Saclay
[2] Université de Paris VII
[3] Physics Department, Purdue University

Abstract

We have searched the Asgat 1990-1991 data base on the Crab pulsar for pulsed emission without success. We estimated the upper limit to VHE emission from the pulsar to be $4.6 \times 10^{-12} cm^{-2} s^{-1}$ (99 % CL, $E_\gamma \geq 600 GeV$, 10% duty cycle).

1 Asgat: a Fast Timing Atmospheric Čerenkov Telescope

The Asgat atmospheric Čerenkov telescope (Goret *et al.* 1991a), located in the Southern French Pyrenees ($1700m$ a.s.l.), consists of seven 7-meter diameter light collectors placed at the apexes and center of an approximate hexagon of side $\approx 60m$. At the focus of each collector a cluster of seven $12cm$ photomultipliers gives a field of view of $\approx 3°$ FWHM. Analog signals from the pmt's in each cluster are summed and preamplified before being sent through programmable delay generators to constant fraction discriminators whose thresholds are set to ≈ 45 photoelectrons. An event trigger is generated whenever $N \geq 4$ discriminators fire in a $40ns$ coincidence gate.

The cosmic primary's incident direction is reconstructed from the relative time of arrival of the Čerenkov light front at the telescopes. With a lever arm of $\approx 60m$ and a time resolution of $\approx 1ns$, the array can locate the direction of arrival of a plane Čerenkov front to $\approx 0.2°$ r.m.s.. However, Monte Carlo simulations show that a $500GeV\gamma$ ray generates a Čerenkov front which is far from plane, more nearly resembling a sphere of a few kilometers radius. Sampling this spherical front over a few points spread at the most over a few tens of meters does not allow the recognition of the curvature: this introduces a further uncertainty (parallax) of $\approx 0.7°$ r.m.s. in the reconstructed arrival direction. The angular resolution (σ_θ) of the array is then $\approx 0.8°$ r.m.s.. The arrival time of an event is recorded with a resolution of $20\mu s$ and an absolute accuracy of $\pm 0.5ms$ by means of a France Inter radio receiver.

2 Data

Between 1990 December and 1991 February the Crab nebula and pulsar were tracked for ≈ 35 hours under clear sky conditions. The target was offset by 1° East with respect to the center of the field of view. Under these conditions the effective energy threshold for γ rays is $\approx 600GeV$, for a collection area $\approx 2 \times 10^4 m^2$. This data base was shown to contain a steady signal from the nebula/pulsar at the $\approx 3\sigma$ level

(Goret *et al.* 1991b). Here we report on the search for emission pulsed at the $\approx 33ms$ pulsar period.

3 Analysis

Among all the events recorded we select the $N_6 = 166118$ events ($\approx$ 25% of the total) that triggered at least the outer 6 telescopes of the array. For these, in fact, the reconstructed arrival direction bears the least uncertainty. A plane is then fitted to the relative trigger times of the individual telescopes, and events yielding a χ^2 goodness of fit probability of less than 10% are discarded. We are left with 122216 events ($\approx$ 75% of N_6).

The arrival time of each event is corrected to the barycenter of the Solar System using the JPL's DE200 Planetary Ephemeris Program (Standish 1982); the corresponding pulsar phase computed using the Jodrell Bank Crab Pulsar Monthly Ephemeris (Lyne and Pritchard 1992).

Events are then sorted in two classes: ON, *i.e.*, those whose reconstructed arrival direction lies within 1.2° ($1.65\sigma_\theta$) of the pulsar (22459 events), and OFF (all the others). We show the ON light curve in Figure 1: there is no evidence of any periodic emission.

We derive upper limits to the pulsed flux from PSR0531+21 for three possible types of light curve:

1. one peak in the phase range $(-0.04, 0.06)$, coincident with the first peak of the COS-B light curve as observed in the $150MeV - 3GeV$ energy range (Clear *et al.* 1987);
2. for a 10% duty cycle, typical of most radio pulsars;
3. one peak in the phase range $(-0.02, 0.02)$, coincident with the peak observed by Dowthwaite *et al.* (1984) at $\approx 1TeV$.

The limits, computed as a 2.33σ (99 % CL) excess above the average, are summarized in Table 1.

A search on a night-by-night basis also failed to reveal evidence of sporadic pulsed emission.

TABLE 1

Upper limits (99 % CL) to VHE emission from PSR0531+21

Phase range	Flux $10^{-12}cm^{-2}s^{-1}$	Comments	Reference
-0.04, 0.06	≤ 4.6	COS B 1st peak	Clear *et al.* 1987
10 % duty cycle	≤ 4.6		
-0.02, 0.02	≤ 2.8	one narrow peak	Dowthwaite *et al.* 1984

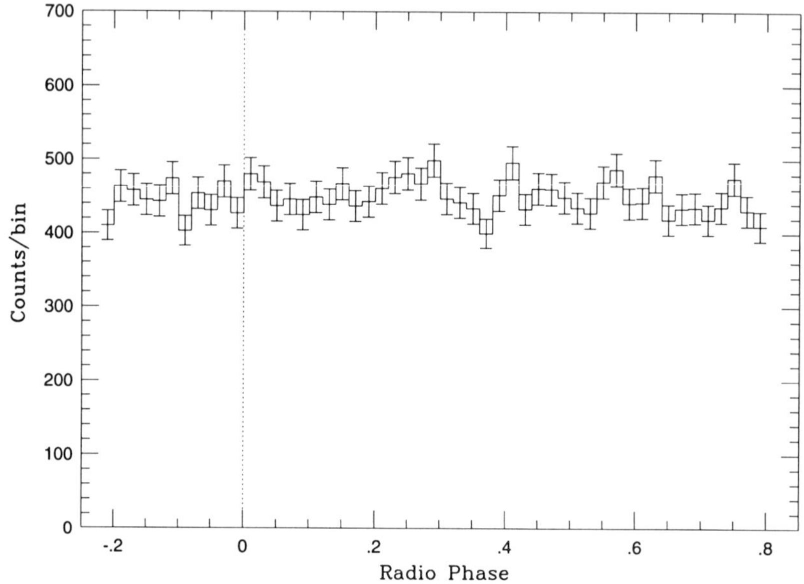

Fig. 1—Light curve obtained folding the ON data set with the frequency of PSR0531+21 ($\chi^2_{49} \approx 57$). Phase 0 coincides with the first radio peak.

4 Discussion

Over the past 25 years several detections of VHE γ rays from the Crab pulsar have been claimed in the literature (Weekes 1988). The most sensitive limits have been reported since 1988 and do not seem to confirm emission at the level reported previously (*e.g.*, Dowthwaite *et al.* 1984, Tümer *et al.* 1985). Figure 2 shows some detections and upper limits reported in the literature to compare to our own.

It is seen that recent limits to the pulsed TeV emission fall short of the extrapolation of the Egret pulsed spectrum by at least two orders of magnitude. This confirms the outer gap model prediction of a high-energy cut off of the pulsed emission above a few GeV (Cheng, Ho, and Ruderman 1986). The polar gap model (Daugherty and Harding 1982) does not apply to the production of TeV γ rays since, in the high magnetic field found above the polar caps, photons are very efficiently absorbed by pair production and cannot escape the magnetosphere.

Positive detections reported before 1985 (Figure 2 and Weekes 1988) claim pulsed fluxes larger than the more sensitive limits reported afterwards. It may be argued that the pulsar was active until the early 1980's and that pulsed TeV emission subsequently subsided. In the outer gap model curvature radiation from the primary $e^{\pm}$ pairs, accelerated to $\approx 10^{13} eV$, is limited to $\approx 10 GeV$. Photons of much higher energy could result from inverse Compton scattering of ambient optical photons on these primary pairs. However, the apparent stability over many years of the MeV to GeV

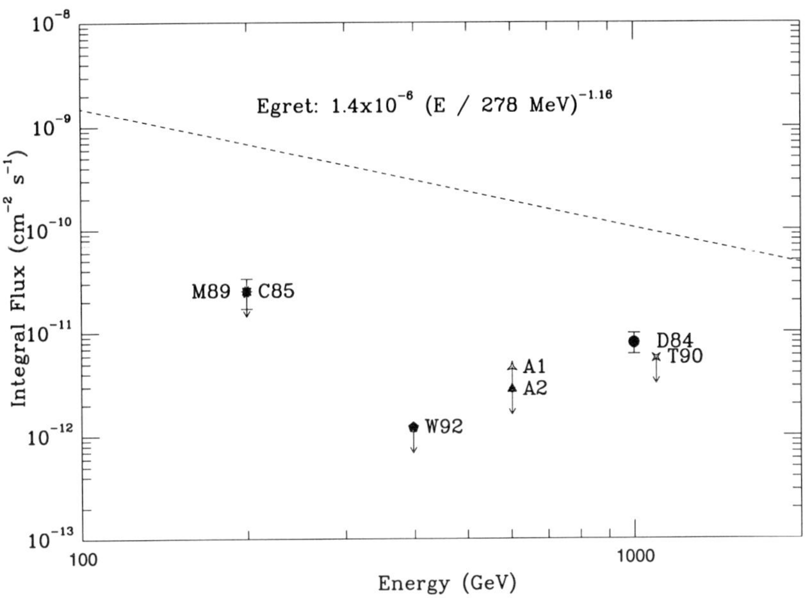

Fig. 2—Some detections (D84: Dowthwaite *et al.* 1984; C85: Tümer *et al.* 1985) and upper limits (M89: Ackerlof *et al.* 1989; T90: Bhat *et al.* 1990; W92: Punch *et al.* 1992; A1: this work, 10 % duty cycle; A2: this work, narrow peak) on PSR0531+21. Published upper limits have been rescaled to the 99 % CL.

flux, measured by COS B (Clear *et al.* 1987) and EGRET (Nolan *et al.* 1992), and of the optical pulsed flux (Middleditch, Pennypacker and Burns 1987; Kristian 1978) does not support the idea of dramatic changes in the pulsar environment after 1985. Nevertheless, it should be pointed out that the absence of contemporaneous optical and γ-ray data does not allow us to firmly constrain the strength of any possible TeV inverse Compton emission in the early 1980's.

GV acknowledges the support of the taxpayers of the European Economic Community.

References

Ackerlof, C. W. *et al.* 1989, Proc. Workshop *Gamma-Ray Observatory Workshop* (Greenbelt), Ed. W. N. Johnson, p. 4-49
Bhat, P. *et al.* 1990, *Proc.* 21st *ICRC* (Adelaide), **2**, 148
Cheng, K. S., Ho, C., and Ruderman, M. 1986, *Ap J.*, **300**, 522
Clear *et al.* 1987, *Astron. Astrophys.*, **174**, 85
Daugherty, J. K., and Harding, A. K. 1982, *Ap J.*, **252**, 337
Dowthwaite, J. C. *et al.* 1984, *Ap J.*, **286**, L35
Nolan, P. *et al.* 1992, *Bull. AAS* 23, **4**, 1439
Goret, P. *et al.* 1991a, *Proc.* 22nd *ICRC* (Dublin), **2**, 630
Goret, P. *et al.* 1991b, *AIP Conf. Proc.* **220** *High Energy Gamma-ray Astronomy*, Ed. J. Matthews, p. 65
Kristian, J. 1978, *Bull. AAS*, **10**, 425
Lyne, A. G., and Pritchard, R. S. 1992, private communication
Middleditch, J., Pennypacker, C. R., and Burns, M. S. 1987, *Ap J.*, **315**, 142
Punch, M. *et al.* 1992, These Proceedings
Standish, M. 1982, *Astron. Astrophys.*, **114**, 297
Tümer, O. T. *et al.* 1985, *Proc* 19th *ICRC* (La Jolla), **1**, 139
Weekes, T. C. 1988, *Phys. Rep.*, **160**, 1

Ultra High Energy Gamma Rays and Observations with CYGNUS/MILAGRO

Daniel D. Weeks,[1,2] and Gaurang B. Yodh[3]

[1]The University of New Mexico
[2]Los Alamos National Laboratory
[3]The University of California at Irvine

Abstract

This talk discusses high-energy observations of the Crab pulsar/nebula and the pulsar in the X-ray binary, Hercules X-1, and makes the case for continued observations with ground-based γ-ray detectors. The CYGNUS Air Shower Array has a wide field of view and monitors several astrophysical γ-ray sources at the same time, many of which are prime objects observed by the Compton Gamma Ray Observatory (GRO) and air Čerenkov telescopes. This array and the future MILAGRO Water Čerenkov Detector can perform observations that are simultaneous with similar experiments to provide confirmation of emission and can measure source spectra at a range of high energies previously unexplored.

1 Introduction

In the last several years, important scientific questions related to the production of γ rays by compact astrophysical sources have emerged. In the MeV/GeV energy range (referred to here as "low-energy"), γ rays are detected using satellites and balloon-borne experiments. At higher energies (above a few hundred GeV), the flux of γ rays is smaller and detectors with larger collection areas are required. These γ rays initiate Extensive Air Showers (EAS) of relativistic charged particles and photons in the Earth's atmosphere. The Čerenkov light emitted by the shower particles as they pass through the atmosphere can be detected in the Very High Energy (VHE) range with mirror telescopes, whereas the direct detection of the secondary particles in the Ultra High Energy (UHE) range uses an array of scintillation detectors on the ground.

The conventional view is that high-energy radiation is produced at, or near, the remnants of supernovae (white dwarfs, neutron stars, and black holes). The detected radiation must be neutral in order that it be undeflected by the interstellar magnetic field. The indirect detection of the primary γ rays requires careful consideration of the background cosmic rays which are random in time and direction. While the most likely candidate for the carrier of the high-energy signal is the photon, the possibility that the radiation is an unknown particle (but still neutral) has been recognized (Nagle, Gaisser, and Protheroe 1988).

There is evidence at high-energies (above ~200 GeV) that γ-ray emission from compact astrophysical objects occurs and is in some cases sporadic on time scales of less than one day. The sporadic emission may be due to rapid changes in the physical

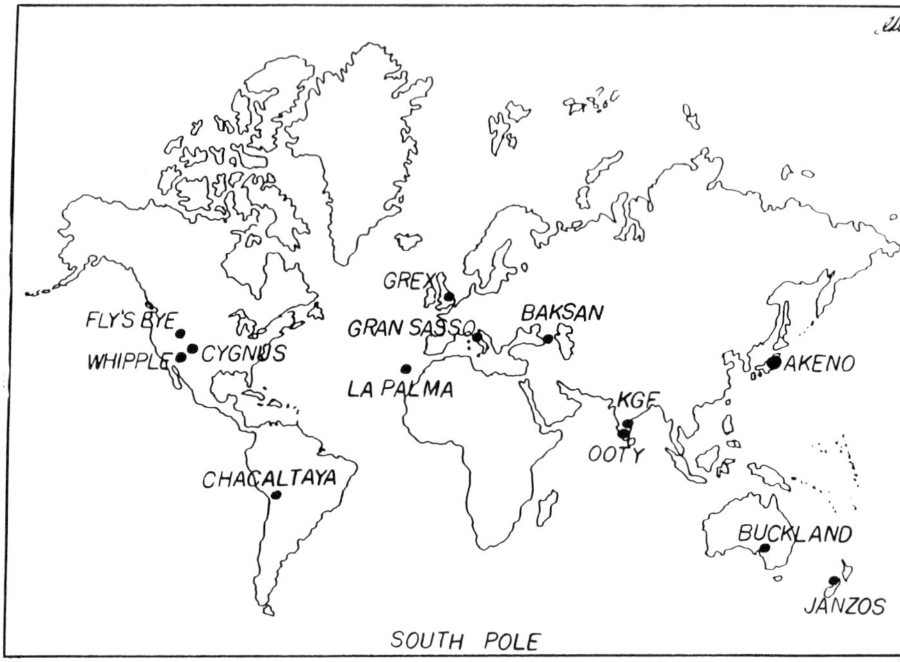

Fig. 1—Geographical positions of selected ground-based UHE air shower arrays and VHE telescopes. The plot is superposed onto a Mercator projection of the World.

structure or environment of the compact object characterized by an increase in pulsed radiation. Recently, claims of sporadic UHE emission have become common, whereas there is no evidence of long-term steady UHE emission.

Detections of sporadic γ-ray emission resulting from independent experimental observations of the same source can lend credence to an initial claim, especially if the emission has some specific "signature"(*i.e.*, pulsar periodicity or correlation with flares at other wavelengths, etc.). Simultaneous observations allow the statistical significances of multiple detections to be combined, resulting in an improved, overall significance. Figure 1 shows the geographical locations of some world-class, ground-based γ-ray observatories, many of them capable of observational overlap with other observatories. This capability is discussed in more detail later in this talk.

2 Results at High Energies

Although many claims exist for the detection of high-energy γ-ray emission from astrophysical sources in general, most are not on a firm base (2-3σ is not compelling). The most commonly employed analytical techniques are 1) tests for episodic increases in shower rate due to γ-rays, 2) directional excess, and 3) periodicity studies using known source dynamics. At high energies, the statistics are poor (small signal-to-noise ratio) which makes γ-ray source detection difficult. Additionally, the possibility that the sources emit γ rays sporadically further confuses the issue. There are, however, a

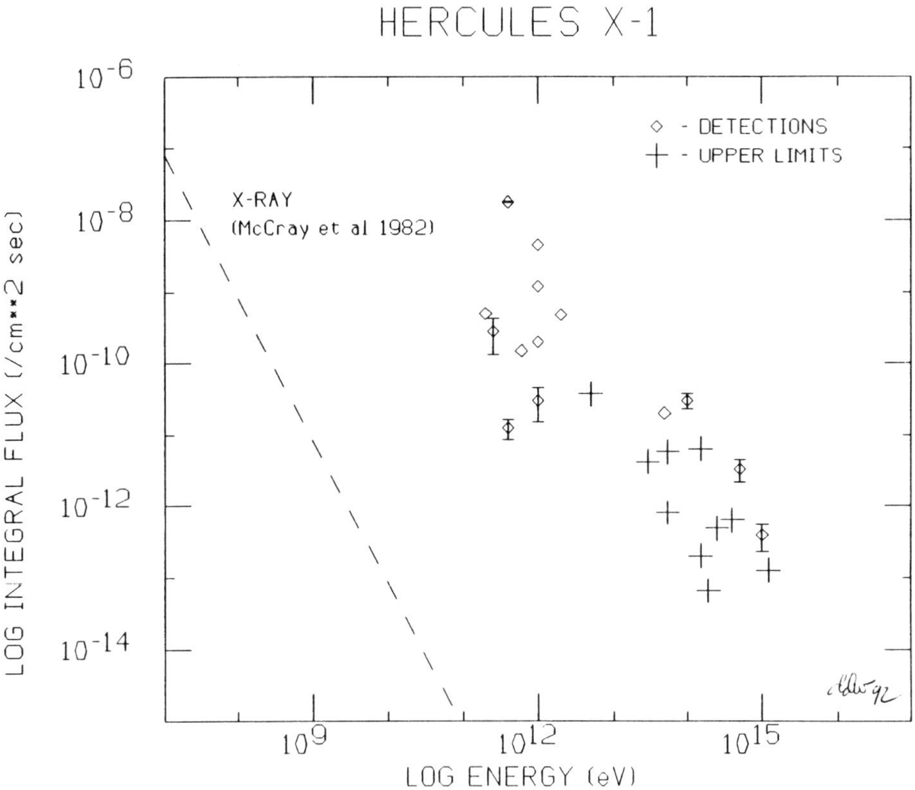

Fig. 2—The high-energy source spectrum for the X-ray binary Her X-1. The X-ray spectrum for Her X-1 (McCray *et al.* 1982) is not expected to extrapolate to the VHE/UHE observations due to different photon production mechanisms. In general, the detections correspond to sporadic emission, and the flux-limits correspond to long-term studies. The sensitivity of ground-based γ-ray detectors increases with increased observation duration (proportional to $\sqrt{t}$). Qualitatively, the results displayed in Fig. 2 give a spectral index that is $\simeq$-1.

couple of compact astrophysical objects that are established γ-ray emitters: Her X-1 and the Crab.

2.1 The Pulsar in Hercules X-1

The pulsar in the X-ray binary system was detected in 1986 data using the CYGNUS Array (Dingus 1988). Figure 2 shows the experimental γ-ray fluxes and limits measured for Her X-1, including most of the short- and long-term steady emission as well as periodic emission. The X-ray spectrum for Her X-1 (McCray *et al.* 1982) is shown in the figure, and is not expected to extrapolate to the VHE/UHE observations due to different photon production mechanisms. In general, the detections correspond to sporadic emission, and the flux-limits correspond to long-term studies. The sensitivity of ground-based γ-ray detectors increases with increased observation duration (proportional to $\sqrt{t}$). Qualitatively, the results displayed in Figure 2 give a spectral index that is $\simeq$-1.

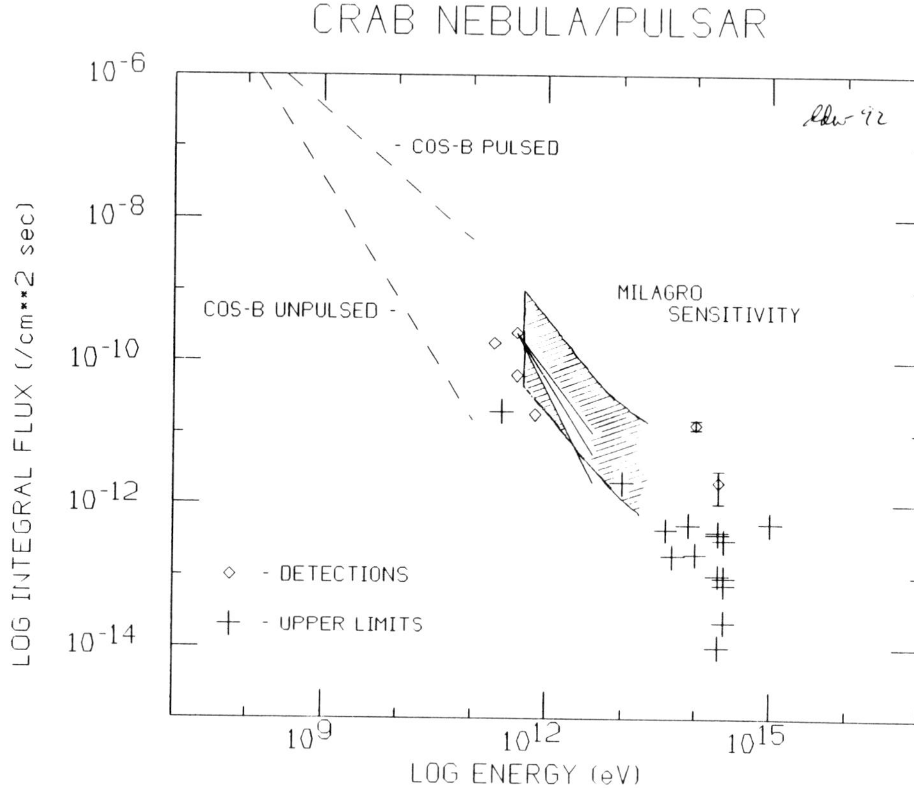

Fig. 3—The source spectrum for the Crab pulsar/nebula. The COS-B satellite results are shown as extrapolated fluxes for pulsed and unpulsed emission. In general, the limits correspond to long-term observations. The VHE results of the Whipple group (Vicanti, *et al.* 1991) are also shown to indicate the range of empirically determined errors on the spectral index. The MILAGRO sensitivity is shown and fills a gap between VHE and UHE observations. The upper- and lower- bounds of the hatched region correspond to one-day and one-year detections of the Crab Nebula at the 5σ level, respectively.

2.2 The Crab Pulsar/Nebula

The Crab Nebula/Pulsar has emerged as a VHE γ-ray standard candle - the Whipple VHE telescope has detected steady γ-ray emission at a high significance level. The result of the Whipple group (Vicanti, *et al.* 1991) is shown along with other less significant detections of the Crab in Figure 3. The Whipple data indicates how the source spectrum extrapolates to higher energies using their empirically determined spectral index. A model has been proposed at this workshop to explain how inverse Compton scattering can promote γ rays to high energies with a cut-off in energy around 10^{14}eV to 10^{16}eV (Harding, *et al.* 1992). It thus becomes important to conduct observations at these energies (UHE) to determine if such a cut-off exists.

2.3 Curiosities and Questions

High-energy observations of compact astrophysical objects have exposed several seemingly anomalous effects within the framework of our understanding of the nature of γ-rays, EAS's, and sources in general. Some of the curiosities and implications are:

- Detections of sporadic emission from Her X-1 have yielded the surprising fact that the γ rays are pulsed at a frequency shorter (by about ~0.16%) than the expected X-ray frequency (Lamb, *et al.* 1988, Resvanis, *et al.* 1988, Dingus 1988, Dingus, *et al.* 1988). One common characteristic of these four detections is that the emission occurs during short intervals of thirty minutes or less. It is a remarkable source, indeed, that can exhibit such a wide range of γ-ray pulsar periods, so that, if taken as real, the theoretical challenge of understanding the nature of the particle production at the source becomes more difficult. A-posteriori models have been devised to describe a process that could produce frequency shifted pulsar emission (Slane and Fry 1989, Cheng and Ruderman, 1989, Ahronian and Atoyan 1991). However, the situation calls for definitive studies that probe this source over a wide range of energies.
- A puzzling detail of the detection of Her X-1 is that the UHE signal showers detected with the CYGNUS experiment contain about the same number of muons expected for background cosmic ray induced showers (Dingus, *et al.* 1988). Apparently, the interactions of γ-ray photons at these energies are not fully understood, or a new type of neutral, cosmic radiation has been detected.
- A significant VHE detection of unpulsed emission from the Crab Nebula using the Whipple atmospheric Čerenkov photon imaging technique has been reported (Vicanti, *et al.* 1991). The BATSE and EGRET γ-ray detectors aboard the GRO have detected pulsed emission from the Crab pulsar in the low-energy range, while the Whipple experiment detects steady emission at the TeV energy range. These detections would imply that the γ-ray production mechanisms are different. However, at this meeting we heard preliminary evidence that EGRET detects an unpulsed signal above about 5 GeV, and that the Whipple telescope detected a sporadic burst that is periodic (Punch 1992). It is important to continue observing the Crab at all energies to help determine the nature of the γ-ray emission from the Crab Nebula/Pulsar.
- The same imaging technique applied to previous Her X-1 sporadic/periodic detection from the Whipple experiment does not yield a signal (Reynolds, *et al.* 1991), suggesting that the radiation from Her X-1 is not γ-ray radiation.

3 Correlative Studies

The reliability of many of the claims of VHE/UHE detections could be strengthened if other detectors can verify a specific source signature, or if the event excesses in one detector are substantiated by contemporaneous event excesses measured with other ground-based and satellite detectors. The observational sky coverage of ground-based γ-ray detectors is shown in Figure 4. The CYGNUS array (denoted by a "C") is seen

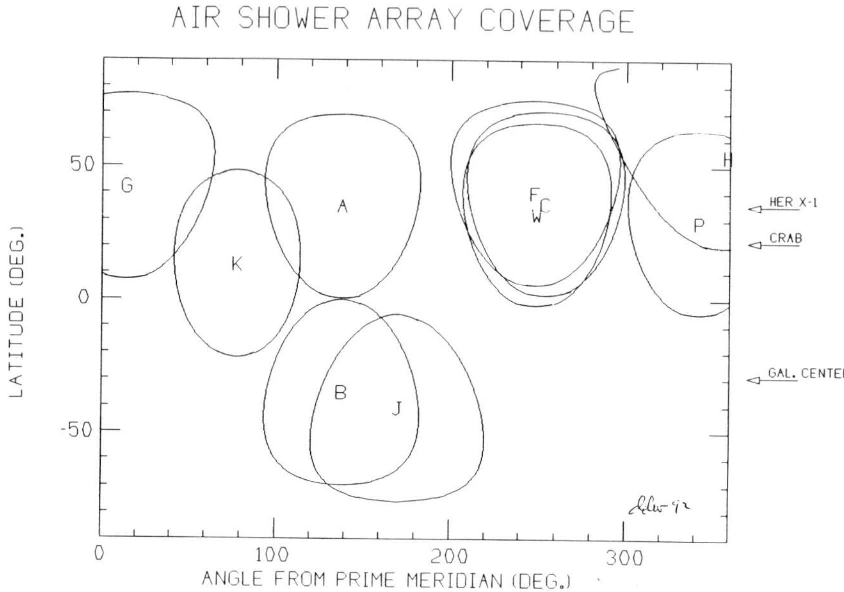

Fig. 4—Mercator projection of the celestial sky overlap for ground-based VHE and UHE gamma-ray detectors. The field of view is ±35 degrees from zenith for each detector. The Cygnus array is located near the center and noted with an "C".

to have observational overlap with the Whipple, CASA, and Fly's Eye (shown together) detectors, indicating the potential for correlation studies of the Crab and Her X-1 by U. S. experiments.

There are two recent examples of reported detections that have found benefit from correlation studies. In one, three air shower arrays (Baksan[Alexeenko *et al.* 1990], KGF [Archaya 1990], and EASTOP at Gran Sasso [Aglietta 1990]) detected excess UHE showers from the direction of the Crab on Feb. 23, 1989 U. T. Individually, the chance probabilities are of the order of a few percent and are not convincing by themselves. However, the fact that all three observations overlap during a 4-5 hour interval on the same day (during source passage) strengthens a claim of γ-ray emission. It should be noted that the CYGNUS (Alexandreas 1991c) and Akeno (Weeks 1990) detectors did not detect event excess on the same sidereal day, implying that this eposodic emission from the Crab began when the source was overhead the KGF detector. In another example, an air Čerenkov experiment at Ooty detected pulsed TeV emission from the Crab at a pulsar phase of 0.0 to 0.1 on Jan. 23, 1985 U. T. (Bhat *et al.*, 1989). The shape of their calculated pulsar light curve is singly peaked and resembles the light curve calculated by Archaya *et al.* (1990) using data taken with an air Čerenkov telescope at Pachmarhi, India, on Jan. 2, 1989 U. T. While these observations are not contemporaneous, they both yield pulsed emission at a common pulsar phase, strengthening each individual claim.

3.1 Ground-Based γ-ray Observations with CYGNUS

The CYGNUS air shower array consists of over two-hundred scintillation detectors and several muon detectors placed on the ground around the LAMPF end-station (Alexandreas *et al.* 1991a) and detects cosmic ray induced air showers above 40TeV with an angular resolution of about 0.75 degrees (as determined by cosmic-ray occultations with the sun and moon [Alexandreas 1991b]).

The CYGNUS event times showing a significant excess of events, as determined by the Li and Ma (1983) method, are tagged as potential intervals during which γ rays might be expected. A more detailed study of a given day's data could be performed to determine if the increased event rate is localized in a time interval that is short compared with the daily source observation time. The distribution of the background event times measured for one sidereal day, with respect to the time of source transit, should follow some expected behavior (the EAS rate being higher near the zenith). A statistically significant excursion, parametrized in terms of the Kolmogorov-Smirnov (KS) test, would be indicative of emission on a time scale of less than one day. If the number of on-source events summed over the entire sidereal day were significant, but the day does not yield a significant KS-test result, this might be indicative of γ-ray emission over an entire day. Furthermore, if both a significant excess of events and a positive KS-test result were obtained, the probability for emission on short time scales compared to the source passage time would be even more likely.

When comparing a CYGNUS detection with a simultaneous detection of another experiment, the combined probability for the correlated observation is determined by (Eadie, *et al.* 1971)

$$P_{total} = P_1 P_2 (1 - \ln(P_1 P_2))$$

where P_1 and P_2 are the individual probabilities of being statistical fluctuations in the number of on-source showers. If a given source has known periodicity, then this information can be used to further test the data.

3.2 MILAGRO for the Future

During the Summer of 1991, five water Čerenkov detectors (about 42 m^2 in area and 2 m in depth) for sampling the air shower were installed within the CYGNUS array. (The water Čerenkov detectors were constructed using commercial swimming pools.) Using seven, 10-inch photomultiplier tubes in each pool the Čerenkov light generated by shower particles, electrons, photons, hadrons, and muons is sampled. The large area and full sensitivity makes it possible to measure the arrival of the front edge of the shower front more precisely than with small area scintillation counters. The angular resolution of the CYGNUS array has improved from its present value of 0.75 degrees to about 0.3-0.4 degrees with the addition of the pools (Sinnis 1992), thus increasing the sensitivity to point gamma-ray sources by a factor of two.

Fig. 5—The MILAGRO detector.

Members of the CYGNUS collaboration have submitted a new proposal (Freedman, *et al.* 1991) for a large water Čerenkov detector (MILAGRO) to be located at Fenton Hill (see Fig. 5), in the mountains behind the city of Los Alamos, at an altitude of 8700 ft. This 5000 m^2 detector would respond to all components of the shower over its entire area, similar to the water Čerenkov detectors described in the previous paragraph. The high altitude and full sensitivity will make it possible to cover the energy range from TeV to PeV in one detector for the first time. The expected sensitivity of MILAGRO is shown in Figure 3 as a 5σ detection of the Crab for one-day (upper-bound) and one-year (lower-bound) observations. To improve the overall capabilities of MILAGRO, the CYGNUS array detectors will be moved to surround the MILAGRO detector.

We are greatful to the conference organizers for providing an excellent environment for this scientific meeting, and appreciate the ongoing efforts of our CYGNUS / MILAGRO collaborators to build and maintain world-class γ-ray detectors. One of us, DDW, acknowledges financial support from the Los Alamos National Laboratory SST-9 division for travel and from the American Astronomical Society for a workstation. This work is supported in part by DOE, NSF, LANL, and IGPP of the University of California. We would also like to thank Cyrus Hoffman for providing improvements to the manuscript.

References

Acharya, B. S., *et al.* 1990, *Proceedings of High Energy Gamma Ray Astronomy AIP* No. **220**, p137.

Aglietta, M. *et al.* 1990, *Proceedings of High Energy Gamma Ray Astronomy AIP* No.220, p145 (1990)

Ahronian, F. A. and Atoyan, A. M., 1991, *Ap. J.*, **381**, 220

Alexandreas, D. E. *et al.* 1991a, *Nucl. Inst. and Methods,* **A311**, 350

Alexandreas, D. E. *et al.* 1991b, *Phys. Rev. D,* **43**, 1735

Alexandreas, D. E. *et al.* 1991c, *Proceedings of the 22nd International Cosmic Ray Conference,* Dublin, Ireland, **1**, 216.

Alexeenko, V. V. *et al.* 1990 *Proceedings of High Energy Gamma Ray Astronomy AIP* No.220, p132.

Bhat, P. N. *et al.* 1989, *Nature,* **319**, 127

Cheng, K. S. and Ruderman, M., 1989, *Ap. J. Lett.*, **337**, L77

Dingus, B. L. *et al.* 1988, *Phys. Rev. Lett.*, **61**, 1906

Dingus, B. L. 1988, PhD Thesis, University of Maryland, Internal report PP No. 88-253.

Eadie, W. T., Drijard, D., James, F. E., Roos, M., and Sadoulet, B. 1971, *Statistical Methods in Experimental Physics,* (North Holland)

Freedman, S. J. *et al.* 1991, MILAGRO proposal submitted to DOE/NSF

Harding, A. *et al.* 1992, these Proceedings.

Lamb, R. C. *et al.* 1988, *Ap. J.*, **328**, L13

Li, T. and Ma, Y., 1983, *Ap. J.*, **272**, 317.

McCray *et al.* 1982, *Ap. J.*, **262**, 301

Nagle, D. E., Gaisser, T. K., and Protheroe, R. J. 1988, *Ann. Rev. Nucl. Part. Sci.*, **38**, 609

Punch, M. 1992, these Proceedings

Reynolds, P. T. *et al.* 1991, *Ap. J.*, **382**, 640.

Resvanis, L. K. *et al.* 1988, *Ap. J. Lett.*, **328**, L9

Sinnis, C. 1992, Pool Status Report, CYGNUS Internal Memorandum

Slane, P. and Fry, W. F. 1989, *Ap. J.*, **342**, 1129

Vicanti, G. *et al.* 1991, *Ap. J.*, **377**, 467

Weeks, D. D. 1990, *Proceedings of High Energy Gamma Ray Astronomy AIP,* No. **220**, p162.

TeV Gamma-Rays from Millisecond Pulsars and the Detectability of Globular Clusters

Ian A. Smith

Department of Astronomy and Astrophysics, University of Chicago

Abstract

The recent discovery of eleven millisecond pulsars in 47 Tucanae indicates that some globular clusters contain several thousand millisecond pulsars. I investigate the detectability of TeV gamma-rays from millisecond pulsars, assuming the energy source that powers this emission is the loss of rotational energy of the spinning neutron star. I find that although individual sources may be too weak to detect, the contribution of many active millisecond pulsars may lead to detectable TeV gamma-ray emissions from globular clusters. Future detections of TeV gamma-rays from globular clusters (or more restrictive upper limits on the fluxes) will lead to better estimates of the number of active millisecond pulsars the clusters contain and to a better understanding of the pulsar emission mechanism. I also derive flux estimates for the following sources: 1) the Crab Nebula, 2) PSR 1937+21, PSR 1953+29, and PSR 1855+09, and 3) nearby millisecond pulsars in the galactic disk.

1 Introduction

The Crab Nebula and several millisecond pulsars have been reported to be sources of TeV (10^{12} eV) gamma-rays over the past two decades [see Weekes (1992) for a review]. However, it was not until recently that a credible, statistically significant result had been obtained: a steady signal at the 20σ level for the Crab Nebula was detected at the Whipple Observatory (Vacanti *et al.* 1991), corresponding to a flux of $7.0\ (\pm\ 0.4) \times 10^{-11}$ photons $cm^{-2}s^{-1}$ above its effective energy threshold of 0.4 TeV. This exciting discovery leads to the question of the detectability of TeV gamma-rays from other sources. In particular, globular clusters are potentially interesting. The recent discovery of eleven millisecond pulsars in 47 Tucanae (Manchester *et al.* 1991) indicates that some globular clusters contain many thousands of active millisecond pulsars. Although individual millisecond pulsars may be too weak to detect, the contribution of many of them in a globular cluster may lead to a detectable TeV gamma-ray emission. Chen (1991) has recently estimated the 100 MeV gamma-ray emission from millisecond pulsars in globular clusters, which may be detectable by the Compton Gamma Ray Observatory.

I examine the question of the detectability of TeV gamma-rays from millisecond pulsars, only assuming that the energy source is the loss of rotational energy of a spinning neutron star. This assumption leads to an estimate of the gamma-ray flux in terms of the observable pulsar parameters as outlined in §2. As an example, I calculate the efficiency for the conversion of spin-down energy into TeV gamma-rays that is required to explain the observed flux from the Crab Nebula. In §3 I calculate the predicted fluxes for globular clusters, in particular 47 Tucanae. In §4 I consider the pulsars PSR 1937+21, PSR 1953+29, and PSR 1855+09, which have been claimed to be TeV gamma-ray sources by Bowden *et al.* (1991), and other nearby millisecond pulsars in the galactic disk. I give a short summary in §5.

2 Theoretical Flux Calculation

I assume that the energy source that powers the TeV gamma-ray emission is the rotational energy loss of the slowing down pulsar, which is taken to be a spherical neutron star of mass M, radius R, spin period P, and period derivative $\dot{P}$. The rotational energy loss rate is given by (Michel 1991)

$$\dot{E} = -\frac{8\pi^2}{5}\frac{MR^2\dot{P}}{P^3}. \tag{1}$$

Although there is a range of values for P and $\dot{P}$ obtained from the radio observations of the millisecond pulsars, using the typical values of $P = P_{-2} \times 10^{-2}$ s and $\dot{P} = \dot{P}_{-18} \times 10^{-18}$ s s^{-1} in equation (1) gives

$$\dot{E} = -3.2 \times 10^{34}\ (M/M_\odot) R_6{}^2 \dot{P}_{-18} P_{-2}{}^{-3}\ \mathrm{ergs\ s^{-1}}, \tag{2}$$

where $M_\odot$ is the mass of the Sun and $R = R_6 \times 10^6$ cm.

A fraction of the spin-down energy may go into particle acceleration that eventually leads to the emission of TeV gamma-rays. If a fraction α of the spin-down energy goes into TeV gamma-rays at the low energy threshold E_γ of the gamma-ray detector, then the number of observable gamma-rays emitted per second by the pulsar is $\dot{N} \sim \alpha|\dot{E}|/E_\gamma$. This estimate crudely assumes that all the gamma-rays are created with energy E_γ, which requires the gamma-ray spectrum to be steep enough that the lower energy threshold dominates: for the Crab Nebula, the measured source spectrum (number flux per unit energy) was $N(E) \propto E^{-2.4}$ (Vacanti *et al.* 1991).

For simplicity, I assume the TeV gamma-rays are emitted isotropically, giving a photon flux at Earth $F = \dot{N}/4\pi d^2$, where d is the distance to the source. This assumption will not be valid if the TeV gamma-rays are produced close to the surface of the neutron star, because some of them will be lost to electron-positron pair production in the magnetic field of the neutron star. The attenuation of photons in a non-rotating dipole magnetic field was studied by Ho, Epstein, and Fenimore (1990). As an example, a 0.4 TeV gamma-ray emitted essentially parallel to the local magnetic field line at the surface of a neutron star of dipole magnetic field strength 2.5×10^8 G will escape the magnetosphere if it is emitted from within a polar cap of angular size $\sim 20°$. A more isotropic emission of the TeV gamma-rays can be obtained if they are generated at a distance of several neutron star radii. Using equation (2), the TeV gamma-ray flux is

$$F = 1.6 \times 10^{-10}\ (M/M_\odot) R_6{}^2 \dot{P}_{-18} P_{-2}{}^{-3} \alpha (E_\gamma/1\ \mathrm{TeV})^{-1} (d/1\ \mathrm{kpc})^{-2}\ \mathrm{cm^{-2} s^{-1}}. \tag{3}$$

Although I have assumed that the energy source is the rotational energy loss of the slowing down pulsar, I have not used any particular model for the generation of the TeV gamma-rays to obtain the flux in equation (3). Instead, except for the efficiency α, the terms on the right hand side of equation (3) are either directly observable (P, $\dot{P}$, E_γ, d), or can be estimated from standard neutron star models (M, R). If a source is detected, it is possible to invert equation (3) to find the efficiency α, leading to constraints on the models of the particle acceleration and gamma-ray production (Usov 1983; Cheng and de Jager 1990). Note that α is the efficiency for the conversion of spin-down energy to the *observed* TeV gamma-rays; the efficiency at the source will be higher than this if some of the TeV gamma-rays are absorbed near to the source or en route to the observer.

As an example, I apply equation (3) to the Crab Nebula. (Note that for the Crab, the observed flux is steady, rather than pulsed; the presence of the nebula may require a different source model, and thus a different efficiency, than for other millisecond pulsars). Using the Whipple observed flux of 7.0×10^{-11} photons $cm^{-2}s^{-1}$ above the effective energy threshold $E_\gamma = 0.4$ TeV (Vacanti *et al.* 1991), the 1982-1987 radio observations of Lyne, Pritchard, and Smith (1988) to obtain a characteristic $P = 33$ ms and $\dot{P} = 4.2 \times 10^{-13}$ s s^{-1}, the distance to the Crab Nebula $d = 2$ kpc (Trimble 1968), and a neutron star mass $M = 1.4M_\odot$ and radius $R_6 = 1$ in the flux equation (3) gives a conversion efficiency $\alpha = 4 \times 10^{-5}$. This calculation shows that the loss of rotational energy is easily able to power the observed TeV gamma-ray emission. The conversion efficiency is quite small; note that the efficiency for converting spin-down energy into radio emission in the Crab is only 10^{-9}, which is extremely inefficient compared with other pulsars (Michel 1991).

3 Globular Clusters

The exciting recent discovery of eleven millisecond pulsars in 47 Tucanae (Manchester *et al.* 1991) indicates that some globular clusters contain thousands of active millisecond pulsars. Although individual pulsars in the globular cluster may be too faint to detect, the contribution of many millisecond pulsars might make the globular cluster a strong TeV gamma-ray source. The current TeV gamma-ray detectors have angular resolutions that are tenths of a degree, at best (Vacanti *et al.* 1991). These detectors will be unable to distinguish individual sources in globular clusters, whose core radii are typically a few arc minutes or smaller; for example, the angular size of the core of 47 Tucanae is $0.47'$ (Meylan 1988). Instead, the observed TeV gamma-ray flux from a globular cluster will be the sum of the fluxes from all the individual, active millisecond pulsars.

I estimate the TeV gamma-ray emission by assuming there are N identical, active pulsars in the globular cluster. This assumption is simpler than the one used by Chen (1991), who convolved the individual pulsar flux with the distribution of the pulsars with respect to the pulsation period; I have chosen not to do this, because there are currently too few pulsars known to form a representative distribution. Also, the period derivatives of the observed pulsars are not currently known; these are needed to give a better estimation of the TeV gamma-ray flux. Multiplying equation (3) by N gives the flux from a globular cluster:

$$F_{GC} \sim 1.6 \times 10^{-10} \left(\frac{M}{M_\odot}\right) R_6{}^2 \dot{P}_{-18} P_{-2}{}^{-3} \alpha_{-3} \left(\frac{E_\gamma}{1 \text{ TeV}}\right)^{-1} \left(\frac{d}{1 \text{ kpc}}\right)^{-2} N_3 \text{ cm}^{-2}\text{s}^{-1}, \tag{4}$$

where $\alpha = \alpha_{-3} \times 10^{-3}$ and $N = N_3 \times 10^3$. (I have not used the efficiency calculated for the Crab, since the emission mechanisms for the Crab and for millisecond pulsars need not be the same). It can be seen from equation (4) that the flux from a globular cluster may be detectable at flux levels comparable to the Whipple Crab detection provided there is a large enough number of reasonably efficient millisecond pulsars in it. As the efficiency of the individual pulsars goes down, more pulsars are needed to give a detectable signal.

For 47 Tucanae (declination $\delta = -72°21'$), assuming $d = 4.7$ kpc, $M = 1.4M_\odot$, $E_\gamma = 0.4$ TeV and the rest of the variables in equation (4) are unity gives a predicted flux of $F_{GC} \sim 3 \times 10^{-11}$ cm^{-2}s^{-1}; this is slightly lower than the current 3σ upper limit on the > 0.4 TeV gamma-ray flux of 4.4×10^{-11} cm^{-2}s^{-1} from 47 Tucanae (Bowden *et al.* 1991). However, all the millisecond pulsars detected so far in 47 Tucanae have periods $P < 6$ ms (Manchester *et al.* 1991). Because the flux from an individual source varies as P^{-3}, if all the millisecond pulsars in the cluster have similarly short periods, the total TeV gamma-ray flux will be higher than the above estimate. In particular, PSR 0021-72K ($P = 1.79$ ms) may be similar to PSR 1937+21 ($P = 1.56$ ms), see §4. If PSR 0021-72K has a period derivative $\dot{P} \sim 10^{-19}$ s s^{-1}, and $\alpha_{-3} \sim 1$, the flux from this pulsar alone is 4.4×10^{-13} cm^{-2}s^{-1} (for a detector threshold of 0.4 TeV). Using the upper limit to the flux quoted above implies that there are $\lesssim 100$ active millisecond pulsars like PSR 1937+21 in 47 Tucanae.

Although it is further away ($d = 9.7$ kpc), another potentially interesting source is M15 (declination $\delta = 11°57'$). Future detections (or more restrictive upper limits on the fluxes) of TeV gamma-rays from these or other globular clusters will help to estimate the number of active millisecond pulsars the clusters contain and will lead to a better understanding of the pulsar emission process.

4 Galactic Disk

The millisecond pulsars PSR 1937+21, PSR 1953+29 and PSR 1855+09 have also been claimed to be TeV gamma-ray sources by Bowden *et al.* (1991). The observed fluxes and detector threshold energies are given in Table 1, which also lists the pulsars' P, $\dot{P}$ (Rawley, Taylor, and Davis 1988), d (Johnston and Bailes 1991), and the required efficiency α using equation (3), assuming $M = 1.4M_\odot$ and $R_6 = 1$. Because of the uncertainty in the distance to PSR 1855+09, I have used the two values $d = 0.3$ kpc and $d = 1.0$ kpc in Table 1, with the latter being the currently favored value (Johnston and Bailes 1991). It can be seen from Table 1 that the efficiencies required to give the observed fluxes are high, and for PSR 1953+29 (and PSR 1855+09 if it is at a distance of 1 kpc), $\alpha > 1$ is needed. These results are difficult to explain using the spin-down model. However, it seems unlikely that these remarkably high fluxes will be confirmed by future observations; indeed, at this meeting, the Whipple group announced that their upper limit to the flux from PSR 1855+09 above 0.4 TeV is 3.4×10^{-12} $cm^{-2}s^{-1}$ (Punch *et al.* 1992), which is ~ 100 times lower than the flux observed by Bowden *et al.* (1991).

TABLE 1
Detected Galactic Millisecond Pulsars

Pulsar	F ($cm^{-2}s^{-1}$)	$\dot{P}_{-18}$	P_{-2}	E_γ (TeV)	d (kpc)	α
1937+21	3×10^{-11}	0.105	0.156	1.0	5.0	0.1
1953+29	3×10^{-11}	0.030	0.613	1.0	2.7	7.5
1855+09	4×10^{-10}	0.017	0.536	0.4	0.3	0.6
1855+09	4×10^{-10}	0.017	0.536	0.4	1.0	6.5

Assuming there are n millisecond pulsars uniformly distributed through the galactic disk of radius r_g and thickness h, an estimate of the mean pulsar separation, and hence the distance to the closest one from Earth is given by

$$\lambda \sim \left(\frac{\pi r_g{}^2 h}{n}\right)^{1/3} = 0.19\ (r_g/15\ \mathrm{kpc})^{2/3}(h/1\ \mathrm{kpc})^{1/3}(n/10^5)^{-1/3}\ \mathrm{kpc}. \quad (5)$$

Using the distance from equation (5) in equation (3), provided there are ~ 10^5 millisecond pulsars in the galactic plane and the conversion efficiency $\alpha \gtrsim 10^{-3}$, implies there may be several galactic millisecond pulsars detectable at TeV gamma-ray energies. Future TeV gamma-ray observations may provide estimates of the galactic number density of millisecond pulsars. However, the simulations of Johnston and Bailes (1991) suggest that the number of millisecond pulsars in the Galaxy with luminosities similar to PSR 1937+21 and PSR 1855+09 need not exceed ~ 60 and ~ 2500 respectively. This might make PSR 1937+21 and PSR 1855+09 the only nearby detectable TeV gamma-ray sources. However, more thorough searches for millisecond pulsars at radio wavelengths are required, particularly at high galactic latitudes, to obtain an accurate determination of n.

5 Summary

I have investigated the detectability of TeV gamma-rays from millisecond pulsars, assuming the energy source that powers this emission is the loss of rotational energy of the spinning neutron star. This investigation led to flux estimates for the following sources:

1) The Crab Nebula. The highly statistically significant signal from the Crab can be easily explained in terms of the spin-down model, and the required efficiency for the conversion of rotational energy into TeV gamma-rays is low, $\alpha = 4 \times 10^{-5}$.
2) Globular Clusters. The contribution of many active millisecond pulsars in a globular cluster may lead to a detectable TeV gamma-ray emission; if there are 10^3

active millisecond pulsars in 47 Tucanae, with a conversion efficiency $\alpha = 10^{-3}$, the predicted flux is close to the current detector sensitivities. The number of pulsars required to give a detectable signal can be an order of magnitude lower than this if they are all similar to PSR 1937+21. Future detections of TeV gamma-rays from globular clusters will lead to better estimates of the number of active millisecond pulsars the clusters contain.

3) PSR 1937+21, PSR 1953+29, PSR 1855+09. The large observed fluxes that are claimed for these sources are difficult to understand in terms of the spin-down model, requiring conversion efficiencies $\alpha \gtrsim 1$.
4) Galactic Disk. If there are $\sim 10^5$ millisecond pulsars in the galactic disk, some of the closer ones may be detectable TeV gamma-ray sources. Future TeV gamma-ray observations may provide estimates of the galactic number density of millisecond pulsars.

I thank M. Ruderman for initially asking me at the Neutron Stars ASI (Crete, 1990) whether galactic millisecond pulsars could be detectable sources of TeV gamma-rays. I also thank Don Lamb for focusing an earlier version of this manuscript. This work was supported at the University of Chicago by grants NAGW-830, NAGW-1284, and NAG-51454.

References

Bowden, C. C. G., *et al.* 1991, in *High Energy Gamma-Ray Astronomy*, ed. J. Matthews (New York: AIP), p. 75.
Chen, K. 1991, *Nature*, **352**, 695.
Cheng, K. S., and de Jager, O. C. 1990, *Nucl. Phys. B (Proc. Suppl.)*, **14A**, 28.
Ho, C., Epstein, R. I., and Fenimore, E. E. 1990, *Ap. J. (Letters)*, **348**, L25.
Johnston, S., and Bailes, M. 1991, *M. N. R. A. S.*, **252**, 277.
Lyne, A. G., Pritchard, R. S., and Smith, F. G. 1988, *M. N. R. A. S.*, **233**, 667.
Manchester, R. N., *et al.* 1991, *Nature*, **352**, 219.
Meylan, G. 1988, *Astr. Ap.*, **191**, 215.
Michel, F. C. 1991, *Theory of Neutron Star Magnetospheres* (Chicago: University of Chicago Press).
Punch, M., *et al.* 1992, these Proceedings.
Rawley, L. A., Taylor, J. H., and Davis, M. M. 1988, *Ap. J.*, **326**, 947.
Trimble, V. 1968, *A. J.*, **73**, 535.
Usov, V. V. 1983, *Nature*, **305**, 409.
Vacanti, G., *et al.* 1991, *Ap. J.*, **377**, 467.
Weekes, T. C. 1992, *Space Sci. Rev.*, **59**, 315.

Relativistic Shock Waves and the Excitation of Plerions

Jonathan Arons,[1,2] Yves A. Gallant,[1,2] A. Bruce Langdon,[2,3] and Tser-Yuan Brian Yang[1]

[1]Theoretical Astrophysics Center, University of California at Berkeley
[2]Institute of Geophysics and Planetary Physics, Lawrence Livermore National Laboratory
[3]X-Division, Lawrence Livermore National Laboratory

1 Introduction

The excitation of diffuse, nonthermal astrophysical synchrotron sources by energy lost from central compact objects has long been a puzzle in high-energy astrophysics. Pulsars and their surrounding plerionic nebulae form the nearest at hand and best studied examples of this problem. Other examples include the excitation of extragalactic radio sources by jets, and possibly the emission from some active galactic nuclei themselves. The Crab Nebula is the best studied of the plerions, and we will take the physical problems it poses to be characteristic of the class. It is typical of what we shall call "diffuse" plerions, in which the synchrotron radiation and adiabatic expansion dominate the energy losses from relativistic electrons and positrons. The termination of the outflow from a pulsar in a compact binary, as the relativistic wind encounters the mass lost from the companion star, may form an example of a "compact" plerion, the difference from the diffuse case being in the competition between synchrotron losses and the acceleration process itself.

Relativistic, magnetohydrodynamic (MHD) models of the coupling between pulsar and nebula are the most successful at the present time. Building on earlier ideas (Piddington 1967; Rees and Gunn 1974; Kundt and Krotscheck 1980), Kennel and Coroniti (1984a,b) showed that such a wind from the pulsar, terminated by a standing, transverse magnetosonic shock wave, could give a good account of the nebular dynamics and of the nebular spectrum of high-energy photons (near infrared, optical, X-rays and gamma rays), which are an instantaneous probe of the pulsar's outflow. The prominent results of their model are several. 1) The post shock flow fits into the slowly expanding nebula only if the upstream flow is almost entirely dominated by kinetic energy, with

$$\sigma \equiv \frac{B_1^2}{r\pi m N_1 \gamma_1 c^2}.$$

Here B_1 is the upstream magnetic field, N_1 is the upstream density, and γ_1 is the Lorentz factor of the flow. 2) If the upstream flow is entirely composed of pairs (an assumption), and if the post-shock energy distributions are power laws with $N(E) \propto E^{-s}$ (another assumption), then the synchrotron emission from the downstream flow can be fit to the observed X- and γ−ray spectra of the Nebula, taken as an unresolved object, if $s \approx 2.3$. Once this fit is made, the optical and near infrared spectra of the nebula (again taken as an unresolved object) are *predicted* (more or less) correctly by the model.

This theory is sufficiently successful as a macroscopic explanation of the phenomenology to provide a setting for investigation of the microscopic physics. Several questions are implied by the macroscopic flow model. Does a relativistic, magnetosonic shock in a pair plasma actually create the assumed power law distributions downstream? If so, what is the mechanism of acceleration in such shocks? Are there other observable consequences of this mechanism, other than the desired particle acceleration, which allow one to test the shock hypothesis directly? We emphasize that one cannot appeal to the ever popular ideas of diffuse shock acceleration proposed for particle acceleration by interplanetary and nonrelativistic interstellar shocks (Bell 1978; Quenby and Lieu 1989) even as extended to relativistic shocks with magnetic field parallel to the flow (*e.g.*, Ellison *et al.* 1990), since the termination shocks of pulsar winds must be fairly close to having the magnetic field perpendicular to the flow upstream when the plerion surrounds the pulsar. Indeed, for the Crab, a laminar Archimedean spiral in the outflow from the pulsar would have Θ_{Bn}, the angle between the flow direction and the magnetic field in the shock frame where we observe the system, differing from 90° only by one part in 10^9! In addition, in relativistic shocks, the particles can stream ahead of the shock and participate in the back and forth motion across the front required in diffusive shock acceleration only when $\Theta'_{Bn} \ll 1/\gamma_1$, where Θ'_{Bn} is the angle between the flow direction and the magnetic field in the proper frame of the upstream fluid. We will argue that $\gamma_1 \sim 10^6$ in the Crab Nebula. Then the magnetic field is still almost exactly transverse even in the upstream fluid frame. Therefore, some other mechanism of shock acceleration must apply.

In relativistic shocks, the mechanisms of high-energy particle acceleration and of basic thermalization can be one and the same - accelerating relativistic particles can be part of the problem of shock structure itself. In fact, the highest energy electrons and positrons in the shock theory we outline below are those which have Larmor radii comparable to the gyro-radii of the heavy ions that control the shock thickness, and are precisely the $e^{\pm}$ that give rise to the highest energy synchrotron photons observed in the *nebular* spectrum. Thus, the problems of shock structure and of nonthermal particle acceleration are one and the same.

2 Shock Structure and Particle Acceleration

Pulsar outflows must have an extremely high Mach number. The Alfven Mach number in the flow is $M_A = \beta_1\gamma_1/\sqrt{\sigma} \cong 10^7$! Under these circumstances, magnetic reflection of particles from the shock front plays an essential role in the shock dynamics. Relativistic, magnetosonic solitary waves with unidirectional flow exist only for $M_A < 1 + 1/\gamma_1$ (Kennel and Pellat 1976, Alsop and Arons 1988, Chiueh 1989). At higher Mach number, incoming particles reflect from the enhanced magnetic field in the wave front and are set into Larmor gyration, leading to a new kind of magnetosonic soliton with reflected particles self-consistently incorporated (Alsop and Arons 1988). This reflection process requires all the momentum of the incoming plasma to be temporarily stored in a magnetic overshoot, with $B_{peak}/B_1 = (1 + \sigma^{-1})^{1/2}$ in the wave frame of a soliton in a pair plasma. The dependence of the overshoot on σ alone is characteristic of relativistic magnetosonic shocks—once $\gamma_1 \gg 1$, the dependence on γ_1 itself is entirely subsumed in the scale of the flow being that of the Larmor radius based on the upstream parameters.

Whereas it is possible to construct solitary wave models with magnetically reflected particles in pair plasmas, unraveling the instabilities and thermalization mechanisms in this highly inhomogeneous environment is a formidable task for analytical theory. We have made use of fully non-linear, fully self-consistent particle-in-cell numerical simulations in order to "experimentally" uncover the relevant physics. The basic method is to integrate the equations of motion of many charged particles in their *self-consistent* fields (Langdon and Lasinski 1976; Birdsall and Langdon 1985). Suppose at some time t_n, one knows the electromagnetic field on a spatial grid, and one knows the positions and momenta of all the particles $\{\mathbf{x}_j(t_n), \mathbf{p}_j(t_n)\}$ (they are not on the grid). One then finds the Lorentz force at each particle's position and uses Newton's law (in relativistic form) to advance the particles to new phase space positions $\{\mathbf{x}_j(t_n+dt), \mathbf{p}_j(t_n+dt)\}$. One then calculates new fields at time $t_n + (dt/2)$ using the charge densities and currents at time t_n, with the particles distributed on the grid with a suitable weight function. These new fields are used to advance the positions of the particles once again, with the time step set by various stability considerations.

We modeled shock waves in pair plasmas by setting up a one-dimensional spatial grid in the computer, with length anywhere from 10 to 40 Larmor radii based on the magnetic field and particle energy in the upstream flow. The initial magnetic field points exactly transverse to the flow velocity. At the initial time, the cold plasma fills the computational box, with flow Lorentz factor $\gamma_1 \gg 1$ and a magnetic field polarized across the flow carried with the plasma at the fluid speed. The ratio of magnetic energy density to flow energy density upstream σ is the only parameter of significance. As expected from theoretical considerations, simulations with the same value of σ but differing γ_1 yield identical results, when $\gamma_1 \gg 1$. We have surveyed parameter space with these spatially 1D models from $\sigma = 13$ down to $\sigma = 0.001$.

At the injection point (the "left wall"), the particles are injected, all with the same upstream velocity, and the magnetic field is carried with the fluid, by requiring the boundary electric field to be that of a perfectly conducting fluid. The opposite end of the grid is represented as a conducting rubber wall - particles bounce off the wall, which acts as a perfect conductor as far as the fields are concerned. It would be better to have outflow boundary conditions on the plasma and radiation, but so far a useful algorithm to implement these in the dense downstream plasma has not been constructed. Because the magnetic field points across the flow, the wall has no influence on the shock structure, once the point of particle reflection is more than a few Larmor radii away from the wall. In the low σ shocks of interest here, the wave energy is never more than a few percent of the thermal and kinetic energy of the plasma, so the conducting wall, which forces the downstream waves to be standing modes, has little influence on the shock once it is well upstream of the wall.

The details of the pair simulations are reported elsewhere (Langdon *et al.* 1988; Gallant *et al.* 1992). The essential results are that one finds an intense, quasi-coherent electromagnetic precursor traveling in front of the shock, while downstream, a few per cent of the total energy is turned into a linearly polarized Rayleigh-Jeans spectrum of electromagnetic waves with temperature close to what one expects from the jump conditions. The incoming plasma stream is reflected from the shock front, followed by rapid thermalization of the Larmor gyration. The downstream spectra of electrons and positrons are almost perfect relativistic Maxwellians. Thus, transverse shocks in a pure pair plasma create a fully thermalized downstream medium, contrary to a basic assumption of the MHD model of plerionic supernova remnants. We have done a few preliminary calculations with the magnetic field oblique to the flow, using what is otherwise the same 1D model. So long as $\Theta'_{Bn} > 50°$, the results are semiquantitatively the same as for the purely transverse case, as one might suspect since the structure is dominated by the reflected particles and the rapid dissipation. Smaller obliquities may be different, but for technical reasons this conclusion is not yet warranted. We have also investigated some aspects of the two-dimensional structure of these shocks, motivated by the observation that the 1D model leaves the plasma with a high temperature in the momenta orthogonal to the magnetic field, while it is still cold in the momenta along B. Such a medium is unstable, most prominently with respect to the Weibel instability. In one 2D shock calculation, we have found this effect does heat the plasma in the direction along B, thus broadening the distribution of gyrational momenta, but the basic shock structure persists, with downstream anisotropy $T_\perp/T_\parallel \sim 2$.

The nature of the thermalization mechanism gives a clue to a shock model which does have nonthermal particle acceleration. As seen in the shock frame, the reflected particles form a gyrating ring with small momentum dispersion in the leading edge of the shock. Such rings are unstable - they form synchrotron masers (Zhelznyakov and Suvorov 1972; Yoon 1990; Hoshino and Arons 1991). Calculations of the growth rate for a homogeneous pair plasma, along with estimates of the maximum harmonic number that can be generated, suggest that the dissipation observed in the pure pair plasmas is due to the formation of synchrotron masers, with consequent plasma heating due to absorption of the maser-generated extraordinary modes.

Now imagine what happens if the upstream flow contains heavy ions, as well as pairs. Since all the species flow into the shock with the same speed (the upstream Larmor radii are small compared to the overall flow scale in which the shock is embedded), $ZN_{1i} + N_{1+} = N_{1-}$. The ions have a rest mass large compared to that of the $e^\pm$. Therefore, they will contribute the dominant kinetic energy in the flow, if $N_{1i} > (m_\pm/m_i)(N_{1+} + N_{1-})$. As the plasma encounters the shock, the lighter pairs, which have Larmor radii a factor $Zm_\pm/m_i$ smaller than the ions, form a leading leptonic shock almost identical to the shock found in a symmetric plasma, if $N_{1i} \ll (N_{1+} + N_{1-})$. The more massive ions have no chance to respond, since the pair shock develops on the pair cyclotron time. They plow on through until they too are magnetically reflected in the enhanced magnetic field of the shock front. An electrostatic field in the direction of flow will be associated with this spatial separation of the ions from the leptons, with electrostatic potential having magnitude $Ze\Phi \sim m_i\gamma_1 c^2$. Once the ions are set into gyration by the reflection from the shock front, they too

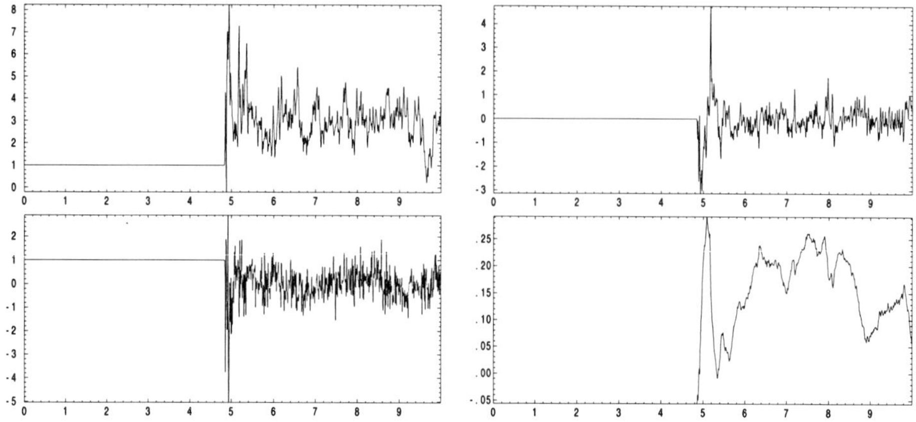

Fig. 1—Electromagnetic structure of a relativistic magnetosonic shock in an electron-positron-proton plasma, at a time when the shock has crossed 50% of the simulation grid. Upper left panel: Magnetic field B_z, with upstream field amplitude equal to unity. Lower left panel: Transverse electric field E_y, with $E_y = (v_1/c)B_z$ at the injection wall. Upper right panel: Electrostatic field E_x. The downstream waves have electrostatic fields with amplitudes roughly one-half the magnitude of the electromagnetic field components, indicating strong elliptical polarization of the post-shock extraordinary modes. Lower right panel: Electrostatic potential in units of $\gamma_1 m_i c^2/e$. Since the potential never exceeds 0.3, the electrostatic field in the shock front acts to decelerate the ion stream, but the reflection is magnetic, as in the shocks in pure pair plasmas. The unit of length is the ion Larmor radius based on upstream parameters.

are synchrotron maser unstable, as they gyrate in the much denser background of shock-heated pairs.

Simulations of shock flows with ions (Hoshino *et al.* 1992) support these conjectures. In Figure 1, we show the electromagnetic structure from a simulation done in the same manner as the simulations of the purely leptonic shocks, but now containing ions with a charge-to-mass ratio appropriate to protons, at $12\omega_{ci}^{-1}$ after the plasma bounced off the simulation wall. In order to keep the computer time finite, the mass ratio is small, $m_i/m_\pm = 20$, and the density ratio is $N_{1i}/(N_{1+} + N_{1-}) = 0.2$, chosen as a compromise to have $N_{1i} \ll (N_{1+} + N_{1-})$ but still have most of the kinetic energy density in the upstream flow. In this case, 80% of the plasma flow energy is in the ions. The upstream flow is almost entirely dominated by the plasma kinetic energy flux - $\sigma = 0.005$. While the choice of γ_1 matters only to the relative scale of the flow, in the interests of making the problem be of direct relevance to plerions and pulsars, we chose $\gamma_1 = 10^6$, corresponding to protons being accelerated through a few percent of the polar cap potential of the Crab pulsar. Note the large overshoot in the magnetic field and the long wavelength oscillations downstream. The first overshoot occurs because as the ions reflect from the shock front, their flow momentum is stored in an increase in the magnetic pressure and in the thermal pressure of the pairs. Behind the overshoot, the ions continue gyrating quasi-coherently, creating compressional enhancements at each turning point, until their gyration finally thermalizes. These oscillations are spaced roughly one ion Larmor radius apart.

In Figure 2 we show zoomed views of the $p_x - x$ projections of ion phase space for a region of the simulation near the shock front itself, at 12 and 13 ion cyclotron times after the simulation began. One sees the quasi-coherent reflection of the ions, followed by their rapid thermalization. The figure also shows the unsteady character of the shock structure. On this scale, the structure of the pair shock, with its initial thermalization to Maxwellian distributions, is infinitesimal, achieved as the ions have barely begun to gyrate. Figure 3 shows the spectra of the downstream electrons and positrons. The positrons are no longer everywhere Maxwellian in energy space, but now have a power law distribution with $N(E) \propto E^{-2}$ at energy/particle exceeding $\gamma_1 m_\pm c^2$, and have energy density comparable to the ions' kinetic energy density in the upstream flow. In this simulation, the electrons also show some sign of power law behavior at energies exceeding $\gamma_1 m_\pm c^2$.

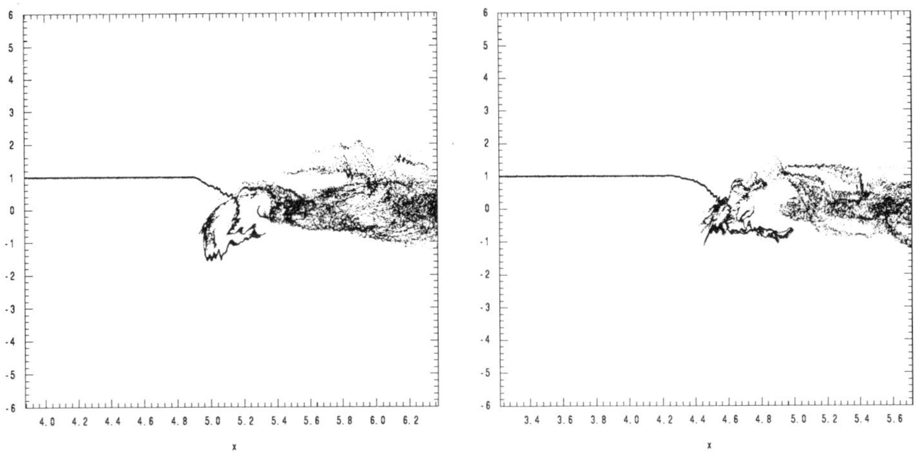

Fig. 2—Ion phase space near the shock front. Left panel: Flow momentum in units of $10^6 m_i c$ versus x for ions near the shock front at $t = 12/\omega_{ci1}$. The unit of length is the ion Larmor radius based on upstream parameters. Note the initial deceleration of the ion stream in the electrostatic field, followed by the rapidly dissolving loop in the magnetic field. The main body of an ion Maxwellian distribution is produced within 2-3 Larmor times after the ions begin decelerating in the shock front. Right panel: ion flow momentum versus x at $t = 13/\omega_{ci1}$ near the shock front. These snapshots illustrate the unsteady character of the shock.

The positron spectrum is Maxwellian at energies below $\gamma_1 m_\pm c^2$ and has upper cutoff at positron energy equal to that of the upstream energy/particle of the protons, $\gamma_1 m_i c^2$. Thus, a relativistic, low σ magnetosonic shock, composed of ions, electrons and positrons, with $N_{1i} \ll N_{1+} + N_{1-}$ but with the upstream flow energy dominated by the ions, can accelerate the downstream positrons into a power law distribution with high efficiency, with the slope of the power law being very close to what is desired for the macroscopic model. This particular simulation is highly resolved, with 64 particles per grid cell per species in the initial state. In other calculations of the shock structure with various values of σ and of γ_1, and in simulations of the synchrotron maser process in isolation in a uniform medium, we have found the slope to be of order 2. Figure 4 shows the spectral slopes found as a function of σ. For $\sigma = 0.0034,\ 0.005$, simulations were done for a variety of numerical parameters (number of particles/cell,

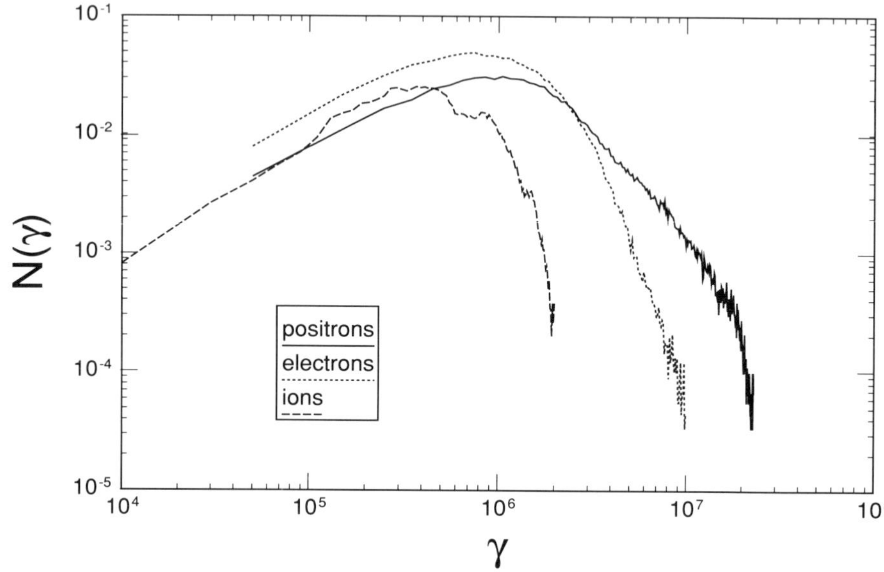

Fig. 3—Post-shock distributions. The positrons have a high-energy power law spectrum with slope $s \approx 2$, whereas the electrons show evidence for a power law supra-thermal distribution with slope $s \approx 3.7$. The upper cutoff occurs where the energy/particle of the pairs equals the upstream energy/particle of the ions, as is expected from cyclotron resonant absorption of extraordinary modes emitted at and above the upstream ion cyclotron frequency.

number of grid cells per ion Larmor radius, ...). The dispersion in the values of s obtained defines the "error" in these numerical experiments in determining the form of the high energy tail of the distribution function. The relatively small mass ratios and the upper limits to the number of particles per cell form the main sources of the variability in s found in the various simulations with the same physical parameters. Figure 4 also shows the fraction of the downstream energy density contained in the nonthermal part of the particle spectra. For all the simulations, this efficiency is between 10% and 20%, with the latter being the value appropriate to the physical conditions inferred in the Crab Nebula.

The prescursor does not appear in this simulation, since the grid scale was equal to $0.2c/\omega_{c1\pm}$ based on upstream parameters, which leaves the precursor unresolved; in separate calculations we have shown that omitting the precursor does not affect the downstream particle acceleration. The electrostatic field in Figure 1 shows a strong double layer structure at the shock front, as the ions are first held back by the pairs frozen to the field, then begin gyrating and pull the pairs forward, while the potential Φ reaches $\sim 0.5\gamma_1 m_i c^2/e$ in the double layer region. Behind the shock, the magnetic field settles down to a mean value of $B_2 \approx 3B_1$, as one would expect from the MHD jump conditions for such a plasma-dominated flow, with intense extraordinary mode noise (Gallant *et al.* 1992). In the simulation frame, the average transverse electric

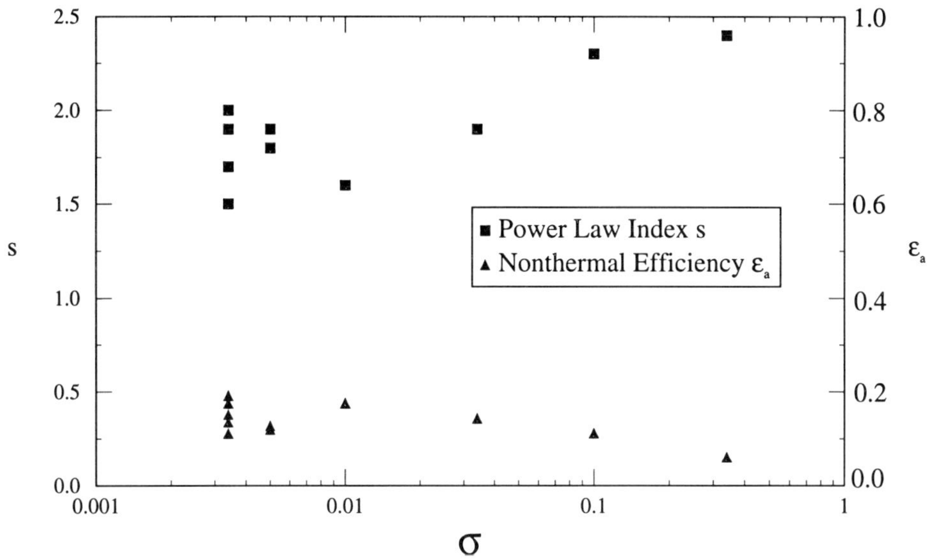

Fig. 4—Nonthermal spectral index and acceleration efficiency as a function of σ. Squares show the index s (left ordinate axis) of the power law part of the downstream positron spectra, as a function of the total σ of the upstream electron-positron-proton flow. The triangles show the nonthermal particle acceleration efficiency of these shocks (right ordinate axis), defined as the ratio of the downstream energy in the nonthermal part of the particle spectrum to the total downstream energy density.

field is noise with a mean of zero, while in the shock frame (only approximately well defined, since the shock velocity is not strictly steady around its mean value of $c/3$), the mean transverse electric field is approximately constant.

It is likely that 2D effects (primarily pitch angle scattering due to Weibel "turbulence") will steepen the spectrum somewhat. In any case, it is clear that these electron-positron-ion shocks do produce downstream positron spectra whose characteristics are highly desirable for the modeling of plerions.

The physics of this acceleration is simple. In this non-symmetric plasma, the extraordinary modes are no longer linearly polarized electromagnetic waves but have a noticeable electrostatic component as they propagate across the magnetic field. For frequencies between the relativistic analog of the lower hybrid frequency and the relativistic cyclotron frequency of the heated pairs as they emerge from the leading leptonic shock, the ion extraordinary modes behave like magnetosonic waves in the pair plasma and have elliptical polarization with the electric vector rotating in the left handed sense with respect to the magnetic field. This is the same sense as the Larmor gyration of the positrons and of the ions. As a result, a positron can linearly absorb the magnetosonic modes at their cyclotron resonance (as well as at higher harmonics) and can gain energy. Because the polarization is elliptical, there is some power in right circularly polarized fields, so the electrons also can gain energy nonthermally from the ions. Because of the small mass ratio used in the simulations, the power in left-handed circularly polarized fields emitted by the ions is much larger than in right-handed circular polarization, so the positron acceleration is much stronger than the acceleration of electrons. Under realistic conditions, we believe the acceleration of electrons and positrons will be much more comparable. It is possible to simulate this flow with mass ratios as high as 100, which gives much larger dynamic range in the particle spectra and should show more comparable electron and positron acceleration; such simulations are in progress.

This mechanism works efficiently only if the collective synchrotron emission of the ions yields extraordinary mode frequencies greater than or equal to the thermal cyclotron frequency of the positrons just behind the leading leptonic subshock. The linear theory shows this is possible only if the upstream thermal momentum dispersion of the ions is small compared to $m_{\pm}/m_i$. This is probably not a severe restriction, because the acceleration of the ions is likely to be by an electromagnetic mechanism, with the ion energy essentially unique on each streamline. In any case, adiabatic expansion in the outflow reduces any initial thermal dispersion drastically by the time the shock is reached. To the degree that this shock acceleration model works successfully, this constraint on the acoustic Mach number of the flow should be regarded as a restriction on any theory of the wind. All of the simulations easily satisfy this requirement.

3 Application to Plerions

Models of astrophysical systems using these results are in the midst of construction. One can get the flavor of these from the following simple considerations. Suppose the magnetosphere of the Crab pulsar emits a dense flow of pairs, with $\dot{N}_{\pm} \sim 10^{38}\dot{N}_{\pm,38}$ coming out in the broad sector around the rotational equator where the pulsar must emit the wind required to feed the torus of X- and gamma-ray emission in the Nebula (Aschenbach and Brinkmann 1975; Pelling *et al.* 1987). A number of magnetospheric models suggest that such a flow is possible. In addition, suppose the magnetospheric electrodynamics causes an outflow of heavy ions to be pulled up from the stellar surface and accelerated to high energy in the outflow around the magnetic equator (Arons 1983 seems to contain the only suggestion of this sort, so far). The ion flux is constrained by the electrodynamics to be close to the Goldreich-Julian flux, $\dot{N}_{GJ} 2\Omega^2\mu/Zec$ for this pulsar, where Z is the charge on the ions. The resulting ratio of the mass densities in the wind is

$$\kappa_i \equiv \frac{m_1 N_{i1}}{m_{\pm}(N_{1+} + N_{1-})} = 1836\frac{A}{Z}\frac{\Omega^2\mu}{ec\dot{N}_{\pm}},$$

where A is the atomic number of the ions, μ is the pulsar's magnetic moment and Ω is its angular velocity. Solving for Z yields $Z \approx 3(A/56)(1/\dot{N}_{\pm,38})(\kappa_i/2)$ in the case of the Crab; the model works for partially ionized iron, the primary constituent of the crust, while if the heavy ions are protons or α particles, an ion flux greatly in excess of the Goldreich-Julian rate would be required.

These ions might be accelerated through a fair fraction of the total polar cap potential $\Phi_{cap} \sim 10^{17}$ Volts, corresponding to $\gamma_1 \sim 5 \times 10^7 (Z/A)(\Phi/\Phi_{cap})$. The equatorial wind with these accelerated ions carries a fraction Φ/Φ_{cap} of the spindown luminosity of the pulsar. If the shock really puts essentially all the energy into the accelerated pairs, as our simulations suggest, and if these pairs radiate essentially all of this in X-rays and gamma-rays, as is indicated by the macroscopic flow models, then one would conclude, from the hard photon luminosity of the nebula, that the ions experience an acceleration potential $\Phi/\Phi_{cap} \sim 0.3/I_{45}$, where I_{45} is the moment of inertia of the neutron star in units of 10^{45} cgs.

We can go much further. From momentum conservation, one expects the shock to occur roughly $10^{17.5}$ cm from the pulsar. Optical observations reveal the presence of the curious "wisps" in the surface brightness in this region, a series of enhancements in the surface brightness elongated in the direction perpendicular to the flow lying to the Northwest of the pulsar; there is also a fainter wisp to the Southeast. The broad band synchrotron emissivity in the magnetic overshoot and downstream magnetic compressions shown in Figure 1 is readily found to be proportional to the $B^{11/3}$, assuming the downstream pitch angle scattering has largely isotropized the pairs. The essential new astrophysical idea is to identify the enhanced surface brightness expected in these compressions with the observed wisps.

From this identification, one can extract a host of conclusions. The details of these will be presented elsewhere. However, several qualitative statements can be made immediately. 1) A reasonable model, with $\sim 10^{38}$ pairs/s flowing out in the wind and 10^{33} iron ions/s with charge $Z = 3$ or 4, yields both a roughly correct optical surface brightness for the thin wisp and wisp 1, the observed spacing between the thin wisp and wisp 1, and the thickness of wisp 1 itself, if $\gamma_1 \sim 10^6$. 2) The series of downstream, large amplitude compressions of the plasma is spaced roughly one ion Larmor radius apart. This spacing is similar to that seen between the first few wisps in van den Bergh and Pritchett's (1989) image if $\gamma_1 \sim 10^6$. 3) The spectrum of wisp 1 should be close to that of Maxwellian synchrotron emission. The spectrum for a uniform medium is well known (Jones and Hardee 1979). With $B \sim 10^{-4}$ Gauss and pair temperature $\sim 10^6 m_{e^\pm} c^2$, the characteristic exponential decline of the spectrum from a relativistic Maxwellian should appear above ~ 1 eV. 4) Successive wisps farther from the ion overshoot should show a progressively more nonthermal spectrum, as the ions transfer their energy to the positrons. 5) Because the pair Larmor radius is very small (~ 1 AU for $B \sim 10^{-4}$ Gauss and $\gamma_1 \sim 10^6$), we identify the observed thickness of the thin wisp [$\sim 10^{16.5}$ cm in the image obtained by van den Bergh and Pritchett (1989)] as being due to the projection of an infinitesimally thin sheet projected on the plane of the sky, as one expects from the flow geometry implied by the Aschenbach and Brinkmann (1975) model of the X-ray emission in the nebula. 6) The flow velocity of the pairs throughout this region is between 0.3c and 0.5c, which provides a Doppler asymmetry to the luminosity just about right to explain the brightness ratio seen between wisp 1 and the one faint wisp observable to the pulsar's southeast. Possible partial anisotropy of the pairs with respect to the local magnetic field enhances this conclusion. 7) The shocks observed in our simulations are unsteady, as is shown in Figure 2. Translated into the shock frame, which is where we observe the Crab Nebula, the unsteady shock velocity becomes aharmonic oscillation of the position of the first ion overshoot (= wisp 1) on a time scale of several months, a phenomenon consistent with time variability of the wisps reported long ago by Scargle (1969). Such variability should be uncorrelated with pulsar glitches and other rotational anomalies. In addition, such unsteady behavior may imply unsteady nebular emission at energies exceeding 1 MeV, as has been occasionally reported. 8) The time variable overshoots can transfer further energy from the gyrating ions to the pairs in the downstream medium, over and above that found in the spatially 1D simulations. Because of the finite rate of pitch-angle scattering present downstream, magnetic pumping can increase the amount of energy transferred to the pairs, and can also act to transform the Maxwellian low-energy pair spectra into non-thermal form, thus allowing the same shock model to explain both the high-energy photons and the radio synchrotron emission from plerions (Arons 1992, in preparation). With this spatially 2D effect appended to the theory, the parameters derived from the wisp model are appropriate to those expected in the diffuse X-ray source formed further downstream. The detailed modeling in progress is designed to test whether these rough numerical conclusions stand up to careful analysis of the flow and predicted surface brightness distribution.

Therefore, we suggest that the structure of a relativistic shock in a relativistic wind composed of pairs plus heavy ions has its spatial structure revealed to us in the detailed morphology of the wisp region in the Crab Nebula. That this may be so is a consequence of the very high rigidity of the outflow from the pulsar, a result in accord with the enormous voltages developed in the magnetosphere combined with the negligible radiation drag on the flow near this photon-poor compact object.

The same ideas can be applied to the millisecond radio pulsars recently discovered in compact binaries (Fruchter *et al.* 1988; Lyne *et al.* 1988), whose relativistic winds interact with the mass lost from the "normal" stellar companion (Kluzniak *et al.* 1988; Phinney *et al.* 1988; Ruderman *et al.* 1989). If one assumes that the same fraction of the potential drop on open field lines goes into the energy/particle of these winds as appears in the wind from the Crab pulsar, one expects $\gamma_1 \sim 10^{5.7}$, whereas if σ is as small as it appears to be in the Crab pulsar's wind, the magnetic field behind the termination shock of the relativistic wind is on the order of 10 Gauss, since the shock occurs less than one solar radius from these pulsars, at least along the line of centers between the pulsar and the star. In this strong field, synchrotron losses limit the energies of the accelerated pairs to values much less than $\gamma_1 m_i c^2$ and convert almost all of the energy of the part of the pulsar's wind shocked within the binary into high-energy radiation. The pairs accelerate to energies on the order of $10^6 m_\pm c^2$ and much of the wind energy is given up as 1-10 MeV synchrotron photons, whose flux at earth might 10^{-6} photons/cm^2-s (Arons and Tavani 1992). Most of the pulsar's wind goes into blowing the nebular bubble observed in Hα emission around the binary (Kulkarni and Hester 1988), but because of the much stronger magnetic field within the close confines of the binary, most of the high-energy emission photon comes from within the binary rather than from the nebular bubble.

The research described here on relativistic shock waves was supported in part by NSF grant AST-9115093 and by IGPP-LLNL grant 92-23, both to the University of California at Berkeley. Part of the work was performed under the auspices of the U.S. Department of Energy at the Lawrence Livermore National Laboratory under contract W-7405-Eng-48.

References

Alsop, D., and Arons, J. 1988, *Phys. Fluids*, **31**, 839.

Arons, J. 1983, in *Proc. Workshop on Electron-Positron Pairs in Astrophysics*, ed. M. L. Burns, A. K. Harding, and R. Ramaty (New York: American Institute of Physics), p. 163.

Arons, J., and Tavani, M. 1992, *Ap. J.*, submitted.

Arons, J. 1992, to be submitted to *Ap. J.*

Aschenbach, B., and Brinkmann, W. 1975, *Astr. Ap.*, **41**, 147.

Bell, A. R. 1978, *MNRAS*, **182**, 147.

Birdsall, C. K., and Langdon, A. B. 1985, *Plasma Physics via Computer Simulation* (New York: McGraw-Hill).

Chiueh, T. 1989, *Phys. Rev. Lett.*, **63**, 113.

Ellison, D. C., Jones, F. C., and Reynolds, S. P. 1990, *Ap.J.*, **360**, 702.

Gallant, Y. A., Hoshino, M., Langdon, A. B., Arons, J., and Max, C. E. 1992, *Ap. J.*, in press.

Hoshino, M., and Arons, J. 1991, *Phys. Fluids B*, **3**, 818.

Hoshino, M., Arons, J., Gallant, Y. A., and Langdon, A. B. 1992, *Ap. J.*, **389**, in press.

Jones, T. W., and Hardee, P. E. 1979, *Ap.J.*, **228**, 268.

Kennel, C. F., and Coroniti, F. V. 1984a, *Ap. J.*, **283**, 694.

Kennel, C. F., and Coroniti, F. V. 1984b, *Ap. J.*, **283**, 710.

Kennel, C. F., and Pellat, R. 1976, *J. Plasma Phys.*, **15**, 335.

Kluzniak, W., Ruderman, M. A., Shaham, J., and Tavani, M. 1988, *Nature*, **334**, 225.

Kulkarni, S. R., and Hester, J. J. 1988, *Nature*, **335**, 801.

Kundt, W., and Krotscheck, E. 1980, *Astr. Ap.*, **83**, 1.

Langdon, A. B., and Lasinski, B. F. 1976, in *Methods in Computational Physics* **16**, ed. B. Alder, S. Fernbach, and M. Rotenberg (New York: Academic Press), p. 327.

Langdon, A. B., Arons, J., and Max, C. E. 1988, *Phys. Rev. Lett.*, **61**, 779.

Pacini, F. 1967, *Nature*, **216**, 567.

Pelling, R. M., Paciesas, W. S., Peterson, L. E., Makashima, K., Oda, M., Ogawara, Y., and Miyamoto, S. 1987, *Ap. J.*, **319**, 416.

Phinney, E. S., Evans, C. R., Blandford, R. D., and Kulkarni, S. R. 1988, *Nature*, **333**, 832.

Piddington, J. H. 1957, *Aust. J. Phys.*, **10**, 530.

Quenby, J. R., and Lieu, R. 1989, *Nature*, **342**, 654.

Rees, M. J., and Gunn, J. E. 1974, *MNRAS*, **167**, 1.

Ruderman, M. A., Shaham, J., and Tavani, M. 1989, *Ap. J.*, **336**, 507.

Scargle, J. D. 1969, *Ap. J.*, **156**, 401.

van den Bergh, S., and Pritchet, C. J. 1989, *Ap. J.*, **343**, L69.

Yoon, P. 1990, *Phys. Fluids B*, **3**, 867.

Zhelznyakov, V. V., and Suvorov, E. V. 1972, *Ap. and Space Sci.*, **15**, 24.

The Crab Nebula's Wisps as Shocked Pulsar Wind

Yves A. Gallant,[1,3] Jonathan Arons,[1,3] and A. Bruce Langdon[2,3]

[1]Theoretical Astrophysics Center, Astronomy Department, U. C. Berkeley
[2]X-Division, Lawrence Livermore National Laboratory
[3]Institute for Geophysics and Planetary Physics, LLNL

1 Introduction: Relativistic, Collisionless Shocks

The Crab synchrotron nebula has been successfully modeled as the post-shock region of a relativistic, magnetized wind carrying most of the spindown luminosity from the central pulsar (Rees and Gunn 1974, Kundt and Krotscheck 1980, Kennel and Coroniti 1984a,b, Emmering and Chevalier 1987). While the Crab is the best-studied example, most of the highest spindown luminosity pulsars are also surrounded by extended synchrotron nebulae, and several additional supernova remnants with "plerionic" morphologies similar to the Crab are known where the central object is not seen (Weiler 1985, Helfand and Becker 1987, Weiler and Sramek 1988). All these objects have nonthermal, power-law spectra attributable to accelerated high-energy particles thought to originate in a Crab-like relativistic pulsar wind. However, proposed models have so far treated the wind shock as an infinitesimally thin discontinuity, with an arbitrarily ascribed particle acceleration efficiency. To make further progress, investigations resolving the shock structure seemed in order.

Motivated by these considerations, we have performed "particle-in-cell" (PIC) simulations of perpendicularly magnetized shocks in electron-positron and electron-positron-ion plasmas (see Arons *et al.*1992). The shocks in pure electron-positron plasmas were found to produce only thermal distributions downstream and are thus poor candidates as particle acceleration sites. When the upstream plasma flow also contained a smaller population of positive ions, however, efficient acceleration of positrons, and to a lesser extent of electrons, was observed in the simulations.

Figure 1 displays the structure of such a electron-positron-ion shock, in the case where most of the upstream energy is in the bulk flow of the ions. The shock structure is characterized by two length scales: the electrons and positrons are first magnetically reflected and thermalized on a length scale on the order of their Larmor radius, $r_{L\pm} = m_e \gamma_1 c^2/(eB_1)$, where B_1 and γ_1 are the upstream perpendicular magnetic field and flow Lorentz factor, respectively. This "pair shock" is entirely analogous to the shock structure observed in pure pair plasmas (Gallant *et al.* 1992), as the background ion flow is essentially unaffected on this length scale.

The ions are then reflected in turn, through a combination of magnetic reflection and electrostatic pull of the charged pair fluid, on a much larger length scale of order the ion Larmor radius $r_{Li} = Am_p \gamma_1 c^2/(ZeB_1)$, where Ze and Am_p are the ion charge and mass, respectively. The downstream dissipation of this ion gyrational energy yields an acceleration of the pairs to power-law distributions, with an efficiency that

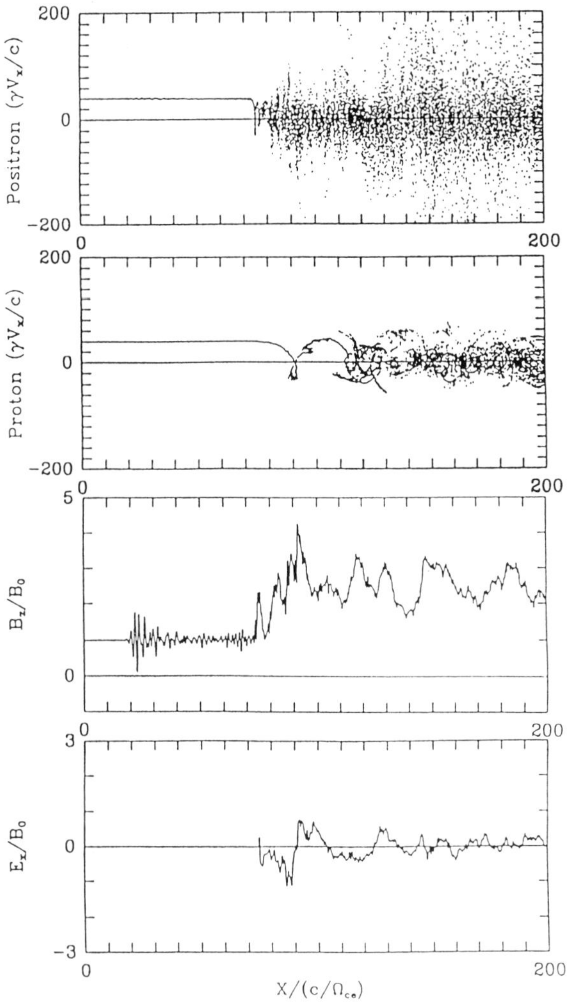

Fig. 1—Structure of an electron-positron-ion shock as revealed by simulations (Hoshino *et al.* 1992). The top two panels display u_x vs x phase space for the positrons and the ions, respectively, while the lower two panels show the magnetic and electrostatic fields. Shock parameters (definitions in §4) were $1/\sigma_i = 2$, $1/\sigma_\pm = 0.9$, and $\gamma_1 = 40$. The unit of length used for the x-axis is $r_{L\pm}$, corresponding to $r_{Li}/20$.

can reach ~ 20% of total wind energy available upstream (Hoshino *et al.* 1992). The increasing number of high-energy positrons downstream can clearly be observed in Figure 1. This acceleration process can be explained as coherent synchrotron emission from the initial ion loop followed by resonant absorption of these waves by the positrons and electrons (Hoshino and Arons 1991).

2 Resolved Shock Structure and the Wisps

One can show that for the parameters of the Crab pulsar's wind, as inferred by Kennel and Coroniti (1984a) or Emmering and Chevalier (1987), the global shock length scale, the ion Larmor radius, is in fact macroscopically large:

$$r_{Li} \approx 0.01 \left(\frac{A}{Z}\right) \left(\frac{\gamma_1}{10^6}\right) \left(\frac{10^{-4}\,\mathrm{G}}{B_1}\right) \mathrm{pc}, \tag{1}$$

where 0.01 pc corresponds to an angular size of 1″ at the Crab distance of 2 kpc. Thus unlike almost all other astrophysical shocks, the shock terminating the Crab pulsar's wind is expected to exhibit an observationally resolvable structure.

The Crab synchrotron nebula has long been known to have structure on precisely this length scale and at the expected distance from the pulsar, namely the "wisps" (Scargle 1969), originally described as "light ripples" (Lampland 1921). A modern, high-resolution CCD image of the wisps, obtained by van den Bergh and Pritchet (1989) at the CFHT, is shown in contour plot form in Figure 2. In addition to the well-known series of bright wisps beginning about 10″ to the northwest of the pulsar, one can clearly distinguish a faint wisp to the southeast of the pulsar, apparently a fainter counterpart of the main wisp to the northwest.

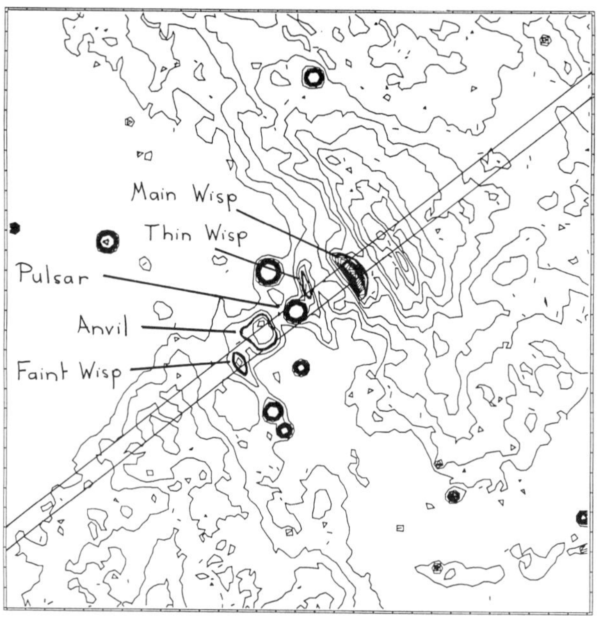

Fig. 2—The central 1′ × 1′ of the Crab Nebula in continuum emission, showing the pulsar and the main wisp features, as identified by Scargle (1969), in his Figure 1*a*, except for the faint wisp. Contours are in equal flux intervals (arbitrary units) starting just above the "underluminous zone" flux level. From the I-band CCD data of van den Bergh and Pritchet (1989).

A cross section along the pulsar-wisps axis of the average surface brightness from this image is shown in Figure 3. It can be seen that the wisps are local brightness enhancements superimposed on the high brightness "plateau" to the northwest of the pulsar. The dashed line represents the estimated foreground and background nebular emission, derived by assuming that the elliptical "underluminous zone" immediately surrounding the pulsar has negligible intrinsic emissivity (Kennel and Coroniti 1984a).

From the toroidal morphology of the X-ray emitting region in the Nebula, identified with the flow downstream of the shock, the wind is inferred to be mostly confined to the equatorial plane of the pulsar (Aschenbach and Brinkmann 1975; Brinkmann, Aschenbach and Langmeier 1985). Thus one would expect the shock to form a ring around the pulsar. The fact that wisps are seen only along the forward and backward flowing directions, *i.e.*, along the two arcs of the ring where the magnetic field is perpendicular to the line of sight, can be partly explained by a combination of Doppler boosting and the anisotropy of the synchrotron emissivity, but probably also

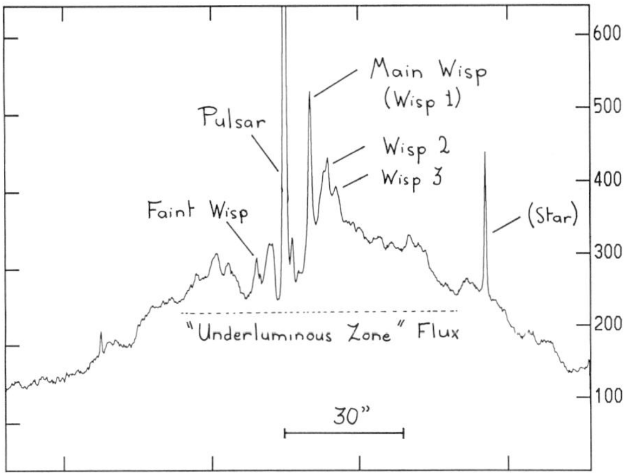

Fig. 3—The brightness profile of the Crab Nebula along the pulsar-wisps axis: averaged flux above sky brightness (arbitrary units) for a strip 2″ wide centered on the pulsar and oriented at a position angle of −53° (drawn in outline across Fig. 2). From the CCD data of van den Bergh and Pritchet (1989).

requires some underlying anisotropy in the synchrotron-emitting particle distribution. Polarization maps of the Crab Nebula (Michel *et al.* 1990) can also be interpreted as evidence for such anisotropy. In this context, investigations are under way to examine the growth and possible saturation of the Weibel instability, thought to be the dominant isotropization mechanism (see Yang *et al.*1992a).

3 Model of Wisp Emission: Synchrotron Radiation from Shock-Heated Pairs

The brightness profile observed in Figure 3 is readily explained in terms of the collisionless shock structure as uncovered by our simulations: beyond the pair shock, the thermalized pairs will radiate synchrotron radiation with critical frequencies corresponding to Lorentz factors on the order of γ_1, *i.e.*, around the optical waveband:

$$\nu_c \sim \frac{1}{2\pi}\frac{\gamma_1^2 e B_1}{m_e c} \approx 3\times 10^{14}\left(\frac{\gamma_1}{10^6}\right)^2\left(\frac{B_1}{10^{-4}\,\mathrm{G}}\right)\ \mathrm{Hz} \tag{2}$$

Moreover, the synchrotron emissivity will correlate very strongly with the magnetic field. Consider the plane-parallel case for illustration: to a good approximation, the pairs are frozen to the field lines, so that their number density $N_\pm$ scales as B. Then, ignoring particle acceleration and radiative losses, average particle energies will be determined by adiabatic compression and expansion, *i.e.*, $\gamma_\pm \propto B^{1/2}$ if the particles gyrate perpendicularly to the magnetic field and conserve their first adiabatic invariant, or $\gamma_\pm \propto B^{1/3}$ if they suffer rapid isotropization. Thus the total synchrotron power $P_{\rm synch}$ will scale as a high power of B:

$$P_{\rm synch} \propto N_\pm \gamma_\pm^2 B^2 \propto B^4 \text{ or } B^{11/3}, \tag{3}$$

where the two choices correspond to the extremes of no isotropization and complete isotropization. From Figure 1, it can be seen that the relativistic shock structure exhibits magnetic field enhancements in the ion reflection regions which can more than account for the observed brightness contrast of the wisps.

The proposed model is also entirely plausible on energetics grounds, as we now demonstrate. The synchrotron luminosity of the main wisp can be roughly estimated as the product of the total number of pairs emitted in the wind per unit time, $\dot{N}_{\pm}$, the time each particle spends in the first wisp region, roughly $r_{\mathrm{L}i}/c$, and the single-particle synchrotron power, $\sigma_T c\gamma^2 B^2/4\pi$. If, in addition, we assume that essentially all the spindown energy of the Crab pulsar goes into the ion-dominated wind energy, *i.e.*, $L_{\mathrm{spindown}} \approx \dot{N}_i m_i \gamma_1 c^2$, we obtain, using a neutron star moment of inertia of $10^{45}\,\mathrm{g\,cm^2}$,

$$L_{\mathrm{wisp}} \approx 5\times 10^{33}\left(\frac{\gamma_1}{10^6}\right)^2\left(\frac{B_1}{10^{-4}\,\mathrm{G}}\right)\left(\frac{\dot{N}_{\pm}/(Z\dot{N}_i)}{1000}\right)\ \mathrm{erg/s}. \tag{4}$$

From this luminosity we can get an estimated flux in the I-band by assuming that the spectrum is roughly flat up to the critical frequency ν_c, and by factoring in the distance to the Crab:

$$F_\nu \sim \frac{5\times 10^{33}\,\mathrm{erg/s}}{4\pi\,(2\,\mathrm{kpc})^2}\,(3\times 10^{14}\,\mathrm{Hz})^{-1} \approx 4\times 10^{-26}\,\mathrm{erg/(cm^2\,s\,Hz)}. \tag{5}$$

The actual total flux from the main wisp, as measured from CCD image of van den Bergh and Pritchet (1989), is $(1.6\pm 0.2)\times 10^{-26}\,\mathrm{erg/(cm^2\,s\,Hz)}$ within the uncertainties due to background subtraction and nebular brightness calibration. Given the very approximate nature of the arguments leading to the above estimate, its agreement with the measured value is remarkably good.

4 Soliton Model for the Shock Structure

So that we may make quantitative comparisons of the observed brightness profile of the wisps with synchrotron emission from shocks with arbitrary values of the upstream wind parameters, we are developing a semi-analytical model of the shock structure on the ion Larmor radius scale. We treat the ions as remaining cold, while the pairs are represented as a hot, magnetized fluid past the pair shock. Since the ions undergo no dissipation, this is analogous to a soliton model, except that we assume dissipation of the pairs' energy by imposing Rankine-Hugoniot conditions on the initial pair shock, which is treated as a discontinuous jump.

The main parameters of the upstream wind are the fractions of the incoming energy respectively invested in bulk flow of the ions, of the pairs, and in frozen-in Poynting flux, which we specify in terms of the two magnetization ratios σ_i and $\sigma_{\pm}$ for the ions and the pairs, respectively:

$$\frac{1}{\sigma_i} \equiv \frac{N_{1i} m_i \gamma_1 c^2}{B_1^2/4\pi} \quad \text{and} \quad \frac{1}{\sigma_{\pm}} \equiv \frac{N_{1\pm} m_{\pm} \gamma_1 c^2}{B_1^2/4\pi}, \tag{6}$$

with the subscript 1 indicating the upstream values. The upstream Lorentz factor γ_1 has negligible influence on the shock structure once $\gamma_1 \gg 1$, provided lengths are measured in units of the Larmor radius $r_{\mathrm{L}i}$.

We seek a self-consistent solution for the ion orbit and the fields as a function of distance x behind the pair shock, with the ion equations of motion given by

$$\frac{d}{d(\Omega_{ci}\tau)}\left(\frac{x}{r_{\rm Li}}\right) = \left(\frac{u_x}{u_1}\right) \tag{7}$$

$$\frac{d}{d(\Omega_{ci}\tau)}\left(\frac{u_x}{u_1}\right) = \left[\left(\frac{\gamma}{\gamma_1}\right)\left(\frac{E_x}{E_1}\right) + \left(\frac{u_y}{u_1}\right)\left(\frac{B_z}{B_1}\right)\right] \tag{8}$$

$$\frac{d}{d(\Omega_{ci}\tau)}\left(\frac{u_y}{u_1}\right) = \left[\left(\frac{\gamma}{\gamma_1}\right) \quad - \left(\frac{u_x}{u_1}\right)\left(\frac{B_z}{B_1}\right)\right], \tag{9}$$

where τ is the proper time, u_1 is the upstream 4-velocity, $\Omega_{ci} = ZeB_1/(Am_pc)$, $r_{\rm Li} \equiv u_1/\Omega_{ci}$, and $E_1 \equiv \beta_1 B_1$.

The fields must satisfy Maxwell's equations including the charge and current densities of the pairs in the source terms. One can show that when the upstream ion flow energy dominates that of the pairs, $1/\sigma_i \gg 1/\sigma_\pm$, these are well represented by bulk $\mathbf{E}\times\mathbf{B}$ drift of the charged pair fluid, and the field equations take the form

$$\frac{d}{d(x/r_{\rm Li})}\left(\frac{E_x}{E_1}\right) = \frac{1}{\sigma_i}\left[\sum\frac{\beta_1\gamma}{|u_x|} - \left(\frac{B_z}{B_1}\right)\right] \tag{10}$$

$$\frac{d}{d(x/r_{\rm Li})}\left(\frac{B_z}{B_1}\right) = -\frac{\beta_1^2}{\sigma_i}\left[\sum\frac{u_y}{|u_x|} - \left(\frac{E_x}{E_1}\right)\right]. \tag{11}$$

For the parameter ranges of interest, it turns out that the soliton structure must include ion reflection, like the solutions found by Alsop and Arons (1988) for the symmetric, pure pairs soliton. The sums in equations (10-11) are over incoming, reflected, and outgoing ion contributions. Solutions to equations (7-11) are obtained by an iterative method, starting with a guess for the fields and gradually improving it by recursively solving equations (7-9) and (10-11).

Figure 4 displays the soliton for the parameters of the shock simulation shown in Figure 1. The soliton can be seen to represent adequately the initial reflected ion loop and its associated magnetic overshoot; a correct representation of subsequent overshoots, however, may require the inclusion of some model of ion energy dissipation. After incorporating the effects of the spherically divergent wind geometry and of Doppler beaming and boosting due to the mildly relativistic bulk motion of the synchrotron-emitting pairs, we should be in a position to fit the observed brightness profile of the wisps as a function of the wind parameters.

A model of the evolution of the pairs' energy distribution, now being developed, which accounts for the pairs' acceleration through a combination of resonant absorption of collective ion cyclotron waves and magnetic pumping in the overshoot structure given by the soliton model, should then provide quantitative predictions about the spectrum of the synchrotron emission from the wisps. A longer-term prospect is that current investigations of the nature and efficiency of the pair isotropization mechanism (Yang *et al.* 1992b) will make it possible to construct complete synthetic brightness maps of the wind shock region, including its azimuthal structure, to be confronted with the full two-dimensional image.

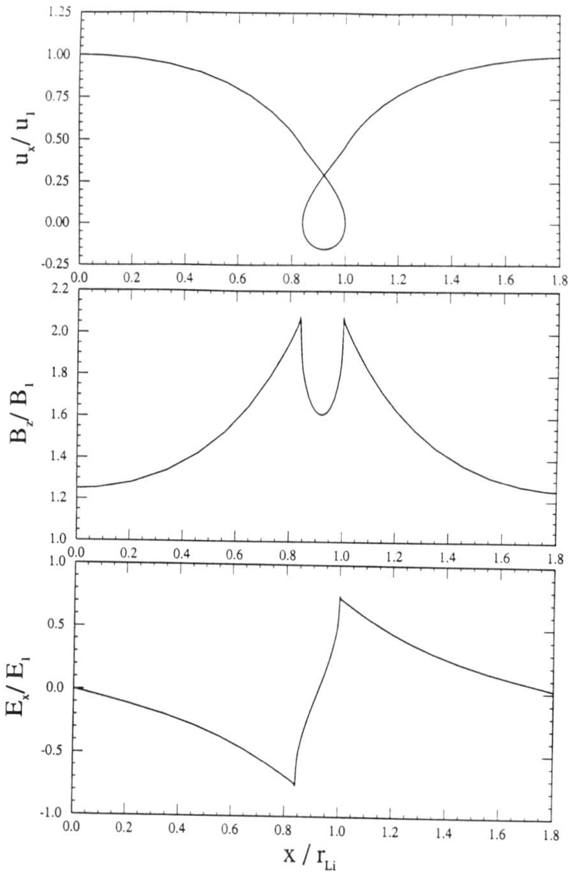

Fig. 4—"Soliton" with $1/\sigma_i = 2$, $1/\sigma_\pm = 0.9$, and $\gamma_1 = 40$. Top panel shows the ion orbits in u_x vs x form, whereas middle and lower panels respectively show the magnetic and electrostatic fields. The reflected ion orbit, doubly peaked magnetic field overshoot, and symmetrically peaked electrostatic field can clearly be recognized as the leading structure of the shock in Fig. 1.

5 Model Predictions: Spectrum and Variability

The relativistic shock structure model makes potentially testable predictions about the spectrum and variability of the wisps, which will be put forth qualitatively here. The wind parameter fitting outlined above, once completed, should allow us to make more quantitative the predictions concerning the wisps' spectrum. The first wisp is expected to have a spectrum close to *Maxwellian* synchrotron (*e.g.* Jones and Hardee 1979), with flux $F_\nu \propto \nu^{1/3}$ roughly up to the thermal critical frequency [eq. (3)] and decreasing rapidly with higher frequencies. With increasing distance behind the shock, and thus with increasing wisp number, the high-frequency part of the spectrum should evolve toward a nonthermal power-law form, as the synchrotron-emitting pairs are gradually accelerated to an energy distribution of the Maxwellian plus power-law type (Hoshino *et al.* 1992).

The other major application of the model concerns the variation of the wisps' position with time. Our shock simulations show that for weakly magnetized flows, shock propagation is intrinsically unsteady, and proceeds in a series "jumps" of the initial reflected loop by distances of order the ion Larmor radius, r_{Li}, on time scales of order the corresponding gyrofrequency. Thus we expect the position of the wisps to vary on time scales of several months, independently of any fluctuations in the outflow from the pulsar. Such variability has indeed been documented by Scargle (1969), although the time resolution between his observations was insufficient to determine the exact behavior of the wisps as a function of time. Earlier observations by Baade suggested that the main wisp forms and slowly moves downstream, eventually to be replaced by a new main wisp (Oort and Walraven 1956). This behavior would be consistent for a weakly magnetized wind, with shock propagation as observed in our simulations. More quantitative predictions about the time-variable behavior will probably become accessible, by means of some form of simulation, once the basic parameters of the wind are better constrained by fitting the time-stationary average structure.

6 Summary and Propaganda

We argue that the structure of the collisionless shock terminating the relativistic wind from the Crab pulsar should be observationally resolvable, and we suggest the identification of the wisps with synchrotron emission from this shock structure. The observed length scale, morphology, and energetics of the wisps all support this interpretation, and the time variability of the wisps' position is also consistent with this picture. The main test of the model will lie in the predicted spectrum of the wisps' continuum emission, which was described qualitatively above, and which modeling in progress should allow us to study more quantitatively. Clearly, modern observations of the spectrum and time-variability of the wisps, in the IR, optical, UV and X-ray wavebands, would prove extremely interesting.

In a meeting largely devoted to high-energy emission from pulsars, it is perhaps not inappropriate to stress that of all the observable manifestations of pulsars, the largest fraction by far of the total spindown luminosity is invested into pulsar winds. Thus relativistic winds are the most important "emission mechanism" to be accounted for by pulsar magnetospheric theory, and the inferences about the pulsar wind parameters and composition expected from our current work should provide detailed observational constraints on this question.

We wish to express our gratitude to S. van den Bergh and C. Pritchet for allowing us to use their CCD image of the Crab Nebula's wisps. The research described here was supported in part by NSF grant AST-9115093 and by IGPP-LLNL grant 92-93 and was performed in part under the auspices of the DOE at LLNL under contract W-7405-Eng-48.

References

Alsop, D., and Arons, J. 1988, *Phys. Fluids*, **31**, 839.
Arons, J., Gallant, Y. A., Langdon, A. B., and Yang, T.-Y. B., 1992, these Proceedings
Aschenbach, B., and Brinkmann, W. 1975, *Astr. Ap.*, **41**, 147.
Brinkmann, W., Aschenbach, B., and Langmeier, A. 1985, *Nature*, **313**, 662.
Emmering, R. T., and Chevalier, R. A. 1987, *Ap. J.*, **321**, 334.
Gallant, Y. A., Hoshino, M., Langdon, A. B., Arons, J., and Max, C. E. 1992, *Ap. J.*, **391**, in press.
Helfand, D. J., and Becker, R. H. 1987, *Ap. J.*, **314**, 203.
Hoshino, M., and Arons, J. 1991, *Phys. Fluids B*, **3**, 818.
Hoshino, M., Arons, J., Gallant, Y.A., and Langdon, A.B. 1992, *Ap. J.*, **389**, in press.
Jones, T. W., and Hardee, P. E. 1979, *Ap. J.*, **228**, 268.
Kennel, C. F., and Coroniti, F. V. 1984a, *Ap. J.*, **283**, 694.
Kennel, C. F., and Coroniti, F. V. 1984b, *Ap. J.*, **283**, 710.
Kundt, W., and Krotscheck, E. 1980, *Astr. Ap.*, **83**, 1.
Lampland, C. O. 1921, *Pub. A. S. P.*, **33**, 79.
Michel, F. C., Scowen, P. A., Dufour, R. J., and Hester, J. J. 1990, *Ap. J.*, **368**, 463.
Oort, J. H., and Walraven, Th. 1956, *Bull. Astr. Inst. Netherlands*, **12**, 285.
Rees, M. J., and Gunn, J. E. 1974, *M.N.R.A.S.*, **167**, 1.
Scargle, J. D. 1969, *Ap. J.*, **156**, 401.
van den Bergh, S., and Pritchet, C. J. 1989, *Ap. J. (Letters)*, **338**, L69.
Weiler, K. W. 1985, in *The Crab Nebula and Related Supernova Remnants*, ed. M. C. Kafatos and R. B. C. Henry (Cambridge: Cambridge University Press), p. 265.
Weiler, K. W., and Sramek, R. A. 1988, *Ann. Rev. Astr. Ap.*, **26**, 295.
Yang, T.-Y. B., Gallant, Y. A., Arons, J., and Langdon, A. B. 1992a, these Proceedings
Yang, T.-Y. B., Gallant, Y. A., Arons, J., and Langdon, A. B. 1992b, in preparation.

Weibel Instabilities in Relativistic Electron-Positron Plasmas

Tser-Yuan Brian Yang,[1] Yves A. Gallant,[2] Jonathan Arons,[1,2] and A. Bruce Langdon[3]

[1] Department of Astronomy, University of California at Berkeley
[2] Department of Physics and Theoretical Astrophysics center, University of California at Berkeley
[3] Lawrence Livermore National Laboratory

1 Introduction

The acceleration of ultra-relativistic particles in astrophysical sources of synchrotron radiation has long been an outstanding puzzle. The nearest at hand and the best studied examples of this problem are pulsars and their surrounding "plerionic" (Weiler 1978) nebulae. Kennel and Coroniti (1984a,b) showed that a relativistic, magnetized electron-positron wind from the Crab pulsar, terminated by a standing, transverse magnetosonic shock wave, could give a good account of the nebular dynamics and of the nebular spectrum of high-energy photons. In a recent theoretical study (Hoshino *et al.* 1992), it has been demonstrated that the relativistic magnetosonic shock wave is an effective mechanism to convert flow energy into highly accelerated positrons with a power-law spectrum $N(\gamma)d\gamma \sim \gamma^{-s}$ (Fig. 1). The efficiency of the energy conversion is about $10 \sim 30\%$ (Hoshino *et al.* 1992).

Wisps have long been observed in the the Crab Nebula (Scargle 1969). Modern observations (van den Bergh and Pritchet 1989) reveal a whole series of wisps to the northwest, and at least one much fainter wisp to the southeast. Recently, a conceptual model of the wisps in the Crab Nebula has been proposed (Hoshino *et al.* 1992) based on the shock structure of a relativistic magnetosonic shock wave in a electron-positron-proton plasma (Fig. 2). The ion overshoot, downstream compressional oscillations, and possibly the pair overshoot can be identified with the wisps.

In this paper, we study the linear stability properties of the Weibel instability (Weibel 1959; Davidson 1983), which are collective electromagnetic oscillations driven by pitch-angle anisotropy of the distribution functions. This instability can cause pitch-angle scattering of the charged particles and the isotropization of the distribution functions. For readers unfamiliar with the Weibel instability, a simple physical picture for the instability in a unmagnetized plasma can be found in the textbook by Chen (1984).

There are a couple of reasons why we think pitch-angle isotropization is relevant to the relativistic magnetosonic shock waves. First, the fact that wisps are observed to be brighter in the region to the northwest of the Crab Pulsar may be explainable by the Doppler beaming effect. However, because the nebular magnetic field wraps around the pulsar in a torus (Aschenbach and Brinkmann, 1975), pitch-angle anisotropy may be required to explain why wisps are not seen in the regions to the northeast and southwest of the Pulsar, for if the particle distribution is completely isotropic in pitch angle, we expect to see ring-like structures. Secondly, the degree of pitch-angle

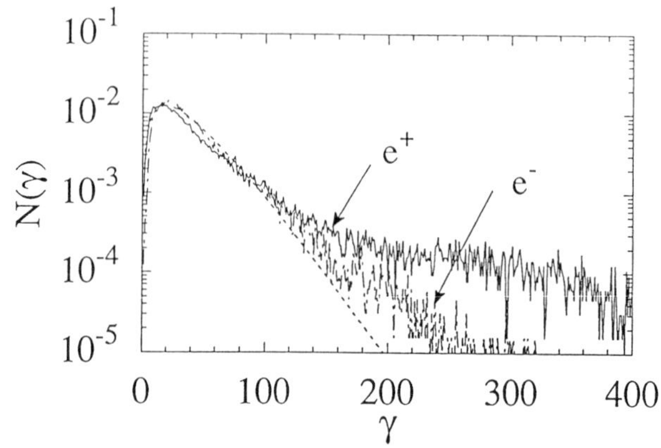

Fig. 1—Energy spectrum for positrons and electrons obtained in a shock simulation (Hoshino *et al.* 1992). The electron spectrum is fitted by a Maxwellian distribution function (the dotted lines).

anisotropy may have strong influence on the energy transfer between the protons and the positrons. For example, the linear theory of the synchrotron maser in a homogeneous medium suggests that the spread in pitch angle reduces the wave emission efficiency. With this in mind we study the Weibel instability in order to answer the following questions:

1. How fast does the pitch-angle isotropization take place?
2. Will the distribution functions completely isotropize, or saturate at a state with finite pitch-angle anisotropy?

2 Basic Characteristics of Electron-Positron-Proton Shocks

The shock structures in electron-positron-proton plasmas (Hoshino *et al.* 1992) (see Fig. 2) have a leading edge similar to shocks in the pure pair plasmas (Gallant *et al.* 1992), in which the downstream energy spectra for both electrons and positrons are well described by relativistic Maxwellians. The leading edge is followed by a region of proton reflection embedded in the shock-heated pair plasma as the more massive protons begin to gyrate behind the leading pair shock. The reflected protons quickly transfer their energy to the positrons through proton emission and the positron absorption of left-handed magnetosonic modes in the shock-heated pairs, the process is the same as that which occurs in synchrotron maser instabilities in a uniform medium (Hoshino and Arons 1991). The energy gained by the positrons appears in the non-thermal, power-law component of their distribution, whereas the electrons retain the Maxwellian spectrum formed in the leading pair shock (Fig. 1).

The reason for the preferential heating of the positrons is that, due to the low mass ratio between the ions and the pairs ($m_i/m_{e_\pm} = 20$) used in the simulations, the electron density is substantially higher than the positron density in order to make the protons the energetically dominant species. This, in turn, makes the emitted magnetosonic waves left-hand polarized and preferably absorbed by the positrons. In a plasma with a realistic mass ratio and a pair density much larger than the number

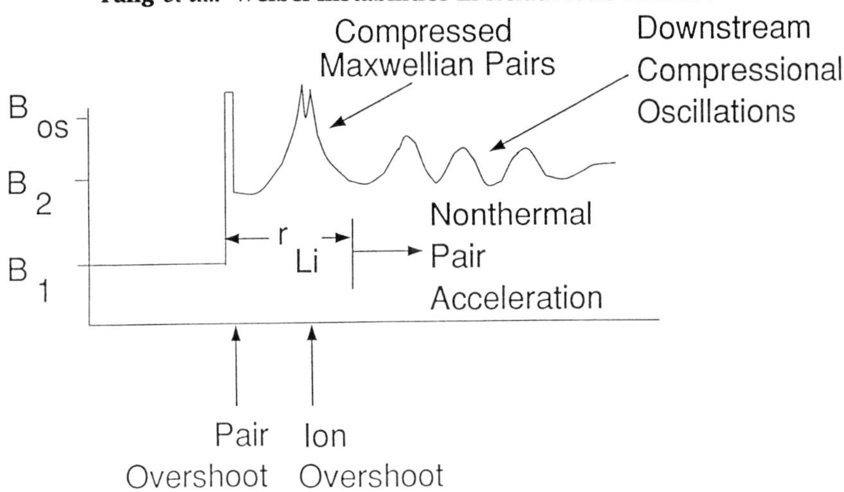

Fig. 2—A cartoon of the mechanical shock structure in the overshoot region.

density of heavy ions, we expect a significant acceleration of nonthermal electrons as well as positrons.

The quasi-coherent gyro oscillations of the protons in the down-stream region form a spatially oscillating transverse current, which causes the ion overshoot and the large amplitude down-stream oscillation of the magnetic field with wavelength comparable to the proton Larmor radius (Fig. 2). As the field increases, the pairs are compressed as an MHD fluid and adiabatically heated. The synchrotron emission from these heated pairs can be identified as the origin of the wisps.

3 The Weibel Instability

3.1 Theoretical Model

The dispersion relation for the waves and instabilities propagating parallel to the background magnetic field has been derived for relativistic magnetized electron-ion plasmas (Sudan 1965). A simple modification readily gives the desired dispersion relation for electron-positron plasmas. It is assumed, in the present analysis, that the electrons and positrons have identical equilibrium distribution functions, i.e., $f^0_{e^-}(\mathbf{x},\mathbf{p}) = f^0_{e^+}(\mathbf{x},\mathbf{p}) = \hat{n}_e F_e(u_\perp^2, u_z)$. Here, $\hat{n}_e$ is the density for the electrons as well as the positrons, u_z is the parallel component of the normalized momentum $\mathbf{u}=\mathbf{p}/mc$, and $u_\perp$ is the magnitude of the transverse component of $\mathbf{u}$. Moreover, $F_e(u_\perp^2, u_z)$ satisfies the normalization condition

$$\int d^3u F_e(u_\perp^2, u_z) = 1. \tag{1}$$

Detailed analysis has been carried out (Yang *et al.* 1992) for the case where the distribution functions of the electrons and positrons have the following form,

$$F_e(u_\perp^2, u_z) = \frac{1}{N_w} H(\gamma_0 - \gamma) H(\gamma_z^0 - \gamma_z), \tag{2}$$

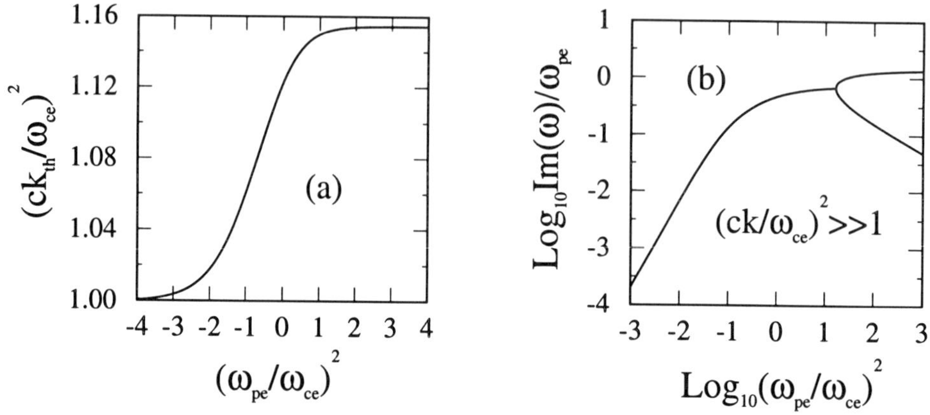

Fig. 3—(a) The normalized threshold wavenumber-squared and (b) the short-wavelength-limit growth rate (measured in units of the relativistic plasma frequency) versus the normalized plasma frequency-squared $(\omega_{pe}/\omega_{ce})^2$, for $\gamma_0 = 100$.

where $H(x)$ is the Heaviside step function defined by

$$H(x) = \begin{cases} 1, & x > 0, \\ 0, & x < 0, \end{cases} \tag{3}$$

N_w is the normalization constant defined by

$$N_w = 2\pi u_z^0 \left(\gamma_0^2 - \frac{\gamma_z^{02}}{3} - \frac{2}{3} \right), \tag{4}$$

$$u_z^0 = (\gamma_z^{02} - 1)^{1/2}, \tag{5}$$

γ^0 and γ_z^0 which satisfy $\gamma_z^0 \le \gamma_0$ are constants. Moreover, γ and γ_z are related to $u_\perp$ and u_z by

$$\gamma = (1 + u_\perp^2 + u_z^2)^{1/2}, \qquad \gamma_z = (1 + u_z^2)^{1/2}. \tag{6}$$

3.2 Stability Properties for Zero Parallel Temperature

The Nyquist analysis (Krall and Trivelpiece 1973) has been performed on the dispersion relation, for the case with zero parallel temperature ($\gamma_z^0 = 1$), to determine the necessary and sufficient conditions for the instability. It was found that there are two unstable (exponentially growing) modes for wavenumbers with magnitudes greater than a threshold wavenumber k_{th}, *i.e.*, $|k| > k_{th} > 0$. Plotted in Figure 3a is the normalized threshold wavenumber-squared $(ck_{th}/\omega_{ce})^2$ versus the normalized plasma frequency-squared $(\omega_{pe}/\omega_{ce})^2$, for $\gamma_0 = 100$. Here, ω_{pe} and ω_{ce} are the relativistic plasma frequency and the relativistic cyclotron frequency, respectively, for the electron and positron with energy $\gamma_0 mc^2$. Figure 3b shows the dependence on the normalized plasma frequency-squared $(\omega_{pe}/\omega_{ce})^2$, of the growth rate Im$\omega$ of the instability in the short-wavelength limit ($ck/\omega_{ce} \gg 1$), for $\gamma_0 = 100$. In Figure 3b, the growth rate is measured in units of the relativistic plasma frequency ω_{pe}. Displayed in Figures 4a and 4b are the normalized growth rates Imω/ω_{ce} versus the normalized wavenumber-squared $(ck/\omega_{ce})^2$ for $\gamma_0 = 100$ and two different values of normalized plasma frequency-squared (a) $(\omega_{pe}/\omega_{ce})^2 = 10^3$ and (b) $(\omega_{pe}/\omega_{ce})^2 = 10^{-3}$.

For ultra-relativistic plasmas ($\gamma_0 \gg 1$), which is the case of interest in the present analysis, detailed analyses of the dispersion relation show that the stability properties depend very weakly on the Lorenz factor γ_0 if the frequencies and wavenumber are measured in units of the relativistic cyclotron frequency $\omega_{ce} = eB_0/\gamma_0 mc$. That is, the normalized threshold wavenumber ck_{th}/ω_{ce} and the curves of ω/ω_{ce} versus ck/ω_{ce} are insensitive to the variation of γ_0 as long as $\omega_{pe}^2/\omega_{ce}^2$ is fixed and γ_0 is sufficiently large. The results in Figures 3 and 4 with $\gamma_0 = 100$ are, therefore, good representatives for general ultra-relativistic plasmas ($\gamma_0 \gg 1$).

From Figure 3a, it is evident that, for a ultra-relativistic electron-positron plasma, the threshold wavenumber, k_{th}, is slightly greater than ω_{ce}/c. This implies that the instabilities occur for wavelengths shorter than the relativistic Larmor radii of the electrons with energy $\gamma_0 mc^2$. For wavenumbers with amplitudes greater than the threshold wavenumber k_{th} (see Fig. 1), there are two unstable modes. For the cases with $(\omega_{pe}/\omega_{ce})^2 = 10^{-3}$, two modes corresponding to a given real wavenumber k have equal growth rates $\mathrm{Im}(\omega)$ and nonzero real frequencies $\mathrm{Re}(\omega)$ of equal amplitudes and opposite signs. This also describes the case with $(\omega_{pe}/\omega_{ce})^2 = 10^3$ when the amplitudes of the wavenumber are sufficiently small.

For the case with $(\omega_{pe}/\omega_{ce})^2 = 10^3$, there are two purely growing modes ($\mathrm{Re}\omega = 0$) with different growth rates (see Fig. 4a) when the amplitudes of the wavenumber are sufficiently large. The growth rate for the lower branch is of the order of the relativistic cyclotron frequency ω_{ce}, and the growth rate for the upper branch increases with increasing amplitude of the wavenumber. At the short-wavelength limit, $(ck/\omega_{ce})^2 \gg 1$, the growth rate of the upper branch is of the order of the relativistic plasma frequency ω_{pe}.

For the case with $(\omega_{pe}/\omega_{ce})^2 = 10^{-3}$, the instabilities show strong resonant features, which include real frequencies close to ω_{ce} ($|\mathrm{Re}\omega/\omega_{ce}| - 1 \ll 1$) and the sharp peak in the growth-rate curve at $(ck/\omega_{ce})^2 \simeq 1$. The resonance is due to the singular shape of the distribution function at $\gamma = \gamma_0$ [see eq. (2)] and will probably disappear for smoother distribution functions.

As evident from Figure 3b, for a weakly magnetized plasma with $(\omega_{pe}/\omega_{ce})^2 > 1$, the maximum growth rate is comparable to the relativistic plasma frequency ω_{pe}. For a strongly magnetized plasma with $(\omega_{pe}/\omega_{ce})^2 < 1$; however, the growth rate in the short-wavelength limit is significantly smaller than ω_{pe}. It is also evident from Figure 3b that purely growing modes ($\mathrm{Re}\omega = 0$) exist only when $(\omega_{pe}/\omega_{ce})^2$ is greater than a minimum value approximately equal to 16.6.

3.3 Stability Properties for Finite Parallel Temperature

For electron-positron plasmas with nonzero parallel temperature ($\gamma_z^0 > 1$), the dispersion relation has been solved numerically over a wide range of system parameters. Plotted in Figures 5a and 5b are the normalized growth rates $\mathrm{Im}\omega/\omega_{ce}$ versus normalized wavenumber-squared $(ck/\omega_{ce})^2$ for $(\omega_{pe}/\omega_{ce})^2 = 10^3$, $\gamma_0 = 100$, and two different values of γ_z^0, *i.e.*, (a) $\gamma_z^0 = 50$ and (b) $\gamma_z^0 = 90$. When the parallel temperature is nonzero ($\gamma_z^0 > 1$), it is evident from Figures 5a and 5b that the instabilities occur in a finite bandwidth $0 < k_{min} < |k| < k_{max}$. Moreover, the bandwidth and

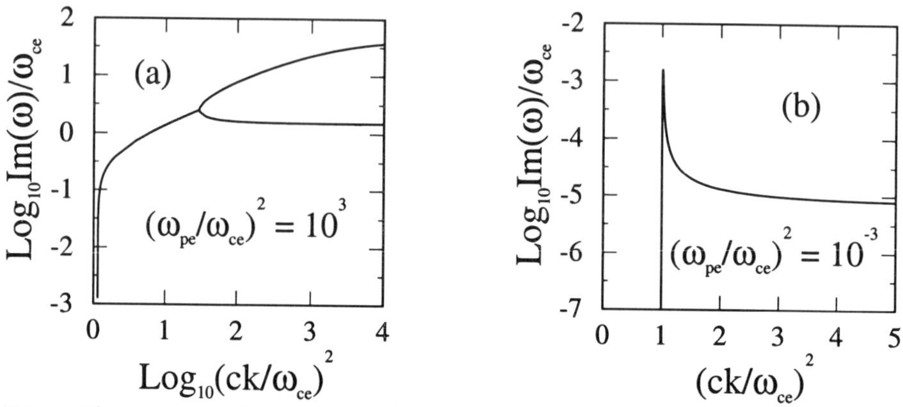

Fig. 4—The normalized growth rate $\mathrm{Im}\omega/\omega_{ce}$ versus the normalized wavenumber-squared $(ck/\omega_{ce})^2$ for $\gamma_0 = 100$ with (a) $(\omega_{pe}/\omega_{ce})^2 = 10^3$ and (b) $(\omega_{pe}/\omega_{ce})^2 = 10^{-3}$.

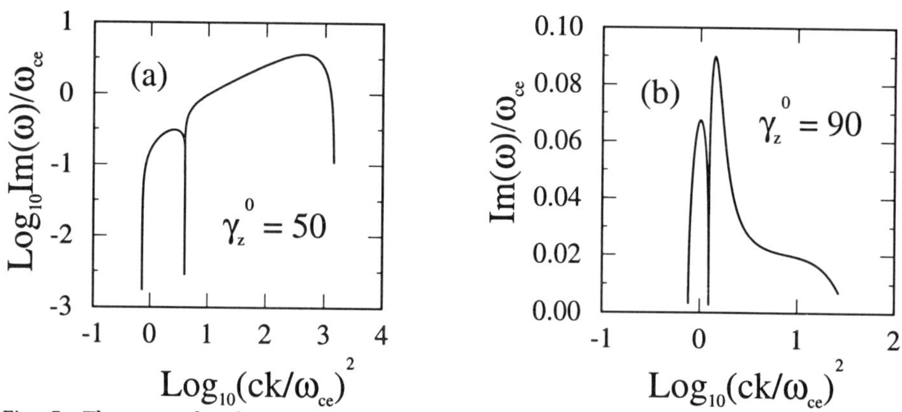

Fig. 5—The normalized growth rate $\mathrm{Im}\omega/\omega_{ce}$ versus the normalized wavenumber-squared $(ck/\omega_{ce})^2$ for $\gamma_0 = 100$ and $(\omega_{pe}/\omega_{ce})^2 = 10^3$ with (a) $\gamma_z^0 = 50$ and (b) $\gamma_z^0 = 90$.

the maximum growth rate of the instabilities decrease with increasing γ_z^0. Although significant decrease in the growth rate can be seen from Figures 5a and 5b, a complete shut-off of the instabilities does not occur for weak, but finite, temperature anisotropy ($0 < \gamma_0/\gamma_z^0 - 1 \ll 1$). Signatures for resonance are observed for the case with $\gamma_z^0 = 90$, corresponding to a weak temperature anisotropy. This resonance is due to the singular shape of the distribution function in equation (2).

4 Summary and Discussions

We studied the linear stability properties of the Weibel instability in a relativistic magnetized electron-positron plasma, with a simple model distribution function [see eq. (2)]. It was found that, for the case with zero parallel temperature, the instabilities occur for wavelengths shorter than the relativistic electron Larmor radius. For weakly magnetized plasmas ($\omega_{pe}^2/\omega_{ce}^2 > 1$), the growth rate increases with the wavenumber and approaches a maximum value comparable to the plasma frequency ω_{pe}. For strongly magnetized plasmas ($\omega_{pe}^2/\omega_{ce}^2 < 1$), however, the growth rate in the short-wavelength limit ($ck/\omega_{ce} \gg 1$) is significantly less than the plasma frequency. In the case of strongly magnetized plasmas, sharp resonance appeared at $ck \simeq \omega_{ce}$ in the plot of the growth rate versus wavenumber. This resonance is due to the singular shape of the model distribution function and will probably disappear when smoother distribution functions are used.

For plasmas with finite parallel temperature ($\gamma_0 > \gamma_z^0 > 1$), the instabilities occur within a finite bandwidth $0 < k_{\rm min} < k < k_{\rm max}$. When γ_z^0 is comparable to, but still less than γ_0, the growth rate decreases significantly from that for $\gamma_z^0 = 1$.

Since a complete shut-off of the instabilities does not occur for weak, but finite, temperature anisotropy ($0 < \gamma_0/\gamma_z^0 - 1 \ll 1$), the distribution function might completely isotropize, if the system were to evolve in a quasi-linear way. However, in some exploratory simulations of a 2-D shock model (Gallant *et al.* 1992), it was observed that a substantial temperature anisotropy was maintained for a period longer than a simple quasi-linear argument would predict. It is our speculation that the saturation of the instabilities and the sustenance of a finite temperature anisotropy are due to other nonlinear effects [*e.g.*, magnetic trapping (Davidson *et al.* 1972)]. The long-term consequences of the Weibel instability and other issues relevant to the relativistic magnetosonic shock waves will be subjects of further investigations.

References

Aschenbach, B., and Brinkmann, W. 1975, *Astr. Ap.*, **41**, 147.

Chen, F. F 1984, *Introduction to Plasma Physics and Controlled Fusion*, Vol. **1**, *Plasma Physics* (New York: Plenum).

Davidson, R. C. 1983, in *Handbook of Plasma Physics.* Vol. **1**, ed. M. N. Rosenbluth and R. Z. Sagdeev, (North-Holland), p. 521.

Davidson, R. C., Hammer, D. A., Haber, I., and Wagner, C. E. 1972, *Phys. Fluids*, **15**, 317.

Gallant, Y., Hoshino, M., Langdon, A. B., Arons, J., and Max, C. E. 1992, *Ap. J.*, in press.

Hoshino, M., and Arons, J. 1991, *Phys. Fluids B*, **3**, 818.

Hoshino, M., Arons, J., Gallant, Y., and Langdon, A. B 1992, *Ap. J.*, in press.

Kennel, C. F., Coroniti, F. V. 1984a, *Ap. J.*, **283**, 694.

Kennel, C. F., Coroniti, F. V. 1984b, *Ap. J.*, **283**, 710.

Krall, N. A., and Trivelpiece, A. W. 1973, *Principles of Plasma Physics*, (McGraw Hill).

Scargle, J. D. 1969, *Ap. J.*, **156**, 401.

Sudan, R. N. 1965, *Phys. Fluids*, **8**, 153.

van der Bergh, S. and Pritchet, C. J. 1989, *Ap. J.*, **338**, L69.

Weibel, E. S. 1959, *Phys. Rev. Lett.*, **2**, 83.

Weiler, K. W. 1978, *Astr. Ap.*, **70**, 419.

Yang, T.-Y. B., Gallant, Y., Arons, J., and Langdon, A. B. 1992, in preparation.

Particle Acceleration and Gamma-Ray Emission by Accreting Neutron Stars

Frederick K. Lamb,[1,2] Russell J. Hamilton,[1] and M. Coleman Miller[1]

[1]Department of Physics, University of Illinois at Urbana/Champaign
[2]Department of Astronomy, University of Illinois at Urbana/Champaign

1 Introduction

The possibility that accreting neutron star systems may produce appreciable fluxes of high-energy electromagnetic radiation is currently of special interest because of ongoing MeV–GeV observations using *GRANAT* and the *Compton Observatory* and the improving sensitivity of ground-based TeV and PeV detectors (see the reports in this volume). Here we summarize the results of our recent investigation (Hamilton, Lamb, and Miller 1992; Miller, Lamb, and Hamilton 1992) of the electrodynamics of magnetic neutron stars accreting from Keplerian disks and the implications for particle acceleration and γ-ray emission in such systems. We argue that the particle density in the magnetosphere is large, making vacuum gaps like those proposed by Cheng and Ruderman (1991) unlikely. Interaction of the disk with the magnetic field of the star produces very large electromotive forces (EMFs), even if the star is slowly rotating. We argue that these EMFs drive very large conduction currents, causing microscopic and macroscopic instabilities in the magnetosphere. Current-driven micro-instabilities may produce relativistic double layers, which can accelerate charged particles to very high energies. Alternatively, twisting of the stellar magnetic field may cause relativistic reconnection events, which can also accelerate particles to high energies. We show that electrons, as well as protons, typically can be accelerated to energies as high as $\sim 10^{12}$ eV in this environment, despite previous claims to the contrary (Katz and Smith 1988; Cheng and Ruderman 1991). Interaction of accelerated particles with X- and γ-rays and accreting plasma may produce high energy radiation detectable by *GRANAT*, the *Compton Observatory*, or ground-based experiments.

2 Disk Star Interaction

Consider a system in which a neutron star with a strong dipolar magnetic field is accreting plasma from a Keplerian disk. For simplicity, assume that the neutron star magnetic and rotation axes are aligned and that the star is slowly rotating. (We expect the qualitative features of the system to be as described below, even if the axes are not aligned.) Assume also that the disk is geometrically thin and situated in the rotational equatorial plane of the neutron star. In analyzing this system, we use cylindrical coordinates (ϖ, ϕ, z) centered on the star. Our analysis focuses on the general features of the system rather than any particular model.

Plasma near the neutron star is highly ionized and has a high electrical conductivity. We therefore expect an electric field $\mathbf{E} \approx -\mathbf{v} \times \mathbf{B}/c$ in this region, which causes the plasma to corotate with the star. The magnitude of the charge density near the star is then (see Goldreich and Julian 1969)

$$|n_+ - n_-| \approx n_{GJ} \approx |\Omega_s \cdot \mathbf{B}|/2\pi ec = 7 \times 10^4 \mu_{30} r_8^{-3} P_s^{-1} \ \mathrm{cm}^{-3} \tag{1}$$

where Ω_s is the stellar angular frequency, μ_{30} is the stellar magnetic moment in units of $10^{30}\,\mathrm{G\,cm}^3$, r_8 is the radius in units of 10^8 cm, and P_s is the stellar rotation period in seconds. We also expect the disk to have a magnetic field that causes the highly conducting plasma near the disk to rotate with an angular velocity close to that of the disk plasma to which it is connected by the magnetic field. The magnitude of the charge density in this region is given approximately by equation (1), with P_s replaced by the local rotation period.

We expect the charge density near the neutron star and the inner disk to be much smaller than the particle density there, for two reasons. First, the electrical potential energy difference between the inner disk and the star is $\sim 10^{15}$ eV (see below), whereas the gravitational potential energy confining protons to the disk is only $\sim 10^6$ eV. Hence, even small fluctuations in the electric field near the disk caused by irregularities in the accretion flow can readily pull protons as well as electrons out of the disk into the magnetosphere. Second, about 10^{40}–10^{42} particles per second flow from the disk to the neutron star. The differing rotation rates of the star and disk plasma twists the magnetic field in this region, creating MHD instabilities that inject plasma into the magnetosphere above the inner disk (see Ghosh and Lamb 1991). Even if only a fraction $\sim 10^{-10}$ of the accreting plasma is injected with a velocity comparable to the local Keplerian velocity, the particle density in the magnetosphere greatly exceeds n_{GJ}. Indeed, X-ray observations of Her X-1 (Brecker *et al.* 1977; Bai 1980) indicate that the particle density above the inner disk is $\gtrsim 10^{11}\,\mathrm{cm}^{-3}$, far greater than n_{GJ}. Formation of a vacuum gap in the magnetosphere requires complete charge separation (see Cheng and Ruderman 1991) and therefore seems unlikely. Our treatment of the interaction between the disk and stellar magnetic field is based on the theory of disk-accreting neutron stars developed over the last fifteen years, which assumes that the particle density in the magnetosphere is large enough to support substantial conduction currents (see Ghosh and Lamb 1991).

The magnetic field of the star can interact with plasma in the disk in a variety of ways. The processes that transport angular momentum in the disk—such as turbulence and magnetic coupling—by their nature allow magnetic flux from the star to couple to the inner part of the disk. The differential rotation of the disk plasma and star generates an EMF that drives field-aligned currents within the magnetosphere. These currents create an azimuthal magnetic field component. The process can be thought of as twisting of the magnetic field due to the differing rotation rates of the disk plasma and star. The resulting magnetic stress on the disk extracts angular momentum from the disk plasma. The Keplerian flow ends at the radius ϖ_0 where the magnetic stress becomes so large that it extracts the angular momentum of the disk plasma in a radial distance much less than ϖ (see Ghosh and Lamb 1991). Inside ϖ_0,

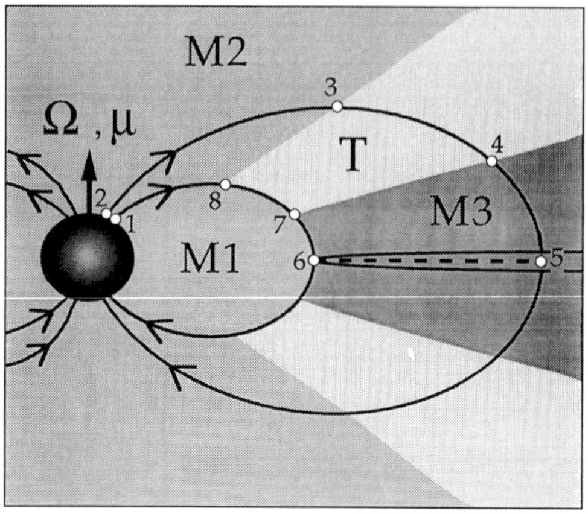

Fig. 1—Schematic diagram of the system analyzed in the text. Plasma on the magnetic field line that connects the inner edge of the disk to the stellar surface (points 6-7-8-1) corotates with the star, as does plasma in regions M1 and M2. Plasma in region M3 orbits the star with the angular velocity of the adjacent disk plasma. We expect velocities in the transition region T to be intermediate between the angular velocity of the star and that of the adjacent disk plasma. The size and shape of region T depends on the particular model.

the angular velocity Ω_d of the disk plasma falls rapidly from the Keplerian value Ω_K, reaching the angular velocity of the star at ϖ_{co}.

3 Electrodynamics

Consider now the electrodynamics of this system. We first calculate the EMF

$$\mathcal{E} \equiv \oint_C \left(\mathbf{E} + \frac{1}{c}\mathbf{v} \times \mathbf{B} \right) \cdot d\boldsymbol{\ell} \tag{2}$$

around the closed path C denoted by the sequence of points 1-2-3 $\cdots$ 8-1 shown in Figure 1. As a first approximation, we assume that the electrical resistance of the matter along this path is negligible.

Noting that $\mathbf{v} \times \mathbf{B} \cdot d\boldsymbol{\ell} = 0$ wherever C follows the poloidal field, we find

$$\mathcal{E} = -\frac{1}{c}\int_A \frac{\partial B_\phi}{\partial t}\, dA + \frac{1}{c}\int_1^2 (\Omega_s \times r)\times\mathbf{B} \cdot d\boldsymbol{\ell} + \frac{1}{c}\int_5^6 (\Omega_d \times r)\times\mathbf{B} \cdot d\boldsymbol{\ell}\,. \tag{3}$$

The first term on the right of this equation comes from the line integral of the electric field around C and describes the back-EMF produced by the inductance of the circuit. The second and third terms come from the line integral of $\mathbf{v} \times \mathbf{B}/c$. We call the sum of the latter two terms the "generated EMF" and denote it $\mathcal{E}_g$.

If the neutron star is slowly rotating, the third term in equation (3) dominates the second and $\mathcal{E}_g \sim \varpi_0^2\, \Omega_K(\varpi_0)\, B_p(\varpi_0)/c \sim 10^{15}\, \varpi_8^{-5/2}\, \mu_{30}\, (M/M_\odot)^{1/2}$ Volts, where $\Omega_K(\varpi_0)$ and $B_p(\varpi_0)$ are the Keplerian angular velocity and the poloidal magnetic field strength at the inner edge of the Keplerian flow. Thus, even if the neutron star is slowly rotating, its interaction with the disk generates very large EMFs.

The large EMFs drive electrical currents between the disk and the star. Although the details of the current distribution depend on the model, all produce current patterns qualitatively similar to the pattern shown in the top panel of Figure 2. The bottom panel shows an equivalent circuit that includes the electrical resistance of the

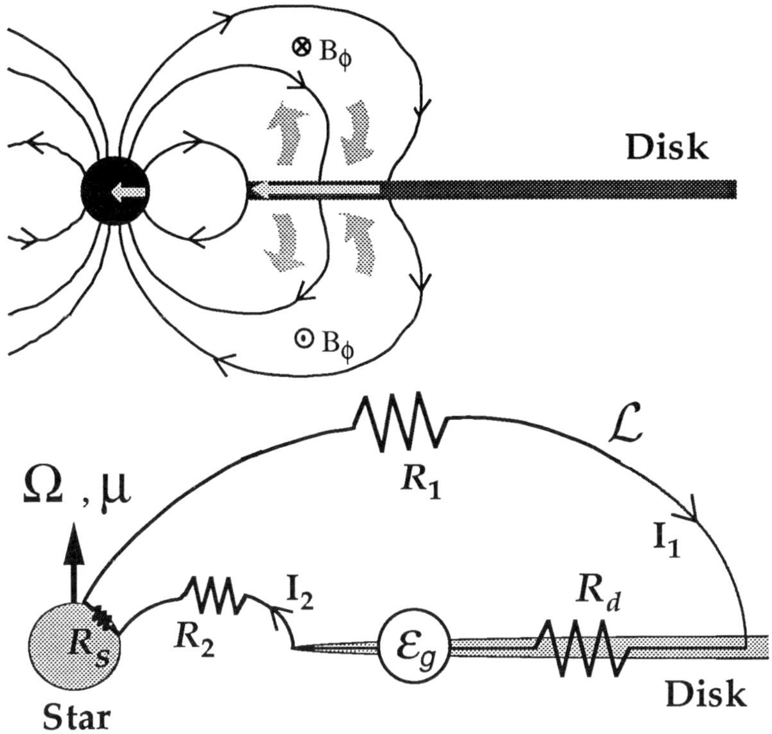

Fig. 2—*Top*: Orbiting motion of the disk plasma generates an EMF that drives cross-field currents (light shaded arrows) within the disk and the neutron star and field-aligned currents (shaded arrows) within the magnetosphere. *Bottom*: Equivalent circuit, showing the generated EMF $\mathcal{E}_g$, the circuit inductance $\mathcal{L}$, and the resistances $\mathcal{R}_s$ and $\mathcal{R}_d$ of the star and disk. Also shown are the resistances $\mathcal{R}_1$ and $\mathcal{R}_2$ and the currents I_1 and I_2 in the two magnetospheric legs of the circuit.

disk, star, and magnetosphere. The time development of the current in this circuit is described by the equation

$$\mathcal{E}_g = \mathcal{L}\frac{dI}{dt} + I\mathcal{R} . \tag{4}$$

The term on the left describes the EMF generated by the disk–star interaction whereas the first term on the right describes the effect of the circuit inductance $\mathcal{L} \sim r^2/c$. The second term on the right describes the effect of the total resistance $\mathcal{R}$ in the circuit.

Equation (4) shows that an initially small current grows linearly with time, corresponding to the linear increase of the azimuthal magnetic field B_ϕ as the field is twisted by the differential rotation of the disk and star. If no other processes intervened, the growth would slow after a time $\sim\mathcal{L}/\mathcal{R}$, and the current would asymptotically approach the value $\mathcal{E}_g/\mathcal{R}$. Our calculations of the resistance $\mathcal{R}$ of the circuit show that it is very small and hence that the asymptotic current would be very large, producing a magnetic pitch $\gamma_\phi \equiv B_\phi/B_p \sim 10^{10}$. We now show that long before the current reaches this value, microscopic and/or MHD instabilities will intervene, interrupting the growth of the current and the twisting of the magnetic field.

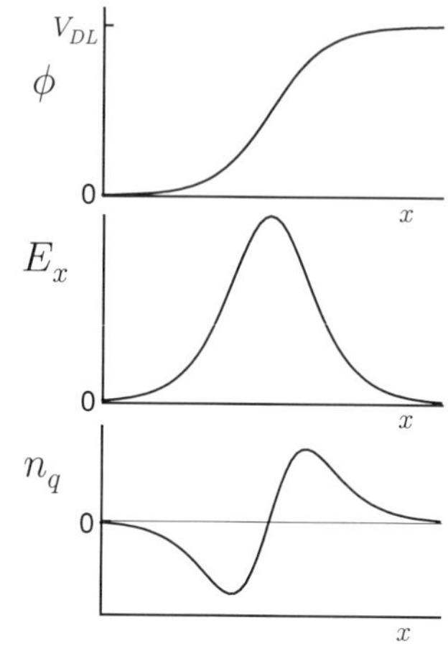

Fig. 3—Example of a stationary double-layer solution, showing the electrical potential ϕ, normal electric field E_x, and charge density $n_q \equiv n_+ - n_-$. Charge separation is maintained by a balance between inertial and electrostatic forces. Although quasi-neutrality is violated at both edges of the layer, the spatially integrated charge is almost zero, and hence the electric field outside the layer is very small. Such a layer produces approximately monoenergetic electrons and protons, since in moving across the layer both attain energies $\sim eV_{DL}$.

4 Instabilities

If the current density in the magnetosphere is high enough, it will drive micro-instabilities. Instability will surely occur if the current density exceeds $j_{\rm micro} = en_e(k_BT/m_e)^{1/2}$, which is many orders of magnitude less than the density of the asymptotic current discussed above; micro-instability may occur at a lower current density. Micro-instabilities may produce anomalous resistance, heating, and, possibly, relativistic double layers (see Spicer 1982). The magnetic field will, as a result, slip through the plasma in the transition region, either steadily or episodically.

Double layers, which are observed in both laboratory and space plasmas, are non-linear BGK-like waves that develop when large currents flow in response to large applied voltages (see Block 1978; Carlqvist 1979). These structures, which consist of two equal but oppositely charged layers, form in unstable plasmas because the ponder-motive force tends to increase the charge separation. An example of a double-layer solution of the Vlasov equation is shown in Figure 3. In the magnetospheres of accreting neutron stars, we expect double layers to be not only strong (voltage $V_{DL} \gg k_BT$), but relativistic ($V_{DL} \gg m_pc^2 \gg m_ec^2$). Although the properties of relativistic double layers are still being explored, several features seem assured.

For self-consistency, the voltage across the layer must be $\approx (\pi m_p cjd^2/e)^{1/2}$, where j is the current density and d is the width of the layer. Electrons and protons crossing the layer are accelerated to energies $\sim eV_{DL}$, and the power that goes into these fast particles is

$$P_{RDL} = \int_V \mathbf{j} \cdot \mathbf{E}\, dV \sim 10^{37} j_{10} V_{14} R_8^2 \text{ ergs s}^{-1} , \qquad (5)$$

where j_{10} is the current density in units of 10^{10} statcoul cm^{-2} s^{-1}, V_{14} is the voltage across the layer in units of 10^{14} Volts, and R_8 is the radius where the layer is located in units of 10^8 cm. As the current density approaches $j_{\rm max} = enc$, the layer will be disrupted. If the layer explodes, the energy ($\sim \mathcal{L}I^2/2$) stored in the circuit will be released in a time $\Delta t \sim R/c \sim$ ms, producing an EMF $\mathcal{E}_{\rm expl} \sim 10^{15}$ Volts and a luminosity in fast particles $\sim 10^{41}$ ergs s^{-1}.

Twisting of the stellar magnetic field will lead to reconnection, if the magnetic pitch becomes sufficiently large ($\gamma_\phi > \gamma_{\rm crit} \sim 10$; see Ghosh and Lamb 1991). The corresponding critical current density is $j_{MHD} \approx (c\gamma_\phi B_p)/(4\pi R_B)$, where R_B is the field-line radius of curvature, many orders of magnitude less than the asymptotic current density. In regions of the magnetosphere where $n \lesssim 10^{10}$ cm^{-3}, B^2 is $\gg 4\pi\rho c^2$ and reconnection events will be highly relativistic, accelerating the bulk flow to relativistic speeds (electron energies $\sim 10^8$ eV and proton energies $\sim 10^{11}$ eV). A small fraction of the particles will reach energies comparable to $e\mathcal{E}_R$, where $\mathcal{E}_R \sim 10^{15}$ V is the EMF generated in the reconnection event.

The rotation of the disk plasma relative to the star continually twists the magnetic field, steadily increasing the field-aligned currents in the magnetosphere. Instead of "shorting out" the EMF and thereby preventing acceleration of particles, as some authors have argued (see, *e.g.*, Wang 1986; Katz and Smith 1988; Mitra 1991), the low internal resistance of the disk–star generator circuit allows these currents to become large enough to trigger processes that accelerate particles. Whether micro-instability or reconnection occurs first depends on the particle density in the most tenuous regions of the magnetosphere. If this particle density is $\lesssim 10^9$ cm^{-3}, field-aligned currents are likely to trigger micro-instabilities and possible double layer formation before reconnection occurs. If instead the lowest particle density is $\gtrsim 10^9$ cm^{-3}, reconnection is likely to occur before the threshold for micro-instability is reached. In some circumstances, the current may continue to grow even after double layers have formed, eventually triggering a reconnection event.

5 Particle Acceleration and Radiation

Suppose there is a strong electric field E in the magnetosphere, produced by a double layer, a reconnection event, or some other process. Relativistic charged particles accelerated by this field will gain energy at a rate $\dot{\epsilon}_{\rm gain} = ecE$. Near the star the magnetic field is $\sim 10^8$–10^{12} G, charged particles lose energy rapidly via synchrotron or curvature radiation, and high energy photons form pairs. For this reason, conditions are more favorable for particle acceleration and escape of high energy photons at distances $\sim 10^7$–10^8 cm from the star, where the field strength is $\lesssim 10^7$ G and the particle density $\lesssim 10^{14}$ cm^{-3}. We therefore consider processes in this region, focusing on results that are independent of model details.

In accretion-powered systems, X-ray photons are copious. A relativistic electron moving through photons of energy ε_γ loses energy via scattering at a rate $\dot{\epsilon}_{\rm loss} \approx n_\gamma c\sigma_{\rm sc}\Delta\varepsilon_\gamma$, where n_γ is the photon number density, $\sigma_{\rm sc}$ is the scattering cross section, and $\Delta\varepsilon_\gamma$ is the average increase in the photon energy in a scattering. Using the Thomson value for $\sigma_{\rm sc}$, Cheng and Ruderman (1991) concluded that this 'inverse Compton drag' prevents electrons from becoming ultrarelativistic. Using instead the exact (Klein-Nishina) expression for $\sigma_{\rm sc}$, we find $\dot{\epsilon}_{\rm gain} \sim 100\,\dot{\epsilon}_{\rm loss}$ for the conditions considered by Cheng and Ruderman. For photon luminosities $\lesssim 10^{38}$ ergs s^{-1}, mean photon energies $\gtrsim 1$ keV, radii $\gtrsim 10^7$ cm, and electric fields $\gtrsim 10^3$ statvolts cm^{-1}, gain exceeds loss at all energies of interest. Stated differently, the probability that an electron will scatter a photon in accelerating across the entire system is less than unity under these conditions. Thus, except in highly luminous systems with weakly magnetic neutron stars, the energies of electrons in the outer magnetosphere are not limited by inverse Compton drag.

Charged particles will follow the magnetic field lines, since the gyroradius is small compared to the radius of curvature of field lines even for particles with the highest energies considered here. The energies of electrons accelerated along the magnetic field are not limited by synchrotron radiation but, under typical conditions, are limited to $\lesssim 10^{13}$ eV by curvature radiation. We conclude that, contrary to previous statements, electrons in the outer magnetosphere can be accelerated to energies $\sim 10^{12}$ eV.

An important question is whether curvature radiation can create a pair avalanche. Typical curvature radiation photons have enough energy to produce pairs if γ_e exceeds 10^6. Relativistic reconnection events do not last long enough for a pair avalanche to develop, but relativistic double layers may produce avalanches of $\sim 10^4$–10^5 pairs, through $\gamma\gamma \to e^-e^+$ and $\gamma e \to e^-e^+e$. As the pair density exponentiates, the current through the double layer also exponentiates, causing the layer to break up or explode.

Double layers can form in both legs of the magnetospheric current loop, accelerating protons toward the disk in one leg and electrons in the other. Thus, we expect them to accelerate both protons and electrons toward the disk, regardless of whether the star's magnetic axis is parallel or antiparallel to its spin axis.

We estimate that curvature radiation by electrons accelerated along magnetic field lines may produce a flux $\sim 10^{-8}$ photons cm^{-2} s^{-1} at 1 kpc. This curvature radiation (and synchrotron radiation, if a pair avalanche develops) might be detectable by the experiments on *GRANAT* and the *Compton Observatory*. Electrons accelerated toward regions of relatively dense plasma, such as the disk, will produce bremsstrahlung. If the spectrum extends up to $\sim 10^{12}$ eV, the flux may be $\sim 10^{-11}$ photons cm^{-2} s^{-1} at 1 kpc. Hence, this radiation may be detectable by VHE experiments.

As discussed by previous authors, ultrarelativistic protons can produce high energy γ-rays via a variety of processes, including $p+\gamma \to \pi_0 \to 2\gamma$ and, in dense plasma, $p+p \to \pi + X$ followed by decay of the π to γ-rays (see Slane and Fry 1989). The flux may be as large as $\sim 10^{-11}$ photons cm^{-2} s^{-1} at 1 kpc. Hence, this radiation may also be detectable by VHE experiments.

The luminosity of these radiations cannot exceed the accretion power at the acceleration radius, which is $\sim GM\dot{M}/\varpi_0$, or $\sim R_s/\varpi_0$ times the luminosity produced near the neutron star surface; here R_s is the neutron star radius. Some previous authors (*e.g.*, Király and Mészáros 1988) have used this bound to argue that acceleration powered by the accretion disk could not explain the high energy γ-ray luminosities being reported at the time. However, more recent reports (Macomb *et al.* 1991; Reynolds *et al.* 1991) suggest that the luminosities of these systems may be much smaller most of the time. If so, a substantial fraction of the luminosity may be produced by the processes discussed here.

This research was supported in part by NSF grant PHY 91-00283 and NASA grant NAGW 1583 at the University of Illinois.

References

Bai, T. 1980, *Ap. J.*, **239**, 328.
Block, L. P. 1978, *Ap. Space Sci.*, **55**, 59.
Brecker, R. H., *et al.* 1977, *Ap. J.*, **214**, 879.
Carlqvist, P. 1979, in *Wave Instabilities in Space Plasmas*, ed. P. J. Palmadesso and K. Papadopoulos (Dordrecht: Reidel), p. 83.
Cheng, K. S., and Ruderman, M. 1991, *Ap. J.*, **373**, 187.
Ghosh, P., and Lamb, F. K. 1991, in *Neutron Stars: Theory and Observation*, ed. J. Ventura and D. Pines (Dordrecht: Kluwer), p. 363.
Goldreich, P., and Julian, W. H. 1969, *Ap. J.*, **157**, 869.
Hamilton, R. J., Lamb, F. K., and Miller, M. C. 1992, in preparation.
Katz, J. I., and Smith, I. A. 1988, *Ap. J.*, **326**, 733.
Király, P., and Mészáros, P. 1988, *Ap. J.*, **333**, 719.
Macomb, D. J., *et al.* 1991, *Ap. J.*, **376**, 738.
Miller, M. C., Lamb, F. K., and Hamilton, R. J. 1992, in preparation.
Mitra, A. 1991, *Ap. J.*, **370**, 345.
Reynolds, P. T., *et al.* 1991, *Ap. J.*, **382**, 640.
Slane, P., and Fry, W. F. 1989, *Ap. J.*, **342**, 1129.
Spicer, D. S. 1982, *Space Sci. Rev.*, **31**, 351.
Wang, Y. M. 1986, *Ap. Space Sci.*, **121**, 193.

Searching for Millisecond Pulsars

John Middleditch,[1] Will Deich,[2] and Shri Kulkarni[2]

[1]Los Alamos National Laboratory
[2]Department of Radioastronomy, Caltech, Pasadena, CA

Abstract

Our understanding of the luminosity function of pulsars, and of millisecond pulsars in particular, depends upon the techniques used to search for them, yet little has been written about the process by which peaks are selected from the Fourier domain. This discussion will detail one such process and present the recent discovery of the 6th and fastest millisecond pulsar discovered in the globular cluster, M15.

1 Introduction

With the exception of a few discoveries of slow and bright periodicities, such as the 71 s periodicity discovered in DQ Her by Walker (1956), and the 685 s periodicity in 4U1820-30 by Stella, Priedhorsky, & White (1987), both of which were revealed by a strip chart recorder, the search for periodicities in astrophysical objects involves the use of very large Fourier transforms and the subsequent testing on the resulting Fourier amplitudes for "significant" patterns. A discussion of this issue has only recently been offered by Kulkarni (1991). This paper will use the same terminology, but will address the problem from a different perspective.

2 Dealing with the Fourier Transform:

2.1 The Response of the Fourier Transform

The major problem that must be addressed is that the response of the Fourier transform is not uniform with frequency, but is "scalloped" with the fine frequency scale (~1/T, where T is the time span of the data which was Fourier transformed. This effect can be seen by taking the equation for a Fourier transform of the data sequence, $(n_j : j = 0, \cdots, N-1)$,

$$A_k = \sum_{j=0}^{N-1} n_j \, e^{-2\pi i \, jk/N}, \tag{1}$$

Substituting for n_j with the inverse transform and rearranging the order of summation gives

$$A_k = 1/N \sum_{\ell=0}^{N-1} A_\ell \sum_{j=0}^{N-1} e^{-2\pi i \, j(k-\ell)/N}. \tag{2}$$

The summation on the far right hand side of equation (2) can be computed, with the result:

$$A_k = 1/N \sum_{\ell=0}^{N-1} A_\ell \, e^{-i\pi(k-\ell)(1-1/N)} \frac{\sin[\pi(k-\ell)]}{\sin[\pi(k-\ell)/N]}. \tag{3}$$

When the total number of equally-spaced data points, N, is large, the expression can be reduced to

$$A_k = \sum_{\ell=0}^{N-1} A_\ell \, e^{-i\pi(k-\ell)} \frac{\sin[\pi(k-\ell)]}{\pi(k-\ell)}. \tag{4}$$

This reduction can also be easily verified by using the Convolution Theorem. With 'k' still an integer subscript in equation 4, the fact that the dominant (and only) contribution to A_k from the sum on the right hand side comes from the local Fourier neighborhood of A_k, is simply trivial. However, there is no reason that 'k' cannot be changed to a continuous subscript, say 'r'. In this case, the dominant contribution to A_r still comes from the locally neighboring Fourier amplitudes, or

$$A_r \sim \sum_{\ell=[r]-m}^{[r]+m} A_\ell \, e^{-i\pi(r-\ell)} \frac{\sin[\pi(r-\ell)]}{\pi(r-\ell)}. \tag{5}$$

In equation (5) the integer 'm' does not need to be large for the expression to hold with decent accuracy; 25 is certainly sufficient, whereas '$[r]$' denotes the nearest integer function on 'r', or some such thing. Equation (5) also points out the fundamental difficulty in detecting periodicities with Fourier analysis, which occurs when the signal frequency, 'r', strays farther and farther from integral values: the significance of any single contributing Fourier amplitude on the right hand side is reduced. The worst case comes when 'r' is half-integral, and then the largest contribution from the points on the Fourier lattice, output by the digital Fourier transform, is down by a factor of $2/\pi$.

2.2 The Use of Interbinning

Nor is it feasible to use equation (5) to "fine bin" by a some factor ≥ 2 with the integer 'm' set to some small number still $\gg 1$ like 7, as the computation is just too expensive (though massively parallel computers might change this). The simplest thing which can be done without too much expense, is to use each consecutive pair of Fourier elements in a two-element interpolation to estimate the Fourier spectrum midway between the lattice frequencies:

$$A_{k+\frac{1}{2}} \propto (A_k + A_{k+1} e^{i\pi * t/T})/(x * \sqrt{2}). \tag{6}$$

The worst case loss of signal-to-noise ratio for this interpolation is less than 10%, much better than the $1 - \frac{2}{\pi}$ fraction lost for no interpolation at all. If it weren't for the fact that most Fourier transforms are done on data sets that have been padded, to some degree, with the mean data value at one end or another, equation (6) would reduce to the simple subtraction of A_{k+1} from A_k. However, two correction factors are needed: the first, t/T, serves for a data duration of the *first* t of time, out of the total time T, spanned by the set actually Fourier transformed. Thus the phase drop from one Fourier element to the next is *less* than 180 degrees (for padding at the *beginning* of the data, it would be *greater* than 180 degrees). The second factor, 'x', is near unity but must increase as t/T gets smaller. In this instance, the data modulation pattern is more widely spaced than the sinc function modulation of equation (5), and the consecutive Fourier amplitudes become more correlated with decreasing t/T.

This technique is named "interbinning" and has been in use since at least the late 1970's (see, *e.g.*, the appendix of Middleditch and Cordova 1982), and by the early '80's was the subject of some literature (see, *e.g.*, Cullers *et al.* 1984), possibly as a response to the use of the more pessimistic, 'non-interbinned', sensitivity by Leahy *et al.* (1983).

Thus peaks are selected, by the "cheap" method of equation (6) for the full "expensive" analysis of equation (5), but after the selection of equation (6) has been made, the further analysis is no longer very expensive because of the vastly reduced number of candidates.

2.3 A Fine-Tuned Method of Search

In addition, equation (5), or rather its resultant value for power, $|A_r|^2$, and phase, $\tan^{-1}(Im(A_r)/Re(A_r))$ may be differentiated *three* times with respect to Fourier frequency ('r') to yield vital information about the nature of the periodicity. The most important of these is the determination of the exact frequency of the peak in the power spectrum, by finding the location of the zero in the derivative of power with respect to frequency. The accuracy of the measurement in frequency is given by

$$\delta f = \frac{3}{\alpha\pi\sqrt{6PT}}, \tag{7}$$

(see, *e.g.*, Middleditch, 1976; Middleditch & Nelson 1976). Here P is the ratio of the power in the peak to the local mean power level, and α is another of the set of the extended parameters, near unity, and defined below ($\alpha T = t$ is the *effective* duration of the *pulsations*, though for padding in the *middle* of the run, α becomes >1). Note that this error is *smaller* than the traditional $0.5/t$ by approximately the significance, σ, where $\sigma \sim \sqrt{2P - \ln(\pi P)}$ is the (equivalent Gaussian) significance of the peak.

The advantage of this technique is that once these more "exact" frequencies are defined for all the events whose lattice, or interbinned, Fourier amplitudes exceed a certain set level, then a search can be made of the power spectrum at frequencies which are *exactly* in harmonic ratios to such frequencies, thanks to the use of equation (5) in the extraction of powers and phases.

Typically, both integral and half integral harmonics are searched, since the strongest peak of a double-peaked pulsar such as the Crab, 1937+21, or 1957+20, is that due to the 2nd harmonic structure ($2\times f$). Such a search is far more effective than simply a blind fold of the Fourier lattice amplitudes, as these folds quickly lose their effectiveness because of the rapid growth in the number of power spectrum elements examined. Yet another important feature of such "interpolated" searches is that the phases of the harmonics are evaluated at frequencies which are in true harmonic ratios. Since most pulsar pulse profiles produce phases in the same harmonic ratio (*e.g.*, 87.40 degrees for the fundamental and 174.69 degrees for the second harmonic), further "tuning" of such searches is also possible, and in fact, desirable.

2.4 The Further Extraction of Information

There are five other details about the nature of the pulsations which can be derived from the first three derivatives of the power and phase with respect to (wrt) Fourier frequency of a peak defined by equation (5) (out of six total we have only used one, the derivative of Power wrt frequency, to accurately define the pulse frequency). The first derivative of the *phase* wrt the frequency gives the time centroid of the pulsations corresponding to a given Fourier peak:

$$c(\equiv centroid) = \frac{-1}{2\pi T}\partial\phi/\partial f; \quad < c >= 0.5 \pm \frac{1}{\sqrt{24P}}. \tag{8}$$

The centroid parameter, 'c', tells one where the pulsations are located within the transformed time series. Pulsations present only in the first half of the series have a centroid of 0.25; those only in the last half would have a centroid of 0.75. This relationship is essentially the famous "drift" equation, familiar to anyone who has synchronously folded pulsar data with a slightly mismatched folding frequency. This parameter and the α parameter as defined below indicate whether subdivision of the time series will result in an increase in signal-to-noise in any of the smaller partitions of data. Thus these enable the user to avoid much useless data hashing.

The second derivative of the Fourier power wrt frequency is a measure of how sharp the Fourier peak is, which, in turn, is a measure of the root-mean-square duration of the pulsations in time. Pulsations that are present near the beginnings and ends of the time series have sharper Fourier peaks than those present only near the center of the time series. A parameter to measure this, α, is defined by

$$\alpha = \frac{1}{\pi T}\sqrt{\frac{3}{2P} \times (-\partial^2 P/\partial f^2)}; \quad < \alpha >= 1.0 \pm \frac{1.}{\alpha\sqrt{10P}}. \tag{9}$$

Since sidelobe peaks have an expectation value for α of $\sqrt{3}$, this parameter is useful in eliminating them from consideration during peak and harmonic searches.

The sixth parameter (counting also power and phase) is derived from the second derivative of the phase wrt frequency. One can prove that this parameter can be used to estimate the actual linear drift of the pulse frequency over the run. This parameter is useful to indicate whether it might be profitable to use the analysis of Figure 1 to search for a moving peak (however, usually by this time, the pulsation will appear at more than one frequency in the transform, and further analysis methods will be called for anyway). The parameter which estimates the $\partial f/\partial t$ is given by:

$$z \equiv T^2\partial f/\partial t \sim \frac{4\pi\alpha^3 T^2}{\partial^2\phi/\partial f^2}\left(1 - \sqrt{1 + \frac{45(\partial^2\phi/\partial f^2)^2}{2\pi^4\alpha^6 T^4}}\right), \tag{10}$$

$$z \sim \frac{-45}{\alpha^3\pi^3 T^2}\partial^2\phi/\partial f^2 \pm \frac{1}{\pi\alpha^2}\sqrt{90/P} \qquad (z \leq 3). \tag{11}$$

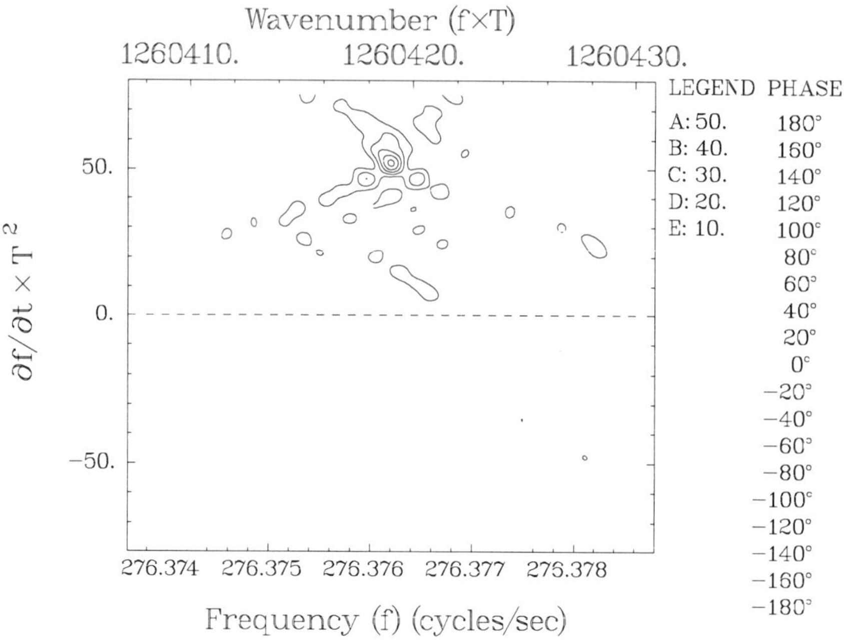

Fig. 1.—The Fourier power from a 75 minute Arecibo Observatory run on the globular cluster, NGC6760, near the 3.6 ms pulsar frequency, 276.37 Hz, is contoured on the frequency-$\partial f/\partial t$ plane. The feature, with about 50 times mean power in the upper half of the frame, is due to a large, apparent *positive* $\partial f/\partial t$ of the 3.6 ms binary pulsar within this cluster when it is very close to its superior conjunction (companion in front of pulsar at the middle of the 1 hr 15 min observation interval). The characteristic "pole" spacing of $6\sqrt{10}/\pi$ in $T^2\partial f/\partial t$ is evident around the peak. These results are from Kulkarni, Deich, Middleditch, Anderson *et al.* 1992.

One can also prove that the *second* derivative of the pulsation frequency wrt time can be estimated using the third derivative of the Fourier power wrt frequency to a precision given by

$$w \equiv T^3\partial^2 f/\partial t^2 \sim \frac{4\pi^2\alpha^4 PT^3}{\partial^3 P/\partial f^3}\left(1 - \sqrt{1 + \frac{945(\partial^3 P/\partial f^3)^2}{\pi^6\alpha^8 P^2 T^6}}\right), \tag{12}$$

$$w \sim \frac{-1890}{\alpha^4\pi^4 PT^3}\partial^3 P/\partial f^3 \pm \frac{6\sqrt{105}}{\pi\alpha^3\sqrt{P}} \quad (w \leq 20). \tag{13}$$

Note that the formulae in equations (10) and (12) for z and w limit their range to maximum (pole) values of $6\sqrt{10}/\pi$ and $12\sqrt{105}/\pi$, beyond which, extrapolation is impossible (one can no longer use purely *local* amplitudes and their derivatives; convolution with other, neighboring Fourier amplitudes is necessary—see Figs. 1 and 2).

Finally, the non-sinusoidal third moment of the pulsation amplitude vs time is believed to be estimated by the third derivative of phase wrt frequency, though systematic effects dependent on α and c need to be removed.

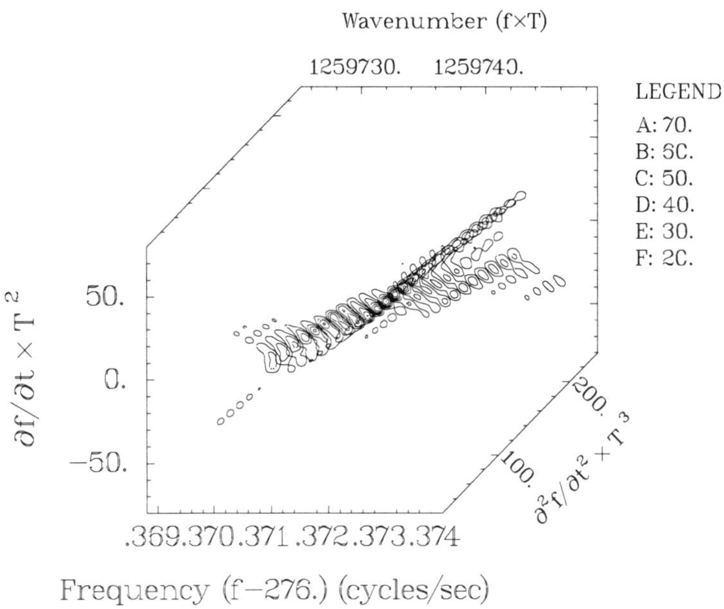

Fig. 2—The Fourier power from the same pulsar (but different orbital phase) as Figure 1, contoured now in the f-$\partial f/\partial t$-$\partial^2 f/\partial t^2$ cube. A peak with more than 70 times mean power is now seen to be well-formed in this space for a value of $\partial^2 f/\partial t^2 \times T^3$ near 130. The large, apparent *positive* $\partial f^2/\partial t^2$ is due to the reaction motion of the 3.6 ms binary pulsar when it is very close to its receding quadrature (at the middle of the 75 minute observation interval). Denser representations of such data are possible when the $\partial^2 f/\partial t^2$ axis (and contour families) are color coded. The characteristic "pole" spacing of $12\sqrt{105}/\pi$ in $T^3\partial^2 f/\partial t^2$is evident for this peak. These results are from Deich, Kulkarni, Middleditch, Anderson, *et al.* 1992.

3 A Pulsar Search

Solitary millisecond pulsars are known to be very stable clocks, with frequency derivatives typically of order 3×10^{-13} Hz/s. Thus the time span over which such pulsars can be observed is the time, T, for which $z \leq 2$, at which point over 10% of the signal-to-noise ratio is lost due to the mismatch of the (uncorrected) time series to the actual pulsar $\partial f/\partial t$. This constraint implies that $T \leq 10^6$ s, or about two weeks.

The best *location* to look for ms pulsars is in the globular clusters, which are known to contain at least many thousands of such objects, if not even more (see, *e.g.*, Lyne *et al.* 1987). Moreover, assuming that a ms pulsar lies within a core radius, or within about 30 arc s from the center of the globular cluster, does not seriously affect such searches, as the error introduced in the ephemeris is not large.

Thus the obvious way to invoke a state-of-the-art pulsar search was to observe decent globular clusters for 1-2 weeks and then Fourier transform the resulting, multi-billion point data set. Billion-point FFT's have been done as early as February 1990 in the search for the pulsar remnant of Supernova 1987A (see, *e.g.*, Perlmutter *et al.* 1990, and Kerrick *et al.* 1991). For the data set taken last October at Arecibo, the time span was 10 days, and the sampling frequency was nearly 2,000 Hz, so 2-billion point FFT's were needed.

The Arecibo data consisted of nine ~ 2-hour transits of data over a 10-night span. A point to note about such long transforms with a substantial percentage of padded data is that the number of degrees of freedom is not given by the number of points spanned by the Fourier transform, but by the number of *actual* data points it contains. Since Arecibo data is also weighted by a factor, $(gain/T_{sys})^2$, which peaks in mid-transit and drops to zero at the ends, the situation is even more extreme. A typical 2-billion point transform of some 140 million or so original data points will really only have about 20 million degrees of freedom, so that by the time a Fourier peak rises above 20 times the mean local power, it is probably caused by some real source.

Thus, when all of the M15 data taken in October of '91 was correctly connected together and Fourier-transformed, in addition to the first five known pulsars discovered in this cluster at 9.03, 17.815, 32.768, 208.21, and 214.99 Hz, a sixth pulsar, the fastest yet discovered in this cluster, appeared at 248.3 Hz, with nearly 29 times mean power. The section of output with this discovery is included below in Figure 3 to give an example of how the Fourier parameters discussed above appear in a real sample (α is labeled as "purity").

```
1  bin number   apparent frequency  power/ sigma  bin 1 phase  pulse frc  centroid purity of  est. df/dt 3rd mom    ddf/dtt     local
   with signal  of this signal and   local   in    at start +   of total  of pulse pulsation  cyc/run**2            cyc/run**3  poisson
  in transform   its error (hertz)   power  peak   error (deg)  + error    +error  (1=const)   est err     est err   est err    levels

    194018508.015  247.3819222.000000   12.94  4.71    47.1 11.3 .000001.191 .478.057 0.903.097 1.04 1.03 14.1  6.7  31.7 5.4  0.0229
    194755101.710  248.3211107.000000   17.47  5.56   -24.5  9.7 .000001.195 .522.049 1.102.069 1.31 0.60 25.5  3.9   0.5 4.7  0.0247
    194755110.796  248.3211223.000000   20.55  6.08   -38.8  8.9 .000001.194 .511.045 1.103.063 1.18 0.55 25.2  3.5   8.8 4.3  0.0247
  % 194755119.883  248.3211339.000000   23.34  6.51$ -53.4  8.4 .000001.194 .502.042 1.101.059 1.07 0.52 24.7  3.3  12.2 4.1  0.0247
  % 194755128.973  248.3211455.000000   25.69  6.86$ -68.4  8.0 .000001.193 .496.040 1.095.057 0.99 0.50 24.3  3.2  13.2 3.9  0.0247
  % 194755138.067  248.3211571.000000   27.46  7.10$ -84.0  7.7 .000001.193 .493.039 1.088.055 0.92 0.49 23.8  3.2  12.2 3.7  0.0246
  % 194755147.163  248.3211687.000000   28.58  7.26$-100.0  7.6 .000001.192 .492.038 1.080.055 0.86 0.48 23.5  3.1  10.9 3.7  0.0246
  % 194755156.263  248.3211803.000000   28.97  7.31$-116.8  7.5 .000001.192 .493.038 1.071.055 0.77 0.49 23.1  3.2   8.0 3.6  0.0245
  % 194755165.366  248.3211919.000000   28.62  7.26$-134.1  7.6 .000001.192 .495.038 1.063.056 0.68 0.50 22.8  3.2   6.4 3.7  0.0245
  % 194755174.472  248.3212035.000000   27.53  7.11$-151.9  7.7 .000001.192 .497.039 1.054.057 0.57 0.52 22.5  3.4   5.0 3.7  0.0245
  % 194755183.582  248.3212151.000000   25.90  6.89$-170.4  8.0 .000001.193 .498.040 1.046.059 0.44 0.54 22.2  3.5   4.5 3.8  0.0245
  % 194755192.694  248.3212267.000000   23.88  6.59$-189.3  8.3 .000001.193 .498.042 1.042.062 0.30 0.57 22.0  3.7   4.8 4.0  0.0245
  % 194755201.807  248.3212383.000000   21.60  6.24$ 152.1  8.7 .000001.194 .497.044 1.043.065 0.16 0.60 22.0  3.9   4.8 4.2  0.0245
    194755210.917  248.3212500.000000   19.10  5.84   134.3  9.3 .000001.194 .495.047 1.046.069 0.04 0.63 22.0  4.1   3.2 4.5  0.0245
    194755220.021  248.3212616.000000   16.39  5.37   117.9 10.0 .000001.195 .492.050 1.051.074-0.07 0.68 22.0  4.4  -0.8 4.8  0.0245
    194755229.118  248.3212732.000000   13.50  4.82   103.4 11.0 .000001.197 .489.056 1.053.082-0.17 0.74 21.9  4.8  -6.5 5.3  0.0245
    197188233.024  251.4234575.000000   13.00  4.72   -83.5 11.2 .000001.194 .503.057 1.246.070-0.27 0.54 31.5  3.5  30.4 5.4  0.0235
```

Fig. 3—A section from the output of the 2-billion point Fourier transform which detected PSR2127+11F at 248.3 Hz. The 9.1-cycle spacing of "equivalent" candidate frequencies is due to the nature of the observing "window" at Arecibo Observatory, *i.e.*, 9×(~2-hours "on" and ~ 22-hours "off") + one "on".

This work was carried out under the auspices of the Department of Energy.

References

Cullers, D. K., Oliver, B. M., Day, J. R. & Olsen, E. T. 1984, *SETI Science Working Group Report* NASA Tech. Report 2244.

Kerrick, A. *et al.* 1991, "Search for Pulsed Gamma-Rays > 0.75 Mev from SN1987A," in *Proceedings, 22nd International Cosmic Ray Conference*, Dublin, August 1991.

Kulkarni, S. R. 1991, *in preparation.*

Kulkarni, S. R., Deich, W. T. S., Middleditch, J., Anderson, S. B. *et al.* 1992, in preparation.

Leahy, D. A., Darbro, W., Elsner, R. F., Weisskopf, M. C., Sutherland, P. G., Kahn, S. & Grindlay, J. E. 1983, *Ap. J.*, **266**, 160.

Lyne, A. G., Brinklow, A., Middleditch, J., Kulkarni, S. R., Backer, D. C. & Clifton, T. R. 1987, *Nature* **332**, 45.

Middleditch, J. & Cordova, F. A. 1982, *Ap. J.* **255**, 585 (Appendix).

Middleditch, J. & Nelson, J. E. 1976, *Ap. J.* **208**, 567.

Middleditch, J. 1976, *Ph. D. Thesis Appendix*, LBL Report No. 3639, (more extensively edited copies available from JM upon request).

Perlmutter, S., *et al.* 1990, "Search for a Near-infrared Pulsar in SN1987A," at the ESO/EIPC Workshop, SN1987A and other Supernovae, Marciana Marina, Isola d'Elba, France, Sep. 17-22.

Stella, L., Priedhorsky, W. & White, N. E. 1987, *Ap. J. (Lett.)*, **312**, L17.

Walker, M. F. 1956, *Ap. J.*, **123**, 66.

CGRO Timing Analysis: BATSE-Specific Effects

Robert B. Wilson,[1] Mark H. Finger,[2] Gerald J. Fishman,[1]
Charles A. Meegan,[1] William S. Paciesas[3]

[1]Space Science Laboratory, NASA/Marshall Space Flight Center
[2]Computer Sciences Corporation, NASA/MSFC
[3]Dept. of Physics, University of Alabama in Huntsville

Abstract

The methods of processing BATSE data for pulsar analyses are discussed, along with systematic effects that occur for large area, uncollimated detectors.

1 Introduction

The analysis of BATSE data for detection and measurement of pulsars uses the standard approach as described in Lyne and Graham-Smith (1990) of correcting the time of detection of the interacting photons to the time they would have arrived at the Solar System barycenter (SSB), before searching for a periodic signal. There are at least two complications in this process that are intrinsic to large area, uncollimated detectors of the BATSE type: counting rates too high to permit individual photon time-tagging, and scattering of source photons from the Earth's atmosphere, which introduces a time delay for some of the interacting photons. The high count rates necessitate binning of the data for times short compared to the source period, or folding the data on-board at the approximate period. The magnitude of the phase dispersion produced by the binning and the scattering effects are the subject of this paper.

2 Analysis Sequence

Briefly, the processing of the BATSE folded-on-board data consists of the following sequence of operations:

- Obtain the start time, T_s, of an on-board accumulation (latched GRO clock).
- Compute the midpoint time, $T_{mid} = (T_s + P_f \times N_f)/2$, where P_f is the folding period used on-board, specified to the nearest μs and set to the mean Doppler-shifted observed period, and N_f is the number of periods folded on-board before readout, always an integer.
- Compute the arrival time at the SSB of an event occuring at T_{mid}, using the source position as determined from timing analysis of the radio pulsations.
- Compute the SSB times of the 64 bin edges of the central "fold" of the accumulation.
- Using the source ephemeris parameters, compute the phases in the SSB corresponding to the bin edges.

- Deposit the counts from the current integration in phase bins fixed in the SSB, splitting counts linearly between those bins based on the relative overlap.

For pulsars with periods longer than ~8s, BATSE Large Area Detector (LAD) data exist in each received packet with 1.024s resolution (4 energy channels), or 2.048s (16 energy channels), so that no advance planning is required to allow pulsed signal searches. These data are processed for an epoch-folding analysis as above, except that each "bin" is treated individually, and T_{mid} is the center of its accumulation time.

3 Binning Effects

The binning of the counts into bins each T_{bin} = 1/64 of the folding period produces timing errors uniformly distributed between $\pm 0.5 \times T_{bin}$. The mean value of this phase error is $1/64 \times (12)^{-1/2} = 0.00451$. The line marked "BINNING" in Figure 1 represents that contribution. It is shown only for periods greater than 4.096 ms, since the controller does not function properly for all periods shorter than that value. These periods are handled by folding at a multiple n of the true period.

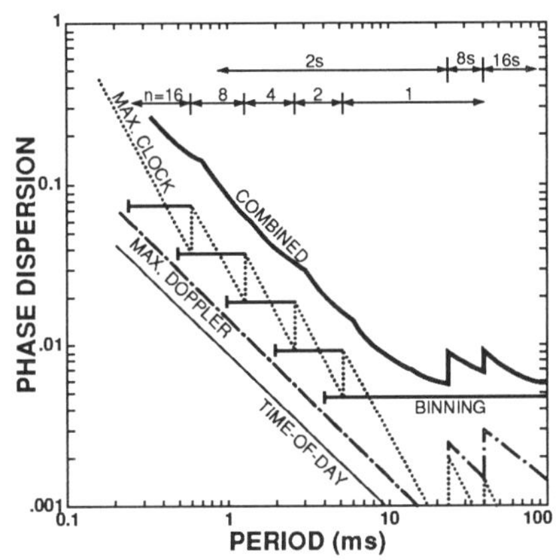

Fig. 1—Contributions to phase dispersion for BATSE folded-on-board data types. Sources with periods greater than that shown are affected only by the intrinsic resolution loss due to the binning, shown by the line labeled "BINNING". See text for a detailed discussion

The boundaries between different values of n are shown in the upper part of the figure. The family of horizontal lines which change value at those boundaries are the corresponding values of dispersion, increasing with n since fewer bins divide each period.

The pulsar controller has a period resolution of 1 μs. The phase dispersion caused by summing the data over the indicated integration times (2 s, 8 s, or 16 s) with a 0.5 μs offset from the desired period is shown by the line "MAX CLOCK". The sawtooth steps in that line below 20 ms are produced by folding the data at a multiple of the true period.

The Doppler effect due to the Earth's motion is never present in the period offsets, since the folding periods chosen are set to the mean observed period. However, the motion of GRO in its orbit becomes a significant contributor to period offsets for periods less than 10 ms, if a fixed period is used over the entire orbit. The line labeled "MAX. DOPPLER" represents the contribution to phase dispersion for a source lying

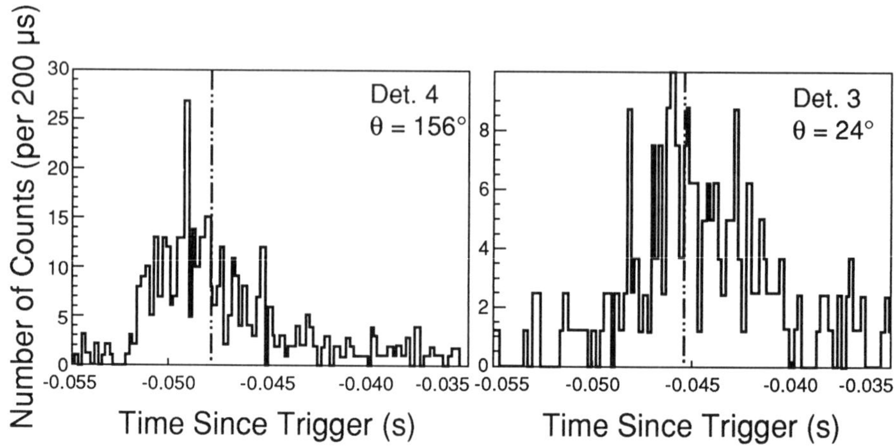

Fig. 2—Time history for GB910711, as seen by two different BATSE LAD detectors. The angle θ is the angle between the detector normal and the geocenter. The dotted line is the median arrival time of the source photons, occurring at different times in each detector due to different proportions of direct and scattered flux.

in the orbit plane, averaged over the portion of the orbit when the source is above the horizon.

The time that marks the start of each on-board accumulation is read from the GRO clock, digitized to an accuracy of 1/64 ms. The line labeled "TIME-OF-DAY" represents the contribution to dispersion which occurs due to the use of this sampled clock when summing successive integrations together.

Adding the above contributors in quadrature, since they are due to uncorrelated effects, one obtains the curve labeled "COMBINED". It is seen that for pulsars with periods greater than about 10 ms, the basic 64-bin resolution can be maintained by suitable choice of operating mode, if the source ephemeris is known well enough to limit the period discrepancy to 1 part in 10^4.

4 Scattering Effects

4.1 Direct Observation of Time Delay

A clear demonstration that a significant number of source photons are scattered off the Earth's atmosphere is provided by a time history of the shortest cosmic γ-ray burst yet observed by BATSE, as shown in Figure 2, taken from Bhat *et al.* 1992. The BATSE LAD detector normals are parallel to the faces of a regular octahedron, placed at the corners of GRO. Each detector thus has a different exposure geometry to both the source and the geocenter, so that they detect different amounts of direct and scattered flux. Some detectors will see *only* scattered flux, as does detector 3 for this event. The time delay of the rising edge of the event in the Earth-facing detector is about 2.5 ms, consistent with the additional path traveled by the scattered photons. The delay is not as large for data collected for pulsars, since only detectors whose normals face within ~60° of the source are used, not detectors which see only scattered flux.

4.2 Effects on Pulse Profiles

Data collection for the Crab has normally been scheduled for about 15 minutes per orbit, starting when GRO traverses closest to the vernal equinox. This position is shortly after source rise, when the angle to the source above the limb of the Earth is between ~15°–60°. This range is a function of the right ascension of the ascending node of the GRO orbit, which precesses with a period of ~50 days.

In order to search for intrinsic pulse profile variability and to detect possible systematic effects due to this changing scattering contribution, data for the Crab pulsar for the entire mission have been epoch-folded into 1/2 day intervals, for which livetimes of ~4000s are usually obtained. The pulse profiles in the energy range 40–100 keV were fit to a template formed by minimizing χ^2 bin-by-bin between the template and the set of 1/2 day background-subtracted profiles, over a range of days which utilized the same source ephemeris information from Taylor *et al.* (1992). Figure 3 shows the variation of the goodness of fit with time. Since the phase bins are not independent, due to the splitting of counts between bins, the mean value expected for the χ^2 distribution per degree of freedom is approximately 2/3. Features separated by the orbit precession period of 50 days may be due to scattering effects. Similar features at the same times are also observed in 20–40 keV and 100–235 keV with somewhat smaller amplitude, but not at higher energies. These are the energy ranges in which atmospheric scattering is expected to be most important.

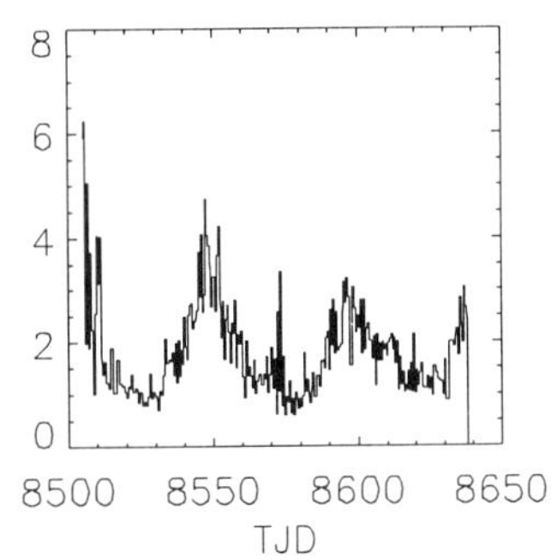

Fig. 3—Values of χ^2 per degree of freedom for each 1/2 day interval, for departures from a best-fit template derived for the data, for deposited energies in the range 40–100 keV.

The contributions to χ^2 that cause the excursions peaking at TJD 8550 and 8600 come mainly from the vicinity of both rising and falling edges of the main peak, and the falling edge of the interpulse, the sharpest features in the profile. Since the scattered photons should produce broadened and time-delayed contributions to the profile, they would produce deviations from the mean template shape in those phase intervals.

5 Conclusions

For periods greater than ~10 ms, the ability of BATSE to observe short-period isolated pulsars is not significantly affected by the need to fold and bin the data on-board. Below that period, data with less phase resolution can be obtained to detect the shortest periods expected for neutron stars.

Scattering into the uncollimated BATSE detectors produces detectable effects on the measured pulse profiles. For sufficiently bright sources, data may be selected to exclude geometries which produce large scattered contributions in order to obtain the best pulse profiles.

References

Bhat, P. N., Fishman, G. J., Meegan, C. A., Wilson, R. B., Brock, M. N., and Paciesas, W. S. 1992, submitted to Nature.

Lyne, A. G., and Graham-Smith, F. 1990, *Pulsar Astronomy* (New York: Cambridge University Press).

Taylor, J. H. 1992, public database: anonymous login via **ftp** to **puppsr.princeton.edu**.

Timing Analysis of EGRET/CGRO Data

David J. Thompson

NASA Goddard Space Flight Center

Abstract

The Energetic Gamma Ray Experiment Telescope (EGRET) on the Compton Gamma Ray Observatory carries out pulsar timing on the basis of analysis of individual gamma rays with energies above about 20 MeV. Although the basic calculation is similar to those used for ground-based observatories, the low photon detection rate requires long observations and limits the possibilities of searching for unknown periodicities.

1 Introduction

The Energetic Gamma Ray Experiment Telescope (EGRET) has been described briefly in an earlier paper in these proceedings (Thompson, et al., 1992). Operating in the energy range above about 20 MeV, EGRET is the high-energy telescope on the Compton Gamma Ray Observatory, detecting gamma rays which interact through pair production.

1.1 The EGRET Data

The basic unit of EGRET data is the event, a single spark chamber picture of a γ-ray interaction plus associated information. The CGRO spacecraft system provides some of this information, including: the location of the spacecraft, the pointing direction of the observatory, and the Universal Coordinated Time (UTC) (but see the note below). For scientific analysis, other information about the event is derived from the EGRET subsystems: the arrival direction of the gamma ray, the γ-ray energy, and the current operating mode of the telescope. Bertsch *et al.* (1989) give additional details about the data and analysis procedures.

Pulsar timing for the EGRET data uses an adaptation of the Princeton TEMPO program and the Jet Propulsion Laboratory DE200 solar system ephemeris, which is the J2000 ephemeris also used by the spacecraft for aspect information. Table 1 summarizes the calculations needed to extrapolate the γ-ray arrival time at EGRET to an equivalent arrival time at the solar system barycenter.

The pulsar phase ϕ of a gamma ray with barycentric arrival time T_b is determined from the pulsar rotational frequency f and its first and second derivatives $\dot{f}, \ddot{f}$ and epoch T_0:

$$\phi = \phi_0 + T \times f + \frac{1}{2} \times T^2 \times \dot{f} + \frac{1}{6} \times T^3 \times \ddot{f}\,, \tag{1}$$

TABLE 1
Arrival Time Corrections

Description	Form (approximate)
Projection on the line of sight to the pulsar of the vector from the solar system barycenter to the Earth center.	Annual sinusoid of magnitude up to 500 seconds.
Difference between atomic time and UTC.	Step function with one-second jumps tabulated.
Projection on line of sight of geocenter to EGRET vector.	Once-per-orbit sinusoid of magnitude 25 milliseconds.
Relativistic correction to rate of Earth-based clocks	Annual sinusoid of magnitude 1.6 milliseconds.
Propagation delay due to the gravitational field of the Sun (Shapiro delay)	Annual, up to 110 microseconds, peaked strongly when source is in the direction of the Sun.

where $T = T_b - T_0$ and ϕ_0 is the phase offset.

2 Effects of a Satellite

Pulsar timing from a satellite involves several differences from ground-based measurements:

- The stability of the on-board clock must be maintained in a relatively harsh environment. The GRO clock has a stability within approximately 50 microseconds over periods of months. EGRET has no difficulty coherently adding photon arrival times which span nearly 100 million pulses from the Crab pulsar.
- The absolute time (UTC) must be determined remotely. The clock on the Compton Observatory has been found to be offset from UTC by an amount whose magnitude has yet to be precisely determined. Until that offset has been measured, absolute timing with GRO instruments is not possible.
- The satellite motion requires frequent updates to the calculation of the barycenter vector. At 7.5 km/s, the Compton Observatory motion produces up to 25 microseconds of change in light travel time per second.

3 EGRET Counting Statistics

Despite the large area of EGRET, the relatively low fluxes of high-energy gamma rays give low counting rates even for bright sources. For the brightest high-energy γ-ray source, the Vela pulsar, the average time between gamma rays detected from the source is about 3 minutes. For the Crab, the average separation between EGRET-observed gamma rays is 10 minutes, or about 18,000 pulsar rotations. Acquiring enough statistics requires observations long enough that the period derivative can not generally be neglected.

An additional timing restriction is intrinsic to the EGRET data system. Because each γ-ray interaction is transmitted as a set of coordinates derived from the magnetic core spark chamber readout, the time needed to complete the analysis of one event has a minimum of about 110 milliseconds. EGRET cannot, therefore, ever observe two consecutive pulses of the Crab, Vela, or similar fast pulsars.

These timing limitations make searches for unknown short periods extremely difficult. Techniques such as those described by Buccheri, *et al.* (1987) and Middleditch (1992) can be applied to EGRET data, but making an unambiguous identification of an unknown period and period derivative appears to be a serious challenge. If an approximate period and period derivative are known, a search becomes more feasible.

4 Summary

The Compton Gamma Ray Observatory high-energy telescope (EGRET) conducts pulsar studies on a photon-by-photon basis, applying the timing corrections to the solar system barycenter for each detected γ-ray. The data are valuable for known pulsars, but the counting statistics limit searches for completely unknown periods.

References

Bertsch, D. L., *et al.* 1989, *Proceedings of the Gamma Ray Observatory Science Workshop* (Greenbelt, MD, GSFC), 2-52.

Buccheri, R., Özel, M. E., and Sacco, B. 1987, *Astr. Ap.*, **175**, 353.

Middleditch, J. 1992, these Proceedings.

Thompson, D. J., *et al.* 1992, these Proceedings.

Pulsars with Planets?

F. Curtis Michel

Rice University

Abstract

Timing noise could be the source of the "planets" reported around PSR1257+12.

1 Introduction

Some time ago, following the discovery that the Goldreich-Julian model could not work as advertised (Michel 1980), Alex Dessler and I (Michel and Dessler 1981) suggested that perhaps a disk of matter might be left in orbit around pulsars, with pulsar action arising from electromagnetic interaction between the disk and the neutron star (Michel 1983, 1991). Although our only intention was to try to save the day for pulsar theory, this model provoked amazing hostility in reviewers, who viewed having any matter whatsoever left around pulsars as preposterous. Anonymity prevented us from ever discovering the basis for this wisdom. In the succeeding years, a more permissive outlook has evolved and now disks have even been invoked to explain planets around pulsars. What was outrageous in a theory becomes obvious when needed to save the appearances. Indeed, I would be quite happy to find just the planetary masses suggested by the PSR1257+12 detections (Wolszczan and Frail 1992), because their total masses are precisely of the order of magnitude predicted 12 years ago. Unfortunately, we have been disappointed by previous claims of discovery of planets. The figurative "evaporation" of PSR 182910 (Lyne and Bailes 1992) is only the latest incident; one of the earliest reports was of a planet around the Crab pulsar (Richards *et al.* 1970).

2 Timing Noise

The spurious planet in the Crab pulsar was due to the large amplitude timing noise in that pulsar. The origin of timing noise is not clear, but it is plausibly associated with the interaction between two components of the pulsar interior, one of which is slowed by magnetic torques and the other only by (normally) weak coupling to the first. This picture is the same one that is posited to account for the glitch phenomenon. The effect of timing noise is to make the pulsar an inaccurate clock, which more or less randomly runs slower and faster. When the secular slowing down of the pulsar, as well as glitches, are removed, this quasi-random component tends to give quasi-periodic oscillations in the residuals. Whether these oscillations are "real" or not is an open question: they could be akin to sunspot cycles or they could just be a feature of quasi-random noise. To my knowledge the latter possibility has not been excluded. Observations of the Crab timing residuals (after removing secular slowing down and large glitches) for over more than a decade show a rough periodicity with a time of slightly over a year, and a power spectrum shows 5 spikes, two on either side of a major spike at this frequency (Lyne, private communication at this conference). Thus the Crab could have 5 planets, but no one is tempted to interpret the data that way any more.

3 Noise in Millisecond Pulsars

The possibility of timing noise was raised and dismissed (Wolszczan and Frail 1992) on the basis that timing noise has not been observed in millisecond pulsars. However, for a long time, pulsars with stellar companions had not been observed either. At this point it would be no more surprising finally to detect timing noise in a millisecond pulsar than to find actual planets. In this regard, we should remember that we have no way of knowing how young a given millisecond pulsar might be, and properties that might depend on age would be dominated by the predominantly old population of these objects.

Another set of planets that turned out to be due to noise were those attributed to the non-existent half-millisecond pulsar in SN 1987A. Indeed, there the same claim (Wolszczan and Frail 1992) was made that there were "two planets and perhaps a third." It seems unreasonable to many people that noise can masquerade as periodic modulation, but it must be remembered that this is "red" noise not random noise (the phase, which is presumably locked to the crust, cannot jump discontinuously to new values), and we already have two examples (Crab timing noise and SN 1987A TV monitor noise) where that was exactly the case. Notice also that the amplitudes of the excursions are similar, with the Crab accumulating a deviation of something like one rotation in a year, quite similar to the amplitude of the PSR 1257+12 deviations.

4 Conclusions

One cannot very well conclude from such comments that the planets are actually timing noise. And a colleague questioned why I was casting doubt on a discovery that would support earlier ideas that I was involved in. I guess the answer is that we are (supposedly) taught to be skeptical about things in science, particularly those things that we would really like to believe in. Also, planets have a lot of angular momentum, which must be transferred even from a millisecond pulsar to a nearby disk with almost 100% efficiency to move the matter out to planetary distances. In the specific case of PSR 182910 there was additionally the suspicious coincidence that the period was almost exactly half that of the Earth year (Helfand and Hamilton 1991). The 3:2 "resonance" in periods for PSR 1257+12 is similarly not particularly reassuring. Gil and Jessner (1992) attribute this resonance and the planets to free precession, while in a timing noise picture, the resonance would arise from the harmonics needed because the quasi-periodicities are actually random and need to be modulated with additional harmonics.

References

Gil, J. A., and Jessner, A. 1992, preprint.
Helfand, D. J. and Hamilton, T. T. 1991, *Nature*, **352**, 481.
Lyne, A. G., and Bailes, M. 1992, *Nature*, **355**, 213.
Michel, F. C. 1980, *Ap. Space Sci.*, **72**, 175.
Michel, F. C. 1983, *Ap. J.*, **266**, 188.
Michel, F. C. 1991, *Theory of Neutron Star Magnetospheres* (Chicago: University of Chicago Press), p. 302.
Michel, F. C., and Dessler, A. J. 1981, *Ap. J.*, **251**, 654-664.
Richards, D. W., Pettengill, G. H., Councelman, C. C. III, and Rankin, J. M. 1970, *Ap. J. Lett.*, **160**, L1.
Wolszczan, A., and Frail, D. A. 1992, *Nature*, **355**, 145.

Daily Glitch Monitoring of 35 Pulsars

David J. Nice

Joseph Henry Laboratories and Department of Physics, Princeton University

Abstract

We began a program of near-daily pulsar timing observations in February 1989. Currently thirty-five pulsars are observed. Six glitches have been detected, one in the Crab pulsar, two in the Vela pulsar, and three in PSR 1737−30. The Crab pulsar glitch was followed by a large permanent step in rotation spin-down, a feature seen in one previous Crab pulsar glitch and not seen in other pulsars. The two Vela pulsar glitches were only three months apart; the later glitch was much smaller than the earlier one. This repeats a pattern seen in a pair of glitches in 1971.

1 Observations

We have implemented an automated system for the collection and analysis of pulsar timing data using an 85′ radio antenna at NRAO/Green Bank. The telescope is available for this project approximately 80% of the time, and observes a fixed daily schedule, currently including thirty-five pulsars. Observations are made at 610 MHz. (A second system, operating at 327 MHz, is under development.)

Data are collected using the Princeton Mark III pulsar data acquisition system (Stinebring *et al.* 1992). Signal-averaged pulse profiles are accumulated for one to eight minutes, depending on source strength. Several consecutive profiles are collected from each pulsar, with total observing time for a single source varying from five minutes to one hour. The data are automatically reduced to a series of pulse arrival times by a computer at the Green Bank site, and the results are sent to Princeton by electronic mail for analysis on a daily basis. Timing accuracy is determined by the signal-to-noise ratio and is typically 1 ms, although the exact value varies substantially from pulsar to pulsar and day to day.

2 Glitch Analysis: PSRs 0531+21, 0833−45 & 1737−30

The accumulated time series are fit to a standard pulsar spin-down model; free parameters are position, pulse phase, rotation frequency Ω, and spin-down rate $\dot{\Omega}$. Timing noise is evident in many of the residual series, but the dominant deviations from the model are several large (macro) glitches in PSR 0531+21 (in the Crab nebula), PSR 0833−45 (in the Vela supernova remnant), and PSR 1737−30. All three of these pulsars are known to glitch repeatedly (*e.g.*, Lyne & Pritchard 1987; Cordes, Downs & Krause-Polstorff 1988; McKenna & Lyne 1991).

The residual series for the three glitching pulsars were analyzed in subsets bound by the dates of discrete events. Each data subset was independently fit with a rotation frequency and spin-down rate (as well as an arbitrary pulse phase ϕ_0). An additional term representing a single exponential decay in phase (with amplitude a and time constant τ) was added for the large glitches in the Crab and Vela pulsars. Thus the barycenter-reduced pulse phase was modeled as

$$\phi(t) = \phi_0 + \Omega_0(t - t_0) + (1/2)\dot{\Omega}_0(t - t_0)^2 + ae^{-(t-t_0)/\tau}$$

with the last term omitted for all but the two largest glitches. The rotation frequency is the derivative of equation (1):

$$\Omega(t) = \Omega_0 + \dot{\Omega}_0(t - t_0) - (a/\tau)e^{-(t-t_0)/\tau}.$$

The changing rotation frequency as a function of time is shown in Figure 1. The effect of the exponential term after large glitches is demonstrated in Figure 2.

Table 1 lists changes in pulsar parameters across glitches. The epoch range listed in the Table is the interval between the last observation before a glitch and the first observation afterward. (Pulse numbering ambiguities across the glitches prevent us from calculating the exact time of glitch in most cases.) The "long-term" parameter changes are calculated after removing the exponential term in equation (1). The "short-term" jumps are additional changes in Ω and $\dot{\Omega}$ due to the exponential term only. The actual short-term values depend on the precise epoch of the glitch; we have calculated lower limits by assuming the latest possible epoch.

The step in spin-down rate for the Vela pulsar may be indicative of a second, longer-timescale exponential decay, as has been observed after previous glitches (Cordes & Downs 1988). The step in spin-down rate for the Crab cannot be explained by a second exponential, however, as that model would not allow the pulsar's post-glitch rotation frequency to fall below its pre-glitch trend (as it is seen to do in Fig. 1). The permanent step in frequency derivative is indicative of a transfer of moment of inertia between various components of the pulsar or of a change in the coupling between them; it cannot be explained by a simple two-component angular momentum transfer.

A similar permanent change in spin-down occurred after a 1975 glitch (Gullahorn *et al.* 1977). These steps in spin-down add a slight contribution to the braking of the Crab pulsar. The braking index is defined as (*e.g.*, Manchester & Taylor 1977)

$$n = \Omega\ddot{\Omega}\dot{\Omega}^{-2}.$$

We can estimate $\ddot{\Omega}$ due to discrete jumps as $\ddot{\Omega} \simeq \delta\dot{\Omega}/\delta t$ where δt is the interval between jumps. Using measured values of $\Omega = 29.95\ \mathrm{s}^{-1}$, $\dot{\Omega} = -3.78 \times 10^{-10}\ \mathrm{s}^{-2}$, and $\delta\dot{\Omega}/\dot{\Omega} = 3 \times 10^{-4}\ \mathrm{s}^{-2}$, and taking the time between large Crab glitches as 14 years (the interval between the 1975 and 1989 glitches), the braking index n due to jumps alone is approximately 0.05, small compared to $n \simeq 2.5$ calculated between glitches.

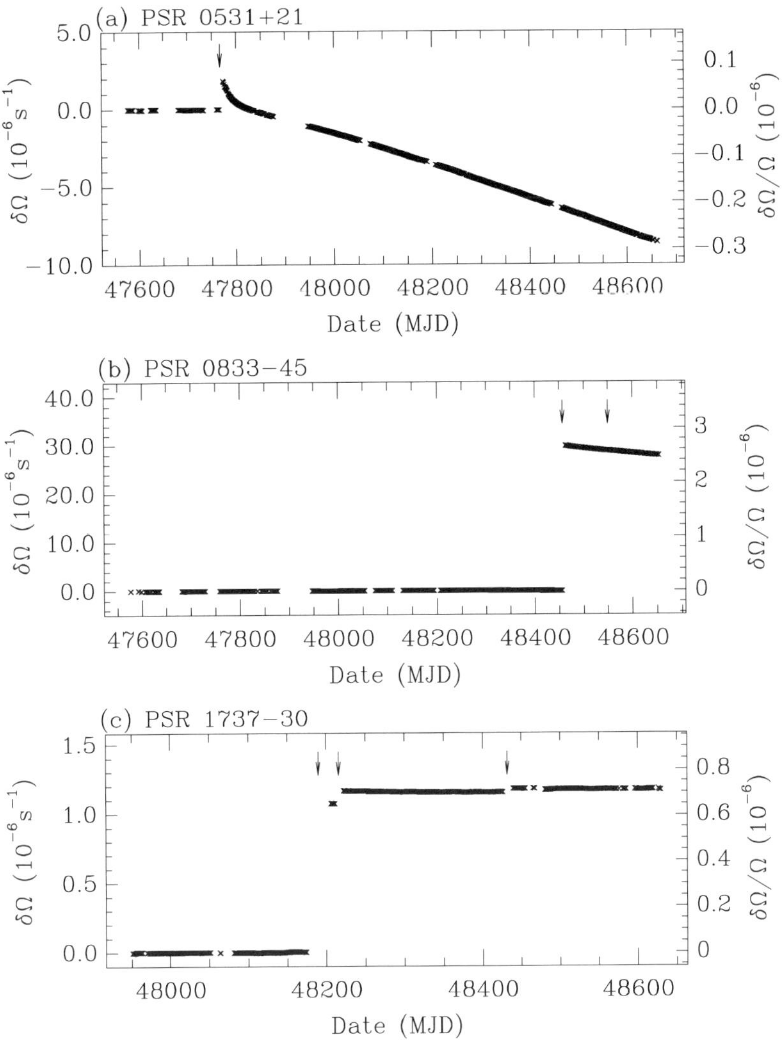

Fig. 1—Rotation frequency as a function of time for (a) PSR 0531+21, (b) PSR 0833-45, and (c) PSR 1737-30. Frequencies were calculated using overlapping 7-day data spans (14-day spans for PSR 1737-30). A pre-glitch trend has been removed in each case. Gaps in the data indicate times for which we have no data. The glitches are indicated by arrows. (The second glitch in the Vela pulsar, only 1% the size of the first, is difficult to see in this plot.)

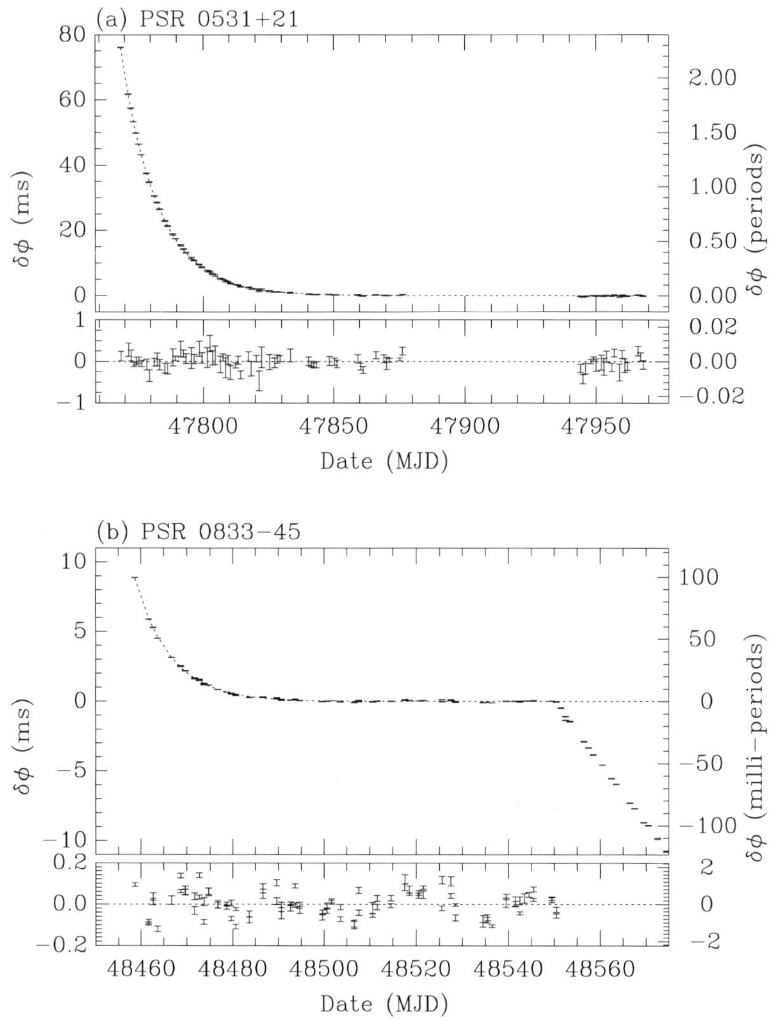

Fig. 2—Residual pulse phases are shown as functions of time for (a) PSR 0531+21 and (b) PSR 0833-45 immediately after large glitches in each. A model including one exponential decay term and a cubic polynomial (to allow for spin-down) has been fit to the data. The lower plots show the residual phases when the full model has been removed. The upper plots show the effect of the exponential term on the arrival times of the pulses. For PSR 0531+21, the 200-day span shown was the largest that could reasonably be fit without significant timing noise. For PSR 0833-45, the data span used for the fit is bracketed by glitches on both sides; the latter glitch is shown on the plot at day MJD 48550.

TABLE 1
Parameters of Macroglitches[a]

PSR 0531+21	Epoch (MJD)	47766.54 – 47768.54
	Long-term $\delta\Omega/\Omega$	$1.87(2)\times10^{-8}$
	Long-term $\delta\dot{\Omega}/\dot{\Omega}$	$3.0(2)\times10^{-4}$
	Short-term $\delta\Omega/\Omega$	$> 6.2(2)\times10^{-8}$
	Short-term $\delta\dot{\Omega}/\dot{\Omega}$	$> 4.1(2)\times10^{-3}$
	Decay Timescale	14.0(2) days
PSR 0833−45	Epoch (MJD)	48456.75 – 48458.73
	Long-term $\delta\Omega/\Omega$	$2.647(1)\times10^{-6}$
	Long-term $\delta\dot{\Omega}/\dot{\Omega}$	$7.14(1)\times10^{-3}$
	Short-term $\delta\Omega/\Omega$	$> 1.32(4)\times10^{-8}$
	Short-term $\delta\dot{\Omega}/\dot{\Omega}$	$> 1.30(6)\times10^{-2}$
	Decay Timescale	7.7(2) days
	Epoch (MJD)	48549.51 – 48551.48
	$\delta\Omega/\Omega$	$3.7(1)\times10^{-9}$
	$\delta\dot{\Omega}/\dot{\Omega}$	$1.5(3)\times10^{-4}$
PSR 1737−30	Epoch (MJD)	48187.82 – 48194.81
	$\delta\Omega/\Omega$	$6.55(1)\times10^{-7}$
	Epoch (MJD)	48214.74 – 48218.74
	$\delta\Omega/\Omega$	$5.5(1)\times10^{-8}$
	Epoch (MJD)	48431.17 – 48432.15
	$\delta\Omega/\Omega$	$1.61(1)\times10^{-8}$

[a]Numbers in parentheses are errors in smallest digit quoted.

3 Limits on Glitches in Other Pulsars

To test for less obvious glitches in the other time series, the data sets were searched for short-term changes in rotation frequency. The method was simply to calculate rotation frequencies in two adjacent $n/2$-day spans and take the difference. This analysis resulted in the limits listed in Table 2 for n=14, 28, and 42 days. The limits on glitches improve as the span is increased, but it is possible that particularly sharp events could be smeared out as n increases. (It should be pointed out that "microglitches," short-term changes in pulsar rotation patterns, may not strictly be changes in frequency. Thus the limits presented here are technically limits on "small macroglitches.")

TABLE 2

Limits on Macroglitches for 32 Pulsars

Pulsar	maximum $\delta\Omega/\Omega$ (10^{-9})		
	14-day-spans	28-day-spans	42-day-spans
0136+57	2.434	0.743	0.255
0329+54	2.184	0.759	0.205
0355+54	1.516	0.304	0.196
0525+21	9.711	2.989	0.616
0628+21	14.927	3.814	1.410
0736−40	11.037	1.923	0.670
0740−28	0.735	1.187	1.255
0818−13	8.751	1.712	0.534
0823+26	2.205	0.883	0.299
0835−41	3.230	0.259	0.144
0950+08	1.308	0.445	0.187
1133+16	1.853	0.506	0.232
1237+25	2.726	0.476	0.228
1508+55	3.937	0.983	0.783
1556−44	2.013	0.575	0.412
1641−45	5.936	1.175	1.630
1642−03	1.248	0.540	0.447
1749−28	0.931	0.362	0.516
1818−04	1.696	1.025	0.392
1831−03	5.640	1.934	0.756
1839+56	25.513	6.197	1.595
1859+03	9.583	2.889	1.653
1911−04	2.458	1.266	0.675
1919+21	3.670	0.657	0.228
1929+10	2.390	0.969	0.487
1933+16	0.671	0.326	0.197
1946+35	1.733	1.140	0.734
2016+28	2.139	0.475	0.276
2020+28	1.477	0.634	0.149
2045−16	2.736	0.650	0.327
2111+46	7.414	1.928	0.788
2217+47	1.184	0.416	0.296

The National Radio Astronomy Observatory is operated by Associated Universities, Inc., under contract from the National Science Foundation. Further support for this project has come from the U.S. Naval Observatory, the Dudley Observatory, and the N.S.F. Numerous individuals have contributed to the 85′ pulsar observing project, in particular V. M. Kaspi, D. R. Stinebring and J. H. Taylor.

References

Cordes, J. M., Downs, G. S., and Krause-Polstorff, J. 1988, *Ap. J.*, **330**, 847.
Gullahorn, Gordon E., Isaacman, Richard, Rankin, J. M., and Payne, R. R. 1977, *A. J.*, **82**, 309.
Lyne, A. G. and Pritchard, R. S. 1987, *MNRAS*, **229**, 223.
McKenna, J. and Lyne, A. G. 1990, *Nature*, **343**, 349.
Manchester, R. N. and Taylor, J. H. 1977, *Pulsars* (San Francisco: Freeman).
Stinebring, D., Kaspi, V. M., Nice, D. J., Ryba, M. F., Taylor, J. H., Thorsett, S. E., and Hankins, T. H. 1992, *Rev. Sci. Instr.*, in press.

The Distribution of Old Neutron Stars in the Galaxy

R. S. Foster,[1] K. S. Wood,[2] P. Hertz,[2] and P. Michelson[3]

[1]Center for Advanced Space Sensing, Naval Research Laboratory
[2]Space Sciences Division, Naval Research Laboratory
[3]Physics Department, Stanford University

Abstract

The population of old neutron stars in the Galaxy might be observable as two distinct classes of astrophysical sources. A class of old low-luminosity nearby millisecond-period radio pulsars may dominate the local neutron star distribution. A second population of old neutron stars that have moved far into the halo of our Galaxy could be the source of gamma-ray burst events. In this paper, we explore the galactic distribution of old neutron stars and the means by which they can apparently be detected by different processes in both the radio and gamma-ray bands of the electromagnetic spectrum. If the millisecond pulsars and gamma-ray burst sources represent different manifestations of old non-accreting neutron stars with different evolutionary histories, then we can make several predictive tests for the properties of gamma-ray burst statistics and the distribution of high latitude millisecond pulsars.

1 Introduction

The distribution of neutron stars in the Galaxy has become an important issue for understanding the distribution of millisecond pulsars at high galactic latitude and possibly for explaining the distribution of gamma-ray burst (GRB) events. Both populations, GRB sources and millisecond pulsars, may represent different manifestations of the end state of neutron stars.

The recently announced nearly isotropic distribution of GRB sources observed with the BATSE instrument by the Compton Gamma Ray Observatory (Meegan *et al.* 1992) implies that the source population must be at a distance beyond 50 kpc (Brainerd 1992). A $\langle V/V_{max} \rangle$ test on the flux density of the 153 GRBs indicates an excess of strong bursts with $\langle V/V_{max} \rangle = 0.348 \pm 0.024$ (Meegan *et al.* 1992). Similar limits on 89 GRB events from Ginga give a value for $\langle V/V_{max} \rangle = 0.35 \pm 0.035$ (Ogasaka *et al.* 1991). A value less than 0.5 indicates that sources are not uniformly distributed within the region of space sampled; for instance, the observations may be sensitive beyond the edge of the source distribution. If GRB events are from neutron stars, as indicated from the observed first and second harmonics of cyclotron absorption features (Murakami *et al.* 1988), and the presence in several bursts of broad spectral features, interpreted as red-shifted 511 keV emission (*e.g.*, Mazets *et al.* 1981), then there is a clear discrepancy between the population of neutron stars that produce GRBs and the old galactic neutron star population inferred from radio pulsar observations.

The discovery of a high-latitude ($b = 74°$) millisecond pulsar has raised questions about the local distribution of millisecond pulsars in the galactic field (Wolszczan 1990). Simple arguments indicate that (1) there exists a local low-luminosity population of millisecond pulsars; (2) that between 150 and 500 millisecond pulsars should be visible in the sky above a limiting luminosity of 1 mJy kpc^2; (3) the total number of millisecond pulsars in the Galaxy exceeds 6×10^4; and (4) at least half of all local millisecond pulsars visible from the Earth should be found above 15 degrees in galactic latitude. Further high-latitude searches for millisecond pulsars will aid in enhancing the robustness of these estimates.

2 High-Latitude Millisecond Pulsars

Comparisons between the observed local "old" neutron star population and models for the distribution of old neutron stars are presented in an attempt to use the modeled distributions to discern the detection probability of discovering a high-latitude millisecond pulsar within 1 kpc of the Sun.

For purposes of this simulation we have used a code developed by Frei, Huang, and Paczynski (1991) to compute the expected z-height and velocity probability distributions of neutron stars. The model assumes (1) an initial scale height of young radio pulsars, (2) an initial velocity distribution inferred from observed proper motions of radio pulsars, and (3) a model for the galactic potential in the local neighborhood.

Paczynski (1990) and Hartmann, Epstein, and Woosley (1990) point out that *the slow velocity component of the population determines the scale height of old neutron stars in the Galaxy.* The high-velocity objects move up into the galactic halo and stay bound within about 30 kpc of the disk due to the presumed presence of dark matter. The estimated escape velocity from this disk is near 500 km s^{-1}. The high-velocity objects escape to the outer halo but contribute negligibly to either determining the scale height of the population or the local millisecond pulsar population distribution. The velocity of the young neutron star population has been measured for over 75 radio pulsars to have an average speed near 200 km s^{-1} (Cordes 1986 and Lyne, Anderson, and Salter 1982).

Using a velocity distribution determined by Paczynski (1990) that follows a Maxwellian form, the differential velocity function takes the following form:

$$p(u)du = \frac{4}{\pi}\frac{du}{(1+u^2)^2}, u = \frac{v}{v*}, v^* = 270\text{km s}^{-1}, \qquad (1)$$

after fitting the data of 26 pulsar velocities reported by Lyne, Anderson, and Salter (1982). The model uses an initial distribution of neutron stars born within 50 to 75 pc of the galactic plane. The galactic potential used is given by Paczynski (1990) and accounts for a variation in the potential as a function of galactic radius.

The program convolves the initial distributions in scale height and velocity to determine a distribution of energies. The distribution of total energies is calculated for a steady state population as a function of height above the galactic plane. Hence the resulting distribution is the distribution of all old neutron stars born over the lifetime of the Galaxy.

A geometrical factor is applied to the generated distribution to select only those objects closer than 1 kpc. These pulsars were then sorted as a function of galactic latitude in 5 degree steps and are tabulated in Table 1. The Sun is assumed to be at a galacto-centric distance of 8.0 kpc. The model gives an old neutron star scale height of 400 pc. Half of all local millisecond pulsars should be found above a galactic latitude of 15°.

TABLE 1
Simulated Local Millisecond Pulsar
Distribution Closer Than 1 KPC

Galactic Latitude Range (degrees)	Cumulative Probability	Interval Probability
0 – 5	26.64	26.64
5 – 10	44.49	17.85
10 – 15	56.11	11.62
15 – 20	64.84	8.73
20 – 25	71.77	6.94
25 – 30	76.80	5.03
30 – 35	81.19	4.39
35 – 40	84.84	3.65
40 – 45	88.10	3.26
45 – 50	91.00	2.89
50 – 55	93.48	2.48
55 – 60	95.43	1.95
60 – 65	96.83	1.40
65 – 70	98.09	1.26
70 – 75	98.78	0.69
75 – 80	99.56	0.78
80 – 85	99.91	0.34
85 – 90	100.00	0.09

With five millisecond and binary pulsars identified as a result of undirected surveys, a projected surface density on the sky can be inferred. From initial surveys at least 780 square degrees (Segelstein *et al.* 1986, Fruchter 1989, and Nice 1991) of sky have been covered below a galactic latitude of 30 degrees with three millisecond pulsars discovered. Above 30 degrees more than 150 square degrees have been surveyed (Wolszczan 1991) with one millisecond and one binary pulsar discovered. The sensitivity of the deepest surveys, all using the Arecibo telescope have approached 1.5 mJy. Using these preliminary statistics, the projected surface density of low-luminosity nearby millisecond pulsars appears to range from 1 per 75 to 1 per 300 square degrees of sky surveyed. Following this line of reasoning we expect from 150 to 500 millisecond pulsars visible in the sky above a limiting radio luminosity of about 1 mJy kpc^2. Assuming the total galactic volume of stars is 800 kpc^3, then we infer the galactic content of millisecond pulsars to be between 6×10^4 and 2×10^5 objects.

The total number of local neutron stars, where local is within about 0.5 kpc of the Sun, has been estimated by Paczynski (1990) to be

$$n_{NS} \simeq 0.0014 \text{ pc}^{-3} \left(\frac{R}{8 \text{ kpc}}\right)^{\frac{-3+4z}{1 \text{ kpc}}} \times \left[1 + \left(\frac{z}{0.2 \text{ kpc}}\right)^2\right]^{-1}, \tag{2}$$

where R is the distance from the galatic center and z is the height above the galactic plane. The above number assumes that there are 10^9 neutron stars in the Galaxy and corresponds to a local density of 1.4×10^6 kpc^{-3} neutron stars born over the life of the Galaxy. Paczynski's local density implies that there are 5×10^6 neutron stars in the local 1 kpc volume or 100 times more than the number inferred from local millisecond pulsars. If the mean age of a neutron star is 10^9 years and the mean age for a "normal" radio pulsar is 10^7 years, then only 1% of the neutron stars are currently observable as radio pulsars. This means that the number of millisecond pulsars in the Galaxy should be roughly comparable to the number of "normal" radio pulsars in the Galaxy. Due to the steep $L^{-2}dL$ distribution of pulsar luminosities, there is a strong selection bias against discovering distant radio pulsars. Most radio pulsars are intrinsically weak sources and are therefore most likely to be found within a few kiloparsecs proximity to the Sun. Yet, the distribution of old neutron stars extends into the galactic halo; almost all radio pulsars that exist there are too weak to be easily detected with current search sensitivities.

3 Neutron Star Models of Gamma Ray Burst Distributions

Explaining the apparently large scale height associated with the nearly isotropic distribution of GRB sources seen by the BATSE detector on the Compton Observatory may actually resolve an old problem in neutron star models. The problem in question concerns the time scale for an electromagnetic torque to spin-down a strong 10^{12} gauss neutron star.

It has long been recognized that only a few candidate periods have ever been identified in the light curves of GRB events (Wood *et al.* 1981). When periodicities have been identified, they have generally fallen in the range of 1 to 10 seconds so that it was impossible to observe a large number of cycles within the duration of the burst. Most identified periodicities are considered questionable. The total fraction of sources with any indication of periodicity is less than a few percent. Yet, if GRBs come from neutron stars, then the gamma-ray burst neutron star (GBNS) population clearly has an underlying spin period distribution, whether or not we know what that distribution is. If a population of GBNS exists, then the population has a mean period, $\langle P \rangle$, greater than a few seconds. This will ensure that the fraction of bursts that show any periodicity will only be a few percent, given the known distribution of burst durations.

Reaching a $\langle P \rangle$ of 10 to 100 seconds from an initial spin period $<< 1$ s by means of electromagnetic torques using a neutron star magnetic field strength of 10^{12} gauss takes of order 10^9 years. In the past the paradigm has been that neutron star fields decay in only 10^7 years. The focus of research in the past has been on finding mechanisms to produce non-electromagnetic torques to spin down the neutron star population. Accretion mechanisms have been considered, but they need either a dense local interstellar medium and very low stellar velocities or a companion star to provide the material for the accretion torques.

The population of "normal" radio pulsars has been assumed to undergo a field decay during their expected 10^7 year radio lifetime (Ostriker and Gunn 1969). Continuity arguments from an analysis of the radio pulsar distribution in the P-$\dot{P}$ diagram indicate that the longer period pulsars also tend to have smaller magnetic field strengths. Yet, mechanisms for producing field decay on a relatively short 10^7 year time scale are hard to justify. An electromagnetic torque acting on the system in 10^9 years will give an older pulsar population a mean spin-period of tens of seconds and perhaps a few percent of these objects will show periodicity.

It is interesting to note that most radio pulsars with observed velocities are traveling at speeds near or above 200 km s^{-1} and moving into or out of the galactic plane. A neutron star born in the plane will travel up to $\sim$ 60 kpc above the plane in 10^{8-9} years. In the same amount of time the source will also be able to spin-down from a fraction of a second period to perhaps 10 to 100 seconds. Is it a clue or just a coincidence that these two time scales appear to be very similar?

If old neutron stars are the sources of GRBs, then GBNS are a slowly rotating, non-accreting, old halo population of stars with a scale height above 50 kpc. Furthermore, the largest source of energy available for burst production in the system appears to be in the form of rotational kinetic energy at about 10^{46} ergs. (For a point of comparison, the magnetic energy available is $< 10^{42}$ ergs). A mechanism must then exist for producing GRBs, such as the "starquake" scenario (*e.g.*, Ruderman 1991a,b). At an arbitrary start time let the star be near the equilibrium configuration for its initial spin period, $P(0)$. The star spins down to $P(0)+dP$ without changing configuration. After the deviation from equilibrium reaches a critical value, the star relaxes to a new equilibrium configuration appropriate to its new spin period in a starquake kind of event, giving off gamma-ray photons with a total energy of 10^{41} ergs per burst, in 10^{4-5} bursts over the lifetime of the neutron star.

The characteristics that define an underlying neutron star population as a source for GRB would be as follows. The number of sources as a function of flux, S, would naturally saturate at some S_{max} because of the limitation on scale height implied by spin-down. Stars with spin periods $>>$ 100 seconds no longer have enough rotational energy to power the GRB phenomenon, hence the GRB source population cannot persist to scale heights greater than that reached in the source spin-down lifetime. A simple $\langle V/V_{max}\rangle$ test would have a value less than 0.5. The number of sources distributed as a function of galactic latitude needs to be nearly isotropic to satisfy the observed GRB distribution. The fraction of GRB sources showing periodicity is less than a few percent. The mean period of the population is of order a few seconds. No steady X-ray flux should be associated with any GRB source. Similarly, there should be no optical counterparts or archival optical flashes associated with GRB sources.

Given the above characteristics of a hypothetical GBNS population, we can now define a suitable population and given some characteristics of the distribution. Most of the following values are roughly uncertain by a factor of $10^{\pm 0.5}$. The age of GBNS is roughly $10^{8.5}$ years. The scale height of the GBNS population must range from 30-100 kpc. The magnetic fields of "normal" neutron stars are fixed at 10^{12} gauss and do not decay. The total energy budget available for GRBs is $W = (1/2)I\Omega^2 \sim 10^{46}$ ergs. The energy of a single burst is about $E_B = 10^{40-41}$ ergs. The birth rate of the sources is $R_B = 10^{-2}$ yr^{-1}. The total number of GRBs from an individual neutron star ranges from $N_B = 10^{4-5}$. Finally, a rough estimate of the energy efficiency of old neutron stars to produce GRBs can be estimated as $\eta = N_B E_B / W = 10$ %.

4 Discussion

Ways to test the idea that GRBs come from an underlying population of old neutron stars can be divided into two classes: phenomena that one expects to observe and phenomena that are not expected to be observable. The conclusive association of steady X-ray counterparts, archival plate optical flashes, or optical counterparts with a GRB source will rule out the above scenario. If neutron stars are the source for GRB events, then one expects occasional cyclotron features (that might not even last the duration of the bursts) or find evidence for periodicity in the burst light curves. Another possible observational test is that the sources that show occasional periodicity may be found predominantly at low galactic latitudes where the mean period is shorter and hence closer to the mean duration of the GRB events. In this model GRBs should *not be* associated with significant gravitational radiation, unlike the situation in some cosmological scenarios involving neutron star coalescence (Mao and Paczynski 1992).

Further analysis working out the detailed distributions for the number of sources at a given flux, $N(S)$, the number of source in a given volume compared with the number expected, given the brightest observed source, $\langle V/V_{max}\rangle$, and the number of sources at each galactic latitude, $N(b)$ is in preparation (Wood *et al.* 1992). In order for this model to work the distributions must conform in a sufficiently detailed manner to the neutron star ejection history and geometry, including the initial source distribution, initial velocity distribution, and possible selection effects associated with beaming in the bursts.

It is interesting and intriguing to speculate that two very different phenomena, radio emission from millisecond pulsars and gamma-ray bursts from neutron stars, may in fact be coming from old neutron stars in our Galaxy. In one case we are observing a very local population, while in the other case we are detecting primarily objects high in the galactic halo.

R.S.F. is supported by an National Research Council/Naval Research Laboratory Cooperative Research Associateship. Additional support for attending this meeting and preparing this manuscript was provided by Los Alamos Laboratories. We also acknowledge several fruitful discussions with Roger Romani.

References

Brainerd, J. J. 1992, *Nature*, **355**, 522.

Cordes, J. M. 1986, *Ap. J.*, **311**, 183.

Frei, Z., Huang, X., and Paczynski, B. 1991, submitted.

Fruchter, A. S. 1989, *Ph.D. Thesis*, Princeton University.

Hartmann, D., Epstein, R. I., and Woosley, S. E. 1990, *Ap. J.*, **348**, 625.

Lyne, A. G., Anderson, B., and Salter, M. J. 1982, *Mon. Not. R. Astron. Soc.*, **201**, 503.

Mao, S. and Paczynski, B. 1990, *Ap. J. Lett.*, **388**, L45.

Mazets, E. P., Golenetski, S. V., Aptekar', R. L., Yu, A., Gur'yan, A., and Ill'inskii, V. N., 1981, *Nature*, **290**, 378.

Meegan, C. A., Fishman, G. J., Wilson, R. B., Paciesas, W. S., Pendleton, G. N., Horack, J. M., Brock, M. N., and Kouveliotou, C. 1992, *Nature*, **355**, 143.

Murakami, T. *et al.*, 1988, *Nature*, **335**, 234.

Nice, D., 1991, personal communication.

Ogasaka, Y., Murakami, T., Nishimura, J., Yoshida, H., and Fenimore, E., 1991, ISAS Research Note 484.

Ostriker, J. P. and Gunn, J. E. 1969, *Ap. J.*, **157**, 1395.

Paczynski, B. 1990, *Ap. J.*, **348**, 485.

Ruderman, M. 1991a, *Ap. J.*, **382**, 576.

Ruderman, M. 1991b, *Ap. J.*, **382**, 382.

Segelstein, D. J., Rawley, L. A., Stinebring, D. R., Fruchter, A. S., and Taylor, J. H. 1986, *Nature*, **322**, 714.

Wolszczan, A. 1990, *I.A.U. Circular*, **5073**.

Wolszczan, A. 1991, personal communication.

Wood, K. S., Byram, E. T., Chubb, T. A., Friedman, H., Meekins, J. F., Share, G. H. 1981, *Ap. J.*, **247**, 632.

Wood, K. S., Foster, R. S., Hertz, P., and Michelson, P. 1992, *Ap. J.*, in preparation.

Diffuse Galactic Gamma-Ray Emission from Pulsars

Dieter H. Hartmann,[1] Lawrence E. Brown,[1] Donald D. Clayton,[1]
Neil Schnepf,[1] James M. Cordes,[2] and Alice Harding[3]

[1] Department of Physics and Astronomy, Clemson University
[2]Astronomy Department, Cornell University
[3]Laboratory for High Energy Astrophysics, NASA Goddard Space Flight Center

Abstract

We investigate the contribution of young, isolated pulsars to the diffuse galactic γ-ray glow of our Galaxy. The Crab, Vela, and Circinus pulsars have been unambiguously identified in the high energy band, and γ-ray emission from other pulsars could be detectable with Compton Observatory. The unique source Geminga and perhaps some of the ~20 unidentified COS-B sources might also involve neutron stars. To improve our understanding of the nature and evolution of pulsars, high energy γ-ray observations provide a powerful diagnostic. Although detection of individual pulsars is important, the integrated γ-ray output from all Galactic pulsars is an additional tool for the study of high energy physics in pulsar magnetospheres. Employing realistic radio pulsar kinematics, we perform Monte Carlo simulations to determine the contribution of unresolved γ-ray pulsars to the diffuse Galactic background. From COS-B data we calculate photon probability maps to constrain the total pulsar contribution. Just as the application of angular information to individual pulsars can enhance their detection probability in the γ-ray band, we expect 2D-maps to provide better constraints than latitude- or longitude-integrated flux profiles.

1 Introduction

Although more than ~ 500 pulsars have now been observed in the radio band, our understanding of their magnetospheric physics is still incomplete (Arons 1992). This is partly because only a small fraction of the total emitted power derived from spin down emerges as pulsed radiation. Prior to Compton Observatory (GRO), pulsed γ-rays had been detected unambiguously from PSR 0531+21 (Crab) and PSR 0833-45 (Vela). If the γ-ray emission of Crab and Vela is representative of young pulsars, observations in the γ-ray band are likely to hold an important key for the understanding of rotation powered pulsars. Searches for additional γ-ray pulsars in data from SAS-2, COS-B, HEAO-1, and the MPI Compton Telescope have not been successful (Schönfelder 1985), but the recent GRO discovery (Wilson *et al.* 1992) of PSR 1509-58 (Circinus) justifies the hope for additional discoveries (Schönfelder 1989; Thompson 1989; Thompson and Kniffen 1989; Bucceri & Schönfelder 1989; Zhao *et al.* 1989). We emphasize that even improved upper limits on pulsed γ-ray emission are desirable from the theoretical point of view. Because no generally accepted γ-ray pulsar model exists, reliable predictions as to which radio pulsars are the most promising candidates cannot be made. Prior to the launch of GRO, radio astronomers initiated extensive observing campaigns to provide the gamma-ray community with the pulsar timing measurements needed to extract the faint γ-ray pulses from the background. A successful search for pulsed γ-rays from radio pulsars relies on the availability of accurate radio data and γ-ray photon time-tagging to better than $\sim$ 1 ms. Fortunately, the required radio observations can be performed for a large number of pulsars without excessive use of telescope time. The status of these coordinated observations is summarized in two recent conference proceedings (Taylor 1989a,b).

2 Gamma-ray pulsars

The energy loss rate due to pulsar spin down is given by

$$L \sim 4 \times 10^{31} \, I_{45} \, \delta_{-15} \, P^{-3} \text{ergs s}^{-1}, \tag{1}$$

where I_{45} is the moment of inertia in units of 10^{45} g cm^2, P the pulsar period (in s), and δ_{-15} the period derivative divided by 10^{-15}. Observations in the radio band recover only a small fraction ($\sim 10^{-6}$) of this energy. A much higher fraction ($\sim 10^{-3}$) emerges in the form of pulsed γ-rays and X-rays, making high energy observations of pulsars a promising avenue to a better understanding of pulsar magnetospheres. Arons (1992) argues that pulsars lose the bulk of their energy in the form of a plasma dominated MHD wind composed of electron-positron pairs and a small fraction of heavy nuclei that carry the bulk of the energy. These winds excite emission from the surrounding medium, observed as incoherent synchrotron emission from the SNR. Models of the magnetospheric γ-ray luminosity invoke either particle acceleration well above the neutron star surface, the so-called outer gap model (*e.g.*, Cheng, Ho, and Ruderman 1986a,b), or near the polar cap (*e.g.*, Daugherty & Harding 1982; Harding 1981a,b). Observation of phase-resolved spectra should eventually allow us to distinguish between "cap" and "gap" models. Chiang and Romani (1992a,b) recently extended polar cap model calculations to the class of recycled (millisecond) pulsars, which may produce a detectable γ-ray flux from globular clusters if they indeed contain as many as 10^4 pulsars (Kulkarni, Narayan, & Romani 1990). The γ-ray efficiency is commonly defined as that fraction of the spindown energy that is emitted in the γ-ray band, *e.g.*,

$$\eta_\gamma^{\text{gap}} = L_\gamma / E \propto L_\gamma(P, \delta, ...) \, P^2 \, \tau \, , \tag{2}$$

where $\tau = P/2\delta$ is the characteristic age of the pulsar. Ruderman and Cheng (1988) estimate that η_γ is proportional to $P^4 \, \delta^{-1}$ for periods less than 0.05 s, and $\eta_\gamma \propto (P\delta)^{-1}$ for longer periods ($P \leq 0.1$ s). Spindown past this period will further increase the efficiency until $\eta_\gamma = 1$ at $P \sim 0.125$ s. At still larger periods, γ-ray emission is quenched. Cheng and Ruderman suggested that this increase for post-Vela type pulsars might be associated with the unidentified sources in the COS-B database. Observations suggest normalization to η_γ(Crab) = 4×10^{-4} and η_γ(Vela) = 6×10^{-3}. On the other hand, calculations by Harding (1981a) suggest

$$\eta_\gamma^{\text{cap}} \propto P^{1.8} \, \delta^{-0.5} \, . \tag{3}$$

In addition to the spectra, an estimate of the beaming fraction, f_b, is essential for determining the overall flux normalization. Empirical models for the beaming fraction (pulse width) at radio wavelengths suggest $f_b \propto P^{-1/2}$ (Rankin 1990), as predicted by the polar cap model. Based on these emission characteristics one can estimate the pulsar contribution to the diffuse Galactic γ-ray glow by folding luminosity and spectrum with a model of the time evolution of the pulsar population (Bailes and Kniffen 1992). There are two distinct aspects of this task, 1) one must determine the evolution of the pulsar γ-ray luminosity function, and 2) one needs to consider pulsar kinematics to determine the apparent brightness distribution on the sky. In this contribution we emphasize the latter.

Before discussing integrated maps, let us consider some individual sources. Although distance estimates to pulsars are rather uncertain, we can apply the above luminosity scaling laws to identify promising γ-ray pulsar candidates in the current set of radio pulsars (*e.g.*, Hartmann *et al.* 1992). In the cap model of Harding (1981) all pulsars emit below the Crab-Vela flux level, but the Cheng *et al.* model suggests two pulsars above that level. Indeed, searches for pulsar emission in the COS-B database (Bennett *et al.* 1990) resulted in a possible detection of γ-ray pulses from PSR 1951+32 at the 10^{-3} confidence level, supporting the earlier claim by Li *et al.* (1987) which was criticized by D'Amico *et al.* (1987). Clearly, a few additional pulsar detections in the γ-ray band would greatly aid in modeling their flux evolution. Detection of pulsed γ-ray emission from faint sources is hampered by the strong Galactic background. To enhance the signal-to-noise ratio for such sources, and thus to enhance γ-ray pulsar detection probabilities, Brown *et al.* (1992) have utilized the angular distribution of individual photons received from the vicinity of the pulsar. Employing the COS-B point-spread function (PSF) to calculate a weighted photon lightcurve, they showed that the detection significance is greatly increased. Here we apply the same technique to the recently detected Circinus pulsar (Wilson *et al.* 1992). We analyzed all COS-B data of observing periods 7 and 61 (2-3/76 and 7/81). The PSF per steradian was re-derived from Vela pulsar data. Figure 1a shows the lightcurve derived using the ephemeris of Manchester *et al.* (1985) and a value of $-1.699\ 10^{-23}\ s^{-1}$ for the second period derivative (Ulmer *et al.* 1991). The background was determined from the counts in the phase range 0.6-1. The significance, σ, plotted is counts minus background divided by the standard deviation in the background. Integrating counts with phases from 0.325 to 0.350, we determine the Circinus spectrum and plot it relative to the flux from the Crab pulsar in Figure 1b. The trend of spectral hardening with increasing photon energy suggested by the BATSE observations obviously does not continue to COS-B energies. The post-facto COS-B detection of the Circinus pulsar reported here is only marginal, so that we interpret our spectral estimates as upper limits. Observations with EGRET should provide a detection with greater significance.

3 Mapping the diffuse glow

Detection of new γ-ray pulsars will, of course, be essential for modeling γ-ray emission mechanisms in rapidly rotating isolated pulsars. We know that only a few gamma-loud pulsars are detectable by current instruments, but we attempt to extract

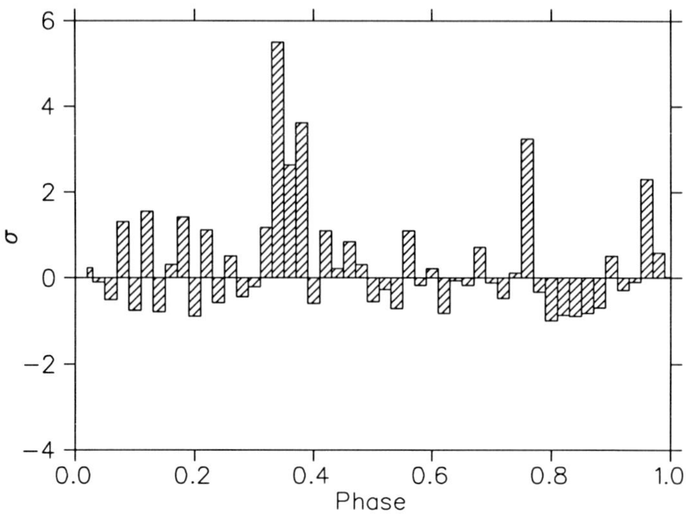

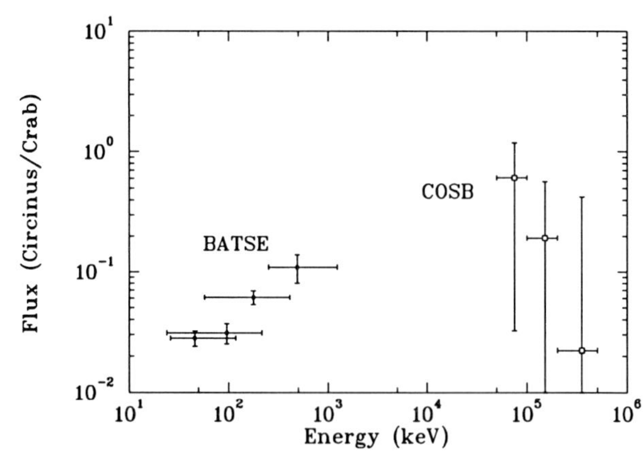

Fig. 1—Light curve for the Circinus pulsar using our weighting technique for COS-B data. (b) Circinus spectrum, normalized to the Crab pulsar flux, for BATSE and COS-B observations.

the integrated signal of all sources below threshold to constrain high energy emission from radio pulsars. To estimate the diffuse Galactic glow due to the superposition of γ-rays from pulsars, one must consider selection effects inherent in radio pulsar surveys. Early work along these lines (Higdon & Lingenfelter 1976; Harding 1981a,b) suggested that perhaps up to 20% of the diffuse emission above 100 MeV observed with COS-B or SAS-2 might be due to unresolved pulsars. However, recent fits to the diffuse galactic emission derived from the COS-B database using detailed HI and CO emission maps (Strong *et al.* 1988) have shown that interactions of cosmic rays (CR) with interstellar gas and inverse Compton scattering of star light off CR electrons could be responsible for most of the diffuse emission between 100 MeV and a few GeV. There doesn't appear to be enough residual emission to allow a 20% contribution from unresolved pulsars.

Using both cap and gap models, Bailes and Kniffen (1992) performed extensive Monte Carlo calculations for the LMT (Lyne *et al.* 1985) and EC (Emmering & Chevalier 1989) models of pulsar evolution. Their simulations confirm that for both LMT and EC models the bulk of the pulsar γ-ray emission is due to currently unknown radio pulsars. However, while pulsars in the EC model hardly contribute to the total γ-ray emission from the Galactic plane, the LMT model could account for up to 100%. Bailes and Kniffen also show that the results are very sensitive to the treatment of pulsar periods and field strengths. For pulsars to dominate the diffuse γ-ray glow, one requires the mean field strength to be below the canonical value, 10^{12} Gauss, and typical periods to be shorter than 20 ms. The question of field evolution is under debate, but recent population synthesis strongly suggests that field decay time scales are much longer than average pulsar life times (Bhattacharya *et al.* 1992). Another important variable is the assumed radial birthfunction for pulsars (*e.g.*, Narayan 1987; Hartmann, Epstein, & Woosley 1990), since it directly affects the distance distribution of pulsars and thus their fluxes. Bailes and Kniffen (1992) calculated latitude-integrated flux profiles and compared them to SAS-2 data. The quality of the CR fits (Strong *et al.* 1988) indicates that one should employ full sky distributions instead. Any map of the diffuse pulsar glow must be based on a realistic treatment of pulsar positions on the sky. Bailes and Kniffen (1992) assumed a simple linear trajectory connecting initial and final position for pulsars born uniformly over the past 25 Myrs. We calculate pulsar orbits in a realistic Galactic potential (Hartmann, Epstein, & Woosley 1990) and compare the resulting sky distribution with that derived from linear trajectories (Figure 2). To demonstrate the differences between these two calculations more clearly, we assume that γ-ray pulsars are standard candles. It is clear from this figure that a detailed comparison of sky flux maps must be based on a realistic treatment of Galactic dynamics for the evolving pulsar population.

It is our pleasure to thank the organizers of this workshop for providing an environment for stimulating discussions with many of the experts on the physics of isolated pulsars. This work was supported in part by NASA grant NAG 5-1578 (DH).

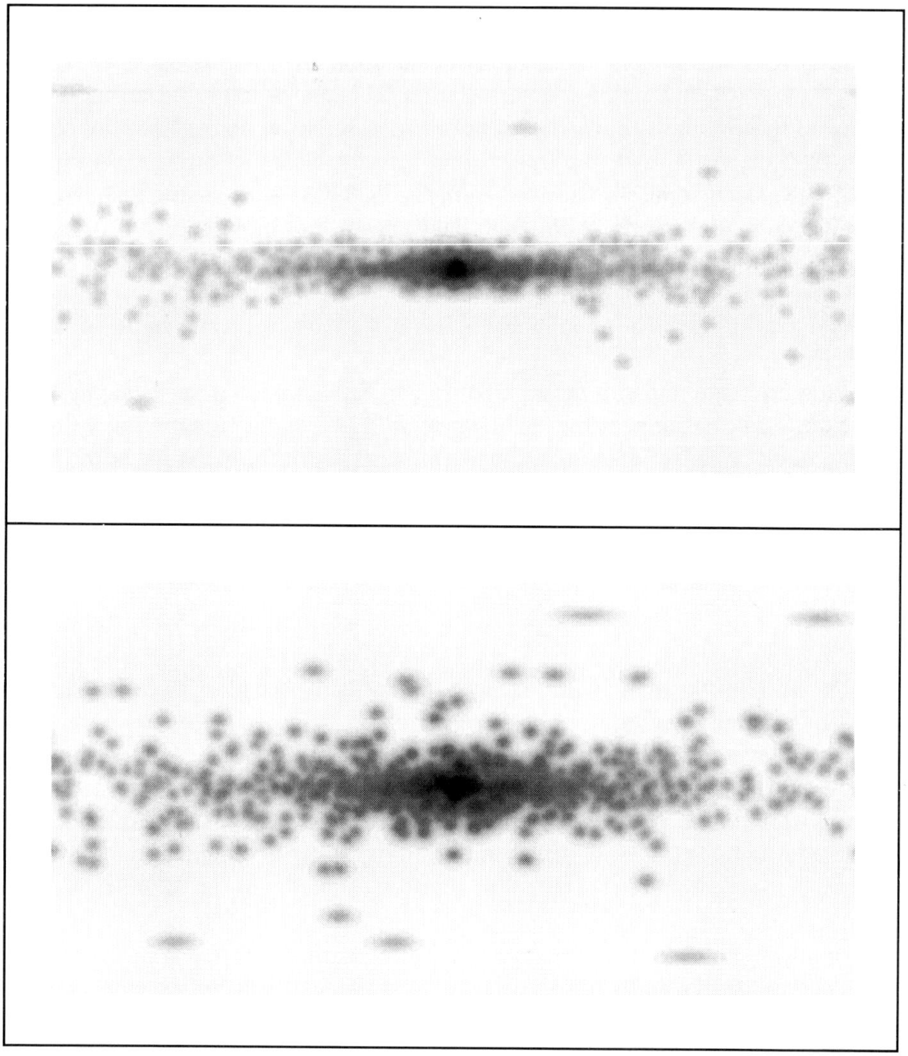

Fig. 2—Sky brightness distribution for γ-ray pulsars younger than 5×10^7 years. The maps are based on one thousand stars, assumed to be standard candles. The top figure shows a distribution based on stellar orbits in a realistic Galactic potential. In contrast, the bottom figure shows the same stars assuming linear trajectories.

References

Arons, J. 1992, in *The Magnetospheric Structure and Emission Mechanisms of Radio Pulsars*, Proceedings of IAU Coll. No. 128, ed. T. Hankins, J. Rankin, and J. Gil, (Pedagogicl Univ. Press: Zielona Góra Poland), pg. 56.

Bennett, K., *et al.* 1990, 21st ICRC, ed. R. J. Protheroe, OG 3.3-1, pg. 181 ApSS, 146, 205

Bhatttacharya, D., Wijers, R. A. M. J., Hartman, J. W., and Verbunt, F. 1992, A&A, in press

Brown, L. E., Clayton, D. D., and Hartmann, D. H. 1992, *Compton Observatory Science Workshop*, NASA Conf. Proc. No. 3137, ed. C. R. Shrader, N. Gehrels, and B. Dennis, pg. 267

Bucceri, R., and Schönfelder, V. 1989, in *Timing Neutron Stars*, ed. H. Ögelman and E. P. J. van den Heuvel, NATO ASI Ser. (Kluwer: Dordrecht), p. 419

Cheng, K. S., Ho, C., and Ruderman, M. 1986a, *Ap. J.*, 300, 500

Cheng, K. S., Ho, C., and Ruderman, M. 1986b, *Ap. J.*, 300, 522

Chiang, J., and Romani, R. W. 1992a, *Compton Observatory Science Workshop*, NASA Conf. Publ. No. 3137, ed. C. S. Shrader, N. Gehrels, and B. Denns, pg. 237

Chiang, J., and Romani, R. W. 1992, *Ap. J.*, submitted

D'Amico, N., *et al.* 1987, IAU Circ. No. 4507

Daugherty, J. K., and Harding, A. K. 1982, *Ap. J.*, **252**, 337.

Emmering, R. T., and Chevalier, R. A. 1989, *Ap. J.*, **345**, 931

Harding, A. K. 1981a, *Ap. J.*, 245, 267

Harding, A. K. 1981b, *Ap. J.*, 247, 639

Hartmann, D. H., Epstein, R. I., and Woosley, S. E. 1990, *Ap. J.*, **348**, 625

Hartmann, D. H., Liang, E. P., and Cordes, J. M. 1992, in *High Energy Gamma-Ray Astronomy*, ed. C. Akerlof, J. Matthews, and J. van der Velde, AIP, in press

Higdon, J. C., and Lingenfelter, R. E. 1976, *Ap. J. Lett.*, 208, L107

Kulkarni, S. R., Narayan, R., and Romani, R. W. 1990, *Ap. J.*, **356**, 174

Li, J. D., *et al.* 1987, IAU Circ. No. 4492

Lyne, A. G., Manchester, R. N., and Taylor, J. H. 1985, MNRAS, 213, 613

Manchester, R. N., *et al.* (1985), *Nature*, **313**, 374

Narayan, R. 1987, *Ap. J*, **319**, 162

Rankin, J. M. 1990, *Ap. J.*, **352**, 247

Ruderman, M., and Cheng, K. S. 1988, *Ap. J.*, 335, 306

Schönfelder, V. 1985, *Naturwissenschaften*, 72, 133

Schönfelder, V. 1989, in *GRO Science Workshop*, ed W. N. Johnson, pg. 3-24

Strong, A. W., *et al.* 1988, A&A, 207, 1

Thompson, D. J. 1989, in *The EGRET Science Symposium*, ed. C. Fichtel, S. Hunter, P. Sreekumar, and F. Stecker, NASA Conf. Publ. No. 3071, 115

Thompson, D. J., and Kniffen, D. A. 1989, in *GRO Science Workshop*, ed. W. N. Johnson, 4-150

Taylor, J. H. 1989, in *GRO Science Workshop*, ed. W. N. Johnson, 4-143

Taylor, J. H. 1989, in *The EGRET Science Symposium*, ed. C. Fichtel, S. Hunter, P. Sreekumar, and F. Stecker, NASA Conf. Publ. No. 3071, pg. 81

Ulmer, M. *et al.* (1991), *Ap. J.*, **369**, 485

Wilson, R. B. *et al.* 1992, IAU *Circular* No. 5429.

Zhao, Y., Lu, T., Huang, K., Lu, J., and Peng, Q. 1989, *A&A*, **223**, 147

On the X-Ray Luminosity-Power Relation for Young Plerions

Y. Chen, Q. Y. Qu, and Z. R. Wang

Department of Astronomy, Nanjing University

1 Introduction

Crab-like supernova remnants (plerions) are the kind of SNR which has a central neutron star or pulsar. It is generally believed that the central pulsar drives the evolution of the plerion (Chevalier 1976, Weiler 1985, Weaver and Woosley 1980).

A statistical relation between the soft X-ray (0.2-4 keV) luminosity L_X of the synchrotron nebula (including the pulsar) and the pulsar rotating energy loss rate $\dot{E}$

$$\log L_X = \alpha \log \dot{E} + \beta \tag{1}$$

was found by Wang and Seward after reanalyzing the *Einstein* data (Wang and Seward 1988a, 1988b, Seward and Wang 1988), where α is in the range 1.3 to 1.4, $\beta - 13.2$ to -16.6, and L_X, $\dot{E}$ are both in units of erg s^{-1}.

Reynolds and Chevalier (1984, henceforth RC) proposed a theoretical model on the evolution of the plerions: The ejected material of Type II supernova is made up of two components—a fast moving shell and slower expanding core material. The pulsar continuously injects some combination of relativistic particles and magnetic field into the bubble, with energy input given by

$$L(t) = \frac{L_0}{(1 + t/\tau)^p} \tag{2}$$

where L_0 is the initial injected power, $p \equiv (n+1)/(n-1)$ with n the braking index equal to $\Omega\ddot{\Omega}/\dot{\Omega}^2$, and τ is a characteristic slowing-down time, generally $\sim 300 - 700$ yr. The bubble loses energy by synchrotron radiation and adiabatic expansion. Its evolution is divided into four phases. In Phase I the bubble expands into core material, either until it sweeps the latter all up or $t = \tau$. If the core is swept up first, Phase II succeeds, as the constant-mass shell continues to gain energy and accelerate. When $t = \tau$, pulsar energy input starts to decline sharply, and Phase III begins. The bubble enters Phase IV at $t = t_r (\sim 1.3 \times 10^4$ yr) when the reverse shock passes. In the nonequipartition case, they fitted the radio properties such as $\Sigma - D$ relations of five plerions with little difficulty, while fits of X-ray luminosity and spectrum were unsuccessful with the parameters used for fits of the radio data.

After modeling the particle injections as double flows and adopting RC's evolution theory for the luminosity and the particle distribution, the $L_X - \dot{E}$ relation of the young plerions can be approximately obtained. Moreover we predict that these remnants have steep radiation spectra with indices $\sim 1.3 - 1.4$ in high energy wavelengths.

2 Double-Flow Model and High Energy Radiation

Considering that the injection of a single power-law flow of particles cannot explain simultaneously the luminosities and spectra of both radio and X-ray emissions, we adopt double-flow model on RC's suggestion. One flow of particles is accelerated near the pulsar, which produces radio radiation, whereas the other is accelerated in the pulsar wind zone or at the wind shock, with a spectrum index ~ 2 or somewhat larger, which is energetically unimportant and produces X-ray radiation. Under the circumstances that the detailed acceleration mechanism for the particles is unclear, we, following Pacini and Salvati (1973) and RC, make two assumptions:

i) The first flow of injected particles has an energy distribution

$$J_I(t,E) = J_I^{(0)} \left[\frac{L(t)}{L_0}\right] E^{-(2z_r+1)} \mathrm{erg}^{-1}\mathrm{s}^{-1}$$
$$\left(m_e c^2 < E < E_{Imax}, 0 < z_r < \frac{1}{2}\right) \qquad (3)$$

and the second has

$$J_{II}(t,E) = J_{II}^{(0)} \left[\frac{L(t)}{L_0}\right] E^{-2z_x} \mathrm{erg}^{-1}\mathrm{s}^{-1}$$
$$(E_{IImin} < E < E_{IImax}, 2z_x \gtrsim 2)\ . \qquad (4)$$

According to RC's result of fitting the radio data, $E_{Imax} \sim 1$ erg, so that the synchrotron radiation of the first flow is cut off at optical wavelength.

ii) In the nonequipartition case, the injected relativistic electrons and magnetic field have a constant ratio of power, $R_{els}\,(\lesssim 1)$: $L_{els} = R_{els}L_B$.

The energy distribution and luminosity of both the flows should obey the rules of evolution derived by RC, and the dynamical evolution of the bubble is the same as the single-flow case.

Some important remnants which Wang and Seward (Wang and Seward 1988a, 1988b; Seward and Wang 1988) used for the establishment of statistical relation (1) are at the ages between τ and t_r, namely in Phase III. For example, the logarithms of the characteristic ages of Crab, LMC 0540, MSH 15-52, and Vela are 3.09, 3.22, 3.19, and 4.05 (Seward 1989), respectively; and CTB 80 is aged less than 10^4 yr (Strom, Angerhofer, and Dickel 1984; Wang and Seward 1984; Blair, Kirshner, Fesen, and Gull 1984; Strom and Blair 1985; Ögelman and Buccheri 1987). In view of the fact that the samples in Phase III are comparatively complete among the remnants observed by *Einstein* and *EXOSAT* observatories, we discuss only the plerions in Phase III.

The break energy E_b is defined as the energy at which the radiative-loss time scale equals the dynamical time-scale, $E_b = (c_1 B^2 t)^{-1}$, where $c_1 \equiv 4e^4/9m_e^4c^7$. Hence the break frequency is $\nu_b = 1.82 \times 10^{18} E_b^2 B$. For Crab Nebula, $B \sim 10^{-3}$ gauss (Fazio, Helmken, O'Mongain, and Weekes 1972) and $t = 937$ yr, we find $\nu_b \sim 10^{12}$ Hz, much lower than the frequencies of soft X-ray. Assume that the second flow of relativistic electrons forms an energy distribution of number density as $N_{II}(E,t) \propto E^{-(2z_x+1)}$

due to synchrotron loss, where $E > E_{IImin} > E_b$. In Phase III the X-ray luminosity is formulated as (RC)

$$_{X\nu} = A_X t^{l_X} \nu^{-z_X} \tag{5}$$

where

$$l_X = b(1 - z_X) - p \tag{6}$$

$$A_X = \frac{4\pi c_j J_{II}^{(0)} \tau^{p-2b} A_B^{z_X+1}}{c_1 (2z_X - 1) B^2(\tau)} \tag{7}$$

where b and A_B come from the magnetic field's evolution law $B(t) = A_B t^{-b}$, $c_j \equiv 1.35\times10^{-22}(6.26\times10^{18})^{z_X} a_{2z_X+1}$, and a_{2z_X+1} can be evaluated by consulting ordinary formulae of synchrotron radiation.

The X-ray luminosity of the second flow of electrons in the frequency range $\nu_1 - \nu_2$ is

$$L'_X = A_X \left(\int_{\nu_1}^{\nu_2} \nu^{-z_X} d\nu \right) t^{l_X} . \tag{8}$$

On the other hand, because $t > \tau$ in Phase III, equation (2) can be turned into

$$\dot{E} = L(t) \simeq L_0 (t/\tau)^{-p} \tag{9}$$

From equations (8) and (9), we obtain

$$\log L_X \simeq \log L'_X = \left[1 + \frac{b}{p}(z_X - 1) \right] \log \dot{E} + k \tag{10}$$

where

$$k = \log A_X + \log \int_{\nu_1}^{\nu_2} \nu^{-z_X} d\nu - [p + b(z_X - 1)] \log(L_0^{\frac{1}{p}} \tau) . \tag{11}$$

L'_X in equation (8), which refers to the luminosity of the bubble, dominates the X-ray emission of the plerion. For example, the pulsed X-ray emissions of the Crab Nebula, Vela, 1055-52, etc. are two or more orders of magnitude less than the emissions of the whole plerions. So it could be taken to be true that $L_X \simeq L'_X$.

Comparing equation (10) with (1), we have

$$1 + \frac{b}{p}(z_X - 1) \sim \alpha . \tag{12}$$

The average value of α could be taken as 1.35 (Wang and Seward 1988a, 1988b, Seward and Wang 1988). And in Phase III, $b = 2$ (Reynold and Chevalier 1984). For a pure dipole radiator, $p = 2$, the known observed values of p are also around 2 (Manchester and Peterson 1989). Therefore from equation (12) we obtain the spectrum index $z_X \sim 1.3 - 1.4$. The indices of some plerions in Phase III have been measured as follows. The Crab Nebula: 1.05−1.15, 3c58: 1.04−1.56, Vela X: 1.09−1.34 (Davelaar and Smith 1985), MSH 15-52: 1.5±1.3 (Seward, Harnden, Szymkowiak, and Swank 1984; Chiappetti and Bell-Burnell 1982), CTB 80: 0.9±0.5 (Angelini, *et al.* 1988) or $1.7^{+0.2}_{-0.6}$[11]. These values and their arithmetic mean are all around 1.3, consistent with our result. And the shock-accelerated particles might be expected to present a spectrum $\sim E^{-2}$

or somewhat steeper (Reynold and Chevalier 1984). This means that $2z_X \gtrsim 2$, also in coincidence with the result.

By comparison between equations (10) and (1), we also know that k is approximately a constant: $k \sim \beta$. In (11), the spectrum index $z_X \sim 1.3 - 1.4$ is approximately a constant, $b = 2$ and ν_1, ν_2 are all constants, so that the establishment of the statistical relation (1) indicates that the parameters p, τ, L_0, A_X, etc. have little deviation from their own typical values. Therefore, with regard to the plerions at the ages of $10^2 - 10^4$ yr, the theoretical relation (10) corresponds just to the statistical relation (1).

Consequently an inference or prediction is obvious: if these young remnants are observed in the frequency range corresponding to the second flow but different from (0.2–4) keV, the $\log L_X - \log \dot{E}$ curves obtained would have the same slope as that of Wang and Seward's, only being translated up or down from the latter. The reason is that equation (12) which gives slope α has nothing to do with the observational bands.

Whether or not the second flow of electrons affects the dynamical evolution of the bubble depends on the energy density of the electrons. On the preceding assumptions, the electron density is given by (RC)

$$N_{II}(E,t) = \frac{3J_{II}^{(0)}\tau^{p-2b}}{4\pi c_1(2z_X - 1)B^2(\tau)A_R^3} E^{-(2z_X+1)} t^{2b-p-3} \mathrm{erg}^{-1}\mathrm{cm}^{-3} \qquad (13)$$

where A_R is found by the radius of the bubble $R = A_R t$. Hence the energy density of the electrons is

$$\begin{aligned} U_{IIe} &= \int_{E_{IImin}}^{E_{IImax}} N_{II}(E,t)EdE \\ &\simeq \frac{3J_{II}^{(0)}\tau^{p-2b}}{4\pi c_1(2z_X - 1)^2 B^2(\tau)A_R^3} E_{IImin}^{1-2z_X} t^{2b-p-3}\,. \end{aligned} \qquad (14)$$

The presupposition that the second flow of electrons is energetically unimportant requires that $U_{IIe} \ll U_B = B^2(t)/8\pi$, so we can place a restriction on E_{IImin}:

$$E_{IImin} \gg \left[\frac{6J_{II}^{(0)}\tau^{p-2b}t^{2b-p-3}}{c_1(2z_X - 1)^2 B^2(\tau)B^2(t)A_R^3}\right]^{\frac{1}{2z_X-1}}. \qquad (15)$$

Our supposition for the second flow $E > E_{IImin} > E_b$ can satisfy the above restriction.

3 Conclusions

1. In the case of double-flow electrons, with regard to the plerions at the ages of $10^2 - 10^4$ yr, the $L_X - \dot{E}$ relation found by Wand and Seward from the X-ray observation is reproduced approximately from RC's theory.
2. It is completely possible that there exists the second flow of electrons which has no influence on the dynamical evolution of the bubble.

3. It is predicted that these young plerions have steep radiation spectra with indices $\sim 1.3 - 1.4$ in other high energy wavebands, and the $\log L_X - \log \dot{E}$ curves would have the same slope ~ 1.35.

This project was supported by the National Natural Science Foundation of China.

References

Angelini, L., White, N. E., Parmar, A. N., Smith, A., and Stevens, M. A. 1988, *Ap. J.*, **330**, L43.
Blair, W. P., Kirshner, R. P., Fesen, R. A.. and Gull, T. R. 1984, *Ap. J.*, **282**, 161.
Chevalier, R. A. 1976, *Ap. J.*, **207**, 872.
Chiappetti, L. and Bell-Burnell, S. J. 1982, *M.N.R.A.S.*, **201**, 1025.
Davelaar, J. and Smith, A. 1985, *The Crab Nebula and Related Supernova Remnants*, M. Kafatos and R. Henry, Ed. (Cambridge Univ. Press, London), 219.
Fazio, G. G. Helmken, H. F., O'Mongain, and E., Weekes, T.C. 1972, *Ap. J.*, **175**, L117.
Ögelman, H. and Buccheri, R. 1987, *Astron. Astrophys.*, **186**, L17.
Manchester, R. N. and Peterson, B. A. 1989, *Ap. J.*, **342**, L23.
Pacini, F. and Salvati, M. 1973, *Ap. J.*, **186**, 249.
Reynold, S. P. and Chevalier, R. A. 1984, *Ap. J.*, **278**, 630.
Seward, F.D. 1989, *Space Sci. Rev.*, **49**, 385.
Seward, F. D., Harnden, F. R., Jr., Szymkowiak, A., and Swank, J. 1984, *Ap. J.*, **281**, 650.
Seward, F. D. and Wang, Z.-R. 1988, *Ap. J.*, **332**, 199.
Strom, R. G., Angerhofer, and P. E., Dickel, J. R. 1984, *Astron. Astrophys.*, **139**, 43.
Strom, R. G. and Blair, W.P. 1985, *Astron. Astrophys.*, **149**, 259.
Wang, Z.-R and Seward, F.D. 1984, *Ap. J.*, **285**, 607.
Wang, Z.-R. and Seward, F.D. 1988a, *Vistas in Astron.*, **31**, 271.
Wang, Z.-R. and Seward, F.D. 1988b, *Acta Astron. Sinica*, **29**, 33.
Weaver, T. A. and Woosley, S. E. 1980, *Ann. NY Acad. Sci.*, **336**, 335.
Weiler, K. W. 1985, *The Crab Nebula and Related Supernova Remnants*, M. Kafatos and R. Henry, Ed. (Cambridge Univ. Press, London), 265.

Spectrum Synthesis of Gamma-Ray Burst in Atmosphere of a Neutron Star

Jerzy Madej

Astronomical Observatory, Warsaw University

Abstract

This paper presents model atmospheres of a magnetized neutron star, which momentarily increases its effective temperature from $T_e = 10^7$ K to 10^8 K due to an impulsive energy release below depth of thermalization. The latter T_e exceeds the Eddington temperature by a factor $10^{1/2}$. Hydrostatic equilibrium is maintained by the magnetic pressure gradient, which develops when the model atmosphere stretches during the initial phase of a burst. Theoretical spectra are presented for two surface gravities, $\log g = 15.0$ and 15.5 (cgs units) and are compared with the "standard" spectrum of soft gamma-ray repeaters.

1 Introduction

Gamma-ray bursts are frequently interpreted as results of violent energy release in the envelope or surface layers of magnetized neutron stars. There exist a number of papers discussing the origin of bursts, energy requirements, and tentative theoretical spectra of bursts (see review by Higdon and Lingenfelter 1990). However, spectra predicted by various available models are practically always given as frequency dependent *emission coefficients*, which can be set to the monochromatic fluxes of radiation only in a very particular case of optically thin emitting medium. This is not relevant to the case of impulsive energy release in deep layers of a neutron star.

Gamma-ray bursts are clearly super-Eddington events on the surface of a neutron star. For some reasonable parameters of a neutron star, including chemical composition dominated by hydrogen and/or helium, the limiting Eddington effective temperature $T_{Edd} \approx 3 \times 10^7$ K ≈ 2.5 keV. This value corresponds to the total luminosity exceeding 10^{38} erg s^{-1} (effects of general relativity are neglected). For higher effective temperature, gravity cannot maintain hydrostatic equilibrium against the radiation pressure gradient in a nonmagnetic stellar atmosphere. This is the case in soft gamma-ray repeaters, which exhibit temperatures order of magnitude higher or more (Mazets *et al.* 1982; Atteia *et al.* 1987; see also Norris *et al.* 1991 and references therein). Temperatures of classical gamma-ray bursters, if connected with neutron stars, exceed that limit by several orders of magnitude.

This paper presents an attempt to construct a model atmosphere of a magnetized neutron star atmosphere with homogeneous field B_0, where impulsive energy release at its bottom increases its effective temperature T_e above the Eddington limit T_{Edd}.

However, expansion of the model is quickly stopped by the magnetic stresses, at least for some short time during initial phases of the burst. Requiring hydrostatic equilibrium, one can determine such distortion of the initial magnetic field, which added to gravity allows for determination of the equilibrium gas pressure in the model, which is possible for the initial B_0 being larger than some limiting value. The final step is spectrum synthesis of the model atmosphere. However, the effects of the magnetic field are not included in the radiative transfer.

Assumed parameters of the burst are relevant to the case of soft repeater. The nature of the energy release is not discussed here. Preliminary results of such an approach, reported here, suggest that soft repeaters and their spectra can be interpreted as super-Eddington events in magnetized neutron star atmospheres being both in hydrostatic and radiative equilibrium.

2 Setting the Model

Consider the plane-parallel atmosphere of a neutron star with homogeneous horizontal magnetic field B_0. When its effective temperature T_e does not exceed the Eddington temperature T_{Edd}, the atmosphere remains both in radiative and hydrostatic equilibrium. Magnetic field B_0 with zero gradient has no impact on gradients of gas pressure. If the magnetic field intensity is low enough, then its influence on both the electron velocity distribution and the radiative transfer can be neglected.

An explosion in the envelope (comet impact or star quake) causes a momentary increase of T_e to very high values (*e.g.*, above T_{Edd}). Increase of the radiation pressure gradient causes expansion of the atmosphere with the magnetic field lines, since (in case of infinite electric conductivity of gas) diffusion of the magnetic field lines across plasma is not possible. This develops nonzero magnetic forces which add to gravity, and they suppress extensive drawing of the field. In case of perfect freezing of the field in ionized matter, this can quickly stop expansion and ensure hydrostatic equilibrium also for $T_e \gg T_{Edd}$.

In the present 1-dimensional model, magnetic forces develop only in connection with stretching of the atmosphere under the influence of the increased radiation pressure gradient. More realistic geometry of that event should include also magnetic forces due to the curvature of magnetic field lines.

2.1 Equations of Model Atmosphere

Standard equations which define plane-parallel stellar atmosphere in both hydrostatic and radiative equilibrium (Mihalas 1978) are supplemented by a single algebraic equation describing stratification of the magnetic field. Thomson scattering optical depth τ is chosen as the independent variable, and $d\tau = +\sigma_T \rho dz$. The depth of explosion (at the lower boundary of the model) is denoted by τ_E and remains a fixed number.

1. Hydrostatic equilibrium: in presence of the magnetic field B

$$\frac{dP_g}{d\tau} = \frac{g}{\sigma_T} - \frac{dP_r}{d\tau} - \frac{dP_m}{d\tau} = \frac{GM}{R^2\,\sigma_T} - \frac{1}{c}\int_0^\infty (\kappa_\nu + \sigma_\nu)\,F_\nu\,d\nu - \frac{d}{d\tau}\left(\frac{B^2}{8\pi}\right) \tag{1}$$

2. Radiative equilibrium:

$$\frac{dF_{bol}}{d\tau} = 0 \tag{2}$$

3. Freezing of the magnetic field in gas:

$$\frac{d\Phi_m}{dm} = \frac{d\Phi_m}{dz}\frac{dz}{dm} = \frac{B(m)}{\rho(m)} = \text{invariant}\,, \tag{3}$$

where Φ_m and $dm = +\rho dz$ denote magnetic flux and mass variable, respectively. Equation (3) represents the assumption, that magnetic field lines cannot diffuse in matter, since the electric resistivity of gas vanishes in the model.

4. Equation of state: as for ideal gas

$$P_g = (N_a + N_e)kT, \tag{4}$$

where N_a and N_e denote the number of atoms and electrons per unit volume, respectively.

Quantities describing monochromatic radiation field both in the equilibrium equations (1) - (2) and in the final spectra have been determined from the following radiative transfer equation, which includes free-free absorption (κ_ν) and Compton scattering from fully relativistic thermal electrons (σ_ν):

$$\mu\frac{\partial I_\nu}{\partial\tau} = \eta_\nu\,(B_\nu - I_\nu) - \zeta_\nu I_\nu - \zeta_\nu J_\nu \int_0^\infty \Phi(\nu,\nu')\,\frac{c^2}{2h\nu'^3}J_{\nu'}\,d\nu' + \tag{5}$$
$$+\,\zeta_\nu\left(1+\frac{c^2}{2h\nu^3}I_\nu\right)\int_0^\infty \Phi(\nu,\nu')J_{\nu'}\left(\frac{\nu}{\nu'}\right)^3 \exp\left[-\frac{h(\nu-\nu')}{kT}\right]d\nu'\,,$$

where $\eta_\nu = \kappa_\nu/\sigma_T$ and $\zeta_\nu = \sigma_\nu/\sigma_T$. Variables I_ν and J_ν denote intensity and mean intensity of radiation, respectively. Auxiliary function Φ denotes angle-averaged, normalized, Compton scattering redistribution function, which includes also scattering events proceeding with large energy exchange between relativistic thermal electron and high-energy photon. Details of determination of the redistribution function Φ and numerical treatment of the transfer equation are presented in Madej (1989; 1991).

Therefore, equation (5) is well suited to solve transfer of gamma-ray photons also in case, when $h\nu$ approaches or exceeds $m_e c^2$, *i.e.*, when the diffusion (Kompaneec) approximation fails. However, no effects of the magnetic field on the radiative transfer are included here.

2.2 Method of Solution

Sample atmosphere with no magnetic field and with $T_e < T_{Edd}$ is computed as the initial model (Madej 1991). Such a model is identical with the one which includes the uniform magnetic field B_0, and therefore yields the run of $B_0/\rho_0(m)$. Then, the final model of high $T_e \gg T_{Edd}$ is iterated, where expansion of the initial model under the influence of strongly increased radiation pressure gradient generates gradients of the magnetic pressure with the opposite direction. New distribution of the magnetic field B, gas pressure P_g and density ρ is then determined from equations (1), (3), and

(4). In this paper it is assumed, that the intensity of the magnetic field at the top of the high T_e model remains equal to B_0. Consequently, the condition of hydrostatic equilibrium causes monotonic decrease of B with increasing optical depth. Equation (2) determines the run of temperature T in both low- and high-luminosity models. Iterations involve frequent replacements of various independent variables: τ, m and geometrical depth z.

The equation of radiative transfer is solved assuming the Eddington approximation in angular distribution of the radiation field. Computations presented in this paper are based on the Feautrier technique (Mihalas 1978). Model atmospheres are computed on a large grid of optical depth points ($nd = 40$) equally spaced in logarithmic scale, ranging from $\log \tau = -4.0$ to $+2.5$. The spectrum of radiation inside the model atmosphere was approximated by $nf = 74$ discrete frequencies, and the transfer of radiation was solved in all those points simultaneously.

3 Results

Two low-luminosity models have been computed according to the above procedure, both with $T_e = 10^7$ K, and surface gravities $\log g = 15.0$ and 15.5, respectively (cgs units). The assumed chemical composition included hydrogen and helium with the solar number abundance $N_{He}/N_H = 0.11$. A uniform magnetic field $B_0 = 5 \times 10^{10}$ G is present in both atmospheres, but it does not, however, alter the structure of the model atmospheres since both monochromatic opacities (free-free absorption and Compton scattering) do not include magnetic field effects by an assumption. Moreover, at the present stage, models include no general relativistic effects.

An impulsive release of energy just below the thermalization depth raises the effective temperature to 10^8 K. A neutron star atmosphere without magnetic field would be immediately accelerated outward by the radiation pressure gradient. However, in presence of the initially homogeneous magnetic field B_0, a model atmosphere of infinite electrical conductivity just stretches, and new distribution of the magnetic field B exhibits nonzero gradient. Joint action of the gradient of B and gravity enables the existence of hydrostatic equilibrium in the hot (super-Eddington) model, and it is possible when the initial intensity B_0 exceeds some minimum value, tentatively estimated here as 2 to 3 $\times 10^{10}$ G. The total magnetic flux across model atmospheres remains fixed in both considered cases.

Table 1 and Figures 1 and 2 yield some information on the physical properties of the model atmospheres of such a hot magnetic neutron star. Third column of Table 1, denoted by HGT, gives the total geometrical height (in cm) of computed atmospheres, between the uppermost level of $\log \tau = -4.0$ and the lower boundary at $\log \tau = +2.5$, where the explosion occurs. It is very interesting to note, that atmospheres of neutron stars are geometrically very thin, which is obviously due to very high surface gravity. Moreover, an impulsive energy release which raises T_e by an order of magnitude and is interpreted as a soft repeater event later in this paper, increases the thickness of the models by about 1 centimeter only! On the other hand, gradients of B which appear in the hot model are enormously large (compare numbers in the last column of Table 1).

TABLE 1

Model Atmosphere Properties

$\log g$	T_e	HGT	B
15.0	10^7	26.10	5.000+10
	10^8	27.45	3.622+10
15.5	10^7	8.13	5.000+10
	10^8	8.74	3.312+10

[a]All values are in cgs units.

This column gives values of B at the bottom of each model, $\log \tau = +2.5$, while in all 4 cases we keep $B = B_0$ at the top.

Figures 1 and 2 present the run of temperature $T(\tau)$ in all models as a function of the Rosseland optical depth τ_R. Both initial low-temperature models exhibit non-monotonic run of $T(\tau_R)$ (Fig. 1), with the effects of cooling by free-free absorption in the intermediate layers, and the Compton heating in the uppermost layers. A sudden increase of T_e reduces absorption significantly, leaving effects of the Compton heating in their full scale (see Fig. 2). As the result, shortly after the beginning of explosion, the atmosphere of a neutron star gets almost isothermal, and everywhere the local temperature $T(\tau_R)$ gets much higher than the effective temperature T_e (Madej 1991).

Figure 3 displays outgoing spectra of both models with super-Eddington effective temperature $T_e = 10^8$ K, compared with the blackbody curve of the same temperature. The high-energy branches of the photon number spectra practically overlap in models of both initial surface gravities and exhibit exponential decrease, the latter being interpreted in some papers as due to free-free emission (see Mazets *et al.* 1982).

In both models high-energy branch of the spectra is distinctly separated from the analogous branch of the blackbody curve. This separation is the characteristic feature of strongly scattering atmosphere, that the temperature guessed from the position of exponential branch of its outgoing spectrum is significantly higher than the effective temperature (see, *e.g.*, Madej 1991). Presented models are too cool to fit precisely the exponential standard spectrum of soft repeaters (Atteia *et al.* 1987; Higdon and Lingenfelter 1990), but they suggest that the temperatures which can be inferred from free-free fit to observations are much greater than the corresponding T_e of bursting neutron star atmospheres.

Figure 3 does not exhibit photon deficiency at low energies, but this can be attributed to not fully adequate radiative transfer considered in this paper.

4 Summary

The equations of magnetized atmosphere of neutron star in radiative and hydrostatic equlibrium are presented in case where T_e increases from a low initial value to above the Eddington temperature. Still, equilibrium is maintained by the magnetic

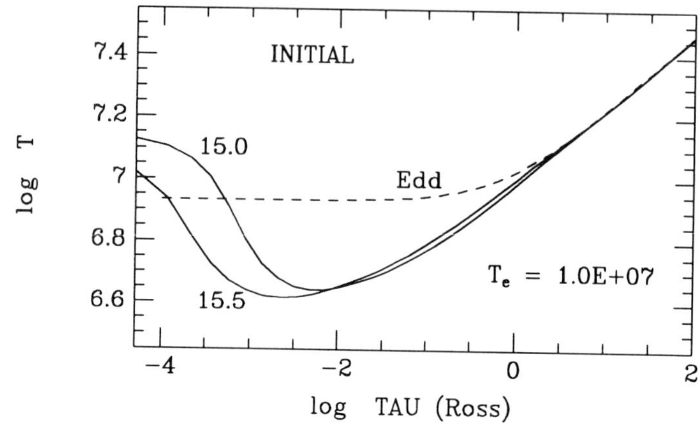

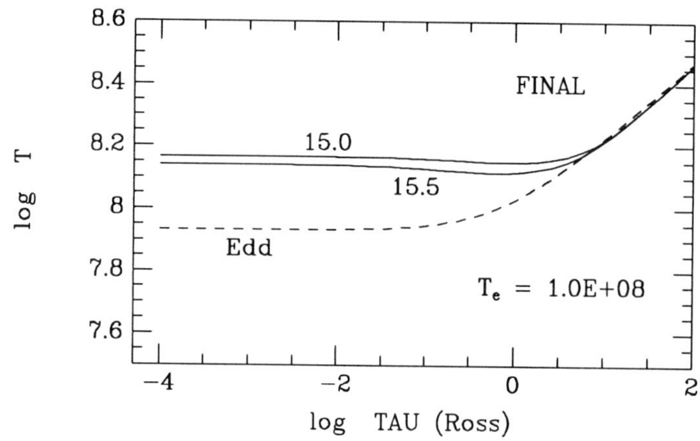

Figs. 1 and 2—Gas temperature distribution vs. the Rosseland optical depth in low-temperature models (upper figure), subsequently heated to temperature 10 times higher (lower figure). Both effects of Compton heating and cooling by free-free true absorption are well pronounced in models with low $T_e = 10^7$ K. The same models heated to 10^8 K become almost isothermal and are everywhere much hotter than T_e. The dashed line denotes the run of temperature given by the Eddington model atmosphere.

pressure gradients, which develop during slight expansion of the atmosphere during the initial phases of a gamma-ray burst. Hydrogen–helium model atmospheres are computed for the case of impulsive rise of the initial $T_e = 10^7$ K to 10^8 K, and $\log g = 15.0$ and 15.5 (cgs units). Intensity of the initial uniform magnetic field $B_0 = 5.0 \times 10^{10}$ G.

Theoretical spectra computed for both models cover the energy range slightly exceeding 100 keV, and exhibit exponential decrease of the high-energy branch. The position and shape of this branch are similar to that exhibited by the standard spectrum of soft gamma-ray repeaters (see Higdon and Lingenfelter 1990).

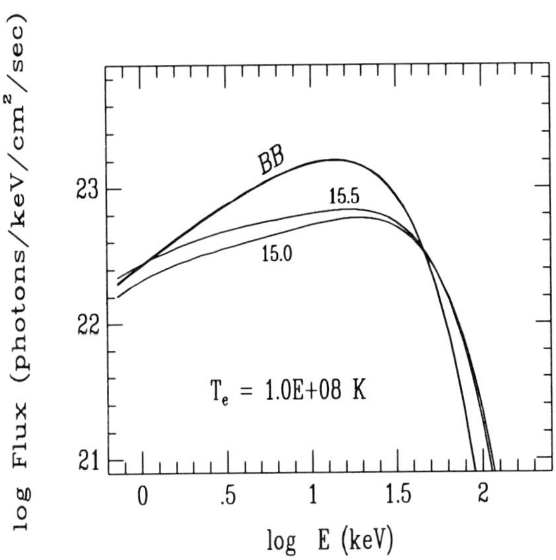

Fig. 3—Spectra of hydrogen-helium models computed for $T_e = 1.0 \times 10^8$ K and two surface gravities $\log g = 15.0$ and 15.5, respectively. Thick solid curve (BB) denotes the spectrum of a blackbody at the temperature 10^8 K.

Neutron star atmospheres with strong surface magnetic field and effective temperature much higher than T_{Edd} seem to represent SOFT REPEATERS—a class of gamma-ray burst sources of short duration. Such a model can therefore replace existing interpretation of their spectra as due to optically thin free-free emission.

This work has been supported by grants KBN-2 1177 91 01 and KBN-2 1213 91 01. I would like to thank Drs. R.I. Epstein, C. Ho, and B. Paczyński for financial support to my attendance in the 1992 Taos Meeting.

References

Atteia, J.-L., Boer, M., Hurley, K., Niel, M., Vedrenne, G., Fenimore, E.E., Klebesadel, R.W., Laros, J.G., Kuznetsov, A.V., Sunyaev, R.A., Terekhov, O.V., Kouveliotou, C., Cline, T.L., Dennis, B., Desai, U., and Orwig, L. 1987 *Ap. J. (Letters)*, **320**, L105.

Higdon, J. C., and Lingenfelter, R. E. 1990, *Ann. Rev. Astron. Astrophys.*, **28**, 401.

Mazets, E.P., Golenetskii, S.V., Guryan, Yu. A., and Ilyinskii, V.N. 1982 *Astrophys. and Space Sci.*, **84**, 173.

Norris, J.P., Hertz, P., Wood, K.S., and Kouveliotou, C. 1991 *Ap. J.*, **366**, 240.

Madej, J. 1989 *Ap. J.*, **339**, 386.

Madej, J. 1991 *Ap. J.*, **376**, 161.

Mihalas, D. 1978 *Stellar Atmospheres* (Freeman: San Francisco).

Dead Isolated Pulsars as Emitters Of Gamma-Ray Bursts

Igor G. Mitrofanov

Institute for Space Research, Moscow

Abstract

Observational evidence is presented that dead isolated pulsars (DIPs) are probable sources of cosmic gamma-ray bursts (GRBs). Statistical observational data on GRBs (Sky positions, time of recurrence, hardness/brightness correlation, and V/V_{max} tests) are discussed, as the phenomenology for the modeling of DIPs' population. Alternatives are compared for population of GRBs sources (evolution, large range of intrinsic luminosities, double-population case) on the base of three astronomical models: Thick Galactic Disc (TGD), Solar Galactic Vicinity (SGV), and Extended Galactic Corona (EGC).

1 Introduction

The origin of cosmic gamma-ray bursts (GRBs) is still unclear after more than 20 years from their discovery. To date there is no direct observational indication on any particular objects responsible for GRBs—both galactic and cosmological hypotheses are considered. For galactic models neutron stars (NSs) are thought to be the best candidates.

Galactic population of lone NSs has been studied mainly using the radio pulsar data. However, the pulsar stage is known to be relatively short, about several million years, with respect to the galactic age of 10^{10} years. After the "active stage," lone NSs become unobservable objects—dead isolated pulsars (DIPs). Assuming the identification of GRBs with sporadic outbursting of DIPs, one might add much of GRBs' data to the total amount of the NS phenomenology, to explore them after the relatively short "active stage."

NSs are the only known galactic objects constantly emitting gamma-rays in the stage of "young active pulsars" (Crab and Vela). One might assume that in the "post-active stage," when NSs are DIPs, they also could emit gamma-rays, provided their magnetosheres are activated by some sporadic internal or external mechanisms (Ruderman and Cheng 1988, Mitrofanov 1992a).

Strong magnetic field of NS provides the favorable physical conditions for emission of gamma-rays, provided some mechanism damps energy inside the magnetosphere: it might be either rotation or vibration of global magnetic field, or MHD-waves, or radiative ejection. A lot of specific gamma-ray emission processes could occur in the strong magnetic field $> 10^{12}$ Gs of NS: the synchrotron cooling is the most efficient one, the e^+/e^- pair creation might lead to electromagnetic avalanche, and Compton up-scattering on relativistic outflowing particles may also be important, provided enough soft X-rays are emitted from the surface of NS.

2 GRB Statistics as a Phenomenology For Modeling Of DIPs

The sensitivity of the BATSE instrument aboard the Compton Gamma-Ray Observatory (Fishman 1991), presently the best one, is about $I_{min} \sim 10^{-7}$ erg/cm^2. The largest known intensities of the strongest events are about $I_{max} \sim 10^{-3}$erg/cm^2. This gives the intensity range of detected GRBs as large as 10^4. The first important fact is the homogeneous positions of all these GRBs on the Sky.

Assuming the "standard candle" hypothesis for GRBs sources, one might suppose that the difference of intensities comes mainly from that of the distances, and therefore, the ratio between distances to "distant" and "close" objects is

$$\frac{d_{max}}{d_{min}} \sim 100 \left(\frac{I_{max}}{10^{-3}\text{erg/cm}^2} \right)^{1/2} \left(\frac{10^{-7}\text{erg/cm}^2}{I_{min}} \right)^{1/2} \tag{1}$$

Assume that in the Thick Galactic Disk (TGD) in the vicinity of the Sun the number density of NSs, which may generate GRBs, is $n_{GRB} = \delta_{GRB} n_{NS}$, where δ_{GRB} is the fraction of outbursting NSs and $n_{NS} \sim 10^{-3}\text{pc}^{-3}$ is the number density for all of them (Paczynski 1990). Assuming that for all estimates of d_{min} the number density n_{GRB} is homogeneous, d_{min} could be estimated as 4 $\delta_{GRB}^{-1/3}$ pc. The scale height of the TGD of DIPs is about 500 pc (Hartmann, Epstein and Woosley 1990). Therefore, for $\delta_{GRB} \ll 1$ the value of $d_{max} \sim 400\ \delta_{GRB}^{-1/3}$ pc seems to contradict the observed isotropy of GRBs on the Sky.

The known distribution of active pulsars by peculiar velocities v shows that for several percents of them $v \gg 300$ km/s. These objects have to escape from the Galaxy Plane to populate the Extended Galactic Corona (EGC) (Jennings 1984, Shklowskii and Mitrofanov 1985). Assuming the model of EGC for GRBs, one might easily explain the isotropy on the Sky. The scale distances of this population should be $d_{min} \sim 1$ kpc and $d_{max} \sim 100$ kpc.

Another statistical approach is associated with the $< V/V_{max} >$-test (Schmidt 1968), where V is the volume of the sphere containing a source, and V_{max} is the maximum volume accessible for the given instrument sensitivity. For homogeneous non-cosmological distribution of any sources with "standard" parameters, the average value of V/V_{max} equals to 0.5. Different experiments led to different estimates of $< V/V_{max} >$, but all are well below this value (*e.g.*, see Mitrofanov *et al.* 1992).

The $< V/V_{max} >$ test would agree with the isotropy on the Sky if the number density of sources, which are "standard candles," decreased with increasing distances from an Observer. On the other hand, models with "non-standard" sources might easily fit observed $< V/V_{max} >$ value, provided they have either different burst-rates, or different mean luminosities, or different mean energy spectra at different distances.

The next statistical issue is the estimation of the mean recurrent time t_{rec} of GRB sources. If they are "standard candles," it may be assumed that GRBs are emitted by each source as a Poissonian sequence at a constant (for each source) rate. Although about $N_{obs} \sim 10^3$ of "classical" GRBs were recorded during the total time $t_{obs} \sim 3$ yrs of observations with high sensitivity, no recurrent events were found among them. If the total number of observable sources is N_{tot}, then one might expect $N^{(1)}$ and $N^{(2)}$ cases of 1-burst and 2-burst registrations, where $N^{(1)} = (t_{obs}/t_{rec})\exp(-t_{obs}/t_{rec})N_{tot}$ and $N^{(2)} = 1/2(t_{obs}/t_{rec})^2\exp(-t_{obs}/t_{rec})N_{tot}$, respectively. Taking into account, that $N^{(1)} \sim N_{obs} \sim 10^3$ and $N^{(2)} < 1$, one estimates $t_{obs}/t_{rec}N_{obs} < 1$ and, therefore one obtains the mean time of recurrency of a particular source of GRBs $t_{rec} > 3 \times 10^3$ yrs and the total number of those outbursting sources $N_{tot} = 10^6(t_{rec}/3 \times 10^3 \text{ yrs})$.

To check the validity of "standard sources" assumption, the database had been developed, which contains GRBs detected in the APEX experiment on the PHOBOS-2 spacecraft. Among others it provides the possibility to compare the average energy spectra for "strong" and "weak" events with $0 < V/V_{max} < 0.3$ and $0.3 < V/V_{max} < 1.0$, respectively. "Strong" GRBs were found to have energy spectra which are harder than "weak" ones (Mitrofanov *et al.* 1992). The evidence of the "brightness/hardness" correlation of GRBs points out that sources should further be considered as "non-standard candles." The following three alternatives could be proposed in this case:

1. GRBs are emitted by a single population whose members have different distances to Observer according to eq. (1). Mean energy spectra of radiated GRBs with increasing distance, become softer.
2. GRBs are emitted by a single population, whose members have comparable distances to Observer $d_{min} \sim d_{max}$, but very large range of outbursting energy $\sim 10^4$. In this case "strong" GRBs with harder energy spectra correspond to $E_{GRB}^{(max)}$, and "weak" ones with softer spectra—to $E_{GRB}^{(min)}$.
3. GRBs are emitted by two different astronomical populations, which are responsible separately for "strong" and "weak" GRBs (Lingenfelter and Higdon 1992).

In case (A) the "brightness/hardness" correlation is associated with the difference between close and distant sources (respectively, with "strong/hard" and "weak/soft" events). If DIPs are considered, their evolution might go together with the motion in space from the close point to the distant one. Therefore, the conception of "evolution+motion" might relate close DIPs (which are thought to be "young" and "strong") with distant DIPs (which are thought to be "old" and "weak").

In case (B) the "brightness/hardness" correlation results from the different energization of outbursting DIPs. Finally, in case (C) one might suppose totally different origin of "strong" and "weak" events, even if DIPs are accepted, as candidates for both types.

Taking into account these three alternatives A, B, and C, different models of DIPs as GRBs emitters are further considered, which are associated with Thick Galactic Disk (TGD), Solar Galactic Vicinity (SGV) and Extended Galactic Corona (EGC).

3 Modeling of GRBs with DIPS in Thick Galactic Disk

NS are known to be born in the spiral arms of the galaxy. The motion of DIPs depends on the peculiar velocity which it had obtained at the Supernova explosion. The escape velocity is known to be about $v_0 \sim 200$ km/s at the Solar Galactic Vicinity

(SGV). DIPs with $v < v_0$ oscillate around the plane with the maximum deviation $z_{max} \sim$ 500 pc (Hartmann, Epstein and Woosley 1990). The Thick Galactic Disk (TGD) might be associated with these objects, and the alternative (A) could be considered, taking into account the evolution.

During the total life time of $t_{age} \sim 10^4$ Myrs a DIP with $v \sim v_0$ covers a long path about 2000 kpc. It corresponds to several oscillations near the plane and several turns around the center. If the evolutionary timescale of t_{ev} were about t_{age}, DIPs on different stages of evolution could be found at any z. In this case no evolution effect would be seen when close objects with low z and distant objects with high z were compared.

Therefore, for the TGD model, the evolution of outbursting DIPs with $v \leq$ 200 km/s has to occur early in their lives, when ages < 5 Myrs and z-deviations < 1 kpc are related by direct proportionality. After that age, DIPs have to lose their ability to generate GRBs. Therefore, the number δ_{GRB} of outbursting DIPs might be as small as $\delta_{GRB} = t_{ev}/t_{age} \sim 10^{-3}$.

Assuming the homogeneous number density of DIPs about $n_{GRB} = \delta_{GRB} n_{NS}$, one might estimate the radius of sphere d_{max}, which contains the necessary number of outbursting objects $N_{tot} = 10^6 (t_{rec}/3 \times 10^3 \text{ yrs})$:

$$d_{max} = 600 \delta_{GRB}^{-1/3} (t_{rec}/3 \times 10^3 \text{ yrs})^{1/3} \text{ pc} \tag{2}$$

For $\delta_{GRB} \sim 10^{-3}$ one obtains $d_{max} \sim 6$ kpc. It is too much, in comparison with the scale height of TGD. Another alternative (B) of single population with broad range of luminosities and $d_{min} \sim d_{max}$ is fully incompatible with the model of TGD.

According to the two population alternative (C), one might assume that DIPs from TGD are responsible either for "strong" or for "weak" GRBs. In the first case, one has to take into account that "strong" events, considered separately, show the isotropy on the Sky (Fishman 1992). Therefore, they should be within the scale height of TGD. However, more distant similar objects, which also should exist according to the TGD model, would be seen like the "belt" concentrated in the Galactic Plane, which in fact is not observed.

In the second case, TGD could be associated with the population of "weak" GRBs. In this case they could be considered as "standard candles." The absence of the evolutional "age/distance" relation permits us to assume that $\delta_{GRB} \sim 1$ and, therefore, $d_{max} \sim 600$ pc. However, there are two conflicting observational limits in this case. To explain the absence of the "belt" on the sky, the distance d_{max} must be assumed below 600 pc. In this case the deep survey of the Center of the Galaxy by the Einstein Observatory would give a significant number $\geq 10(600 \text{ pc}/d_{max})$ of observable X-ray events associated with distant GRBs sources (Mitrofanov 1981), which in fact was not found (Helfand and Vrtilek 1983). Therefore, the alternative (C) with TGD population for "weak" events does not seem to be acceptable either.

4 Modeling Of GRBs With DIPs In Solar Galactic Vicinity

Inside the TGD population the age of outbursting DIPs should be restricted by $\sim 5-10$ Myrs, provided the "age/distance" evolution would be accepted. The number density of those sufficiently young NSs throughout the Galaxy might be strongly increased around those regions, where a "flare" of star formations and Supernova explosions had occurred. One might assume that $\sim$ 10 Myrs ago this "flare" happened in the Solar Galactic Vicinity (SGV).

The mean birth rate of NSs in the Galaxy is $R_{NS}^{G} \sim 10^{-12} M_{\odot}^{-1}$ yr^{-1}. If the total mass of the SGV is $M_{SGV} \sim 10^{8} M_{\odot}$, then during $\sim$ 10 Myrs the total number of NSs created in this region is about $N_{NS}^{SGV} \sim 10^{3} \times A^{SGV}$, where A^{SGV} is the enhancement coeficient of the birth rate of NS in SGV in comparison with the total Galaxy. If A^{SGV} is about 10^3, then about 10 Mys ago the total of 10^6 NSs had been created in the SGV with all possible peculiar velocities.

Therefore, in accordance with those hypotheses $\sim 10^6$ DIPs are outflying now from the SGV to all directions along the distance scale about 0.5–1 kpc. It is just that number of sources of GRBs, which are necessary to provide observed number of events. So, one might assume, that those DIPs from SGV are responsible for observed GRBs (Mitrofanov 1992b). In this case the isotropy of GRBs around the Sun corresponds to the symmetry of outflying DIPs from the birth place. The mean value of $<V/V_{max}> < 0.5$ represents the decreasing number density of outflying objects. Finally, the "hardness/brightness" correlation might correspond to the evolution of outrunning DIPs during the time scale about 10 Myrs.

Assuming that "close" DIPs are responsible for "strong" GRBs (alternative A), one might estimate the average total energy emitted in one GRB as

$$E_{GRB} \sim 10^{37} \left(\frac{I_{max}}{10^{-3}\ \mathrm{erg/cm^2}}\right) \left(\frac{d_{min}}{10\ \mathrm{pc}}\right)^2 \mathrm{ergs} \tag{3}$$

In this case $t_{ev} \sim t_{age} \sim 10$ Myrs, and each DIP radiates at least $Q = t_{ev}/t_{rec} \sim 3 \times 10^3$ events. So the total energy, which each DIP has to radiate in all bursts should be about

$$E_{tot} = E_{GRB} Q \sim 3 \times 10^{40} \left(\frac{I_{max}}{10^{-3}\ \mathrm{erg/cm^2}}\right) \left(\frac{d_{min}}{10\ \mathrm{pc}}\right)^2 \left(\frac{t_{ev}}{10^7\ \mathrm{yrs}}\right) \left(\frac{3 \times 10^3 \mathrm{yrs}}{t_{rec}}\right) \mathrm{ergs} \tag{4}$$

This energy is rather small in comparison to the total energy of NS rotation, $E_{rot} \sim 10^{46} (1\ \mathrm{sec}/P)^2$ ergs, which could be considered as a possible source of outbursting activity.

All mechanisms of DIPs outbursting in gamma-rays are related with the strong magnetic field (*e.g.* Mitrofanov 1992a). Therefore, assumed evolution of DIPs might be resulting from the magnetic field decay. The time scale of the magnetic field decay should be about 10 Myrs in this case. After this time DIPs are thought to lose the ability to generate GRBs. Therefore, the model of SGV is compatible with alternative (A). It might agree also with alternative (B) provided the outbursting activity of DIPs occurs at $d_{max} \sim 1$ kps from the Sun when they are 10 Myrs old. The radiated energies E_{GRB} and E_{tot} for "strong" events should be as high as 10^{41} ergs (see eq. (3)) and

3×10^{44} ergs (see eq. (4)) respectively. Model of SGV might also correspond either to 'strong'-bursts population or to 'weak'-burst one for the alternative (C).

5 Modeling Of GRBs With DIPs In Extended Galactic Corona

A fraction of DIP with large peculiar velocities $v > 300$ km/s might escape to the Extended Galactic Corona (EGC). Taking into account those objects, one should assume in accordance with alternatives (A) and (B), that all detected GRBs are radiated in the Corona (see Shklowskii and Mitrofanov 1985). Thus, the total number of outbursting objects in the Corona should be about $N_{GRB} \sim 10^6$.

Assuming the distance scale of the Corona $d_{max} \sim 100 - 300$ kpc, the age of NS, which could cover this distance, should be about 300–1000 Myrs (see Expression (1)). Assuming the evolution scenario with the "time/distance" correlation, one has to suppose that $t_{ev}/t_{age} \sim 0.03 - 0.1$. The fraction of high-velocity NS is about 0.01 or less. Therefore, the fraction of outbursting galactic NSs could be estimated about $\delta_{GRB} \sim 3 \times 10^{-4} - 10^{-3}$. For all $\sim 10^9$ NSs in the Galaxy this estimation agrees with the necessary number of outbursting DIPs in EGC.

For the evolutionary scenario (A) the distance-scales about $d_{min} \sim 1$ kpc and $d_{max} \sim 100$ kpc for the EGC model result in the following estimate of the mean radiated energy in one GRB:

$$E_{GRB} = 10^{41} \left(\frac{I_{max}}{10^{-3}\mathrm{erg/cm}^2} \right) \left(\frac{d_{min}}{1\mathrm{kpc}} \right)^2 \mathrm{ergs} \tag{5}$$

The total number of GRBs radiated by one source is about $Q = t_{ev}/t_{rec} = 3 \times 10^5$ and leads to the total energy store for outbursting activity as large as

$$E_{tot} = E_{GRB} Q = 3 \times 10^{46} \left(\frac{I_{max}}{10^{-3}\mathrm{erg/cm}^2} \right) \left(\frac{d_{min}}{1\mathrm{kpc}} \right)^2 \left(\frac{t_{ev}}{10^9\mathrm{yrs}} \right) \left(\frac{3 \times 10^3\ \mathrm{yrs}}{t_{rec}} \right) \mathrm{ergs} \tag{6}$$

The model of EGC might agree with the alternative (B) provided both "strong" and "weak" GRBs could be radiated from $d_{max} \sim 100$ kpc.

For two population alternative (C) both estimates for E_{GRB} and E_{tot} should be the same as eqs. (5) and (6), provided only "weak" bursts are associated with the Corona. For "strong" burst sources at the distance $d_{max} \sim 100$ kpc, the estimate of radiated energy $\sim 10^{45}$ ergs should be valid for both alternatives (B) and (C).

This very large energy is the main problem for the EGC model for GRBs. It is comparable to the total rotational energy of DIPs, and therefore no mechanism could be responsible for GRB emission, which converts the rotational energy to gamma-rays. The energy E_{tot} is about 1% of the total kinetic energy of NSs which have a velocity about 300 km/s, $E_{kin} = 10^{48}$ ergs. One might assume that during the Supernova explosion some physical process works, which supplies to NS a large kinetic energy about 10^{48} ergs, and provides a compatible internal resources for outbursting activity. The evolution of DIPs should be connected in this case with some internal triggering mechanism (NS starquakes?), which during 1000 Myrs would convert this energy to GRBs with the efficiency about 1%.

The simplest verification of the EGC model could be the observation of GRBs from the similar Corona around the Andromeda Nebula, which could be seen in the future with more sensitive instruments. Besides, the Center/Anticenter anisotropy could be expected for the strongest GRBs when a sufficient number of them will be detected.

6 Conclusions

1. The model of the Thick Galactic Disk is not compatible with the present observational data on GRBs.
2. Observations of GRBs agree with models of the Solar Galactic Vicinity or Extended Galactic Corona, provided the evolution of DIPs is assumed (alternative (A)). In the first case, outbursting energy $E_{tot} \sim 3\times10^{40}$ ergs may come from NS rotation and its evolution during $\sim$ 10 Myrs should be finished by loss of outbursting activity. In the second case, a much larger internal energy resource $E_{tot} \sim 3\times10^{46}$ ergs is necessary, and the evolution of DIPs corresponds to $\sim 10^3$ Myrs.
3. Alternative (B), such that the observed range of GRBs corresponds to difference $\sim 10^4$ in radiated energy, implies that only those DIPs can produce bursts which are located either at the outer boundary of SGV population ($d_{mean} \sim$ 1 kpc) or at the outer boundary of EGC population ($d_{mean} \sim$ 100 kpc). The problem in this case is that the objects with the largest luminosities that would normally produce "strong" events could also be detected as "weak" events when located at the distances as large as 10–100 kpc (for SGV model) or 1–10 Mpc (for EGC model). The absence in the sky distribution of "weak" GRBs of any evidence of galactic or/and local intergalactic structure with those distance scales suggests that alternative (B) should be rejected.
4. Alternative (C) might be associated with the combination when "weak" events come from SGV and "strong" ones—from EGC, or vice versa: when "strong" events come from SGV and "weak" ones—from EGC. In both cases the population responsible for "strong" GRBs might contribute also to "weak" events number (see point 3) with well-pronounced angular structure. Therefore, the possibility (C) seems to be also quite improbable.

The author thanks the Workshop organizers for their support and hospitality.

References

Jennings, M. C. 1984, in *Proc. of High Energy Transients in Astrophysics*, ed. S. Woosley (New York: AIP), 412.

Helfand, D.J., Vrtilek, S.D. 1983, *Nature* **304**, 41.

Hartmann, D., Epstein, R. I., and Woosley, S. E. 1990, *Ap. J.*, **348**, 625.

Fishman, G. J. 1991, in *Proc. of the Huntsville Workshop on Gamma-Ray Bursts*

Lingenfelter, R. E. and Higdon, J. C. 1992, *Nature*, in press.

Mitrofanov, I. G. 1981, *Astrophys. Sp. Sci*, **80**, 303.

Mitrofanov, I. G. 1992a, in *Proc. of the Los Alamos Workshop on Gamma Ray Bursts*, ed. Cheng Ho, R. Epstein and E. Fenimore (Cambr. Univ. Press), 37.

Mitrofanov, I. G. 1992b, in preparation.

Mitrofanov, I. G. *et al.* 1992, in *Proc. of the Los Alamos Workshop on Gamma-Ray Bursts*, ed. Cheng Ho, R. Epstein and E. Fenimore (Cambr. Univ. Press), 203.

Paczynski, B. 1990, *Ap. J.*, **348**, 485.

Rudermann, M. and Cheng, A. 1988, *Ap. J.* **335**, p.306.

Schmidt, M. 1968, *Ap. J.*, **151**, 393.

Shklowskii, I. S. and Mitrofanov, I. G. 1985, *MNRAS*, **212**, 545.

Planets Around Pulsars

Menashe Banit, Mal Ruderman, and Jacob Shaham

Physics Department and Columbia Astrophysics Laboratory, Columbia University

Abstract

We discuss possibilities for the formation of planets in circular orbits around pulsars. We conclude that the formation of such planets may occur in a very late phase of Low-Mass X-Ray Binary (LMXB) evolution: the Binary-Millisecond-Pulsar (BMP) phase. Evaporation of the companion star in that phase may supply matter to a circumbinary excretion disk in which planets similar to those observed in PSR1257+12 may form.

1 Introduction

Measurements of pulse arrival time delays of the millisecond pulsar PSR1257+12 can be fit by assuming two planets orbiting around the pulsar (Wolszczan and Frail 1992). The fit uses masses that correspond to $m\sin i$ of 2.8 and 3.4 earth masses and periods of 98.2 and 66.6 days respectively. Here, m is a planet mass and i is the inclination angle of the orbital axis to the line of sight. Fitting of the orbital parameters gives an eccentricity lower than 0.03 for both planets. Future pulse arrival times measurements, which should exhibit the expected orbital parameters variation due to interplanetary gravitational interaction, can confirm the planets' presence (Rasio *et al.* 1992).

Generally, pulse arrival time delays may also result from other effects, such as free precession (Pines and Shaham 1972) or Tkachenko oscillation (Tkachenko 1966; Ruderman 1970). Apart from the difficulty in understanding the appearance of two modes of oscillation, there are additional constraints on these models. In the free precession model, the pulse arrival delay amplitude is limited to a quarter of the pulse period (Nelson, Finn and Wasserman 1990) while for PSR1257+12 the delay amplitude is about half of the pulsation period. As for Tkachenko oscillations, no single fundamental oscillations of this kind were measured for any other pulsar. Moreover, it is not clear how such a large amplitude of motion of magnetic flux tubes relative to vortex lines in the neutron-star's superfluid core can remain undamped.

On the other hand, the planets' low orbital eccentricities challenge the usual neutron star formation theory. Neutron stars are born during the core collapse of massive stars $8 < \frac{M}{M_\odot} < 20$, in a supernova (SN) explosion. The progenitor star is usually a Red-Giant (RG) with an extended envelope of radius larger than $1AU$ (Woosley and Weaver 1986; Meader and Meynet 1988). Whether planets that are at a smaller orbital radius could survive this phase without spiraling into the RG's center is not yet clear.

More importantly, since more than half of the system's mass is ejected in the SN explosion, the planets will unbound if the explosion is symmetric. The gravitational binding energy increase of the collapsing core alone is expected to produce an eccentricity of about 0.1. Pulsar proper motions at velocities of 10^2km s^{-1} are, in part, attributed to an asymmetry in the SN explosions (Narayan and Ostriker 1990). A well designed neutron star kick velocity (Hills 1983) could keep a given planet in a bound circular orbit. However, if the planetary orbits are circular before the SN explosion, the probability that *both* observed planets will remain bound with such low eccentricities is nil. For this reason, a model in which the planets pre-exist has problems.

Below we first consider the proposal that the planetary orbits after the SN explosion were, in fact, eccentric, but were circularized (Field, Roger and Rybicki 1992) by planetary wind fueled by pulsar radiation (Ruderman, Shaham and Tavani 1989). We show that such a process may not be efficient enough. We then discuss briefly other models and, finally, suggest a new mechanism by which planets might form in circular orbits around a millisecond pulsar (Banit and Shaham 1992).

2 Wind-induced Circularization

Processes that tidally lock planetary spin to orbital motion are ineffective. Planetary spin may, however, play a role in the circularization of an orbit through planetary wind-delay effects (Field, Roger and Rybicki 1992): at each point of its orbit, the planet's inner face is illuminated, but wind emission is somewhat delayed because of thermal inertia effects. The planet's recoil due to its wind has therefore both a radial component and a component tangential to the orbit. The latter induces a change in angular momentum. When the wind mass loss time scale is much longer than the orbital period (the "adiabatic" limit), the radial component of the recoil can be ignored altogether to the lowest order in eccentricity: it does not affect the angular momentum, and to that order its influence on the kinetic energy of the planet averages out to zero over one period. With no wind delay, the recoil is, of course, purely in the radial direction.

We considered two limits of wind-emission mechanisms: one in which the pulsar's illumination far exceeds the planetary Eddington luminosity (which we term the super-Eddington limit (spE)); and a second limit in which the illumination is below the planetary Eddington luminosity (which we term the sub-Eddington limit (sbE)). If l is the radiation flux at the planetary surface and μ is the wind mass flux, then $\mu \propto l^{\frac{1}{3}}$ in the spE limit and $\mu \propto l$ in the sbE limit (Banit *et al.* 1992).

The instantaneous rate of change of the planetary angular momentum per unit mass, **j**, with orbital phase, ϕ, due to the tangential wind recoil component, depends on the instantaneous values of the radius vector r, the wind velocity and the mass flow rate. If the wind delay is independent of the illumination and the planetary spin is normal to the orbital plane, then $| \frac{dj}{d\phi} | \propto r^{\frac{5}{3}}$ (spE) $\propto r$ (sbE). The corresponding changes in the planetary specific energy per unit mass are $\propto r^{-\frac{1}{3}}$ and $\propto r^{-1}$. For prograde spin, j will increase; for retrograde, it will decrease. Similarly, the energy per unit mass, e, will increase when the spin is prograde and will decrease when the spin is retrograde. For a fixed value of e, the circular orbit has the highest angular momentum. Therefore, the sign of $\frac{d\epsilon}{dt}$, where ϵ is the eccentricity, will depend on the magnitude of

$$\Sigma \equiv -\left\langle \frac{d\log|\, e\,|}{d\log j} \right\rangle = \left\langle \frac{v_\phi{}^2}{|\, e\,|} \right\rangle ,$$

where v_ϕ is the azimuthal planetary velocity and where the $\langle\ \rangle$ average is taken over a full planetary period with the weight function

$$\frac{d\log j}{d\phi} .$$

One needs $\Sigma < 2$ for circularization with prograde rotation and $\Sigma > 2$ when rotation is retrograde. For example, when the tangential recoil force has a power law form, $f_\phi \propto r^{-\gamma}$, the above condition translates to $\gamma < \frac{3}{2}$ for circularization with prograde rotation and $\gamma > \frac{3}{2}$ for retrograde rotation (Chen and Eardley 1992).

If we use $f_\phi = \dot{m}w_\phi$, where w_ϕ is the azimuthal wind velocity component relative to the planet, then

$$\frac{\delta e}{\delta\phi} \simeq v_\phi w_\phi \frac{1}{m}\frac{\delta m}{\delta\phi}$$

and

$$\frac{\delta j}{\delta\phi} \simeq \frac{w_\phi}{v_\phi}\frac{1}{m}\frac{\delta m}{\delta\phi} .$$

These equations assume that the planet has a negligible size. $\delta m/\delta\phi$ is a constant in the sbE case. For a constant wind angle (*i.e.*, constant $\frac{w_\phi}{w}$ where w is the wind speed relative to the planet) this leads to a power-law force with $\gamma = 2 > \frac{3}{2}$; hence in the sbE case, circularization requires a *retrograde* spin (Banit *et al.* 1992). The opposite is true for the spE case, where $\gamma = \frac{4}{3} < \frac{3}{2}$.

However, the spin orientation itself is not independent of the process we are considering, and circularization may not occur at all in the sbE case. The angle between the spin and the angular momentum direction satisfies the inequality

$$\delta(\hat{\Omega}\cdot\hat{\mathbf{j}}) \equiv \left(\frac{1}{j}\right)\hat{\Omega}\cdot[\delta\mathbf{j} - \hat{\mathbf{j}}(\hat{\mathbf{j}}\cdot\delta\mathbf{j})] > 0,$$

for *most* possible configurations. Here $\hat{\Omega}$ is the spin direction and $\hat{\mathbf{j}}$ and $\delta\mathbf{j}$ are the direction and variation of the orbital angular momentum, respectively. For $\gamma > \frac{4}{3}$ the inequality holds for any $\epsilon < 0.92$, and it always holds for the *retrograde* regime, *i.e.*, when $\hat{\Omega}\cdot\hat{\mathbf{j}} < 0$. This shows that the process that we are considering, variation of orbital parameters induced by planetary wind, tends to align the orbital angular momentum with the spin: a retrograde orientation of the planetary spin is unstable. The ratio between the timescales of orbital angular momentum alignment with the spin and the orbital angular momentum change is of order unity. Thus, when the change in specific orbital angular momentum is large, the spin orientation becomes prograde. As we shall see below, the circularization processes in question here indeed require large changes in specific angular momentum. Therefore, circularization is unlikely to occur in the sbE limit unless the post-supernova planetary spin happens to be very close to retrograde.

3 The Rate of Circularization

In the adiabatic limit, if $f_\phi \propto r^{-\gamma}$ then

$$\left(\frac{\epsilon}{\epsilon_o}\right) = \left(\frac{a}{a_o}\right)^{\frac{1}{2}(\gamma-\frac{3}{2})},$$

to lowest order in eccentricity (Banit *et al.* 1992). Here a is the semi-major axis, ϵ is the eccentricity and a_o and ϵ_o are their initial values respectively.

Thus, in the sbE case

$$\left(\frac{\epsilon}{\epsilon_o}\right) \simeq \left(\frac{a}{a_o}\right)^{\frac{1}{4}},$$

and in the spE case

$$\left(\frac{\epsilon}{\epsilon_o}\right) \simeq \left(\frac{a}{a_o}\right)^{-\frac{1}{12}}.$$

These relations mean that in both limits, as long as the adiabatic approximation holds, a reduction of the eccentricity from 0.1 down to the claimed planetary eccentricities (Wolszczan and Frail 1992) requires a change of more than two orders of magnitude in the orbital semi-major axis. Such an evolution would have to be finely tuned. Since the adiabatic approximation is likely to hold, this alone leads us to believe that wind-induced circularization is not likely to be the process that results in the observed low planetary eccentricities.

For completeness we have run computer simulations that verified our analysis in the adiabatic approximation, and also simulated cases in which the adiabatic approximation breaks down. The latter correspond to a pulsar spin down timescale shorter than one orbital period: circularization may now be possible when the rotation is close to retrograde, depending upon the initial orbital phase. We note that breakdown of the adiabatic approximation would require a high pulsar magnetic field and a very short initial pulse period.

In another wind-related model for planet-pulsar formation (Krolik 1991), the circular orbit is guaranteed. In this model, the planet is the residual of an ablated (single) companion in a BMP. The model does not explain the presence of two planets.

In yet another model (Wijers *et al.* 1992) circularization occurs via drag in the envelope of a Thorne-Zytkow (TZ) star (Thorne and Zytkow 1977). If we assume that the envelope is a spherically symmetric, with a density distribution profile $\rho(r) \propto r^{-n}$, then one can show in the adiabatic approximation and in the limit of Keplerian orbital motion that

$$\left(\frac{\epsilon}{\epsilon_o}\right) \simeq \left(\frac{a}{a_o}\right)^{\frac{n+1}{2}}$$

to lowest order in eccentricity (Banit *et al.* 1992). For any reasonable index n, this mechanism seems to lead to efficient circularization, because the dragging force is proportional to the square of the planetary velocity: this leads to a strong deceleration at periastron and makes the circularization efficient. (In the wind-induced circularization process, the eccentricity variations on opposite sides of the orbit tend to cancel each other out). If the planetary spiral-in timescale is shorter than the TZ object's envelope lifetime, circularization may be possible. The difficulty with this model is that it is hard to understand how PSR1257+12 could spin up to less than its present period of 6.2 ms in a TZ object: to do that, the accretion rate will have to be an order of magnitude higher than the Eddington rate (we assume that the object lifetime is $\simeq 10^4$yrs).

4 A New Mechanism

PSR1257+12 is an old ($\simeq 10^9$ yrs), low magnetic field ($\simeq 10^9$ G) millisecond ($P \simeq 6$ ms) pulsar. Such pulsars are believed to have been spun-up to such short periods by accretion in LMXBs (Alpar *et al.* 1982). In later evolution accretion stops, possibly because of the strong pulsar radiation flux (Iliarionov and Sunyaev 1975; Ruderman, Shaham, and Tavani 1989). That, in turn, may ablate a very low mass companion (Ruderman *et al.* 1989). This scenario is how an isolated millisecond pulsar may form.

Observations of the BMPs systems 1957+20 (Fruchter, Stinebring and Taylor 1988) and 1744-24A (Lyne *et al.* 1990) provide evidence for the ablation of companions in such an evolutionary phase. The orbital period of the BMP1957+20 system appears to be decreasing on a very short timescale: $\frac{P_{orb}}{|\dot{P}_{orb}|} \simeq 3 \times 10^7$ yrs (Ryba and Taylor 1991). A possible reason for both the large eclipse size and the rapid orbital shrinking is that the wind is relatively cold and slow (Banit and Shaham 1992) ($\frac{V_{wind}}{V_{escape}} < 1$, V_{wind} is the wind velocity at the sonic point and V_{escape} is the escape velocity from the companion). The wind gets away from the binary by the combined effects of pulsar radiation heating and pressure. Gravitational interaction of the particles in such a wind with the binary system can produce a torque that reduces the system's angular momentum. Particles are delayed in the vicinity of the second Lagrangian point, behind the companion, and their specific angular momentum can increase by a factor $Q \simeq 1.3 - 1.4$ for BMP1957+20 (Banit and Shaham 1992). Computer simulations of 3-dimensional ballistic trajectories produced the eclipse shape and the orbital shrinking time-scale (Banit and Shaham 1992). This model fits the observed parameters if the companion is bloated to its RL dimension, as has been independently suggested by recent observations (Fruchter and Goss 1992).

In winds like that the particle specific energy is, on the average, below escape energy. One therefore expects matter to move out via an excretion disk (Webbink 1976; Pringle 1991), which for 1957+20 would begin at $\simeq 3 \times 10^{11}$cm, about twice the system's orbital separation.

Below, we show that the physical conditions in such disks can lead to planetary formation (Banit and Shaham 1992). This circumbinary disk is expected to be locally Keplerian and will have angular momentum transport on a timescale $\tau_\nu \simeq \frac{r^2}{\nu} \simeq 10^6 - 10^7$yrs (Pringle 1981) where r is the orbital radius and ν is the kinematic viscosity. This timescale is shorter than the orbital shrinking timescale for 1957+20. The amount of matter that is evaporated is more than two orders of magnitude higher that the required planetary mass, and may be relatively heavy in chemical composition as it originates in the core of an evolved white dwarf. At a distance of about $0.5AU$ the black-body temperature in the evolving disk may be less than 1000^oK (using the spin-down luminosity of 1957+20, and neglecting the heat generated in the disk).

Such a temperature can be considered an upper limit since the high energy electron-positron component of the pulsar wind may not be efficiently absorbed in such a disk. In such conditions dust grains in the disk can form (Takenori 1987), concentrate in the midplane of the disk (Weidenschilling 1987), and, by gravitational instabilities (Goldreich and Ward 1973), form planetesimals, coalesce (Safronov 1969) and lead to planetary formation on a timescale shorter than $\simeq 10^6$yrs (Wetherill 1980; Wetherill 1989; Lin, Woosley and Bodenheimer 1991). These planets are likely to have low eccentricities, similar to the eccentricities observed in 1257+12.

In *accretion* disks matter that moves inward accretes onto the compact object. Here, by contrast, matter cannot find itself inside the binary because radiation pressure drives it out. The inner boundary of the resulting *excretion* disk lies just outside of the binary at a specific relative Keplerian angular momentum that is somewhat below Q, and shrinks with it. From this inner boundary, angular momentum is transported outwards by the radially expanding disk. At any given moment, the angular momentum lost by the binary up to then is stored in that expanding excretion disk (unless part of it is blown off by the radiation pressure); the angular momentum balance determines the surface density profile of the disk.

It is interesting to compare this model with that proposed by Lin, Woosely and Bodenheimer (1991). There, a fraction of the SN explosion debris forms a disk around a newly born neutron star. The disk's typical initial radius is about 10^8cm, which corresponds to more than one order of magnitude less specific angular momentum than in the circumbinary disk model. Thus, in the latter model, surface density at $\simeq$1AU would take a shorter time to build up, and the limit on the fraction of the original disk's mass that can carry the angular momentum of the formed planets is much higher. It seems, therefore, that planet formation can proceed more efficiently in a circumbinary disk of a "windy BMP" than in such a SN debris disk.

We gratefully acknowledge support from NASA (grants NAGW-1618 and NCC 5-37) and the NSF (grant AST89-01681).

References

Alpar, M. A., Cheng, A. F., Ruderman, M. A., and Shaham, J., 1982, *Nature* **300**, 728.
Banit, M., Ruderman, M., Shaham, J., and Applegate, J., 1992, *In preparation.*
Banit, M., and Shaham, J., 1992, *Ap. J. Lett.*, **388**, L19.
Chen, K., and Eardley, D. M., 1992, *Preprint.*
Field, R. G., Roger R. D., and Rybicki G. B., 1992, *Preprint.*
Fruchter, A. S., and Goss, W. M., 1992, *Ap. J. Lett.*, **384**, L47.
Fruchter, A. S., Stinebring, D. R., and Taylor, J. H., 1988, *Nature* **333**, 237.
Goldreich, P., and Ward, W. R., 1973, *Ap. J.* **183**, 1051.
Hills, J. G., 1983, *Ap. J.* **267**, 322.
Iliarionov, A. F., and Sunyaev, R. A., 1975, *Astron. and Astrophys.* **39**, 185.
Krolik, J. H., 1991, *Nature* **353**, 829.
Lin, D. N. C., Woosley, S. E., and, Bodenheimer P. H., 1991, *Nature* **353**, 827.
Lyne, A. G., 1990, *et al. Nature* **347**, 650.
Meader, A., and Meynet, G., 1988, *Astron. Astrophys. Suppl. Ser.* **76**, 411.
Narayan, R., and Ostriker, J. P., 1990, *Ap. J.* **352**, 222.
Nelson, R. W., Finn, L. S., and Wasserman, I., 1990, *Ap. J.* **348**, 226.
Pines, D., and Shaham, J., 1972a, *Phys. Earth and Planet. Inter.* **6**, 103.
—— 1972b, *Nature Phys. Sci.*, **253**, 43.
—— 1972c, *Comm. Ap.*, **6**, 37.
Pringle, J. E., 1991, *M. N. R. A. S.* **248**, 754.
Pringle, J. E., 1981, *Ann. Rev. Astron. Astrophys.* **19**, 137.
Rasio, F. A., Nicholson, P. D., Shapiro, S. L., and Teukolsky, S. A., 1992, *Nature* **355**, 325.
Ruderman, M., 1970, *Nature* **225**, 619.
Ruderman, M., Shaham, J., and Tavani, M., 1989, *Ap. J.* **336**, 507.
Ruderman, M., Shaham, J., Tavani, M., and Eichler, D., 1989, *Ap. J.* **343**, 292.
Ryba, M. F., and Taylor, J. H., 1991, *Ap. J.* **380**, 557.
Safronov, V. S., 1972, *Evolution of the Protoplanetary cloud and the Formation of the Earth and Planets* (Nauka, Moscow, 1969); (Engl. trans.) NASA TTF-667.
Takenori, N., 1987, *M. N. A. R. S.* **224**, 107.
Thorne, K. S., and Zytkow, A. N., 1977, *Ap. J.* **212**, 832.
Tkachenko, V. K., 1966, *Sov. Phys. JETP* **23**, 1049.
Webbink, R. F., 1976, *Ap. J.* **209**, 829.
Weidenschilling, S. J., 1987, *Gerlands Beitr. Geophys.* **96**, 21.
Wetherill, G. W., 1980, *Ann. Rev. Astron. Astrophys.* **18**, 77.
Wetherill, G. W., and Stewart, G. R., 1989, *Icarus* **77**, 330.
Wijers, R. A. M. J., van den Heuvel, E. P. J., van Kerkwijk, M. H., and Bhattacharya D, 1992, *Nature* **355**, 593.
Wolszczan, A., and Frail, D. A., 1992, *Nature* **355** 145.
Woosley, S., and Weaver, T. A., 1986, *Ann. rev. Astron. Astrophys.* **24**, 205.